The Properties of Electrodeposited Metals and Alloys

The Properties of Electrodeposited Metals and Alloys

A Handbook

William H. Safranek
Battelle's Columbus Laboratories
Columbus, Ohio

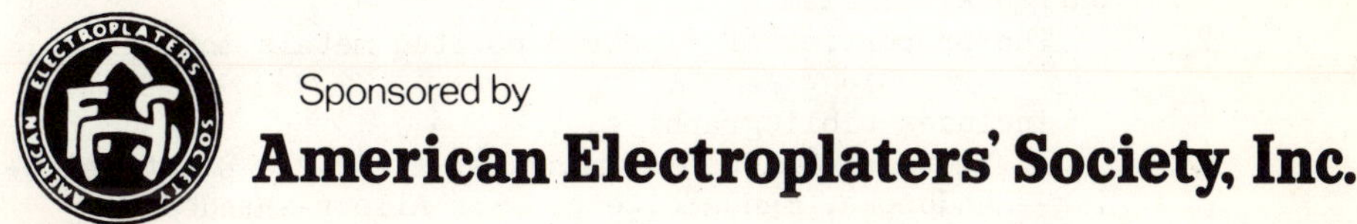

Sponsored by
American Electroplaters' Society, Inc.

AMERICAN ELSEVIER PUBLISHING COMPANY, INC.
NEW YORK LONDON AMSTERDAM

AMERICAN ELSEVIER PUBLISHING COMPANY, INC.
52 Vanderbilt Avenue, New York, N.Y. 10017

ELSEVIER PUBLISHING COMPANY
335 Jan Van Galenstraat, P.O. Box 211
Amsterdam, The Netherlands

International Standard Book Number 0-444-00140-9

Library of Congress Card Number 73-9214

Library of Congress Cataloging in Publication Data

Safranek, William H
The properties of electrodeposited metals and alloys.

Includes bibliographies.
1. Metals--Handbooks, manuals, etc. 2. Electroplating--Handbooks, manuals, etc. 3. Alloys--Handbooks, manuals, etc. 4. Metal coating--Handbooks, manuals, etc.
I. Title.
TA459.S23 620.1'6 73-9214
ISBN 0-444-00140-9

factured in the United States of America

Contents

Preface

Expanding interest in the use of electrodeposited metals for resisting corrosion and wear, reducing friction, providing lubricity or good electrical conductivity, or other purposes has inspired the development of considerable property data in the past 20 years. This compilation of data on the properties of electrodeposited metals represents the first attempt to organize and correlate the voluminous information published in more than 1500 separate articles that contribute pertinent quantitative data. This book includes all of the useful quantitative data on the physical, mechanical, magnetic, and electrical properties of electrodeposits and is based on a comprehensive search of the English and foreign-language literature. Considerable supplementary information on structure as related to property data has also been included in this book, which is expected to: (a) help design engineers in their task of selecting the appropriate electrodeposited coating for a specific application, (b) provide a single-source guidance for electroplating engineers in adopting process controls for insuring compliance with user specifications that govern applications for electrodeposited metals, and (c) supply a reliable foundation for research engineers to use in investigations aimed at increasing mankind's knowledge of electrodeposits and their properties. Towards these ends, reliable and useful data have been compiled and organized in ready reference form for all of the common and some uncommon metals and alloys that have been electrodeposited. The data are correlated with electrodeposition process conditions because the selection of conditions influence properties appreciably.

Many independent investigators have recognized AES publications as an appropriate forum for disseminating their data on the properties of electrodeposits. Approximately 20 percent of the references cited in the Handbook appeared in *Plating* or the *Technical Proceedings* of the AES. Of course, a large proportion of the other references appeared in foreign language publications or in the journals published in England.

The benefits expected for wide-scale use of data in this Handbook are several. New applications for electroplating and electroforming will occur to product design engineers with a knowledge of the Handbook data. Electroplaters will be able to improve the properties of the electrodeposited coatings applied on specific products, and purchasers will be able to use the Handbook data to assist them in specifying the most appropriate coatings for explicit applications. Increasing awareness of the property requirements for electroplated coatings that are associated with specific applications will be developed. Research engineers will be able to use the Handbook data to provide a foundation for new investigations of properties.

Hopefully, industrial organizations with a financial interest in electroplating will be encouraged to undertake projects aimed at developing supplemental property data. Several obvious gaps exist in the available information. For example, data on the platinum-group metals are sketchy. Many of the industrial processes for depositing other metals are handicapped because of the lack of property data for the deposits obtained with these procedures. As shown in this Handbook, alternative processes for depositing a metal or alloy frequently result in properties that differ considerably. The effects of addition agents on the properties of electrodeposits are usually phenomenal. Thus all new plating processes must be investigated comprehensively in terms of deposit properties and compared with established procedures before they are introduced to industry. An ever-increasing degree of knowledge and sophistication is inevitable for the electroplating industry, as for any other viable business serving mankind.

Several members of the staff of Battelle's Columbus Laboratories assisted the author in preparing some of the chapters in this Handbook. They are Mr. John G. Beach, Mr. John E. Clifford, Dr. David E. L. Dyke, Mr. Carl H. Layer, and Dr. F. R. Morral. Mrs. Nancy Dunn completed the initial literature search and translated some of the Russian language references.

Dr. Fumio Hine, Kyoto University at Uji, Dr. Takeo Ishida, Osaka College of Technology, and Dr. Rihei Tomono, Kyoto City Machine-Industry Consultants, translated many of the Japanese language references.

The author also acknowledges the helpful comments received from the following reviewers: Dr. Abner Brenner, National Bureau of Standards (retired); Dr. George W. Brock, IBM; Dr. Henry Brown, The Udylite Corporation; Mr. John B. Capuano, The Udylite Corporation; Mr. Bennie Cohen, Air Force Materials Laboratory, Wright-Patterson Air Force Base; Mr. Paul B. Croly, The International Nickel Company; Mr. Paul Davis, Tin Research Institute; Dr. Robert Duva, The Sel-Rex Company; Dr. Charles L. Faust, Battelle's Columbus Laboratories; Dr. R. D. Fisher, Ferroxcube Corporation; Dr. D. Gardner Foulke, The Sel-Rex Company; Mr. William B. Harding, The Bendix Corporation; Mr. Christian E. Johnson, National Bureau of Standards; Mr. W. Stuart Lyman, Copper Development Association; Dr. M. P. Makowski, Gould Inc.; Mr. Fielding Ogburn, National Bureau of Standards; Dr. Martin Prager, Copper Development Association;

Dr. Harold J. Read, Pennsylvania State University; Mr. Robert L. Ruleff, Consultant; Mr. Atkin Y. Simonian, The International Nickel Company; Dr. P. K. Subramanyan, Gould Inc.; Dr. Harold J. Wiesner, Lawrence Livermore Laboratory, University of California; Mr. James C. Withers, General Technologies Corporation.

W. H. Safranek
Columbus, Ohio

Chapter 1

Introductory Data

Some data purportedly associated with electrodeposited metals merely repeat previously reported information on the wrought forms of those metals. Because many properties of electrodeposited metals differ appreciably from the properties of castings, forgings or rolled sheet, reference to specific values for the electrodeposits is important when selecting candidates for a specific application. Furthermore, attention must be given to the procedure for electrodepositing the metal or alloy, because alternative procedures frequently produce vastly different properties. Thus property data associated with specific deposition procedures are listed in the individual chapters of this Handbook for the guidance of engineers responsible for selecting materials and procedures.

The properties of electrodeposits that are important for a number of general applications are reviewed in this chapter. These include coatings for a broad spectrum of applications where resistance to corrosion, wear, and/or abrasion must be provided. Electrodeposits providing good electrical conductivity are needed for high-strength steel wire, printed circuits, electrical contacts, and high-frequency radar waveguides. Light or heat reflectance is important in other cases. Reducing friction, preventing galling, providing lubricity, and eliminating stress corrosion or erosion are other purposes of electrodeposition which require consideration of specific properties for specific applications. Dense and structurally stable coatings must be selected for protection against high-temperature oxidation. Coatings with special magnetic properties fill the need for another application for electrodeposited metals, which requires freedom from porosity and inclusions. Electroplating is also used to facilitate the joining of polymeric materials to metal or metal to metal. Examples include brass plating for bonding rubber and tin or tin alloy plating for soldering.

The first major use of electrodeposition was in 1839 and 1840, for fabricating art forms and printing plates by means of electroforming. Some twentieth-century applications are based on the unexcelled fidelity of the electroforming process for reproducing surface textural detail. Electrotypes and phonograph-record stampers are outstanding examples of these applications. Structural shapes fabricated by electroforming since 1950 required a relatively high strength, a high modulus of elasticity, and/or good homogeneity of properties and structure. Electroforming has replaced combinations of deep drawing, stamping or extruding, and welding or brazing for fabricating many shapes, especially those with thin walls, because it combines in a single step strength and homogeneity with good dimensional accuracy. Metal foil and screen are popular products of electroforming because they can be produced with good density, strength, and ductility at a lower cost than casting and rolling or by weaving. Thus physical and mechanical properties are important criteria for many applications of electroforming.

Physical Properties

Density

Some electrodeposits contain voids, which affect their usefulness for protecting other metals from corrosion or high-temperature oxidation. Others are equivalent in density to their cast and forged counterparts. The density of aluminum, chromium, cobalt, copper, gold, lead, nickel, and silver electrodeposits can be equal or nearly equal to that of the corresponding wrought form if the proper electrodeposition conditions are selected. On the other hand, these deposits may contain pores, voids, or impurities that reduce density if appropriate conditions are not adopted. For example, the density of some silver electrodeposits was reduced as much as 24 percent by using inappropriate addition agents that induced inclusions. Density data for electrodeposited metals are compared with the densities of wrought metals in Table 1.1.

Electrical Resistivity

Electrical resistivity depends on purity and density. The best of the cadmium, gold, and silver electrodeposits are equivalent in conductivity to that of the corresponding wrought material. The conductivity of some copper deposits is within 2 percent of that of high-purity, wrought copper. The lowest resistivity value for tin electroplate is only 4 percent higher than the value for its high-purity, metallurgical counterpart. In the case of chromium, lead, nickel, and zinc, the electrodeposit is 8 to 11 percent higher in resistivity, at best, by comparison with the corresponding wrought metal. Cobalt differed by 25 percent and rhodium by 90 percent. Resistivity data are summarized in Table 1.2.

Small concentrations of oxides, hydrates, or salts concentrated at grain boundaries are the chief reasons for the higher-than-expected resistivity values for chromium and nickel deposits. High values for other metals may be due to similar occlusions at grain boundaries. Inclusions induced by some organic impurities greatly increase the resistivity of copper, nickel, and silver. Of course, metallic impurities also increase resistivity. Approximately 0.5 percent cobalt increases the resistivity of nickel about 4 percent, for example. About 0.05 percent sulfur in the form of nickel sulfide increases resistivity by >10 percent; 0.05 to 0.15 percent sulfur prevails in most of the bright nickel electroplates produced with organic brighteners.

TABLE 1.1

Summary of Density Data for Electrodeposited and Wrought Metals

Metal	Density Range for Electrodeposits, g/cu cm[(a)]	Type of Plating Bath for Maximum Density	Maximum Density of Electrodeposit, g/cu cm	Density of Wrought Metal, g/cu cm
Aluminum	2.56 to 2.66	Anhydrous chloride-hydride-ether	2.66	2.71
Chromium	6.9 to 7.18	Chromic acid	7.18	7.19
Cobalt	8.8	Chloride	8.8	8.85
Copper	8.86 to 8.93	Cyanide, fluoborate, pyrophosphate, or sulfate	8.92–8.93	8.96
Gold	to 19.3	Acid cyanide	19.3	19.3
Lead	11.34	Fluoborate	11.34	11.34
Nickel	8.86 to 8.93	Fluoborate, sulfamate, or Watts	8.91–8.93	8.90
Silver	8.0 to 10.5	Cyanide	10.5	10.5
Zinc	7.2 to 7.8[(b)]	Sulfate	7.8[(b)]	7.13

[(a)] No density data were disclosed in the literature on cadmium, iron, the platinum-group metals, and tin.
[(b)] Calculated values.

TABLE 1.2

Summary of Electrical Resistivity Data for Electrodeposited and Wrought Metals[a]

Metal	Resistivity Range for Electrodeposits, microhm-cm[b]	Type of Plating Bath for Minimum Resistivity	Minimum Resistivity of Electrodeposit, microhm-cm	Resistivity of Wrought Metal, microhm-cm
Aluminum	3.1 to 5.8	Anhydrous chloride-hydride-ether	3.1	2.66
Cadmium	6.8 to 8.5	Not disclosed	6.8	6.8
Chromium	14 to 67[c]	Chromic acid	14	12.9
Cobalt	8.4 to 10	Sulfate-chloride	8.4	6.24
Copper	1.70 to 4.6	Fluoborate or sulfate	1.70–1.72	1.67
Gold	2.34 to 4.2	Acid cyanide	2.34–2.44	2.35
Lead	22.9	Fluoborate	22.9	20.7
Nickel	7.4 to 11.5[d]	Cobalt-free Watts	7.4	6.84
Palladium	10.7	Nitrate	10.7	10.8
Rhodium	4.5 to 8.5	Sulfate	4.5	4.5
Silver	1.59 to 1300	Cyanide	1.59	1.59
Tin	11.4 to 18	Fluoborate	11.4	11.0
Zinc	6.6 to 7.5	Cyanide	6.6	5.92

(a) All values relate to measurements at about 20 C.

(b) No resistivity data were disclosed in the literature on iron.

(c) The resistivity of all chromium deposits was reduced to the range of 13 to 16 after annealing at 1200 C.

(d) Annealing at 1000 C reduced resistivity to the range of 7.4 to 8.4, as a rule.

Copper and silver electrodeposits are employed for some electrical connectors because they exhibit a low resistivity. Oxide or sulfide films acquired during service increase contact resistance, however. Gold electrodeposits are becoming increasingly popular because they resist oxidation and because their electrical resistivity is only slightly greater than that for copper. For such applications, porosity in not-too-costly, thin gold deposits must be avoided in order to prevent the growth of substrate-metal oxidation products on the gold surfaces, because such products increase contact resistance appreciably. Gold-plated silver and copper contacts can form high-contact-resistance diffusion compounds if subjected to an elevated temperature. An electrodeposited nickel barrier is frequently employed for minimizing porosity in thin gold overlays and preventing diffusion compounds at elevated temperatures. Chromium is better for preventing diffusion at temperatures as high as 800 C, however.

Annealing sometimes decreases resistivity by dispersing inclusions concentrated at grain boundaries. Such is the case for high-resistivity chromium, copper, or silver. On the other hand, not all high-resistivity nickel deposits respond to such a treatment.

TABLE 1.3

Coefficient of Expansion of Electrodeposited and Wrought Metals

		Coefficient of Thermal Expansion, 10^6/C	
Metal	Type of Plating Bath	Electrodeposit	Wrought Metal
Aluminum	Anhydrous chloride-hydride-ether	24.6[a]	23.6[a]
Cadmium	Not disclosed	16.6[a]	29.8[a]
Chromium	Chromic acid	7.4[b]	6.2[c]
Cobalt	Chloride	14.5[d]	13.8[b]
Cobalt-30% tungsten	Chloride	16.2[e]	—
Copper	Cyanide, fluoborate, or pyrophosphate	16.7[b]	16.5[b]
Copper	Sulfate	17.1[b]	16.5[b]
Gold	Acid cyanide	14.3[c]	14.2[c]
Nickel	Chloride, sulfamate, or Watts	13.6[c]	13.3[c]
Nickel-40% cobalt	Not disclosed	13.1[c]	—
Zinc	Not disclosed	22[a]	39.7[a]

[a] From 20 to 700 C.

[b] From 20 to 200 C.

[c] Near room temperature.

[d] From 0 to 900 C.

[e] 20 to 1000 C.

TABLE 1.4

Permanent Changes in Length and Density of Electroformed Metals from Thermal Treatment to 1050 C

Metal	Plating Bath	Permanent Change in Length, percent[a]	Change in Density, percent[b]
Chromium	Chromic acid (50 C)	−1.1	1.9
Chromium	Chromic acid (80 C)	—	0.7
Cobalt-30% tungsten	Chloride	−0.8	—
Copper	Sulfate with triisopropanolamine	0.33[c]	−17.4
Copper	Sulfate with gelatin	1.2[c]	
Copper	Sulfate with proprietary organic compounds	3.6[c]	—
Nickel	Chloride	—	−1.2
Nickel	Sulfamate at 2.5 pH	1.2	−5.6
Nickel	Sulfamate at 4.3 pH	0.2	−0.6
Nickel	Watts at 2.5 pH	3.8	−7.8
Nickel	Watts at 3.0 pH	—	−0.8
Nickel	Watts at 4.4 pH	2.3	−3.5

[a] A negative sign indicates shrinkage.
[b] A negative sign indicates permanent expansion.
[c] Change from heating to 400 or 500 C.

Many alloy electrodeposits have higher resistivities than their cast counterparts. Examples include alloys of nickel and cobalt, cobalt and tungsten, tin and zinc, and silver plus lead or copper. Heat treatment usually reduces resistivity to values that equal or nearly equal those for the corresponding cast alloys. In the case of the silver alloys, the heat treatment converted supersaturated solutions to two-phase structures. In other cases, annealing probably dispersed small amounts of nonmetallic inclusions during recrystallization.

Thermal Properties

Except for data on expansion coefficients, meager information on thermal properties is available from the literature. The thermal conductivities (at room temperature) of cobalt, gold, and nickel deposits were reported to be nearly the same as the conductivities of their metallurgical counterparts, as follows:

Thermal Conductivity,

cal/sq cm/cm/C/sec

Metal	Electrodeposit	Wrought Metal
Cobalt	0.165	0.165
Gold	0.707	0.710
Nickel	0.20–0.26	0.22

The expansion coefficients for most of the electrodeposited metals are nearly the same as those for the corresponding wrought metal, as shown in Table 1.3. However, these data were developed when the electroformed metals were heated and cooled a second (or third) time. The first cycle of heating and cooling sometimes causes an appreciable expansion or contraction and a corresponding change in density. Table 1.4 summarizes available data on such changes.

Chromium contracts during the first thermal cycle and voids in the deposit are healed. The deposit from an 80 C bath contracts less than one from a 50 C bath, however. An alloy of cobalt containing 30 percent tungsten also contracted upon heat cycling. On the other hand, copper deposited in a sulfate bath containing gelatin or some other addition agents expanded as a result of thermal cycling. Nickel deposited in chloride, sulfamate, and Watts-type baths also expands upon heat treating, but an optimum pH for the sulfamate and Watts-type bath was identified for minimizing this expansion, as shown in Table 1.4, to less than 1.0 percent. Electrodeposits that expand appreciably during thermal cycling must be avoided for any application involving heat treatment or service at high temperatures. Otherwise, porosity introduced during heating will reduce the ability of such deposits for protecting substrates from corrosion or high-temperature oxidation.

Magnetic Properties

The electrodeposited metals and alloys of primary interest for magnetic films and shields include cobalt, binary alloys of cobalt and phosphorus, iron and nickel, nickel and phosphorus, and ternary alloys of nickel, cobalt and phosphorus or of nickel, iron and phosphorus. Cobalt-phosphorus and nickel-phosphorus alloy films applied by electroless plating in solutions containing sodium hypophosphite are also employed for thin films on nonmagnetic substrates for recording information in computers. Some property data for these coatings are listed in Table 1.5.

Cobalt with a low coercivity, from 8 to 25 oersteds, is electrodeposited in chloride or sulfate solutions with a pH <2.0 when the solutions are operated at about 20 C. On the other hand, a coercivity as high as 1200 oersteds has been reported for cobalt electrodeposited in solutions with a pH >4.5, which were heated to at least 40 C. The increases in pH and temperature appear to favor the deposition of a hexagonal-close-packed structure, in place of a mixture of hexagonal-close-packed and face-centered-cubic cobalt. Soft magnetic coatings with a low coercivity are utilized for fast-switching memory devices, whereas thin films with a high coercivity are needed for high-density information storage on drums or disks. Chapter 5, "Cobalt and Cobalt Alloys," details magnetic property data for cobalt and cobalt-alloy electrodeposits and discusses the influence of operating conditions on coercivity, remanent induction, and saturation magnetization. Variations in such conditions, the type and condition of the substrate, and the thickness of the electrodeposit have profound effects on magnetic properties.

Electroless cobalt containing 1 to 1.5 percent phosphorus, which is deposited in alkaline solutions at a high temperature (92 C), exhibits a low coercivity, from 1 to 11 oersteds. On the other hand, a high coercivity ranging from 1000 to 1200

TABLE 1.5

Magnetic Property Data For Electrodeposited and Electrolessly Deposited Metals and Alloys

Metal or Alloy	Type of Plating Bath	Coercivity, oersteds	Ratio of Remanence to Saturation Magnetization (B_r/B_m ratio)
Cobalt	Chloride or sulfate, with a pH <2 (20 or 25 C)	8–25	0.57–0.88
Cobalt	Chloride or sulfate, with a pH >4.5 (40 or 90 C)	600–1200	No data
Cobalt-nickel (3 to 18% Ni)	Chloride, sulfate, or sulfamate	50–140	0.32–0.65
Cobalt-nickel (21 to 35% Ni)	Ditto	100–500	0.7–1.0
Cobalt-nickel-phosphorus (19% Ni and 2% P)	Sulfate	1750	No data
Cobalt-nickel-phosphorus (15% Ni and <2% P)[a]	Alkaline citrate	1400–2000	No data
Cobalt-phosphorus (3% P)[b]	Phosphate-phosphite-tartrate (75 C)	1200–1400	No data
Cobalt-phosphorus (1–1.5% P)[a]	Sulfate (92 C)	1–11	0.93–1.0
Cobalt-phosphorus (5–5.8% P)[a]	Sulfate (50 C)	1000–2100	No data
Iron	Chloride	15–40	No data
Iron-nickel (3 to 5% Ni)	Chloride	1–10	0.8 to >0.95
Nickel	Sulfate-chloride	40–120	0.65 max
Nickel-cobalt-phosphorus (2–7% P)[a]	Alkaline	0.14–4.5	0.1–0.5
Nickel-iron (15–25% Fe)[c,d]	Sulfate and sulfamate	1–5	>0.85
Nickel-phosphorus (3–6% P)[a]	Alkaline	10–80	<0.1
Nickel-phosphorus (7–18% P)[a]	Alkaline	2 max	0.4–0.7

(a) Electroless deposits. Others were electrodeposited.

(b) The coercivity of other electrodeposited alloys containing <3 or >3% phosphorus ranged from 300 to 800 oersteds.

(c) Alloy deposits containing <15 or >35% iron exhibited higher coercivity, ranging to 110 oersteds.

(d) Several ternary alloys containing antimony, arsenic, phosphorus, selenium, or another metal also exhibited low coercivities.

oersteds has been reported for electroless cobalt containing 5 to 5.8 percent phosphorus deposited in solutions maintained at a lower temperature (50 C). A coercivity as high as 2000 oersteds was claimed for an electroless alloy containing 15 percent nickel and < 2 percent phosphorus. Cobalt alloy containing 3 percent phosphorus, which was electrodeposited in a phosphate-phosphite-tartrate solution while alternating current was superimposed on the direct current, also exhibited a high coercivity, ranging from 1200 to 1400 oersteds. Another high-coercivity electrodeposit contained 89 percent cobalt, 19 percent nickel, and 2 percent phosphorus. The coercivity of this alloy film was 1750 oersteds.

A coercivity of < 5 oersteds was reported for nickel-rich electroless films containing a small concentration of cobalt and 2 to 7 percent phosphorus and for cobalt-free nickel films containing 7 to 18 percent phosphorus. Electrodeposited alloys exhibiting low coercivities included an iron alloy with a nickel content of 3 to 5 percent and nickel alloys with an iron content of 15 to 25 percent.

Magnetization of the nickel alloys approaches zero at 15 percent phosphorus, which corresponds to the compound Ni_3P. Only slight magnetization has been detected for alloys containing as much as 9 percent phosphorus, however.

The low-coercivity electrodeposited nickel alloys containing 15 to 25 percent iron show high ratios of remanent induction to saturation magnetization (>0.85). Alloys in this composition range also approach zero magnetostriction. A 19 or 19.5 percent iron content was cited for several films exhibiting zero magnetostriction. Zero magnetostriction also was reported for alloys containing 11 to 20 percent co balt and 10 to 17 percent iron or 10 to 22 percent molybdenum and 23 percent iron.

Mechanical Properties

Strength and Ductility

The tensile strength of individual electrodeposited metals spans broad ranges and depends on the conditions adopted for electrodeposition. At least one procedure has been documented for depositing each of the metals listed in Table 1.6 with a tensile strength exceeding the strength of its annealed, metallurgical counterpart of comparable purity. In several cases, the deposit is two or three times as strong as the corresponding wrought metal. The maximum tensile strength for electrodeposited cobalt is more than four times the strength of annealed, wrought cobalt. Some chromium deposits are nearly seven times as strong as cast or sintered chromium.

A fine grain size is the primary reason for the higher strength of the electrodeposits, as compared with their wrought counterparts. In the case of cobalt, copper, and nickel, grain-size refinement during electrodeposition is more effective for increasing strength than grain-size reduction of the metallurgical material by cold rolling or swaging. The tensile strength of the strongest electrodeposited nickel is about one-third higher than the maximum reported for cold-rolled sheet. The strongest cobalt and copper deposits have tensile strengths about 70 percent higher than the maxima listed for their cold-worked counterparts. High-strength silver de-

TABLE 1.6

Strength and Ductility Data for Electrodeposited Metals

Metal	Plating Bath	Minimum Tensile Strength		Maximum Tensile Strength		Elongation, percent	Wrought Metal[a]		
							Tensile Strength		Elongation, percent
		kg/sq mm	psi	kg/sq mm	psi		kg/sq mm	psi	
Aluminum	Anhydrous chloride-hydride-ether	7.7	11,000	21.7	31,000	2 to 26	9.15	13,000	35
Cadmium	Cyanide	—	—	7.0	10,000	—	7.2	10,300	50
Chromium	Chromic acid	10	14,000	56	80,000	<0.1	8.4	12,000	0
Cobalt	Sulfate-chloride	54	76,500	121	172,000	<1	26	37,000	—
Copper	Cyanide, fluoborate, or sulfate	18	25,000	66	93,000	3 to 35	35	50,000	45
Gold	Cyanide and cyanide citrate	12.7	18,000	21.2	30,000	22 to 45	13.5	19,000	45
Iron	Chloride, sulfate, or sulfamate	33	47,000	109	155,000	2 to 50	29	41,000	47
Lead	Fluoborate	1.4	2,000	1.6	2,250	50 to 53	1.85–2.1	2,650–3,000	42 to 50
Lead-TiO_2 composites	Fluoborate	3.2	4,500	4.1	5,800	7 to 15	3.2[b]	4,500[b]	18[b]
Nickel[c]	Watts, and other type baths	35	50,000	107	152,000	5 to 35	32.2	46,000	30
Silver	Cyanide	24	34,000	34	48,000	12 to 19	16–19	23,000–27,000	50 to 55
Zinc	Sulfate	4.9	7,000	11.2	16,000	1 to 51	9.1	13,000	32

[a] Annealed, worked metal.
[b] Powder compact prepared with 1.5% PbO particles.
[c] Data do not include values for nickel containing >0.005% sulfur.

posits and cold-rolled sheet have about the same strength, but the electrodeposited metal is more ductile.

Coarse-grained electrodeposits have a lower strength than the corresponding metallurgical materials. Some deposits of aluminum, copper, gold, lead, and zinc are weaker than their annealed, wrought counterparts. The conditions adopted for electrodeposition must be selected carefully to avoid those that provoke weak, coarse-grained deposits. The individual chapters on these metals should be consulted for guidance in procedures for depositing fine-grained, strong metals.

High strength is important for many electroforming applications. Resistance to wear and abrasion usually can be improved by adopting conditions to refine grain and increase strength. High-strength, fine-grained deposits usually exhibit high density and freedom from porosity, which is important for using them as corrosion-protective coatings.

Several nickel alloys can be deposited with a higher strength than unalloyed nickel. Tensile strengths exceeding 150 kg/sq mm (214,000 psi) have been reported for nickel alloys containing 40 to 50 percent cobalt or 25 to 40 percent iron. Nickel containing about 0.02 percent sulfur ranges in tensile strength from 149 to 165 kg/sq mm (212,000 to 235,000 psi). This amount of sulfur is introduced in the deposit when organosulfur compounds are added to the deposition solution to reduce stress in the nickel. Strength and ductility data for both nickel and cobalt alloys are summarized in Table 1.7.

Composites of nickel containing boron or tungsten fibers also exhibited high strengths, as shown in Table 1.7. A composite of nickel and silicon carbide whiskers was the strongest electrodeposit reported in the literature. With 3.2 to 3.6 weight percent of such whiskers, tensile strength ranged from 174 to 229 kg/sq mm (249,000 to 327,000 psi).

The tensile strength of notched nickel specimens is about equal to that of unnotched nickel, providing the sulfur content of the nickel does not exceed 0.017 percent. With a higher sulfur content, the ratio of notched and unnotched strength values was reduced to < 0.75.

Boron-fiber-strengthened nickel exhibited a notched to unnotched tensile strength ratio of 1.1 when the fiber content was 6 or 7 percent by volume. An increase in the boron fiber content to 12 percent reduced the ratio to about 0.8.

A composite consisting of nickel and 46 volume percent of silicon carbide filaments retained a high strength of 74 kg/sq mm (106,000 psi) during measurements at 600 C. In comparison, the strength of unalloyed electrodeposited nickel at this temperature is < 25 kg/sq mm ($< 35,000$ psi). No strength data for the composite at higher temperatures were reported.

Cobalt-tungsten alloy is another electrodeposit that retains high strength at high temperatures. Tensile strengths of about 50, 40, and 15 kg/sq mm (71,000, 57,000, and 21,000 psi, respectively) were reported for measurements at 600, 800, and 1000 C, in the case of an alloy deposit containing 28 percent tungsten. Another containing 35 percent tungsten, 3 percent iron, and 1 percent nickel exhibited tensile strengths of about 60, 55, and 40 kg/sq mm (85,000, 78,000, and 57,000 psi) at 600,

TABLE 1.7

Strength and Ductility Data (at 20 C) for Cobalt and Nickel Alloy Electrodeposits

Alloy	Plating Bath	Minimum Tensile Strength		Maximum Tensile Strength		Elongation, percent
		kg/sq mm	psi	kg/sq mm	psi	
Cobalt-nickel (20% Ni)	Sulfamate	—	—	56	80,000	1
Cobalt-phosphorus (1% P)	Chloride	—	—	91	130,000	7
Cobalt-tungsten (20% W)	Alkaline	—	—	110	155,000	<1
Nickel-cobalt (40–50% Co)	Sulfate and sulfamate	140	200,000	192	275,000	2 to 4
Nickel-iron (25–40% Fe)	Sulfate	140	200,000	180	255,000	2 to 3
Nickel-manganese (0.5% Mn)	Sulfamate	117	167,000	131	187,000	5 to 6
Nickel-phosphorus (5 or 6% P)	Electroless chloride (90 C)	39	56,000	48	68,000	2.0
Nickel-phosphorus (8–10% P)	Electroless chloride (90 C)	45	65,000	86	122,000	0.3
Nickel-tungsten (20% W)	Sulfate	78	110,000	110	156,000	~0
Nickel-nickel sulfide (0.02% S)[a]	Sulfamate or Watts	149	212,000	165	235,000	1 to 5
Nickel-boron composite[b]	Sulfamate	112	159,000	144	205,000	—
Nickel-SiC composite[c]	Sulfamate	174	249,000	229	327,000	—
Nickel-tungsten composite[d]	Sulfamate	148	211,000	154	220,000	—

(a) The nickel sulfide was induced with decomposition of organic compounds (such as saccharin) containing sulfur.

(b) A composite containing 24 to 42%, by volume, of 0.1-μm boron fibers.

(c) A composite containing 3.2 to 3.6%, 1-μm-diameter SiC whiskers.

(d) A composite containing 35 to 50%, by volume of 0.025 to 0.1-μm tungsten fibers.

800, and 1000 C, respectively. Figure 5.5 in Chapter 5, "Cobalt and Cobalt Alloys," shows these relationships.

Modulus of Elasticity

High-modulus deposits are desired for electroformed structures. Furthermore, deposits on substrates subjected to an applied load (strain) during postplating fabrication or service should have a high modulus to resist deformation. Some high-modulus metals and alloys are listed in Table 1.8.

A high modulus of elasticity was reported for several composites of nickel and fibers produced by continuous winding of the filaments during electrodeposition. For example, a nickel-tungsten fiber composite had an elastic modulus as high as 24,000 kg/sq mm (34×10^6 psi), which compares with 21,600 and 21,000 kg/sq mm (31 and 30×10^6 psi) for high-purity electrodeposited and wrought nickel, respectively. A composite of nickel and boron fibers had a modulus of 27,600 kg/sq mm (39.2×10^6 psi), about 25 percent greater than the maximum for unalloyed electrodeposited nickel. A modulus of 31,500 kg/sq mm (45×10^6 psi) was reported for a composite of nickel and silicon carbide.

Hardness

Although the usefulness of hardness data for predicting serviceability is questionable, voluminous data have been published for the electrodeposited metals. Some reporters have assumed that strength and hardness increase or decrease proportionally. Yet many exceptions have been cited. Some investigators have reported improved resistance to wear or abrasion as a result of increasing hardness. On the other hand many reports show exceptions, and some of these show better resistance to abrasion for some metals as hardness was reduced. Resistance to friction and galling and structural defects are important supplemental factors that influence wear and abrasion.

Variations in hardness measurement procedures are numerous, and conversions to a standard are either difficult or impractical. Many investigators did not report the load applied to the indentor, which has a profound effect on the hardness value. Despite these measurement problems and the reported inconsistencies in descriptions of attempts to correlate hardness with strength or abrasion resistance, hardness data are often used to differentiate between mediocre and good conditions for depositing metals. Hardness values are particularly useful for evaluating electrodeposited alloys because they indicate changes in composition and/or structure.

The microhardness of nearly all electrodeposited metals spans broad ranges. Figure 1.1 shows ranges for relatively soft deposits and Figure 1.2 shows ranges for harder metals and alloys. Besides chromium, cobalt-phosphorus and nickel-phosphorus alloys are very hard, reaching a maximum hardness of >1200 kg/sq mm after heat treatments that precipitate nickel phosphide. Some iron deposits also are

TABLE 1.8

High-Modulus Electrodeposited Metals and Alloys

Metal	Plating Bath	Modulus of Elasticity			
		Minimum		Maximum	
		kg/sq mm	psi × 10^6	kg/sq mm	psi × 10^6
Chromium	Chromic acid	10,000	15	23,000[a]	33[a]
Cobalt-tungsten (20% W)	Alkaline	—	—	17,000	24
Nickel	Watts and sulfamate	14,900	21	19,100	27
Nickel	Chloride	21,400	30.5	21,600[b]	31[b]
Nickel-cobalt (20% Co)	Sulfamate	—	—	20,300	29
Nickel-phosphorus (8 to 9% P)	Electroless (acid)	12,000	17	20,000	28.5
Nickel-tungsten (25% W)	Alkaline	—	—	17,000	24
Nickel-alumina composite[c]	Watts	—	—	22,000	31
Nickel-boron composite[d]	Sulfamate	22,800	32.6	27,600	39.2
Nickel-silicon carbide composite[e]	Sulfamate	28,400	40	31,500	45
Nickel-tungsten composite[f]	Sulfamate	21,000	30	24,000	34

(a) Increased to 25,000 kg/sq mm (35 × 10^6 psi) by annealing at 1200 C. For comparison, the modulus of elasticity of wrought chromium is 25,000 kg/sq mm (36 × 10^6 psi).
(b) For comparison, the modulus of wrought "A" nickel is 21,000 kg/sq mm (30 × 10^6 psi).
(c) 6 or 9 percent, by weight, of Al_2O_3 particles.
(d) 23 to 50 percent, by volume, of boron fibers.
(e) 40 to 50 percent, by volume, of silicon carbide fibers.
(f) 30 to 50 percent, by volume, of tungsten fibers.

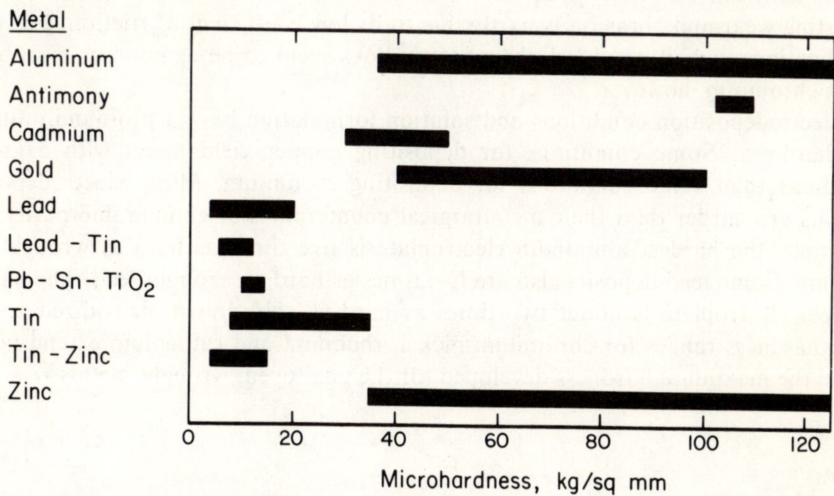

Figure 1.1. Microhardness ranges for relatively soft electrodeposited Metals and Alloys.

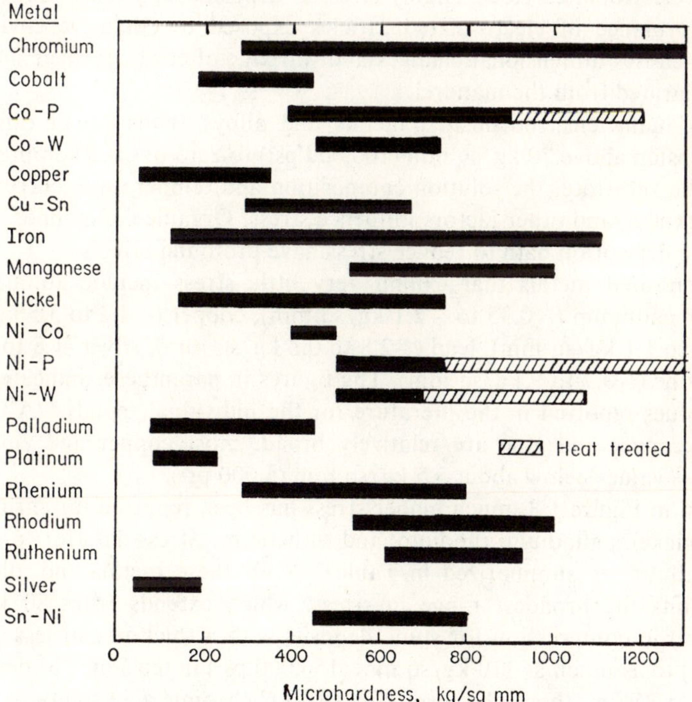

Figure 1.2. Microhardness Ranges for Relatively Hard Electrodeposited Metals and Alloys.

hard, as much as 1100 kg/sq mm. The popularity of chromium electroplate for resisting wear and abrasion is partly due to its low coefficient of friction. For some applications, cobalt or nickel-phosphorus alloys seem to be as good or even better than chromium, however.

Electrodeposition conditions and solution formulation have a profound influence on hardness. Some conditions for depositing copper yield metal with a greater hardness than some conditions for depositing chromium. Many electrodeposited metals are harder than their metallurgical counterparts of comparable purity. For example, the hardest aluminum electroplate is five times as hard as wrought aluminum. Some lead deposits also are five times as hard as wrought lead. The hardest copper electroplate is about two times as hard as cold-drawn, deoxidized copper. The hardness ranges for chromium, nickel, rhodium, and ruthenium extend far beyond the maximum hardness developed for the unalloyed, wrought metals.

Stress

Distortions in the crystal lattice can cause high stress in electrodeposited metals. The resistance of electroplated metals to mechanical stress in service is reduced when the deposits are highly stressed. High tensile stresses reduce the fatigue strength of electroplated steel. Highly stressed deposits reportedly reduce the corrosion performance of electroplated articles exposed to corrosive environments. High stress causes dimensional changes and warpage of electroformed shapes when they are separated from the mandrel.

Stress of many electrodeposited metals and alloys spans broad ranges, from values in tension above 70 kg/sq mm (100,000 psi) to zero or even compressive. The nature of the substrate, the solution composition and temperature, current density, deposit thickness, and other factors influence stress. Organic compounds sometimes added to the deposition bath to reduce stress have profound effects.

Electrodeposited metals that exhibit very little stress include aluminum (0.06 kg/sq mm), cadmium (−0.35 to −2.1 kg/sq mm), copper (−4.2 to 5.5 kg/sq mm), gold (−0.6 to 1.1 kg/sq mm), lead (−2.8 to 0.2 kg/sq mm), silver (1.5 to 2.5 kg/sq mm), and zinc (−0.7 to 9 kg/sq mm). The figures in parentheses indicate the range of stress values reported in the literature for the individual metals.* Although the ranges for copper and zinc are relatively broad, most copper and zinc deposits exhibit stress values below about 3.5 kg/sq mm (5,000 psi).

As shown in Figure 1.3, much higher stress has been reported for chromium, cobalt, iron, nickel, palladium, rhodium, and ruthenium. Stress data for some electrodeposited alloys are summarized in Table 1.9. Of these metals and alloys, chromium exhibits the broadest range in stress, which extends from 80 kg/sq mm (114,000 psi) in compression for some deposits with a thickness of less than 1 μm (<0.04 mil) to as much as 110 kg/sq mm (156,000 psi) in tension. For deposits with a thickness > 25 μm, those obtained in 60 to 70 C chromic acid solutions are highly

* A negative sign denotes compressive stress.

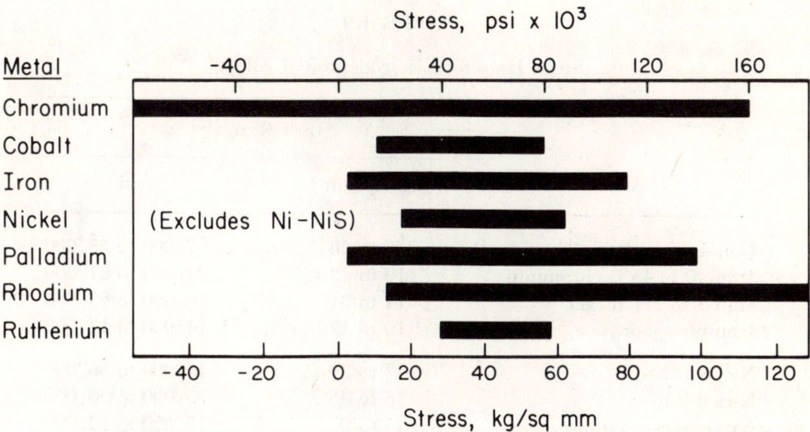

Figure 1.3. Stress Ranges for Highly Stressed Electrodeposited Metals.

stressed in tension, whereas deposits from 75 to 90 C baths exhibit a low tensile stress of < 4.2 kg/sq mm (< 6,000 psi). Some deposits from solutions maintained at 20 to 40 C are stressed in compression, whereas others remain slightly stressed in tension after stress relief by cracking.

Stress in chromium, nickel, and other electrodeposits is reduced as thickness is increased, and usually reaches an equilibrium value when thickness is about 30 μm (1.2 mils). The best solutions for depositing nickel with a minimum stress are sulfamate or sulfamate-fluoborate baths. The addition of chloride or other halide ions to any type of nickel plating bath increases stress. Contamination with ferrous or ferric ions or the addition of cobalt ions increases the stress of nickel deposits. Additions of saccharin and other organosulfur compounds reduce tensile stress or induce a compressive stress. Such deposits can be classed as alloys, since they contain at least 0.01 percent sulfur. Data for alloys are given in Table 1.9.

High stress in chromium, rhodium, and some alloy electrodeposits is usually relieved or partly relieved by cracking. Nearly all hard chromium electrodeposits and most of the rhodium deposits applied for resisting wear contain cracks. The crack density varies, however, as solution formulations and conditions are changed. Compressively stressed chromium contains more cracks than chromium stressed in tension. Such compressively stressed chromium has little or no harmful effect on the fatigue strength of steel or other substrates, whereas chromium deposits highly stressed in tension will reduce fatigue strength at least 22 percent or as much as 73 percent, depending on the strength of the substrate material. Analogously, nickel deposits highly stressed in tension reduce the fatigue strength of steel substrates as much as 59 percent, whereas nickel slightly stressed in tension has only a moderate effect. Compressively stressed nickel has no harmful effect on steels with a fatigue strength that does not exceed 32 kg/sq mm (46,000 psi).

TABLE 1.9

Stress Data for Electrodeposited Alloys

Alloy	Stress Range[a]	
	kg/sq mm	psi
Cobalt-phosphorus[b]	5 to 38.5	7,000 to 55,000
Iron-40 to 45% chromium	19 to 43	27,000 to 61,000
Iron-1 to 2% nickel	14 to 34	20,000 to 55,000
Iron-phosphorus	10 to 12	14,000 to 17,000
Nickel-cobalt	−7 to 39	−10,000 to 56,000
Nickel-iron	16 to 35	23,000 to 50,000
Nickel-nickel sulfide	−10.5 to 9	−15,000 to 13,000
Nickel-phosphorus[b]	−10.8 to 35	−15,400 to 50,000
Palladium-25% nickel	20 to 40	28,500 to 57,000
Tin-35% nickel	−28 to 28	−40,000 to 40,000

[a] A negative sign denotes compressive stress.
[b] Alloys deposited in electroless plating baths.

Structural Features

Cracking occurs in highly stressed deposits, as noted previously. As a rule, these deposits exhibit a banded or lamellar structure, but not all banded structures show cracks. Bright deposits of copper, nickel, and some alloy electrodeposits are generally lamellar. High-strength deposits are fine-grained, whereas low-strength, soft deposits invariably have a coarse, columnar structure. For detailed information on such structures or textures, the individual chapters on specific metals and alloys should be consulted.

Crystal size, determined by X-ray or electron diffraction procedures, varies from $<$ 100 angstroms to at least 1000 angstroms (0.1 μm). Bright or fine-grained deposits consist of the smaller crystals. Examples include cobalt-phosphorus alloys, nickel, and nickel-phosphorus alloy. A minimum crystal size of 120 to 200 angstroms was reported for aluminum, crackfree chromium, and cobalt-zinc-phosphorus alloy (which was deposited in an electroless solution).

Electrodeposited alloys are not infrequently deposited as supersaturated solid solutions. Examples include cobalt alloys of copper, phosphorus or tungsten, iron alloys of manganese, phosphorus, tungsten or zinc, nickel alloys of silver or tungsten, and silver alloys of bismuth, cadmium, copper, lead or zinc. The phosphorus alloys of cobalt and nickel deposited in electroless baths also are deposited as supersaturated solid solutions. Tin-nickel and some tin-copper alloys have a unique, metastable crystal structure which transforms into two phases upon heat treatment. The superstaturated solutions also transform upon heat treatment into two-phase structures.

The crystal faces in many electrodeposits are oriented preferentially. Hexagonal cobalt crystals deposited in sulfate baths show a high degree of preferred orientation, but others deposited in chloride baths are oriented randomly. Preferred orientations were reported for cobalt-nickel-phosphorus and cobalt-tungsten alloys. Copper deposited in sulfate and fluoborate solutions are oriented with (110) planes parallel with the substrate surface, whereas copper deposited in pyrophosphate and high-temperature cyanide baths show (111) and (200) orientations, respectively. Preferred orientation has been reported for some iron and nickel deposits. Tin deposited in alkaline stannate solutions shows preferred orientation with (100) crystal planes parallel with the surface, whereas a (110) orientation was observed for tin from acid baths. The orientation of crystals in tin-nickel alloy depends on the nickel content. Crystal orientation of electroless nickel and cobalt-phosphorus alloys have also been reported. In the case of the cobalt alloy, the degree of orientation depended on the phosphorus content.

Chapter 2

Aluminum and Aluminum Alloys

Some aluminum deposits from organic electrolytes are similar in hardness and strength to annealed, wrought aluminum, but other deposits are much harder and stronger. The hardest and strongest deposits were obtained with periodically reversed current in aluminum chloride - lithium aluminum hydride - diethyl ether baths. Similar tensile strength and hardness values were reported for aluminum deposited with direct current in solutions containing anisole substituted for a part of the diethyl ether solvent. In each case, the high levels of strength and hardness are attributed to a refinement in grain size. Organic additives also refined the grain size and hardened the aluminum, but generally reduced its strength. Such deposits usually contained organic inclusions and were not stable to heating.

Much of the property data for electrodeposits were developed during research on electroforming, which was motivated by a need for lightweight, dimensionally precise radar waveguides, mirrors, and aerospace components. However, electrocladding of nuclear materials provided supplemental interest in the development of property data for electrodeposited aluminum. Relatively thick coatings (>125 μm or > 0.005 inch) were examined during these studies.

Data developed for aluminum deposited in aluminum chloride - lithium aluminum hydride - ethyl ether and mixed ether baths include strength, ductility, hardness, linear thermal expansion, and specular reflectivity. Hardness was also determined for aluminum deposited in other organoaluminum solutions. The tensile strength of aluminum reinforced with boron fibers was reported. These data are rather extensive in view of the limited commercial activity on electrodepositing aluminum.

Thin coatings, less than 25 μm, should be useful for the corrosion protection of steel. With its natural, self-healing oxide film, aluminum provides a durable protective coating for steel. Electrodeposition from organic solutions that must be protected from oxygen, carbon dioxide, and moisture is considered costly for most corrosion-protective applications. Lower cost fused-salt baths have been unsatisfactory for electrodepositing unalloyed aluminum in dense form. (1) However, the aluminum-manganese alloy deposited as fine-grained, dense and bright coatings in a 175 C fused-salt bath exhibits good corrosion-protective characteristics. (2,3) Hardness and electrical resistivity are increased by alloying the aluminum with manganese. (3)

Hardness and Density

The hardness and other properties of electrodeposited (2-11) and wrought aluminum (12) are compared in Table 2-1. Unalloyed electrodeposited aluminum is harder than annealed, wrought, commercially pure aluminum (Alloy 1100). Values for electrodeposits from organic solutions were as high as 138 kg/sq mm, which is five times greater than handbook data for the wrought metal. The harder aluminum deposits are associated with grain refinement induced with additives, (4) (14) or periodically reversed current. The hardness of aluminum alloy containing 20 percent manganese from fused-chloride salt baths (2) (3) reached a value of 450 kg/sq mm with significant grain refinement. (13)

High density, about 98 percent of that of wrought 1100 aluminum alloy, was reported for aluminum deposited in an aluminum chloride - lithium aluminum hydride - diethyl ether (Et_2O) solution. (8) The aluminum deposited in a similar bath that also contained ethoxyethyl trimethyl ammonium chloride was about 4 percent lower in density. (10)

Strength, Modulus, and Ductility

The tensile strength of unalloyed aluminum deposited in the Et_2O solution with direct-current plating was about equal to that of Aluminum Alloy 1100. With periodically reversed current, tensile strength was doubled to 17-22 kg/sq mm (23,500 to 31,200 psi). (8,14) With direct current and a mixed Et_2O-anisole bath, tensile strength reached 15 to 18 kg/sq mm (21,000 to 26,000 psi). (15)

Composite deposits containing nickel-coated boron fibers exhibited higher tensile strengths of 25 and 43 kg/sq mm (35,500 and 61,000 psi). Moduli of elasticity were 11.6 and 15.5 $\times$ 10^3 kg/sq mm, for aluminum with 15 and 25 volume percent boron-fiber content. An ultimate composite strength of 113 kg/sq mm, and a modulus of 19.7 $\times$ 10^{-3} kg/sq mm, were attained with a composite of 40 volume percent fibers, produced by hot pressing of several electroformed single layers.(14) (15)

Elongations as high as 26 percent (8) and 7 percent (14) were reported for aluminum deposited with direct and periodically reversed current, respectively. Organic additives to the aluminum chloride - lithium aluminum hydride - diethyl ether bath generally resulted in harder deposits with lower tensile strength and ductility. (14)

Stress and Thermal Expansion

Stress in unalloyed electrodeposited aluminum was considered negligible, about the limits of measurement readings.(8)

The coefficient of thermal expansion for electrodeposited aluminum, which was $24.5 \pm 1.8 \times 10^{-6}$/C ($-50$ to 93 C), (8) is slightly above the coefficient reported for 1100-0 aluminum, 22.2×10^{-6}/C (-50 to 20 C) and 23.6×10^{-6}/C (20 to 100 C). (12)

TABLE 2.1

Properties of Unalloyed Electrodeposited and Wrought Aluminum

Property	Electrodeposited Aluminum	Wrought Aluminum[a]	Reference Number
Hardness, kg/sq mm	36 to 55[b], 36 to 138[c], 70 to 100[d]	28	4–6, 8–14
Density, g/cc	2.66[b], 2.56[d]	2.71	8, 10, 12
Ultimate tensile strength,			
kg/sq mm	7.7 to 10.0[b], 16.3 to 21.7[e], 14.9 to 18.1[f]	9.15	8, 11, 12, 14, 15
psi × 10^3	11 to 14.4[b], 23.5 to 31.2[e], 21.1 to 25.7[f]	13	
Yield strength,			
kg/sq mm	5.4[b], 12.2 to 14.9[f]	3.45	8, 11, 12, 14, 15
psi × 10^3	7.8[b], 17.6 to 21.4[f]	5	
Elongation, percent	2 to 7[e], 5 to 10[g], 26[b], 4.7 to 10[f]	35	8, 11, 13, 14, 15
Modulus of elasticity,			
kg/sq mm × 10^3	5.6[b]	7	8
psi × 10^6	8.0[b]	10	
Stress,			
kg/sq mm	±0.056[b]	—	8
psi	±80[b]	—	
Electrical resistivity, microhm-cm	3.1 to 5.6[b] and 5.8[h]	2.9	3, 16
Linear coefficient of expansion, 10^{-6}/C	24.6 ± 1.7 (−50 to +93 C)[b]	22.2(−50 to +20 C), 23.6(+20 to 100 C)	8
Specular reflectivity, percent	88(0.3 to 0.9 micron range)[c]	—	8
	95(1.2 to 7 micron range)[b]	—	

[a] Type 1100-0, commercially-pure, annealed aluminum sheet. (12)
[b] Electroformed aluminum, $AlCl_3$-$LiAlH_4$-Et_2O bath. (5)(17)
[c] Electroformed aluminum, $AlCl_3$-ethyl pyridinium bromide-toluene bath. (4)
[d] Electroformed aluminum, $AlCl_3$-$LiAlH_4$-Et_2O-2 ethoxyethyl trimethyl ammonium chloride bath. (10)
[e] Electroformed aluminum, bath (b), using periodic reverse current plating. (14)
[f] Electroformed aluminum, $AlCl_3$-$LiAlH_4$-Et_2O-Anisole bath. (15)
[g] Electroformed aluminum with an iron content of about 600 ppm from bath (b). (11)
[h] Electrodeposited aluminum, $AlCl_3$-KCl-NaCl fused-salt bath. (2)(3)

Specular reflectivity for electroformed mirror surfaces of aluminum was 88 or 95 percent of that for an aluminum mirror standard in the respective wavelength ranges of 0.3–0.9 and 1.2–7.0 microns. (8)

Electrical Resistivity and Structure

The resistivity of electrodeposited 0.5 mil aluminum coatings from fused-chloride-salt baths increased with the codeposition of manganese from 5.8 microhm-cm for unalloyed deposits to 229 microhm-cm for the alloy containing 25 percent manganese. (3) The resistivity for the unalloyed aluminum deposits from the Et_20 solution ranged from 3.1 to 5.6 microhm-cm, (16) compared to 2.9 microhm-cm for commercially pure, wrought 1100 aluminum. (12)

The grain size of the unalloyed electrodeposits was 2700 Å, which was reduced to less than 500 Å by alloying with 4 to 25 percent manganese. (3) The smallest size of 196 Å for a 17 percent manganese alloy corresponds to the bright plating range for aluminum-manganese alloy deposits.

The coarse columnar structure characteristic of direct-current plating in organic baths was modified by periodically reversing the current. A fine grain structure and parallel banding, perpendicular to the direction of crystal growth, was observed in this case. These structural changes are considered stacking faults, in which the regular columnar-growth pattern of the crystal is disturbed.

Purity

The amount of secondary metal constituents in electrodeposited aluminum from the etherate bath (5,7,10) is less than that encountered in commercially pure Aluminum Alloy 1100, (12) as shown in Table 2.2. The lowest secondary-metal content was produced from the etherate bath, made up with reagent-grade chemicals, and operated with high-purity aluminum anodes. (7) Aluminum deposits from "more highly purified" baths were discontinuous. (11) (17) Therefore, a secondary metal appears to be necessary to produce sound thick aluminum electrodeposits (13) in the etherate bath. This secondary metal is possibly iron which is normally introduced as ferric chloride with the aluminum-chloride makeup salts. When 62 mg/l ferric chloride was added to an etherate bath contaminated with 50 ml/l water, good plating characteristics were restored. (11) This bath had not been restored by additions of hydrides, which ordinarily compensate for adverse effects of water.

The iron content of aluminum deposits produced from plating baths made up with commercial-grade aluminum chloride (which contains about 250 ppm of iron) and operated with commercially pure aluminum anodes (which contain up to 0.3 percent iron) was initially 1400 ppm (0.14%), and decreased to about 600 ppm during operation. (13) Sound aluminum deposits containing less than 10 ppm of iron were not reproducibly obtained from etherate baths. (7) (17) Sound aluminum

TABLE 2.2

Composition of Electrodeposited and Commercially-Pure Aluminum

Aluminum Type	Metal Constituents									
	Al[a] %	Fe %	Si %	Cu ppm	Ni ppm	Cr ppm	Ga ppm	Ti ppm	Mn ppm	Pb ppm
Wrought 1100	99.0+	<1	<1	<2000	<500	<500	<500	<500	<1000	<500
Electrodeposited										
$AlCl_3$-$LiAlH_4$-Et_2O bath[6–7]	99.99+	0.001	0.0005	0.5	10	10	10	10	20	<10
$AlCl_3$-$LiAlH_4$-Et_2O bath[11]	99.8+	0.14 to 0.06	(b)	100 to 10	(b)	(b)	(b)	(b)	(b)	(b)
$AlCl_3$-$LiAlH_4$-Et_2O-2-ethoxy-ethyl tri-methyl ammonium chloride bath[10]	99.9+	0.001	0.002	20	<10	<10	<10	<10	<3	<10
$AlCl_3$-ethylene pyridinum bromide-toluene-methyl-butylether bath[4]	99.8+	0.05 to 0.1	0.05 to 0.1	50 to 500	(b)	<50	(b)	(b)	(b)	100 to 1000
$AlCl_3$-$LiAlH_4$-Et_2O-Anisole bath[15]	99.96	<0.002	0.008	1.8	nil	nil	nil	nil	110	≪200

(a) Balance, including oxides which were not determined.
(b) Not determined.

deposits containing 20 to 110 ppm of manganese were reported from the etherate type baths. (6,7,15)

References

(1) Castle, A. W., "An Introduction to the Electrodeposition of Aluminum," *Electroplating and Metal Finishing, 7* (8), 291–294; 305 (1954).

(2) Austin, L. W., Vucich, M. G., and Smith, E. J., "The Effect of Manganese on the Brightness of Aluminum Coatings Electrodeposited from Fused Salt Baths," *Electrochemical Technology, 1* (9/10), 267–272 (1963).

(3) Read, H. J., and Shores, D. A., "Structural Characteristics of Some Electrodeposited Aluminum-Manganese Alloys," *Electrochemical Technology, 4* (11/12), 526–530 (1966).

(4) Safranek, W. H., Schickner, W. C., and Faust, C. L., "Electroforming Aluminum Waveguides Using Organo-Aluminum Plating Baths," *Journal Electrochem. Soc., 99* (2), 53–59 (1952).

(5) Couch, D. E., and Brenner, A., "A Hydride Bath for the Electrodeposition of Aluminum," *Journal Electrochem. Soc., 99* (6), 234–244 (1952).

(6) Brenner, A., "Electrodeposition of Metals from Organic Solutions. I. General Survey," *Journal Electrochem. Soc., 103* (12), 652–656 (1956); Connor, J.H., and Brenner, A., "Electrodeposition of Metals from Organic Solutions. II. Further Studies on the Electrodeposition of Aluminum from a Hydride Bath," *Ibid., 103* (12), 657–662 (1956).

(7) Beach, J. G., and Faust, C. L., "Electroclad Aluminum on Uranium," *Journal Electrochem. Soc., 106* (8), 654–659 (1959).

(8) Schmidt, F. J., and Hess, I. J., "Properties of Electroformed Aluminum," *Plating, 53* (2), 229–234 (1966).

(9) Buschow, A. G., and Esola, C. H., "Commercial Applications of the Aluminum Electrodeposition Process," *Plating, 55* (9), 931–935 (1968).

(10) Beach, J. G., McGraw, L. D., and Faust, C. L., "A Low-Volatility Nonflammable Bath for Electrodepositing Aluminum," *Plating, 55* (9), 936–940 (1968).

(11) Clay, F. A., Harding, W. B., and Stimetz, C. J., "Electrodeposition of Aluminum," *Plating, 56* (9), 1027–1037 (1969).

(12) *Alcoa Aluminum Handbook*, Aluminum Company of America, Pittsburgh (1967).

(13) Battelle's Columbus Laboratories, unpublished information.

(14) Withers, J. C., and Abrams, E. F., "The Electroforming of Composites," *Plating, 55* (6), 605–611 (1968).

(15) Lui, K., Guidotti, R., and Klein, M., "Research and Development of Magnesium/Aluminum Electroforming Process for Solar Concentrators," *Final Report, NASA CR-66427*, Electro-Optical Systems, Inc., under Contract No. NAS 1–6218 (May, 1967).

(16) Heritage, R. J., and Balmer, J. R., "Recent Developments in Aluminum Electrodeposition: A Critical Review," *Product Finishing (London), 10* (3), 54–58 (1957).

(17) Harding, W. B., Bendix Corporation, Kansas City Division, unpublished information.

Chapter 3

Cadmium and Cadmium Alloys

The major use for cadmium plating is for corrosion protection. (1) Thickness usually is in the same range as for zinc plate, i.e., 4 to >25 μm (0.15 mil to > 1 mil) depending on the application and environment. No data were found in the literature on mechanical properties, such as ultimate strength, yield strength or fatigue limit. Apparently, these properties are not significant in uses of interest for cadmium plate.

The characteristics of cyanide baths favor their use for producing dense fine-grained cadmium, with excellent covering power and good plate distribution over complex shapes. Cyanide baths are adaptable to automatic, still, and barrel operations.

Typical cadmium cyanide baths are covered by the following formulas:

	Still Plating	Barrel Plating
Cadmium oxide, CdO	23–35 g/l	17–23 g/l
Sodium cyanide, NaCN	90–120 g/l	70–90 g/l

The cadmium cyanide complex, $Na_2Cd(CN)_4$, and sodium cyanide (free sodium cyanide) to aid anode dissolution are the principal constitutents. Addition agents are used to produce grain refinement and brighten the deposits.

Acid baths are inferior in throwing power and covering power, compared to the alkaline cyanide solutions. Without brightener additives, the acid cadmium deposits tend to be coarsely crystalline and porous. However, cadmium fluoborate baths containing proprietary addition agents are useful for avoiding or minimizing hydrogen embrittlement of steel. Fluoborate solutions provide fine-grained deposits that are considered equal to those obtained from alkaline cyanide baths in protective quality and appearance.

Formulation of a typical cadmium fluoborate both is as follows:

Cadmium fluoborate, $Cd(BF_4)_2$	240 g/l
Boric acid, H_3BO_3	24 g/l
Ammonium fluoborate, NH_4BF_4	60 g/l
Licorice	1 g/l
pH (with fluoboric acid, HBF_4)	3.0–3.5

Although alloys of cadmium with tin (1 to 99 percent), (2,3) zinc (4 to 75 percent), (4) iron (1 to 93 percent), (5) nickel (5 to 23 percent), (6) indium (35 to 65 percent), (7) or copper (8) have been electrodeposited, applications are yet to be developed.

Properties of Electrodeposited Cadmium

Properties of electrodeposited cadmium are compared in Table 3.1 with the properties of electrodeposited zinc and those of the metallurgically prepared wrought metals.

The properties of electrodeposited cadmium are similar to those of wrought cadmium. Hardness values are generally higher for bright, fine-grained deposits than for dull mat deposits (45 to 49 versus 30 to 39 kg/sq mm). Bright cadmium deposits from a cyanide solution with additions of α-naphthylamine sulfate (to saturation) and a wetting agent showed hardness values of 45 to 50 kg/sq mm. (12) A significant increase in hardness was noted for cadmium or zinc deposits produced in a cyanide bath with the use of ultrasonic agitation in place of air agitation: 36 to 84 KHN_{100} for cadmium and 50 to 103 KHN_{100} for zinc. (13)

Stress for cadmium deposits produced in cyanide baths without brighteners or alloying additives is usually compressive (-0.35 to -2.1 kg/sq mm). The internal stress in cadmium deposits (8 and 14 μm thick) from sulfate solutions with additions of naphthalene disulfonic acid and gelation was relatively low (0.5 to 1.5 kg/sq mm) in tension. (18) Brighteners or metallic additives codeposited with cadmium affect hardness, stress, electrical resistance, and structure (normally hexagonal). For example, cadmium-indium alloy deposited from sulfate-sulfamate solutions exhibited tensile stresses of 1.6 to 2.1 kg/sq mm, and microhardness values of 126 to 164 kg/sq mm. (7)

The coefficients of thermal expansion reported (Table 3.1) for electrodeposited cadmium and zinc are lower than those for the polycrystalline wrought form of the metals. The coefficients for the electrodeposited metals are closer to the coefficients measured along the alpha axis of the hexagonal structures, 16.6 versus 18.5 $\times$ 10^{-6}/C for cadmium and 22 versus 15 $\times$ 10^{-6}/C for zinc.

Some electrical resistivity values for electrodeposited cadmium are similar to those for the wrought polycrystalline metal. (19) The higher values reported for electrodeposited cadmium are probably due to differences in structure, intercrystalline continuity, and/or purity of the metal forms.

Solution formulations or operating conditions were not detailed for the deposits with the properties summarized in Table 3.1. In most cases, the electrolyte simply was acknowledged as a cyanide solution.

Cadmium-zinc alloy electrodeposits from cyanide solutions are supersaturated solid solutions with 3 or 4 percent zinc. A face-centered cubic structure with 21 to 38 percent zinc was observed in comparison to the hexagonal form for unalloyed cadmium and zinc electrodeposits. (4) Microhardness of the alloys reached an initial maximum of 42 kg/sq mm with 4 percent zinc and a larger maximum of 110 kg/sq mm at 90 percent zinc, compared to 30 and 90 kg/sq mm, respectively, for the unalloyed cadmium and zinc deposits. Specific resistance increased to 11.5 mi-

TABLE 3.1

Properties of Unalloyed Cadmium and Zinc, Electrodeposited and Metallurgically Prepared

Property	Cadmium		Zinc		Reference
	Electrodeposited	Wrought	Electrodeposited	Wrought	
Hardness, kg/sq mm					2, 4, 9, 10, 11, 12, 13
Dull deposits	30 to 39		35 to 90		
Bright deposits	45 to 50		132 to 148		
Stress[a]					14, 15, 16
psi	−500 to −3,000		−7,000 to +12,900		
kg/sq mm	−0.35 to −2.1		−4.9 to +9.0		
Specific resistance, 20 C, microhm-cm	6.8 to 8.5	6.8[b]	5.9 to 7.5	5.9[b]	4, 17, 18
Coefficient of thermal expansion, 20 to 100 C, 10^{-6}/C	16.6	29.8[b]	22	39.7[b]	17
Ultimate tensile strength,					14, 15
psi	10,000	10,300[c]	7,000 to 18,400		
kg/sq mm	7.0	7.2	4.4 to 12.8		
Elongation, percent	No data	50[c]	Up to 50		
Modulus of elasticity,					
psi	8×10^6	8×10^6 [c]	14×10^6		14
kg/sq mm	5,600	5,600	9,800		15

[a] A negative sign designates compressive stresses.
[b] Polycrystalline metals.
[c] Chill cast sections.

crohm-cm with 4 percent zinc from 8.5 microhm-cm for unalloyed cadmium. With higher zinc content, the specific resistance decreased and approached that of unalloyed zinc (7.5 microhm-cm).

Cadmium-tin alloys (stannate-cyanide solutions) exhibited a tetragonal-hexagonal structure and specific resistivities up to 50 percent higher than their metallurgically prepared counterparts. (2,3) Microhardness decreased from 30 to 13 kg/sq mm with increasing tin content, approaching that of unalloyed tin.

Cadmium-iron alloys (sulfate solutions) showed evidence of a mechanical mixture of the two metals and a body-centered-cubic structure. (5) Cadmium-nickel alloy with 23 percent nickel (sulfamate solution), showed a microhardness of 341 kg/sq mm, about ten times that of unalloyed cadmium. (6) Copper-cadmium alloys (ammonium sulfate—ethylene diamine solutions) were bright, fine-grain silvery deposits with microhardness readings of 330 to 360 kg/sq mm, more than ten times that of unalloyed cadmium from cyanide solutions and more than twice that of unalloyed copper deposits from this type of bath (150 to 160 kg/sq mm). (8)

Hydrogen Embrittlement of Steel

Corrosion protection by cadmium plating, without embrittlement of high-strength substrates, is a matter of major concern for coated structural components. High-strength alloys (tensile strength of 280,000 psi or 200 kg/sq cm, or more) are extremely susceptible to hydrogen embrittlement. Transition metals as the alloying constituents are proved to adsorb the atomic hydrogen discharged during plating. Hydrogen enters the ferritic lattice structure of the steel by means of hydride formation. Atomic hydrogen enters any dislocations and grain boundaries, and forms molecular hydrogen in any cracks and holes. The hydrogen produced at cathode surfaces is in the atomic form because the high cathode overvoltage inhibits its association to form molecular hydrogen. (20)

Baking at 160-230 C for 24 hours after plating is a common practice for reducing hydrogen embrittlement. Hydrogen seems to be easier to remove from steel during baking when dull, relatively coarse-grained cadmium is deposited in cyanide solutions containing no brightener. (21,22) Good ductility is not always restored by baking steel plated with fine-grained, bright cadmium. Baking relieved embrittlement of cadmium-plated SAE 4340 steel of 160 kg/sq cm (230 kpsi) tensile strength, but, no amount of baking was sufficient to enable full recovery of cadmium-plated SAE 4340 steelbars of 200 kg/sq cm (280 kpsi) tensile strength. (23) The damage caused by hydrogen in certain high-strength alloys seems irreversible at temperatures below the melting point of cadmium (321 C).

By cadmium plating in acid-type baths, sulfamate, fluoborate, chloride or perchlorate solutions, the amount of embrittlement is decreased. (24) This is generally attributed to near 100 percent current efficiency for deposition of cadmium from acid baths and little or no hydrogen ion reduction at the cathode. But the coverage of such cadmium deposits from these acid baths is inferior to those of cyanide cadmium.

Corrosion Protection of Steel

Protection by sacrificial action of cadmium coatings (or zinc) on iron, steel, copper, and brass generally prevents corrosion of the substrate metals at exposed areas through scratches or pores in the coating. Because of the high cost of cadmium, the thickness of the deposit is usually less than 25 microns (0.001 inch or 1 mil). Generally, it is applied when a thickness of about 13 microns will suffice.

For equivalent thicknesses, cadmium coatings on steel are considered superior to zinc coatings for use in marine atmospheres, but are considered inferior for use in outdoor industrial atmospheres. Cadmium is not readily attacked by caustic, as is zinc. Cadmium is less susceptible than zinc to "white-powder" corrosion resulting from water vapor condensate.

Cadmium is more readily plated from cyanide baths on malleable and cast iron than is zinc from cyanide baths. Because of its low electrical contact resistance, ease of soldering, resistance to alkali, and self-lubricating properties, in addition to resistance to salt corrosion, cadmium is used rather widely in the electrical and electronic industries. (1)

References

(1) Soderberg, K. G., and Westbrook, L. R., *Modern Electroplating*, ed. 2, edited by F. A. Lowenheim, John Wiley and Sons, Inc., Chapter 4, 64–79 (1963).

(2) Fedot'ev, N. P., Vyacheslavov, P. M., and Andreeva, G. P., "Structure and Properties of Electrodeposited Sn-Cd Alloys," *Journal Applied Chemistry (USSR), 35* (7), 1479–1482 (1962).

(3) Davies, A. E., "The Electrodeposition of Tin-Cadmium Alloys from Fluoride-Fluosilicate Solutions," *Trans. Inst. Metal Finish., 33*, 74–83 (1956). "The Electrodeposition of Tin-Cadmium Alloys from Stannate-Cyanide Solutions," *Ibid., 33*, 85–99 (1956).

(4) Fedot'ev, N. P., and Vyacheslavov, P. M., "The Phase Structure of Binary Alloys Produced by Electrodeposition," *Plating, 57* (7), 700–706 (1970).

(5) Misra, S. S., and Rama Char, T. L., "Simultaneous Deposition of Cadmium and Iron from a Sulfamate Bath," *Metalloberflaeche, 18* (6), 164–168 (1964).

(6) Kudryavtsev, N. T., Tyutina, K. M., Chuan, N. L., and Bogdanovskaya, V. A., "Electrodeposition of a Cadmium-Nickel Alloy from a Sulfamate Electrolyte," *Zashchita Metallov, 4* (6), 707–711 (1968).

(7) Sklyarenko, S. I., Lavrov, I. I., and Mansurova, L. A., "Formation of Electrolytic Coatings of Binary Alloys of Indium with Zinc and Cadmium from Aqueous Sulfamic Acid Solutions," *Journal Applied Chemistry (USSR), 35* (9), 1935–1938 (1962).

(8) Belyakova, L. A., and Gudin, N. V., "Deposition of Bright Coatings of Copper and Its Alloys from Ethylenediamine Electrolytes," *Theory and Practice of Bright Electroplating*, Published by the Israel Program for Scientific Translations for the U.S. Dept. of Commerce and the National Science Foundation, Jerusalem, 161–169 (1965).

(9) Weiner, R., "Mechanical Properties of Several Bright Silver Deposits," *Galvanotechnik, 57* (1), 15–20 (1966).

(10) Weiner, R., and Klein, G., "Examination of the Cathodic Polarization of Bright Plating Solutions," *Metalloberflaeche, 10* (1), 17–21 (1956).

(11) Weiner, R., and Klein, G., "Hardness and Wear Resistance of Electrodeposits," *Metalloberflaeche, 7B* (1), 1–7 (1955).

(12) Sivickis, J., and Visomirskis, R., "Bright Cadmium Plating from Cyanide Electrolytes," *Blestyashchie Komb. Metal. Pokrytiya, No. 2*, 76–82 (1967).

(13) Lanyi, R. J., Lane, D. H., Forbes, C. A., and Ricks, H. E., "The Application of Ultrasonic Energy to Metal Processing," *SAE Trans., 71*, 520–540, 562 (1963).

(14). Such, T. E., "The Physical Properties of Electrodeposited Metals," *Metallurgia, 56* (334), 61–66 (1957).

(15) Newell, I. L., "Stress in Electrodeposited Metals," *Metals Finishing, 58* (10), 56–61 (1960).

(16) Kushner, J. B., "Stress in Electroplated Metals," *Metal Progress, 81* (2), 88–93 (1962).

(17) Meckelburg, E., "Electrolytic Zinc and Cadmium Coatings," *Metall, 14* (8), 798–800 (1960).

(18) Emel'yanenko, G. A., Tat'yanina, S. P., and Vlasov, E. A., "Cadmium Plating from a Sulfate Solution with Additions of Naphthalene Disulfonic Acid and Gelatin," *Zh. Prikl. Khimii, 37* (3), 688–691 (1964).

(19) McCutcheon, F. G., and Musgrave, J. R., *Rare Metals Handbook*, edited by C. A. Hampel, Reinhold Publishing Company, New York, 93 (1954).

(20) Lorcery, P., "Influence of Current Density on the Embrittling Hydrogenation of a High Strength Steel Cadmium Plated in an Alkaline Bath," *Metaux (Corrosion Ind.), 41* (491–92), 261–276; (493), 325–346; (494), 349–390 (1966).

(21) Geyer, N. M., Lawless, G. W., and Cohen, B., "A New Look at the Hydrogen Embrittlement of Cadmium Coated High-Strength Steels," *Tech. Proceedings Am. Electroplaters' Soc., 47*, 143–151 (1960).

(22) Kozan, T. G., "Evaluation of Cadmium Plating Systems for Embrittlement," *Report No. D3-4657, Aerospace Research and Testing Committee Project No. 6-61*, Boeing Company (September 19, 1962).

(23) Brown, B. F., *Naval Research Laboratory Report 4839* (October 1956).

(24) Chilton, J. E., *WDAC Technical Report 57-514, ASTIA Document No. AD 142316 (January 1958).*

Chapter 4

Chromium and Chromium Alloys

Chromium electroplate resists tarnishing and oxidation at temperatures above 300 C. A thin, self-healing oxide film is responsible for this useful property. Deposits ranging in thickness from 0.15 to 0.75 μm (0.005 to 0.025 mil) for preserving a decorative appearance are porous. Thicker coatings often are cracked. Yet chromium with a thickness of 0.05 to 5.0 mil is used for many applications to provide good resistance to wear, abrasion, and friction. In these applications, chromium is deposited directly on aluminum, steel, and many other substrates. Its usefulness on soft substrates is limited because of its poor ductility, however.

Retention of hardness after heating to a temperature as high as 400 or 500 C is another useful property for specific applications. Corrosion resistance in oxidizing atmospheres is excellent, but chromium corrodes rapidly in acid chloride solutions. The crack-free chromium deposited in chromic acid solutions operated at 85 C or with pulsating current (1) provides better protection of basis metals than other deposits.

Impurities, particularly chromic oxide, have an appreciable effect on the properties of electrodeposited chromium. The oxide content is influenced by the electrodeposition conditions. For example, high solution temperatures of 75 or 85 C favor a relatively low oxide content and dense, strong deposits with a relatively low specific resistivity, whereas deposits obtained at low operating temperatures (20-40 C) have a high oxide content and resistivity and a low density.

Chromic acid solutions are usually adopted for chromium plating. Such solutions also contain fluosilicic and/or sulfuric acid. Sodium or calcium hydroxide is added to some solutions to form a tetrachromate salt. Property data for chromium deposited in these solutions are fairly complete. On the other hand, data for chromium deposited in trivalent chromium baths are relatively meager. Yet the trivalent chromium solutions have been the basis for depositing chromium alloys. Some property data for these alloys have appeared in the literature since 1955.

Physical Properties and Purity

The melting point of electrodeposited chromium has been reported as 1890 $\pm$ 10 C (2) and as 1920 C. (3) In comparison, electrolytic chromium ductilized by prolonged heating in hydrogen at 1200 C has a melting point of 1875 C. This high-

TABLE 4.1

Density and Electrical Resistivity Data for Electrodeposited Chromium

Density, g/cu cm		Electrical Resistivity, microhm-cm		Oxygen Content, percent	Plating Bath and Temperature, C		Current Density, amp/sq dm	Reference
Before Annealing	After Annealing[a]	Before Annealing	After Annealing					
6.90	7.12	67	14.3	0.87	(b)	30	5	4
6.94	7.13	62	13.8	0.88	(b)	40	15	4
7.05	7.15	48	13.5	0.40	(b)	50	20	4
7.11	7.17	36	16.1	0.25	(b)	65	40	4
7.17	7.19	21	13.2	0.10	(b)	75	60	4
7.17	7.19	15	13.0	0.05	(b)	85	80	4
7.04	7.16	14	12.9	—	(b)	100	80	4
7.06	7.17	34.5	13.8	—	(b)	50	5	4
7.05	7.15	47.5	13.5	0.40	(b)	50	20	4

7.01	7.14	51.8	13.6	—	(b)	50	30	4
7.01	7.14	54.8	13.6	—	(b)	50	40	4
6.99	7.13	61.6	13.6	0.35	(b)	50	80	4
6.90	—	—	—	0.80	(b)	30	5	5
7.17	—	—	—	0.05	(b)	85	80	5
7.13	7.18	—	—	—	(b)	84	75	6
7.01	—	47	—	0.40	(c)	50	20	4
7.18	7.18	22	13	0.04	(c)	85	20	4
7.16	7.19	—	—	0.11	(c)	85	80	4
6.98	7.14	58	14.1	0.51	(d)	50	30	4
7.14	7.20	29	15.0	0.24	(d)	85	80	4
7.18	7.21	—	—	—	(e)	85	20	4
6.97	7.19	—	—	0.54	(f)	—	—	4
7.14	—	—	—	—	(g)	780	—	7

(a) Heat treated at 1200 C.
(b) 2.5 N CrO_3 with 2.5 g/l H_2SO_4.
(c) 0.5 N CrO_3 with 0.5 g/l H_2SO_4.
(d) 2.5 N CrO_3 with 6 g/l F_2.
(e) 2.5 N CrO_3 with 1.2 g/l F_2.
(f) Trivalent chromium bath.
(g) Fused salt bath containing fluorides.

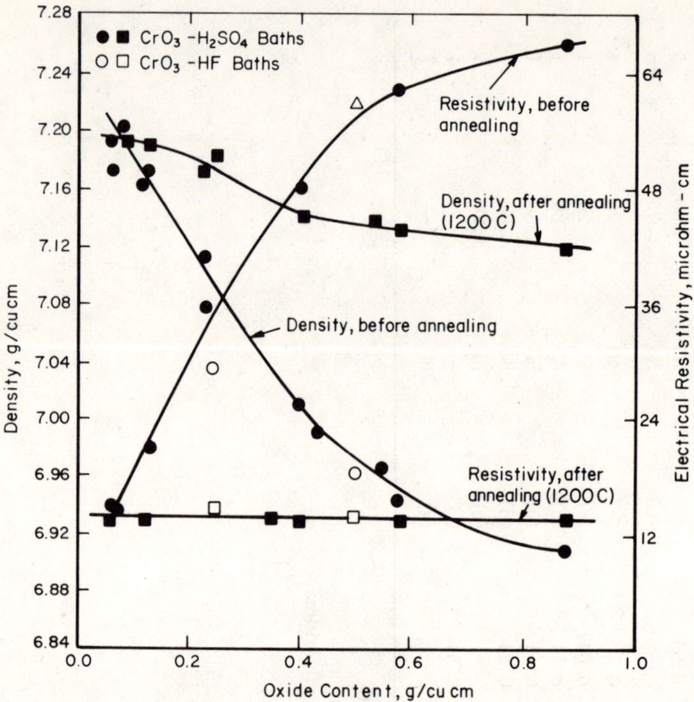

Figure 4.1. Density and Electrical Resistivity of Chromium Electrodeposits as a Function of Oxide Content. (4)

purity chromium contained 0.002 percent oxygen, 0.016 percent nitrogen, 0.00008 percent hydrogen, and 0.000005 percent metallic impurities.

Density values range from 6.9 to 7.2 g/cu cm, depending on the current density and temperature of the electroplating bath, as detailed in Table 4.1. High bath temperatures of 75 or 85 C favored high densities of 7.13 to 7.18 g/cu cm, which compares with a calculated density of 7.20 g/cu cm for body-centered-cubic chromium with a cell parameter of 2.878 angstroms. (4) On the other hand, low densities of 6.90 to 6.94 g/cu cm were reported for chromium deposited in 30 or 40 C solutions. Operation at 50 or 65 C produced chromium with a density of 6.99 to 7.11 g/cu cm.

An increase in the current density slightly reduced the density of typical deposits. Chromium deposited in solutions containing hydrofluoric acid or potassium fluoride instead of sulfuric acid exhibited about the same density as chromium deposited in chromic-sulfuric solutions, when the temperature was identical.

High-density chromium contained 0.03 to 0.05 percent oxygen, whereas the amount of oxygen found by vacuum fusion increased inversely with density (Figure 4-1), up to 0.88 percent. (4) The oxygen in chromium probably exists as hydrous chromic oxide, $Cr_2O_3.nH_2O$, in view of the identification of chromic oxide in films of residue remaining after electrolytic dissolution of the metal. (8)

By employing a high deposition temperature of 90 C and solution flow through a cation-exchange resin, the oxygen content of a 72-hour deposit obtained in a 2.5 N chromic oxide bath was reduced to 0.004 percent. (9) This low oxygen level was not duplicated in other experiments with similar conditions. However, 0.004 percent oxygen also was reported for chromium deposited in an 86 C, 3.9 N chromic acid bath containing 7.4 g/l sulfuric acid. (10) High-purity deposits containing 0.015 to 0.02 percent oxygen, 0.0005 to 0.001 percent hydrogen, and less than 0.0002 percent nitrogen were reported for deposits from conventional solutions purified with ion-exchange resins. (11) Nitrogen and oxygen contents of 0.002 and 0.03 percent, respectively, were reported for chromium deposited in chromic-hydrofluoric acid baths at 100 to 103 C, in comparison with 0.0015 and 0.012 percent nitrogen and oxygen in chromic-sulfuric acid solutions. (12) The addition of a nitrate salt (1 g/l) increased the nitrogen content from 0.002 to 0.1 percent in deposits produced in a chromic-sulfuric-hydrofluoric acid bath. (13) Adding ammonium salts or agitating the bath with nitrogen did not increase the nitrogen content of the deposits.

The hydrogen content of electroplated chromium, which usually ranges from 0.01 to 0.07 percent, is also inversely proportional to density. The average hydrogen content of about 20 specimens was 1/8 of the average oxygen content, which supports the postulation that hydrogen exists in the form of water of hydration associated with chromic oxide inclusions. (4)

Electrical resistivity ranged from 14 microhm-cm for a deposit from a chromic acid bath heated to 100 C to a maximum of 67 microhm-cm for a deposit from a 30 C solution (Table 4.1). (4) Resistivity was directly proportional to the oxide content, as shown in Figure 4.1, and inversely proportional to density. Crystals of high-purity chromium deposited in a fused salt bath which contained < 0.01 percent oxygen exhibited a resistivity of only 3.6 microhm-cm. (15).

Heat treating electroformed chromium obtained in chromic acid solutions operated at 30 C or higher increased density and reduced electrical resistivity appreciably. After annealing at 1200 C for an hour, the density of electroformed chromium was at least 7.12 g/cu cm and frequently as high as 7.15 to 7.19 g/cu cm, (4) as indicated in Table 4.1. Deposits produced in 85 C solutions, which contained a

TABLE 4.2

Contraction of Chromium After Heating

Change in Length, percent	Heat Treating Temperature, C	Plating Bath Composition	Bath Temperature, C	Current Density, amp/sq dm	Reference
1.0	700	Standard chromic acid[a]	About 50	30–40	16
1.1	1050	Ditto	48	30	6
0.8	1200	..	50	20	17
0.2	700	..	85	40–120	16
0.2	1050	..	85	75	6
0.1	1200	..	85	80	17

[a] 250 g/l CrO_3 and 2.5 g/l H_2SO_4.

maximum of about 0.1 percent oxygen, exhibited a density of 7.19 or 7.20 g/cu cm after heat treating. Annealing at 1200 C reduced resistivity to the range of 12.9 to 14.1 microhm-cm, with a few exceptions. A value of 13 microhm-cm was reported elsewhere for electroformed chromium vacuum annealed at 1000 C. (14) The increase in density and decrease in electrical resistivity observed after heat treating is attributed to the healing (sintering) of cracks in the deposit. Chromium contracted during heating. The amount of contraction varied with the temperature of heat treatment and the conditions for depositing the chromium, as detailed in Table 4.2.

The coefficient of thermal expansion derived from expansion data from reheating a chromium electrodeposit varied with the temperature range, as follows: (6)

Temperature Range, C	Coefficient of Linear Expansion, micrometer/meter/C
20–200	7.4
20–400	8.4
20–600	9.2
20–800	9.8
20–1050	11.0

The low coefficient of expansion coupled with the brittleness of conventional chromium is responsible for the formation of expanded cracks visible on the surface when a substrate with a higher coefficient is plated with chromium and heated.

The thermal conductivity of a soft, high-density chromium deposited in a 1550 C, fused salt bath was reported to be 0.16 cal/sq cm/cm/sec/C. (7) Specific heat was 0.107 cal/g/C.

Strength, Modulus of Elasticity, and Ductility

Bright chromium with cracks deposited in 2.5 N chromic acid baths at 30 to 50 C exhibited very low tensile strengths ranging from 10 to 13 kg/sq mm (15,000 to 19,000 psi), as shown in Table 4.3. (4) A low tensile strength also was observed for chromium deposited at 40 amp/sq dm in an 85 C solution, but chromium deposited at 80 or 120 amp/sq dm in an 85 C bath exhibited much higher strengths of 49 and 56 kg/sq mm (70,000 and 80,000 psi). Heat treating at 1200 C raised the tensile strength of all deposits appreciably, as noted in Table 4.3. All strength data relate to electroformed chromium separated from the substrate.

A tensile strength of >140 kg/sq mm (>200,000 psi) was claimed for chromium deposited in a divalent chromium bath prepared with chromous fluoride, sodium fluoride, formic acid, and sodium formate. (19) The bath temperature was > 50 C and its pH was in the range of 1.7 to 4.5. The deposits were reported to be ductile.

A low modulus of elasticity ranging from 10,000 to 14,000 kg/sq mm was reported for chromium deposited in 2.5 N chromic acid baths at 30 to 50 C. (4) Chromium deposited in high-temperature (85 to 100 C) solutions exhibited a modulus of 21,000 to 23,000 kg/sq mm. Heat treating at 1200 C increased the

TABLE 4.3

Tensile Strength and Modulus of Elasticity of Electrodeposited Chromium[4]

Tensile Strength				Modulus of Elasticity						
Before Annealing		After Annealing[a]		Before Annealing		After Annealing[a]				
kg/sq mm	psi	kg/sq mm	psi	kg/sq mm	psi $\times 10^6$	kg/sq mm	psi $\times 10^6$	Plating Bath	Temperature, C	Current Density, amp/sq dm
13	19,000	—	—	13,000	19	—	—	(b)	30	5
11	16,000	—	—	10,000	15	—	—	(b)	30	10
12	17,000	—	—	11,000	16	—	—	(b)	40	15
10	15,000	57	81,000	11,000	16	23,000	33	(b)	50	20
12[c]	17,000	—	—	—[c]	—	—	—	(b)	50	40
13	18,000	—	—	—	—	—	—	(b)	85	40
20	28,000	—	—	22,000	31	—	—	(d)	85	80
25	35,000	43	61,000	22,000	31	—	—	(e)	100	80
21	30,000	62	88,000	14,000	20	26,000	37	(f)	50	30
22	31,000	—	—	21,000	30	—	—	(f)	85	20
34	48,000	—	—	22,000	31	—	—	(g)	85	20
49	70,000	—	—	23,000	33	25,000	35	(b)	85	80
56	80,000	70	100,000	—	—	—	—	(b)	85	120

[a] Heat treated 1 hour at 1200 C.
[b] 2.5 N CrO_3 with 2.5 g/l H_2SO_4.
[c] Chromium deposited at 40 amp/sq dm in a 57 C bath exhibited a tensile strength of 15 kg/sq mm and a modulus of 16,400 kg/sq mm.[18] Heating at 300 C reduced the modulus to 13,000 kg/sq mm.
[d] 2.5 N CrO_3 with 5 g/l H_2SO_4.
[e] 0.5 N CrO_3 with 0.5 g/l H_2SO_4.
[f] 2.5 N CrO_3 with 6 g/l HF.
[g] 2.5 N CrO_3 with 1.2 g/l HF.

modulus of chromium deposited in high- and low-temperature baths to the range of 23,000 to 26,000 kg/sq mm. Chromium deposits heated in vacuum to 1000 C exhibited a modulus of 19,800 kg/sq mm. (14) The modulus increased to 21,000 kg/sq mm at −150 C or 20,500 kg/sq mm at 150 C. Heat treatment at 450 C reduced the modulus of hard chromium deposited in a low-temperature tetrachromate solution from 16,000 to only 3,200 kg/sq mm. (20) By heating at 300 C, the modulus of another chromium deposit was reduced from 16,400 to about 13,00 kg/sq mm. (18)

Chromium deposited in chromic acid solutions exhibits no ductility. From data on tensile strength and modulus of elasticity, an elongation of 0.1 percent was calculated. (4) Elongations from 0 to 0.12 percent were measured for typical hard chromium deposits. (21) Annealing for an hour at 1200 C had no effect on ductility. (4) However, heat treatment for 4 to 16 hours at 1600 to 1650 C was reported to ductilize chromium deposited in a chromic-hydrofluoric acid bath. (22) An elongation of 17 percent and a tensile strength of 20 kg/sq mm (28,600 psi) were obtained for this heat-treated chromium. Measured at 650 C, an elongation of 30 percent and a tensile strength of 10 kg/sq mm (14,500 psi) were reported. (22) At 850 C, elongation increased to 45 percent, whereas tensile strength decreased to about 8 kg/sq mm (11,000 psi). At a low stress level of 2.1 kg/sq mm, an elongation of 35 percent was measured at 980 C during stress-rupture tests on electroformed chromium. (6) At a higher load of 4.2 kg/sq mm, the elongation was only 1.9 percent.

Hardness and Wear Resistance

Hardness values from 126 kg/sq mm for chromium deposited in a high-temperature, fused salt bath (15) to >1200 kg/sq mm have been reported in the voluminous literature on this property of electrodeposited chromium. Deposits with a hardness of > 900 kg/sq mm were reported for chromium from chromic-sulfuric and chromic-sulfuric-fluosilicic acid baths maintained at 30 to 50 C (Tables 4.4 and 4.5). On the other hand, chromium deposited in tetrachromate solutions prepared with chromic acid and sodium hydroxide usually exhibited a hardness of < 700 kg/sq mm (Table 4.6).

The indentation load to derive the numbers listed in Tables 4.4, 4.5, and 4.6 was not always reported. Hence, some variations in hardness may be due to differences in the applied load, from 100 to 500 grams, which varied for a single deposit, as follows: (43)

Load, grams	Hardness, kg/sq mm
100	1150
200	1075
500	900

Hardness values greater than about 1200 kg/sq mm can be questioned because the deposits may have been too thin to avoid the influence of the hardness of the

TABLE 4.4

Hardness of Chromium Deposited in Chromic-Sulfuric Acid Solutions

Hardness, kg/sq mm	CrO_3 Concentration, N[a]	Bath Temperature, C	Current Density, amp/sq dm	Reference
900	0.5	50	20	4
325	0.5	85	20	4
500	0.5	85	80	4
310	0.5	100	80	4
985	2.5	30	5	4
905	2.5	40	15	4
920	2.5	50	20	4
875	2.5	50	80	4
975	2.5	65	40	4
325	2.5	85	20	4
475	2.5	85	60	4
550	2.5	85	120	4
670	2.5	100	80	4
1050	2.5	45	10	23
850	2.5	45	50	23
450	2.5	65	10	23
1000	2.5	65	50	23
1100[b]	2.5	50	~30	24
1082[c]	2.5	52	~50	25
675	2.5	50	30	26
925	2.5	50	50	26
275	2.5	80	30	26
325	2.5	80	50	26
1206	1.5	35	10	27, 28
1186	1.5	55	30	27, 28
1138	2.5	35	20	27, 28
1176	2.5	45	60	27, 28
750[d]	2.5	50	75–100	29
444	2.5[e]	85	20	30
528	2.5[e]	85	40	30
1500–1600[f]	2.5[g]	50	25–60	31
1250	2.5[h]	50	15	32

[a] The ratio of CrO_3 and H_2SO_4 was usually 100:1.

[b] Superimposing alternating current reduced hardness to 630 kg/sq mm.

[c] Adding 0.5 g/l In_2SO_4 increased hardness to 1218 kg/sq mm.

[d] Hardness was increased with ultrasonic agitation to 910 kg/sq mm.

[e] The solution also contained 25 g/l Fe.

[f] Hardness decreased to the range of 1000 to 1300 kg/sq mm after aging at room temperature for 30–40 days.

[g] The solution also contained 5 to 10 g/l Cr_2O_3 and 0.5 g/l powdered opium.

[h] The solution also contained 1 g/l $H_7P(W_2O_7)_6 \cdot 22H_2O$.

TABLE 4.5

Hardness of Chromium Deposited in Chromic-Fluosilicic-Sulfuric Acid Solutions

Hardness, kg/sq mm	CrO_3 Concentration, N	K_2SiF_6 Concentration, N	Bath Temperature, C	Cathode Current Density, amp/sq dm	Reference
1500	2.5	0.08[a]	50	15	33
1200	2.5	0.08[a]	50	45	33
1350	2.5	0.08[a]	50	60–75	34
900	2.5	0.09[a]	30	60	35
1000	2.5	0.09[a]	55	60	35
700	2.5	0.09[a]	80	60	35
1040	2.8	0.09[a]	45	50	36
1060	2.8	0.09[a]	45	100	36
970	2.8	0.09[a]	65	100	36
740	2.8	0.09[a]	75	50	36
840	2.8	0.09[a]	75	100	36
800	2.5	0.09[a]	25	30	37
800	2.5	0.09[a]	25	100	37
820	2.5	0.09[a]	35	30	37
820	2.5	0.09[a]	35	100	37
550	2.5	0.09[a]	55	30	37
850	2.5	0.09[a]	55	100	37
250	2.5	0.09[a]	80	30	37
500	2.5	0.09[a]	80	100	37
1120	5.4	0.04	50	19	38
1110	5.4	0.04	50	56	38
836	5.4	0.13 KF	86	37	38
800	5.4	0.13 KF	86	74	38
900	2.5	0.3 HF	50	30	4
470	2.5	0.3 HF	85	20	4
800	2.5	0.3 HF	85	80	4

[a] $SrCrO_4$ was added to regulate the concentration of soluble SO_4^{2-} ions.

substrate, when indentations were made on polished cross sections. For example, 1275 kg/sq mm was reported for 5 μm of chromium deposited directly on zinc, in comparison with 840 kg/sq mm for 20 μm of chromium on zinc. (44)

Although comparison of data from two or more different reports are subject to the questions raised in the preceding paragraphs, the influence of some operating conditions on hardness can be judged validly because the same trends have been noted by several investigators. For example, an increase in the temperature of the plating bath consistently reduced hardness. However, change in the temperature of chromic-sulfuric acid baths (Table 4.4) has a greater effect than change in the

TABLE 4.6

Hardness of Chromium Deposited in Tetrachromate Solutions

Hardness, kg/sq mm	CrO_3 Concentration, N	Other Constituents, g/l	Bath Temperature, C	Cathode Current Density, amp/sq dm	Reference
350–600	3.8	60 NaOH, 10 Cr_2O_3, 2.5 $MgSO_4$, 1 H_2SO_4	18–25	—	39
360	4.0	20 NaOH, 1.5 sugar, 2.5 H_2SO_4	20	50	40
320	4.0	40 NaOH, 1.5 sugar, 2.5 H_2SO_4	20	50	40
400	4.0	80 NaOH, 1.5 sugar, 2.5 H_2SO_4	20	50	40
600–690	2.5	60 $CaCO_3$, 5–20 $CaSO_4$	—	40	41
800	2.5	50 $CaCO_3$, 2 sugar, 5 $CaSO_4$	20	45	34
1000	3.8	55 NaOH, 2.0 sugar, 2.0 H_2SO_4	20	200	42

temperature of chromic-fluosilicic-sulfuric acid solutions (Table 4.5). The hardness of chromium deposited in chromic-sulfuric acid baths at 80 or 85 C ranges from 275 to 550 kg/sq mm, depending on the current density, whereas chromium from an 80 or 85 C, chromic-fluosilicic-sulfuric acid solution exhibits hardness values from 470 to 836 kg/sq mm. The hardness of deposits from both types of solutions usually exceeded 900 kg/sq mm when the operating temperature was reduced to the range of 30 to 50 C. Chromium deposited in each type of solution was about equal in hardness when temperature and current density were 30 or 50 C and 20 or 40 amp/sq dm, respectively.

A temperature of about 50 C seems to be the optimum temperature for maximum hardness of chromium deposited in chromic-sulfuric acid solutions. (4, 45) Modifications in chromic acid concentration have only minor effects on hardness.

An increase in cathode current density reduces the hardness of deposits obtained in solutions operated at 30 to 50 C. On the other hand, an increase in current density increases the hardness of chromium deposited in solutions heated to 75 to 85 C.

A high trivalent chromium ion concentration above 3.5 g/l reduced hardness, according to one report, (46) but a shift in concentration from 5.5 to 30 g/l had no effect, according to another. (38) Deposits described as machinable chromium, (30) which are produced in an 85 C bath containing 25 g/l ferric ion exhibited a hardness of 444 kg/sq mm.

Some addition agents were reported to increase hardness. They include indium sulfate, (25) alkaloids such as opium, (31) antimony sulfate, (47) silicon tungstate, (47) pyridine, (47) and hydrogen iodide. (47) Ultrasonic agitation increased the hardness of chromium deposited in a chromic-sulfuric acid bath, (29) a chromic-fluosilicic-strontium sulfate solution, (37) and in two baths of an undisclosed composition. (48,49)

Chromium with a hardness of 1000 kg/sq mm or more exhibited a body-centered-cubic structure, whereas softer deposits in the range of 600 to 700 kg/sq mm exhibited a hexagonal structure. (50) A mixture of bcc and hexagonal chromium was observed for deposits with intermediate hardness.

Some investigators reported that wear resistance was independent of hardness. With 0.02 mm or thicker coatings, no difference was observed in deposits with a hardness of 750 to 950 kg/sq mm, for example. (51) Another investigator reported that some soft deposits showed less wear against a revolving nitrited steel drum than harder, bright deposits. (52) Wear resistance was better for deposits with a hardness of 720 kg/sq mm, in comparison with harder chromium (870 kg/sq mm). (53) An increase in hardness reduced the scratch resistance to emery paper of electrodeposited chromium. (54) On the other hand, another investigator reported maximum scratch resistance for hard chromium, in comparison with softer chromium. (55) Increasing the hardness of chromium deposited on steel cylinders or piston rings from 500 to 1050 kg/sq mm reduced wear. (56) The low coefficient of friction exhibited by chromium electrodeposits has been credited for its resistance to wear, in typical applications. For sliding contacts, the coefficient for chromium on chromium was less than one-half the coefficient for steel on steel. (57)

Fundamentally, the hardness of chromium is a result of oxide inclusions, fine grain size, and high internal stress. (4) Chromium with a hardness of 800 kg/sq mm or more contained > 0.2 percent oxygen, whereas softer deposits contained < 0.2 percent oxygen. Heat treating does not reduce the oxide content, although chromium is softened appreciably by heat treating at temperatures above 400 C, as shown in Figure 4.2.

The reduction in hardness by heat treating is attributed to recrystallization and grain growth at 500 to 600 C (62,63) and the agglomeration of oxides. (4) Hydrogen is removed by heat treating at lower temperatures, as indicated in Figure 4.2, so hydrogen does not appear to account for the initial high hardness of chromium deposits. The amount of hydrogen removed during heating is a direct function of the temperature and is about 95 percent complete at 400 C. (4,58) The removal of hydrogen from electrodeposited chromium follows about the same temperature relationship as the removal of water of hydration from hydrous chromic oxide. (4)

Hardness data for chromium at elevated temperatures is shown in Figure 4.3. Hard chromium containing > 0.2 percent oxygen was slightly harder than chromium containing less oxygen during measurements at 600 or 800 C.

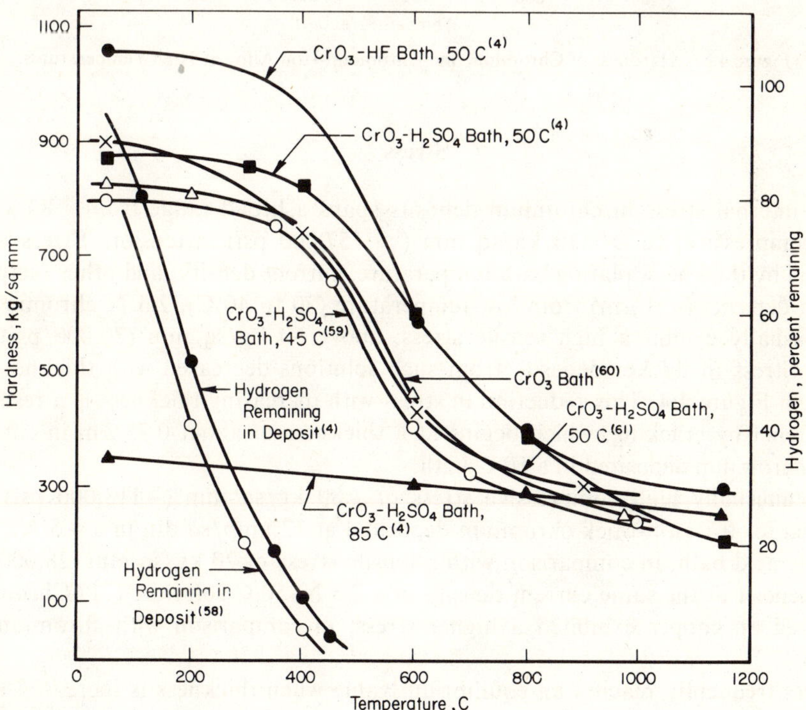

Figure 4.2. Hardness and Amount of Initial Hydrogen Remaining in Deposit after Heat Treating.

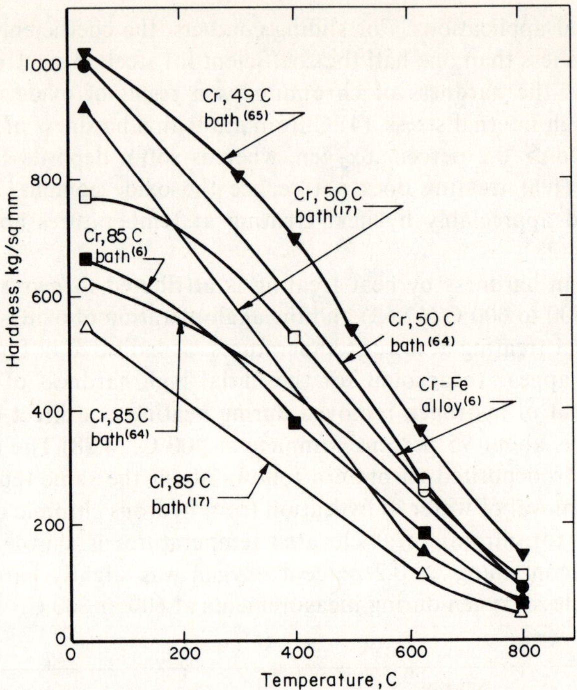

Figure 4.3. Hardness of Chromium and Chromium-Iron Alloy at High Temperatures.

Stress

The internal stress in chromium deposits spans a broad range from −80 kg/sq mm (compressive) to > 110 kg/sq mm (> 157,000 psi) in tension. Stress is influenced by thickness, plating bath temperature, current density, and other factors.

Thin deposits (< 3 μm) from low-temperature (30 to 46 C), 2.5 N chromic acid baths usually exhibit a high tensile stress, above 50 kg/sq mm (70,000 psi). (4, 66-70) Stress in thicker deposits from such solutions decreased with thickness, as shown in Figure 4.4. This reduction in stress with increasing thickness is a result of stress relief by cracking, which begins at a thickness of about 0.75 μm in conventional chromium deposited in a 50 C bath.

The unusually high compressive stress of −80 kg/sq mm (−113,000 psi) was reported for 0.1−μm-thick chromium deposited at 12 amp/sq dm in a 4.5 N, 50 C chromic acid bath, in comparison with a tensile stress of 20 kg/sq mm (28,000 psi) for a deposit at the same current density in a 2.5 N, 50 C solution. (71) Chromium deposited on copper exhibited a higher stress, in comparison with chromium on nickel.

Stress frequently reaches an equilibrium value when thickness is increased to 30 or 50 μm (Figure 4.4). For deposits in this thickness range, increasing the temperature of the plating bath usually increases stress. Examples of this trend appear in Table 4.7 and Figure 4.4. However, low stress values < 5 kg/sq mm (7,000 psi) were reported for chromium deposited at a low current density (10 or 10.8 amp/sq dm) in high-temperature solutions. (74,75) A low tensile stress or a low compressive stress was reported for 30 to 50-μm-thick deposits from 1.0, 2.0, or 2.5 N chromic acid baths, operated at 40 to 45 C. (67,69,70,72) Chromium with a thickness of 25 to 50 μm deposited in a 43 C self-regulating bath containing fluosilicate and sulfate ions exhibited a compressive stress of −8,500 or −10,500 kg/sq mm (−12,000 or −15,000 psi). (69) On the other hand, 2.5 N, conventional chromium baths operated at 20 or 30 C yielded chromium with a high tensile stress of about 21 kg/sq mm. Chromium from solutions maintained at 55 to 65 C also exhibited a high tensile stress (21 to 46 kg/sq mm or 30,000 to 65,000 psi). (5, 68-70,72,73)

The addition of 0.2 g/l selenic acid to a 2.5 N chromic acid solution at 50 C reduced stress to 7.0 kg/sq mm (10,000 psi) when the current density was 15 amp/sq dm. (76) Stress in 5-μm-thick chromium deposited at 5 amp/sq dm in a 50 C, 2.5 N chromium bath was about three times greater than the stress measured for

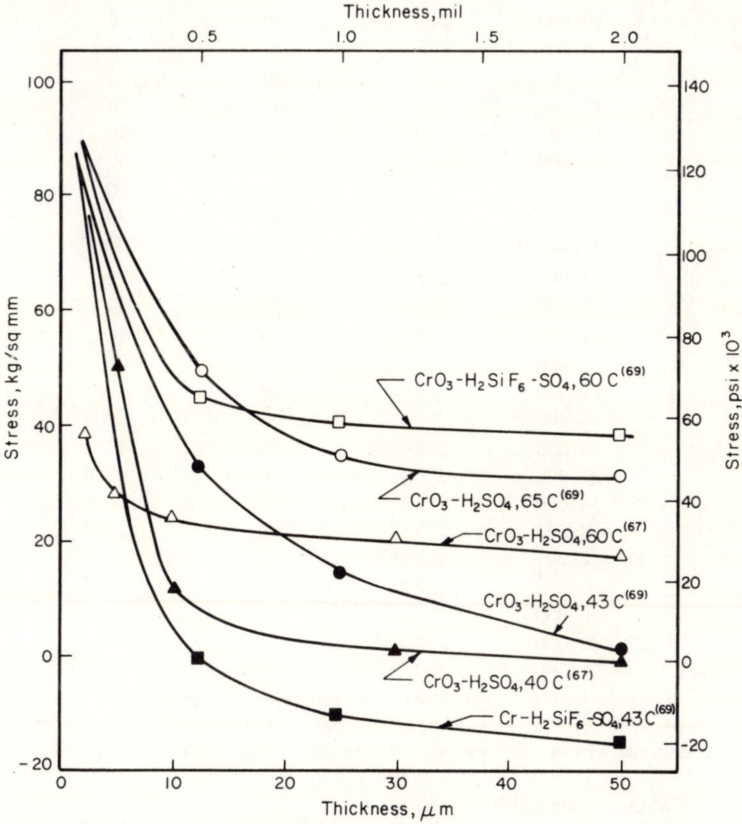

Figure 4.4. Stress in Electrodeposited Chromium as a Function of Thickness.

TABLE 4.7

Stress in Chromium Deposits—Effects of Changing Bath Temperature and Current Density

Stress[a]		CrO_3 Concentration, N[b]	Bath Temperature, C	Cathode Current Density, amp/sq dm	Reference
kg/sq mm	psi				
16.5	23,000	2.5	40	25	5
27.3	39,000	2.5	60	25	5
19.7	28,000	2.5	40	50	5
21.5	30,500	2.5	60	50	5
−2.6	−3,700	1.0	40	20	67
10.0	14,000	1.0	60	20	67
−0.6	−900	1.0	40	40	67
6.0	8,500	1.0	60	40	67
−2.0	−2,800	2.0	40	20	67
18.0	25,500	2.0	60	20	67
−2.0	−2,800	2.0	40	40	67
11.0	15,500	2.0	60	40	67
28[c]	40,000	2.5	55	30	68
46[c]	65,000	2.5	70[d]	30	68
3.5	5,000	2.5	43	40	69
32	45,000	2.5	65	40	69
−10.5	−15,000	1.8[d]	43	40	69
42	60,000	3.2[d]	60	40	69
~21	~30,000	2.5	20–30	20	70
−1.4	−2,000	2.5	40	20	70
~25	~35,000	2.5	55–60	20	70
2.0	2,800	2.5	45	15	72
8.0	11,500	2.5	70	15	72
15.0	21,000	2.5	30	20	73
−2.1	−3,000	2.5	40	20	73
7.7	11,000	2.5	50	20	73
32	45,000	2.5	65	20	73
7	10,000	2.5[e]	25	10.8	74
14	20,000	2.5[e]	35	10.8	74
52	74,000	2.5[e]	50	10.8	74
5.6	8,000	2.5[e]	65	10.8	74
1.4	2,000	2.5[e]	75	10.8	74
4.2	6,000	2.5[e]	85	10.8	74
11.2	16,000	2.5	30	10	75
42.0	60,000	2.5	50	10	75
2.8	4,000	2.5	70	10	75
2.8	4,000	2.5	90	10	75

(a) Stress for deposits with a thickness of 15 μm or more. A negative value designates compressive stress.

(b) The CrO_3:H_2SO_4 ratio was 100:1.

(c) Data for 15-μm-thick deposits, which were nearly crack-free when the plating bath was heated to 75 C.

(d) Self-regulating solution containing $SiF_6^{=}$ and $SO_4^{=}$ ions.

(e) The Cr^{3+} concentration was 0.2 g/l.

chromium deposited in a 20 C tetrachromate bath containing 400 g/l chromic acid and 1.5 g/l sugar. (40) A slight increase in stress was observed when ultrasonic agitation (20 kc/sec) was applied during the electrodeposition of 18-μm-thick chromium. (77) A reduction in the stress in 0.2-μm-thick deposits from a 50 C, 1.5 N chromic acid bath was obtained by superimposing 900 cycles/second alternating current on the direct current. (78)

Chromium with a high tensile stress (e.g., 35 kg/sq mm or 50,000 psi) contains relatively few cracks (< 20 cracks/linear cm), whereas chromium with a low tensile stress (2.1 kg/sq mm or 3,000 psi) or a compressive stress contains >400 cracks/linear cm (> 1,000 cracks/inch). (69,70)

Effect of Stress on the Fatigue Strength of Steel Substrates

Chromium stressed in tension greatly reduces the fatigue strength of the chromium-plated steel. (5,70,73,79–84) Deposits highly stressed in tension reduced fatigue strength to a greater degree than deposits with less stress, as shown in Table 4.8. However, the strength of the steel also affects the amount of reduction in the fatigue strength of the chromium-plated steel. Chromium plating also reduced the fatigue strength of a precipitation-hardened aluminum alloy. (85) Shot peening before plating reduced the loss from 57 to 6 percent, however.

Compressively stressed chromium deposits reduced the fatigue strength of steel substrates very slightly or not at all, depending on the strength of the steel and the degree of compressive stress (Table 4.8). (70,73,79,80) The compressively stressed chromium contained a large number of cracks (500/linear cm, or more), whereas chromium electrodeposits highly stressed in tension contained relatively few cracks (e.g., 40/cm). (69,70) Deposits with a high crack count have little or no harmful effect on the fatigue strength of the steel substrate. (70,86,87)

The relationship between stress (S) in the chromium in tons/sq in., the fatigue limit (F) of the steel in tons/sq in., and the change in the fatigue limit (L) was defined by the following equation: (80)

$$L = 50 - F - 3S.$$

Relationships between stress and the tensile strength of the steel (T) in tons/sq in. and/or hardness (H) of the steel were as follows:

$$L = 5 - T - 3S, \quad L = 50 - H/T - 3S.$$

The reduction in the fatigue strength of steel by chromium plating was attributed to the combined effects of hydrogen pickup during chromium plating, the notch sensitivity of the chromium-plated steel, and high tensile stress, which is incompletely relieved by the formation of cracks. (88)

Steel with a tensile strength of 140 kg/sq mm (200,000 psi), which had been reduced 47 percent in fatigue strength by chromium plating, was reduced only about 10 percent in fatigue strength when it was shot peened before plating. (70) Another

TABLE 4.8

Effect of Stress in Chromium on the Fatigue Strength of Chromium-Plated Steel

Stress[a]		Change in Fatigue Strength of Steel, percent	Chromium Bath		Cathode Current Density, amp/sq dm	Reference
kg/sq mm	psi		Temperature, C	Cr Concentration, N		
42	60,000	−73[b]	—	—	—	79
30	43,000	−34 to −40	60	2.5	25	5
28	40,000	−64[b]	—	—	—	79
17 to 28	24,000 to 40,000	−65[c]	25	2.5	20	73
20	28,000[d]	−65[e]	—	—	—	70
7.7	11,000	−22[e]	50	2.5	20	70, 73
7.7	11,000	−42[d]	50	2.5	20	73
−2.1	−3,000	−33[b]	—	—	—	79
−1.4 to −2.8	−2,000 to −4,000	4[c,e]	50	2.5	20	73
−3.5	−5,000[f]	4	40	2.5	20	70
−12	−17,000	−10.4[b]	—	—	—	79

[a] A negative sign indicates a compressive stress.
[b] The fatigue strength of the unplated steel was 148 kg/sq mm (211,000 psi).
[c] The fatigue strength of the unplated steel was 112 kg/sq mm (160,000 psi).
[d] This chromium deposit contained about 40 cracks/linear cm (100/inch).
[e] The fatigue strength of the unplated steel was 92 kg/sq mm (130,000 psi).
[f] This chromium deposit contained about 500 cracks/cm (1300/inch).

steel with a tensile strength of 112 kg/sq mm (160,000 psi) reduced 40 percent in fatigue strength by chromium plating was reduced only about 5 percent in fatigue strength when it was shot peened before plating. Shot peening induced compressive stress in the surface layers of the steel.

The fatigue strength of steel plated with chromium highly stressed in tension is further reduced by heat treating at a low temperature. Specifically, heat treating at temperatures from 100 to 320 C increased the reduction of fatigue strength caused by chromium plating. (73,80,82,83) However, a temperature of 260 C was effective for restoring the fatigue strength of a normalized steel plated with chromium in an uncommon 6.0 N chromic acid bath containing 3 g/l sulfuric acid. (81)

Heat treating at 400 or 425 C improved the fatigue strength of some chromium-plated steels, (74,79,80) but not of others. (73,79) Heat treating at 525 C, which shifted the stress in chromium plate from tensile to compressive (from 7.7 to about −15 kg/sq mm), eliminated all harmful effects on the fatique strength of steels with a tensile strength of 92 or 112 kg/sq mm (130,000 and 160,000 psi, respectively), whereas heat treating at 425 C was unsatisfactory for restoring good fatigue characteristics for these steel alloys. (73)

To provide resistance to sustained loading for 1000 hours at 90 percent of the ultimate tensile strength of a high-strength steel (140 kg/sq mm) after conventional chromium plating, heat treatment at 190 C for 23 hours or treatment at 535 C for 2 hours was required. (89) However, steel chromium plated at a higher than normal temperature to minimize cracking had to be heated to at least 260 C for 24 hours to resist sustained loads.

Chromium Alloys

Some property data have been reported for chromium alloys containing iron, molybdenum, nickel, phosphorus, or tungsten. Alloys containing 6 or 10 percent iron deposited in chromium-aluminum sulfate baths containing ferrous sulfate were harder after heating to 800 C than conventional chromium (436 or 486 kg/sq mm for the alloys in comparison with 303 kg/sq mm for conventional chromium), although the unalloyed chromium was harder before heating. (6) Alloy containing 6 percent iron retained its original hardness of 600 to 700 kg/sq mm after heating at 600 C. (90) These trivalent chromium plating baths were operated at 54 to 62 C with a current density of 20 to 40 amp/sq dm. Alloy containing 15 percent iron exhibited a hardness of 1000 to 1025 kg/sq mm before heat treating. (91) A chromium alloy containing 60 percent iron deposited in a chromium-potassium sulfate bath containing ammonium sulfamate and sodium fluoride exhibited a high stress of 25 to 27 kg/sq mm (36,000 to 38,000 psi) at a thickness level of 5 μm. (92) These deposits contained many cracks.

The coefficients of expansion for chromium alloys containing 6 and 18 percent iron were slightly higher than the coefficient for chromium. Comparative data are given in Table 4.9. (6).

Chromium alloy containing 5.7 percent iron and 0.9 to 1.2 percent nickel, which

TABLE 4.9

Expansion Coefficients of Chromium and Chromium-Iron Alloys [6]

Temperature Range, C	Coefficient of Thermal Expansion, micrometer/meter/C		
	Chromium [a]	Chromium-6% Iron [b]	Chromium-18% Iron [b]
20 to 200	7.4	8.8	8.8
20 to 400	8.4	9.2	9.1
20 to 600	9.2	9.8	9.9
20 to 800	9.8	10.4	10.4
20 to 1050	11.0	11.7	11.8

[a] Deposited in a 2.5 N chromic acid solution containing 2.5 g/l H_2SO_4.

[b] Deposited in chromium-aluminum sulfate solutions containing ammonium sulfate and ferrous sulfate.

was deposited in a tetrachromate bath prepared with chromic oxide, calcium carbonate, and calcium sulfate, exhibited a hardness of 712 kg/sq mm. (93) By comparison, the hardness of unalloyed chromium deposited in a similar bath containing no iron or nickel was only 480 kg/sq mm.

Chromic acid plating baths containing molybdenum salts have been employed for producing deposits designated as alloys of chromium. Good wear and abrasion resistance were reported for such deposits. (94–97) A deposit containing 3 percent molybdenum exhibited a hardness from 1000 to 1300 kg/sq mm, which was about 10 percent harder than unalloyed chromium. (94) Abrasion resistance was improved 200 to 300 percent by codepositing the molybdenum. With another bath containing 250 g/l chromic acid, 1.5 to 3 g/l fluosilicic acid, and 1 g/l sulfuric acid, hardness increased as concentration of sodium molybdate increased to 100 g/l. (98) Hardness values of 800 kg/sq mm for unalloyed chromium increased to 1100, 1400, and 1550 kg/sq mm with sodium molybdate additions of 50, 75, and 100 g/l. (98) For maximum resistance to wear, which was reported to be three to four times that for hard chromium deposits, electropolishing the chromium-molybdenum alloy was important. (97) Because X-ray diffraction studies showed the presence of molybdenum trioxide, the deposit appears to be a dispersion of molybdenum oxide in chromium. (95)

An increase in hardness in proportion to the phosphorous content was reported for chromium alloy containing up to 15 percent phosphorous. (99) Such deposits were obtained from chromium-ammonium sulfate solutions containing sodium hypophosphite.

A 20 to 25 percent increase in hardness was obtained by adding 0.2 g/l sodium tungstate to a tetrachromate solution prepared with chromic acid, sodium hydroxide, and sulfuric acid. (100) Additions of magnesium oxide (1.0 g/l) increased hardness 10 to 15 percent in deposits described as ductile and lowly stressed. These deposits, like those containing molybdenum, may have been dispersions of the oxides in the chromium matrix. Chromium containing dispersions of zirconia or thoria deposited at high current density in chromic acid plating baths exhibited high hardness and improved wear resistance in comparison with chromium deposits containing no such dispersions. (101) The hardness of chromium deposited at 32 amp/sq dm in a 2.5 N chromic acid bath at 54 C containing 2.5 g/l sulfuric acid was increased from the range of 900 to 1000 kg/sq mm to the range of 1550 to 2300 kg/sq mm by incorporating 1 percent by volume of titanium oxide particles. (102)

Black chromium deposits were described as mixtures of chromium, chromic oxides, and other oxides of chromium. (103) The hardness and wear resistance of these deposits depended on their composition and structure. X-ray diffraction analysis of black chromium deposited in solutions containing urea and/or acetic acid revealed a mixture of hexagonal-close-packed and body-centered-cubic structures. (104) Treatment in hydrogen at 150 C converted the mixture to a bcc form.

Structure

Structural characteristics account for the unique properties of chromium, which is porous in thin layers (0.5 μm or less) and frequently cracked in thicker layers, when electrodeposited in chromic acid solutions. These cracks are filled with hydrous chromic oxide. (8) A structural transition from unstable chromium hydrides to body-centered-cubic chromium during or soon after plating accounts for the cracking observed in most chromium deposits. (105) A low solution temperature, a low concentration of sulfuric acid, or a high current density favors the deposition of hexagonal-close-packed chromium hydride, which has been identified by X-ray diffraction analysis. (105,106) Face-centered-cubic chromium hydride was deposited in a bath containing a high concentration of trivalent chromium hydride. (105) Either form of hydride decomposes during plating or within a few hours after plating to body-centered-cubic chromium and hydrogen. This decomposition involves a volume shrinkage of $>$ 15 percent.

The electrical resistivity of a deposit of chromium hydride produced in a 4.0 N chromic acid bath at 12 to 15 C was increased about ten times when the deposit was heated in a vacuum at 50 C, which also evacuated considerable hydrogen. (107) These results are manifestations of hydride decomposition, volume shrinkage, and the formation of cracks.

Electroplating in high-temperature chromic acid baths favors immediate decomposition of the hydride to bcc chromium, which exhibits less shrinkage or contraction during subsequent heat treating than chromium deposited in solutions maintained at a lower temperature. The contraction during heat treating of chromium deposited in an 85 C chromic acid bath was only 10 to 20 percent that of chromium deposited in a 48 or 50 C solution. (6,16,17)

Figure 4.5 shows a surface view of a typical chromium deposit from a 55 C, 4.0

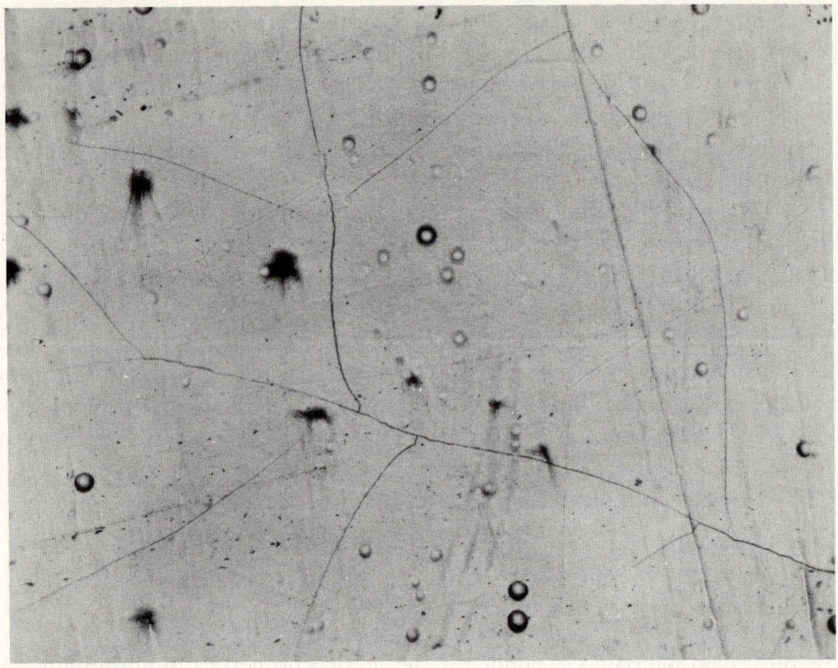

Figure 4.5. Surface View of Conventional Chromium Deposited at 18.8 amp/sq dm in a 55 C, 4.0 N Chromic Acid Bath (100×).

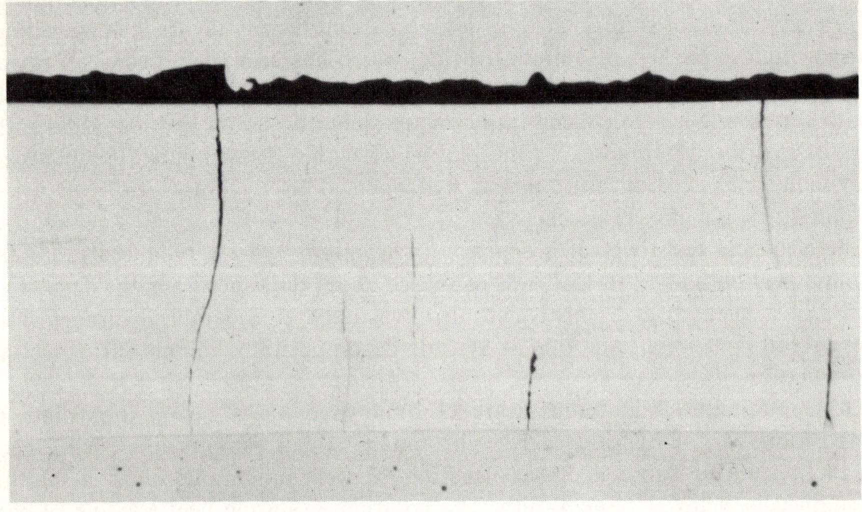

Figure 4.6. Cross Section of Conventional Chromium Deposited at 18.8 amp/sq dm in a 55 C, 4.0 N Chromic Acid Bath (500×).

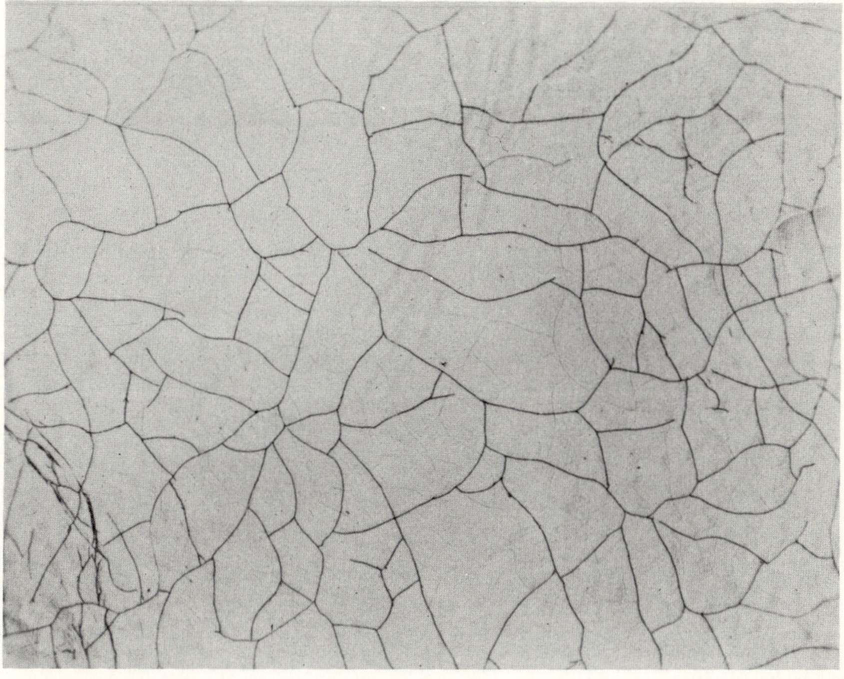

Figure 4.7. Surface View of Chromium Deposited at 18.8 amp/sq dm in a 55 C Chromic-Fluosilicic-Sulfuric Acid Bath (250×).

N chromic acid solution, which exhibited a hardness of about 1,000 kg/sq mm (100-gram load). The cracks and small nodules that appear in this view are typical of hard chromium deposits from chromic-sulfuric acid baths. The cracks frequently penetrate the entire thickness of thick (100 μm) deposits, as illustrated in Figure 4.6. Chromium deposited in 85 C solutions contains fewer cracks, but these cracks are also propagated to the substrate material.

Chromium deposited in chromic-fluosilicic-sulfuric acid solutions contains more cracks than conventional chromium (Figure 4.7), but these cracks are much shorter when cross sections are examined. Figure 4.8 shows cracks with a maximum length of only 5 μm. Deposits like this are frequently preferred for protecting steel from corrosion because corrosive solutions cannot penetrate deposits with a thickness of 25 μm or more to reach the steel substrate.

Plating with pulsating current completely eliminated cracking in chromium deposited in a 60 C, 2.0 N chromic acid bath. (1) Good corrosion protection also was reported for this kind of chromium plate, which exhibited a columnar structure with no evidence of banding. Hardness was in the range of 332 to 389 kg/sq mm with a 100-gram load. X-ray diffraction analysis indicated that chromium produced with pulsating current was either cubic or a mixture of cubic and hexagonal forms. (108, 109)

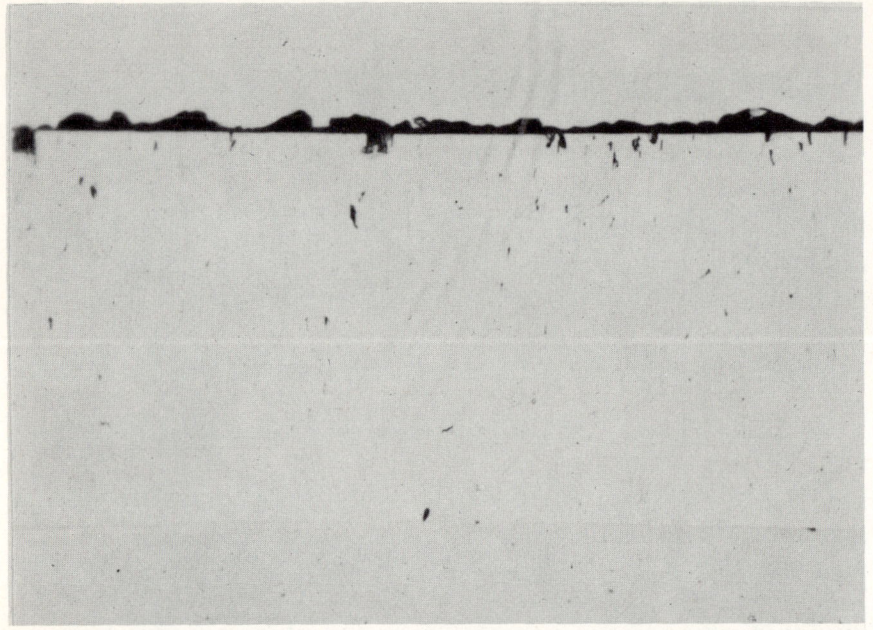

Figure 4.8. Cross Section of Chromium Deposited at 18.8 amp/sq dm in a 55 C Chromic-Fluosilicic-Sulfuric Acid Bath (500×).

Conventional chromium deposits exhibit a banded or lamellar structure when cross sections are etched, but some evidence of columns crossing the bands has been noted. Some cracks terminate at the boundaries between layers. (59,63,68) After prolonged heating at 300 to 500 C, new grains are formed with their major axis perpendicular to the substrate. Large equiaxed grains are formed during prolonged heating at a higher temperature, such as 1100 C. Oxides are agglomerated when the deposits are heated. (4,110)

A grain size of 179 angstroms was reported for crackfree chromium. (111) However, the grain size of conventional bright chromium deposited in a 45 C, 2.5 N chromic acid bath was 1000 angstroms. (59) Annealing at 450 C for an hour increased grain diameters to about 1500 angstroms. After an hour at 1000 C, grains had grown to a diameter of about 20,000 angstroms.

Preferred orientation with (111) planes parallel with the substrate surface has been reported for conventional bright chromium. (63,74,112) This orientation persisted after recrystallization. Microstress in a crackfree chromium deposit was 102 kg/sq mm (148,000 psi) in comparison with a macrostress of 44 kg/sq mm (63,000 psi). (111)

Crackfree chromium-iron alloy deposits from trivalent chromium solutions had a columnar structure, (6) similar to that for chromium deposited with pulsating current. (1) However, the addition of glycine and water glass to a similar solution resulted in a banded structure in chromium-iron alloy deposits of similar composition. (91) Cracking and banding was attributed to the precipitation of ferric salts induced with the addition agents.

References

(1) Faust, C. L., Schaer, G. R., and Semones, D. E., "Electroplating Chromium Directly on Aluminum," *Plating, 48,* 605–612 (1961).

(2) Grube, G., and Knabe, R., "Electrical Conductivity and Phase Diagram of Binary Alloys. XXI. The System Palladium-Chromium," *Z. Elektrochem., 42,* 793–804 (1936).

(3) Smithells, C. J., and Williams, S. V., "The Melting Point of Chromium," *Nature, 124* (3129), 617–618 (1929).

(4) Brenner, A., Burkhead, P., and Jennings, C. W., "Physical Properties of Electrodeposited Chromium," *Proceedings Am. Electroplaters' Soc., 34,* 32–73 (1947).

(5) Fedot'ev, N. P., "Physical and Mechanical Properties of Electrodeposited Metals," *Plating, 53* (3), 309–317 (1966).

(6) Safranek, W. H., and Schaer, G. R., "Properties of Electrodeposits at Elevated Temperatures," *Proceedings Am. Electroplaters' Soc., 43,* 105–117 (1956).

(7) West, P., "Refractory Metals Can Be Plated and Electroformed," *Materials in Design Engineering, 62* (1), 93–94 (1965).

(8) Cohen, J. B., "Film in Chromium Electroplate," *Trans. Electrochem. Soc., 86,* 441–456 (1944).

(9) Black, F. E., Good, P. C., and Asai, G., "Electrodeposition of High-Purity Chromium," *Journal Electrochem. Soc., 106* (1), 43–47 (1959).

(10) Esmore, L., "Electrodeposition of High-Purity Chromium," *Products Finishing, 30* (11), 70–73 (1966).

(11) Gruzensky, P. M., and Block, F. E., "Preparation of High-Purity Electrolytic Chromium," Report Investigation No. 5305, U.S. Bureau of Mines, (1957), 11 pp.

(12) Ryan, N. E., Henderson, F., Johnstone, S. T. M., and Wain, H. L., "Some Properties of Chromium Deposited From an Electrolyte Containing Fluoride," *Nature, 180* (4599), 1406–1407 (1957).

(13) Ryan, N., and Lumley, E. J., "The Source of the Nitrogen Impurity in Electrodeposited Chromium," *Journal Electrochem. Soc., 106* (5), 388–391 (1959).

(14) Fine, M. E., Greiner, E. S., and Ellis, W. C., "Transitions in Chromium," *Trans. AIME, 191,* 56–58 (1951).

(15) Dean, R. S., and McCawley, F. X., "Electrolytic Production of Pure Chromium Crystals and Coatings," U.S. Patent 2,951,794 (September 6, 1960). Assigned to Chicago Development Corporation.

(16) Blum, W., "Summary of Wartime Research on Plating at the National Bureau of Standards," *Proceedings Am. Electroplaters' Soc., 33,* 16-21 (1946).

(17) Lamb, V. A., and Young, J. P., "Chromium Plating of Gun Bores," *Proceedings Am. Electroplaters' Soc., 43,* 260–266 (1956).

(18) Eilender, W., Arend, H., and Schmidtmann, E., "The Elastic Modulus and Strength of Hard Chromium Coatings," *Metalloberflaeche, 3A* (7), 145–147 (1949).

(19) Deyrup, A. J., "Chromium Plating," U.S. Patent 3,069,333 (December 18, 1962). Assigned to E. I. du Pont de Nemours & Company.

(20) Weiner, R., *Die Galvanische Verchromung (Chromium Plating),* Schriftenreihe Galvanotechnik Eugen G. Leuze Verlag, Saulgau/Wurtt. (1961).

(21) Wiegand, H., and Kaiser, H.-R., "Stresses in Hard Chromium Deposits," *Metalloberflaeche, 19* (5), 129–137 (1965).

(22) Brandes, E. A., and Whittaker, J. A., "Ductile Electrodeposited Chromium Tubing and Other Shapes," British Patent 1,081,137 (August 31, 1967). Assigned to Fulmer Research Institute Ltd.; "Production of Ductile Electroforms in Chromium," *Engineer, 220* (5732), 929 (1965).

(23) Fedot'ev, N. P., Vyacheslavov, P. M., and Bardin, V. V., "Hardness of Electrolytic Chromium," *Journal Applied Chemistry (USSR), 29* (3), 521–523 (1956); *Zh. Prikl. Khimii, 29* (3), 476–478 (1956).

(24) Bogoyavlenskii, A. F., Ivanov, B. E., and Khudyakov, V. L., "Chromium Plating of Aluminum With a Direct Current Superimposed by an Alternating Current," *Zh. Prikl. Khimii, 33* (2), 368–372 (1960).

(25) Gabriel, J., "Effect of Additives on the Properties of Chromium Plating Solutions and Chromium Deposits," *Galvanotechnik, 56* (12), 714–721 (1965).

(26) Petrova, O. A., "Wear and Corrosion-Resistant Composite (Two-Layer) Chromium Deposits," *Theory and Practice of Chromium Electroplating,* edited by A. T. Vagramyan and N. T. Kudryavtsev, Akademiya Nauk SSSR. Translated by the Israel Program for Scientific Translations, Jerusalem, 66–73 (1965). TT 65-50001.

(27) Shreider, A. V., "Wear-Resistance and Hardness of Electrolytic Chromium Coatings," *Journal Applied Chemistry (USSR), 29* (1), 77–85 (1956); *Zh. Prikl. Khimii, 29* (1), 73–82 (1956).

(28) Shreider, A. V., "The Influence of the Operating Conditions in Chromium Plating on the Hardness and Wear-Resistance of Chromium Deposits," *Theory and Practice of Chromium Electroplating,* edited by A. T. Vagramyan and N. T. Kudryavtsev, Akademiya Nauk SSSR. Translated by the Israel Program for Scientific Translations, Jerusalem, 51–65 (1965). TT 65-50001.

(29) Vagramyan, A. T., and Solov'eva, Z. A., *Technology of Electrodeposition.* English Translation, Robert Draper Ltd., Teddington (1961), 398 pp.

(30) Gardam, G. E., "The Production of Machinable Cr Deposits," *Journal Electrodepositors' Technical Soc., 20,* 69–74 (1945).

(31) Kozarev, Khr., Marinkov, N., and Mardirosov, N., "Hard Chroming," *Leka Promishlenost (Sofia), 8* (2), 22–24 (1959).

(32) Kubala, F., "New Type of Bath for Hard Chromium Coating," *Korose Ochrana Mater., 8* (3), 66–67 (1964).

(33) Zak, T., "Chromium Plating in a Self-Regulating Bath," *Galvanotech. Oberflaechenschutz, 6* (4), 87–96 (1965).

(34) Pepchuk, P. A., "A Comparative Study of Chromium Plating Baths Used for the Reforming of Automobile and Tractor Parts," *Zap. Leningr. Sel'skokhoz. Inst., 93,* 92–109 (1963).

(35) Kazakbaev, M., and Terminasov, Yu. S., "X-Ray Examination of a Chromium Deposit Obtained in a Self-Adjusting Bath," *Tr. Leningr. Inzh.-Ekon. Inst., 29,* 61–69 (1962).

(36) Antonov, N. M., *et al.*, "Electrolytic-Flow Plant for Rapid Chromium Plating," *Russian Engineering Journal, 46* (10), 67–70 (1966); *Vestnik Mashinostroeniya, 46* (10), 63–66 (1966).

(37) Virolaynen, E. I., and Kaybiyaynen, L. K., "Influence of Ultrasonics on the Structure of Chromium Electrodeposits," *Protection of Metals, 2* (2), 185–189 (1966); *Zashchita Metallov, 2* (2), 221–226 (1966).

(38) Morisset, P., Oswald, J. W., Draper, C. R., and Pinner, R., *Chromium Plating*, Robert Draper Ltd., Teddington (1954), 611 pp.

(39) Drobantseva, N. T., and Saimanova, A. I., "A Comparative Investigation of Chromium Deposits Obtained From Tetrachromate and Standard Electrolytes," *Protective Metallic and Oxide Coatings, Metal Corrosion and Electrochemistry*, Edited by N. P. Fedot'ev, Akademiya Nauk SSSR. Translated by the Israel Program for Scientific Translations, Jerusalem, 26–34 (1968).

(40) Ryaboi, A. Ya., and Shluger, M. A., "Electrodeposition of Chromium From a Tetrachromate Bath," *Journal Applied Chemistry (USSR), 32* (3), 617–622 (1959); *Zh. Prikl. Khimii, 32* (3), 588-595 (1959).

(41) Drobantseva, N. T., and Saimanova, A. I., "Deposition of Chromium From a Self-Regulating Tetrachromate Electrolyte," *Vestn. Khar'kov. Politekh. Inst., No. 13,* 68–71 (1966).

(42) Nud'ga, V. N., and Ginberg, A. M., "Effect of Hydrogen Absorption on the Mechanical Properties of Chromium Coatings Deposited Under Effect of Ultrasound," *Navodorozhivanie Metal. Bor'ba Vodorodn. Khrupkost'yu,* 173–178 (1968).

(43) Howell, R. E., "The Effect of Operating Conditions on the Hardness of Chromium Plate," *Metal Finishing, 49* (3), 50–56, 69 (1951).

(44) Shams El-Din, A. M., "Direct Chrome Plating of Zinc Castings," *Metalloberflaeche, 15,* 241–242 (1961). Cf. CA *55,* 26779g.

(45) Cymbaliste, M., "The Structure and Hardness of Electrolytic Chromium," *Trans. Electrochem. Soc., 73,* 353 (1938).

(46) Greenwood, J. D., "Heavy Electrodeposition for the Engineer. Part 4. Hard Chromium Plating," *Metal Finishing Journal, 12* (138), 260–265 (1966).

(47) Doškár, J., and Gabriel, J., "Improvement in the Properties of Hard Chromium Deposits," *Galvanotechnik, 58* (11), 802–811 (1967).

(48) Müller, Fr., and Kuss, H., "The Effects of Using Cathodes Oscillating at Various Frequencies, Particularly in the Ultrasonic Region, on the Electrodeposition of Metals," *Helvetica Chimica Acta, 33* (1), 217–228 (1950).

(49) Petrov, Yu. N., "Chromium Plating and Iron Plating of Metallic Surfaces for Increasing Their Wear Resistance," *Elektronnaya Obrabotka Materialov, Akad. Nauk Moldavsk. SSR, 4,* 21–28 (1966).

(50) Arend, H., "The Hardness of Hard-Chromium Plates," *Metalloberflaeche, 3B* (5), 72–76 (1951). Brutcher Translation No. HB 3427.

(51) Eilender, W., Arend, H., and Schmidtmann, E., "Hard-Chromium Coatings of Very High Resistance to Wear," *Metalloberflaeche, 3A* (3), 57–59 (1949).

(52) Iitaka, I., "Wear-Resisting Characteristics of Chromium Plate. I and III," *Bulletin Institute Physical and Chemical Research (Tokyo), 22,* 558–570; 187–194 (1943).

(53) Ishida, T., "Wear Resistance of Chromium Plate Against Chromium Plate. Wear Resistance of Chromium Plate (Part I)," *Journal of the Metal Finishing Society of Japan, 14* (1), 21–24 (1963).

(54) Wahl, H., and Gebauer, K., "Wear Testing of Hard Chromium Coatings," *Metalloberflaeche, 2* (2), 25–37 (1948); *Metaux (Corrosion-Ind.), 26* (305), 29–46 (1951).

(55) Hosdowich, J. M., "Scratch Hardness and Abrasion Hardness of Electrodeposited Chromium," *Proceedings Am. Electroplaters' Soc., 36,* 103–125 (1949).

(56) Krylov, K. A., "Hardness of Chromium Relative to Its Wear-Resistance," *Russian Engineering Journal, 39* (12), 6–8 (1959); *Vestnik Mashinostroeniya, 39* (12), 7–10 (1959).

(57) Oswald, J. W., "Hard-Chromium Plating of Press Tools," *Sheet Metal Industries, 33* (354), 691–694 (1956).

(58) Eilender, W., Arend, H., and Schmidtmann, E., "The Influence of Heat-Treatment on the Hardness and Gas Content of Hard Chromium Layers," *Metalloberflaeche, 2A* (7), 143–145 (1948).

(59) Brittain, C. P., and Smith, G. C., "The Influence of Annealing on the Structure and Hardness of Electrodeposited Chromium," *Trans. Inst. Metal Finish., 31,* 146–152 (1954).

(60) Bastien, P., and Popoff, A., "On the Existence of Micro-Fissures in Electrodeposits of Chromium: Their Influence on the Fatigue Limit of Steel Test-Pieces," *Metaux et Corrosion, 23* (277), 191–198 (1948).

(61) Ishida, T., "Hardness of Chromium Plate," *Journal of the Metal Finishing Society of Japan, 13* (10), 427–432 (1962).

(62) Jacquet, P. A., "Structure of Electrolytic Chromium and Its Development in the Course of Annealing," *Bull. Doc., Centre Inform. Chrome Dur* (Feb., 1966), 44 pp.

(63) Snavely, C. A., and Faust, C. L., "Studies on the Structure of Hard Chromium Plate," *Journal Electrochem. Soc., 97* (3), 99–108 (1950).

(64) Brenner, A., "A Microhardness Tester for Metals at Elevated Temperatures," *Plating, 38* (4), 363–366 (1951).

(65) Domnikov, L., "Chromium and Electroless Nickel Deposits. Hardness at High Temperatures," *Metal Finishing, 60* (1), 67–68, 70 (1962). Lozinsky, M. G., and Mirotvorsky, V. S., "Metallurgy and Fuel," *Bulletin Acad. Sci. USSR, Div. Techn. Sci., No. 3* (1959).

(66) Brenner, A., and Senderoff, S., "A Spiral Contractometer for Measuring Stress in Electrodeposits," *Proceedings Am. Electroplaters' Soc., 35,* 53–76 (1948).

(67) Aotani, S., "An Internal Stress of Chrome Plating and Its Relaxation by Heating," *Journal of the Metal Finishing Society of Japan, 16* (1), 7–11 (1965).

(68) Fry, H., "A Study of Cracking in Chromium Deposits," *Trans. Inst. Metal Finish., 32,* 107–127 (1955).

(69) Stareck, J. E., Seyb, E. J., and Tulumello, A. C., "Stress in Chromium Deposits," *Plating, 41* (10), 1171–1180 (1954).

(70) Hammond, R. A. F., "Stresses in Hard Chromium and Heavy Nickel Deposits and Their Influence on the Fatigue Strength of the Basis Metal," *Metal Finishing Journal, 7,* 441–449 (1961).

(71) Cleghorn, W. H., and West, J. M., "Internal Stress and Structure in Thin Chromium Electrodeposits," *Trans. Inst. Metal Finish.*, *45* (2), 43–47 (1967).

(72) Konishi, S., "Stress in Electrodeposition of Chromium," *Metal Finishing, 61* (3), 54–58; (10), 58–64 (1963).

(73) Williams, C., and Hammond, R. A. F., "The Effect of Chromium Plating on the Fatigue Strength of Steel," *Trans. Inst. Metal Finish.*, *32* (1), 85–106 (1955).

(74) Wyllie, M. R. J., "A Semiquantitative Method for Measuring the Ductility of Chromium Electrodeposits," *Trans. Electrochem. Soc.*, *92*, 519–536 (1947).

(75) Wyllie, M. R. J., "The Influence of Internal Stress on the Structure of Electrodeposits," *Journal Chemical Physics*, *16*, 52–64 (1948).

(76) Fry, H., "Chromium Plating," British Patent 798,590 (July 23, 1958). Assigned to British Non-Ferrous Metals Research Association.

(77) Smirnova, A. M., and Kudryavtsev, N. T., "Investigation of the Influence of Ultrasonic Vibrations on the Electrodeposition of Chromium," *Journal Applied Chemistry (USSR)*, *33* (11), 2484–2489 (1960); *Zh. Prikl. Khimii*, *33* (11), 2521–2526 (1960).

(78) Avramenko, V. I., and Popereka, M. Ya., "Internal Stresses in Chromium Coatings Deposited by Alternating Current," *Protection of Metals*, *1* (5), 480–482 (1965); *Zashchita Metallov*, *1* (5), 539–542 (1965).

(79) Stareck, J. E., Seyb, E. J., and Tulumello, A. C., "The Effect of Different Chromium Deposits on the Fatigue Strength of Hardened Steel," *Plating*, *42* (11), 1395–1402 (1955); *Proceedings Am. Electroplaters' Soc.*, *42*, 129–136 (1955).

(80) Hammond, R. A. F., "Some Engineering Aspects of Electrodeposition," *Proceedings Am. Electroplaters' Soc.*, *51*, 9–20 (1964).

(81) Fürstenberg, U.-H., "Chromium Deposits From a High Concentration Electrolyte," *Metalloberflaeche*, *20* (5), 223–228 (1966).

(82) Kocánda, S., "Mechanical Properties of Electrolytically Chrome-Plated Structural Steel Parts," *Przeglad. Mech.*, *16* (7), 279–284; (8), 325–329 (1957).

(83) Logan, H. L., "Effect of Chromium Plating on the Endurance Limit of Steels Used in Aircraft," Research Paper No. 2011, *Journal Research National Bureau of Standards*, *43* (2), 101–112 (1949).

(84) Stareck, J. E., Seyb, E. J., Jr., and Passal, F., "Chromium Plating of Hard Steels," U.S. Patent 2,800,436 (July 23, 1957). Assigned to Metal & Thermit Corporation.

(85) Morgan, C. J., and Brine, F. E., "The Effect of Chromium Plating on the Fatigue Strength of Aluminum Alloy L65," *Trans. Inst. Metal Finish.*, *47* (3), 77–79 (1969).

(86) Stareck, J. E., Seyb, E. J., Jr., and Passal, F., "Chromium Plating of Hard Steels," U.S. Patent 2,800,438 (July 23, 1957). Assigned to Metal & Thermit Corporation.

(87) Stareck, J. E., Seyb, E. J., Jr., and Passal, F., "Chromium Plating of Hard Steels," U.S. Patent 2,800,443 (July 23, 1957). Assigned to Metal & Thermit Corporation.

(88) Wiegand, H., and Kaiser, H.-R., "On the Mechanism of Changes in Strength of Ferrous Metals Following Hard Chromium Plating," *Metalloberflaeche*, *19* (8), 241–252 (1965).

(89) Turns, E. W., "Process Plating Crack-Free Chromium, CF-500, Physical and Chemical Properties, Evaluation of," *Report No. FGT-2732, Contracts AF 33(600)-36200 and AF 33(657)-7248*, General Dynamics/Fort Worth (January 1962). AD 272 106.

(90) Snavely, C. A., Faust, C. L., and Bride, J. E., "Electrodeposition of Chromium and Chromium Alloys," U.S. Patent 2,693,444 (November 2, 1954). Assigned to Battelle Development Corporation.

(91) McGraw, L. D., Gurklis, J. A., Faust, C. L., and Bride, J. E., "Deposition of New Chromium-Iron Alloy Plate of Banded Structure," *Journal Electrochem. Soc.*, *106* (4), 302–305 (1959).

(92) Elsie, P. L., Gowri, S., and Shenoi, B. A., "Iron-Chromium Alloy Deposition," *Metal Finishing*, *68* (11), 52–55, 63 (1970).

(93) Saimanova, A. I., and Drobantseva, N. T., "Electrodeposition of a Chromium-Iron-Nickel Alloy From a Self-Regulating Tetrachromate Electrolyte," *Zh. Prikl. Khimii*, *40* (12), 2722-2727 (1967).

(94) Aotani, K., and Nishimoto, K., "Electrodeposited Alloys. XI. Hardness and Abrasion Resistance of Chromium-Molybdenum Alloy Electrodeposits," *Kinzoku Hyomen Gijutsu (Journal of the Metal Finishing Society of Japan), 21* (7), 356–362 (1970).

(95) Jacquet, P. A., and Lepetit, P., "Conditions for Obtaining, Structure, and Wear Resistance of Chromium-Molybdenum Mixed Deposits," *Bull. Doc., Centre Inform. Chrome Dur*, 1–26 (Sept.-Oct. 1965).

(96) Korovin, N. V., "Alloy Deposition. III. Alloys for Wear Resistance and Low Friction Application," *Electroplating and Metal Finishing, 17* (6), 188–192 (1964).

(97) Jacquet, P. A., Galbrun, J. J., and Popoff, A., "The Influence of the Surface Structure on the Attrition of Hard Chromium and Mixed Chromium-Molybdenum Deposit," *Chrome Dur*, 46–58 (1959–60).

(98) Shenoi, B. A., and Indira, K. S., "Electrodeposition of Chromium-Molybdenum Alloys. An Experimental Study and a New Bath," *Metal Finishing, 63* (5), 56–69; (6), 94–97 (1965).

(99) Bondar, V. V., and Potapov, I. I., "Electrodeposition of Chromium-Phosphorus Alloys From Acid Solutions," *Zashchita Metallov, 5* (3), 346–348 (1969).

(100) Petrov, Yu. N., Dekhtyar, L. I., and Beznosov, A. E., "Effect on Microhardness of the Alloying of Electrodeposited Chromium With Tungsten and Magnesium," *Izv. Akad. Nauk Mold. SSR, 6*, 74–78 (1967).

(101) Hassion, F. X., and Szanto, J., "Occlusion of Various Refractory Materials in Chromium Plate on Steel Substrates," Springfield Armory, U.S. Dept. Comm. AD 621 920 (1965), 43 pp.

(102) Greco, V. P., and Baldauf, W., "Electrodeposition of Ni-Al_2O_3, Ni-TiO_2 and Cr-TiO_2 Dispersion-Hardened Alloys," *Plating, 55* (3), 250–257 (1968).

(103) Yoshizawa, I., Aotani, K., and Chihaya, T., "Colored Chromium Plating. IV. Crystal Structure of Black Chromium Deposit From Plating Baths Containing Fluorosilicic Acid," *Kinzoku Hyomen Gijutsu (Journal of the Metal Finishing Society of Japan), 20* (2), 57–63 (1969).

(104) Okada, H., and Ishida, T., "Hexagonal Modification of Black Chromium Electrodeposit," *Nature, 187* (4736), 496–497 (1960).

(105) Snavely, C. A., "A Theory for the Mechanism of Chromium Plating; A Theory for the Physical Characteristics of Chromium Plate," *Trans. Electrochem. Soc., 92*, 537–575 (1947).

(106) Azhogin, F. F., Garshina, N. V., and Sycheva, V. I., "Hydrogenation of Steel During Electrolytic Chromium Plating," *Zashchita Metallov, 2* (3), 366–368 (1966).

(107) Knödler, A., "Properties of Chromium Hydride," *Metalloberflaeche, 17* (11), 331–337 (1963).

(108) Beckmann, M. E., and Maass-Graefe, F., "Effect of the Wave Form of Rectified Alternating Current on the Growth of Electrodeposits," *Metalloberflaeche*, 5A, 161–169 (1951).

(109) Kakovkina, V. G., and Sysoev, A. N., "Effect of the Form of the Current on the Quality of the Chromium Deposits," *Trudy Kar'kovsk. Politekh. Inst., Ser. Khim.-Tekhnol., 18* (5), 107–115 (1958).

(110) Dale, J. J., "The Structure of Thick Chromium Electrodeposits," *Proceedings of Third International Electrodeposition Conference, Electrodepositors' Technical Society*, 185–194 (1947).

(111) Van Zuilichem, A. G., Reidt, M. J., von Rosenstiel, A. P., and Verbraak, C. A., "Influence of Texture, Internal Stresses, and Undercoatings on the Quality of Crack-Free Chromium Plating," *Metalloberflaeche, 19* (1), 13–17 (1965).

(112) Hume-Rothery, W., and Wyllie, M. R. J., "The Structure of Electrodeposited Chromium," *Proc. Royal Soc. (London), A181*, 331–344 (1943); *A182*, 415 (1944).

Chapter 5

Cobalt and Cobalt Alloys

Electrodeposited cobalt and nickel are similar in many respects, but uses for nickel have surpassed those for cobalt because the market price for nickel was appreciably lower until 1969, when the differential was reduced to less than 20 percent. Prices for cobalt and nickel futures were equalized in February, 1970. Thus future developments in cobalt plating processes are expected to keep pace with developments in nickel plating.

Prior to 1969, applications for cobalt and cobalt alloy plating were centered on electronic gear such as recording tapes, disks, and drums. Thus considerable data have been developed on magnetic properties. Such properties are critically dependent on solution formulation and operating conditions.

A severe shortage in the supply of nickel triggered the adoption in 1969, by at least 100 companies in the United States, of cobalt and cobalt-nickel alloy plating for decorative and protective uses. Many brighteners developed for nickel can be used for cobalt or nickel-cobalt alloy plating. However, the need for a stress reducer to avoid stress cracking of the deposit becomes greater as the cobalt content of the bright electrodeposit is increased. Although reliable data on corrosion performance are meager, there is general agreement that the alloy containing 30 to 50 percent cobalt is similar in behavior to unalloyed nickel, when combined with a decorative chromium overlay.

Some properties of cobalt-nickel alloys have been examined with special attention to electroforming applications. The physical and mechanical property data summarized herein reflect efforts to develop better electroforming processes.

Cobalt-tungsten and cobalt-phosphorus alloys are of interest for their exceptional hardness and resistance to wear. These alloys also exhibit high hardness at high temperatures.

Physical Properties of Cobalt

The density of electrodeposited cobalt is approximately 8.8 g/cu cm, a figure derived by extrapolation from data for cobalt-tungsten alloys. (1) This value is equal to that given for face-centered-cubic cobalt prepared metallurgically, and similar to values reported for electrodeposited nickel (8.8 to 8.9 g/cu cm).

TABLE 5.1

Physical Properties of Electrodeposited Cobalt (1)

Property[a]	Value
Density (extrapolated from data for Co-W alloys)	8.8
Thermal conductivity, cal/sq cm/cm/C/sec	0.165
Expansion coefficient, μ in./in./C (to 900 C)	14.5
Permanent expansion, percent after heating to 950 C and cooling	0.05
Electrical resistivity, microhm-cm @ 20 C	10

[a] Properties for cobalt deposited in a solution containing 120 g/l cobalt and 50 g/l ammonium chloride with a pH of 4.0 and a temperature of 25 C. The current density was 2 amp/sq dm.

Thermal and electrical property data for electrodeposited cobalt are listed in Table 5.1. The coefficient of thermal expansion for cobalt is slightly lower than that for electroplated nickel. However, the amount of occluded salts affecting the thermal stability of the deposits was much lower in a length of electroformed cobalt tube, which increased only 0.05 percent as a result of heating to 950 C and cooling, (1) by comparison with an increase of about 3 percent for electroformed nickel tubes.

Cobalt electroplated in a chloride bath was reported to have an electrical resistivity of 10 microhm-cm, (1) by comparison with only about 6.2 microhm-cm for the metallurgical form. The resistivity of high-purity electroplated nickel ranges from 7.4 to 8.0 microhm-cm. The relatively high resistivity for the electroplated cobalt was associated with a relatively high oxygen content of 0.05 percent. A lower resistivity of 8.4 microhm-cm was reported for cobalt deposited in a sulfate bath containing 60 g/l cobalt ions, 20 g/l boric acid, and 20 g/l sodium chloride, operated with a temperature of 20 C and a current density of 1.0 amp/sq dm. (2)

Mechanical Properties

Strength and Ductility

Electrodeposited cobalt is stronger than soft nickel electroplates, but not so ductile. Strength and ductility data for cobalt deposited in chloride, sulfate-chloride and sulfamate-bromide solutions are summarized in Table 5.2. All three baths produced cobalt with about the same strength (55 to 67 kg/sq mm or 78,000 to 96,000 psi).

An adjustment in the pH of a sulfate-chloride bath from 2.0 to 1.5 increased the tensile strength of the deposit from 67 to 121 kg/sq mm (96,000 to 172,000 psi) and improved ductility slightly. (5) An increase in pH from 2.0 to 4.0 reduced ductility

TABLE 5.2

Mechanical Properties of Cobalt Electrodeposits

Bath	Solution Temperature, C	Current Density, amp/sq dm	Tensile Strength kg/sq mm	Tensile Strength psi	Elongation, percent
Chloride[a]	25	2.0	55	78,200	—
Sulfamate-bromide[c]	49	3.0	63	90,000[b]	1–3
Sulfate-chloride[e], pH = 1.5	60	4.0	121	172,000	0.019[d]
Sulfate-chloride, pH = 2.0	60	4.0	67	96,000	0.017[d]
Sulfate-chloride, pH = 2.0	70	4.0	61	87,000	0.017[d]
Sulfate-chloride, pH = 4.0	70	4.0	54	76,500	0.006[d]
Sulfate-chloride, pH = 4.0	60	4.0	79	112,000	0.010[d]

[a] The solution contained 120 g/l cobalt and 50 g/l ammonium chloride. Its pH was 4.0. The current density was 2.0 amp/sq dm. (1)

[b] The yield strength averaged about 50 kg/sq mm or 70,000 psi. (3)

[c] The solution contained 79 g/l cobalt, 30 g/l boric acid, and 14 g/l cobalt bromide. Surface tension was reduced to 30 dyne/cm. with sodium lauryl sulfate. The pH was 4.0. (3)

[d] Significant strain at fracture for 25-30 μm deposits subjected to a bulge test. (3) A strain value of 0.022 is approximately equivalent to an elongation of 1 percent. (4)

[e] The solution contained 330 g/l $CoSO_4 \cdot 7H_2O$, 45 g/l $CoCl_2 \cdot 6H_2O$, and 30 g/l H_3BO_3. After treatment with activated carbon and cobalt carbonate, the surface tension was adjusted with sodium lauryl sulfate to 35 dyne/cm. (5)

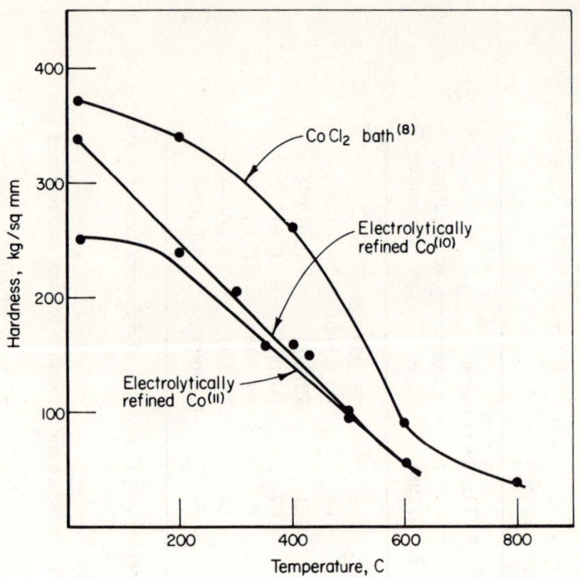

Figure 5.1. Hardness of Electrodeposited Cobalt at Elevated Temperatures.

appreciably for 25 to 30-μm-thick deposits, but strength was changed only 10 percent, or less. A slightly higher tensile strength was measured for cobalt deposited at 60 C, in comparison with deposits produced at 70 C.

Heat treating the deposits from a sulfamate bath at 470 or 630 C increased their tensile strength from about 63 to 85 kg/sq mm (from about 90,000 to 120,000 psi). (3) Yield strength was increased from about 50 to 56 kg/sq mm (70,000 to 80,000 psi). Elongation was unchanged as a result of heat treating.

Hardness

Hardness (resistance to indentation) of electrodeposited cobalt ranged from 180 to 443 kg/sq mm (Vickers), depending on the conditions of deposition and indentation load. The softer deposits (about 180 kg/sq mm) were relatively coarse-grained coatings from 4 N cobalt fluoborate baths at 45 C. (6) Harder deposits (250 to 330 kg/sq mm) were reported for cobalt from a sulfamate bath, (3) and a sulfate solution. (7) Hardness in the range of 330 to 443 kg/sq mm was measured for cobalt deposited in a chloride bath (8) and fluoborate baths operated at 25 or 30 C. (6,9)

Hot-hardness data from three sources are shown as a function of temperature in Figure 5.1. Hardness decreased rapidly at temperatures above 200 C for all three deposits. After heating to 620 C and cooling, cobalt deposited in a sulfamate bath retained a hardness of 350 kg/sq mm. (3) Cobalt deposited in a chloride solution retained a hardness of 290 kg/sq mm after heating to 800 C. (8) By contrast, the hardness of electrodeposited nickel after such a heat treatment is, as a rule, only about 150 kg/sq mm. (3)

Internal Stress

Internal stress is an inherent force in an electrodeposit free from external forces, either compressed or stretched in relation to its normal condition. In the tensilely stressed condition, the deposit has a tendency to contract; in the compressively stressed condition, the deposit has a tendency to expand.

Failures of electroplated and electroformed parts have been attributed to the presence of high tensile stresses, which may cause cracking or a reduction in the fatigure strength of the plated part. High internal tensile stresses may cause warpage or distortion of an electroformed article after separaton from the mandrel. High compressive stresses may cause the electroform to separate from the mandrel during the plating cycle. Other failures sometimes attributed to stress conditions are crazing and blistering. Internal stresses are affected by the following:

(1) Changes in the parameters of the lattice
(2) Changes in the distance between deposit crystallites
(3) Chemical compound formation in the deposit or codeposited impurities
(4) Thickness of the deposit
(5) Operating conditions such as temperature and current density
(6) Cations and anions in the electrolyte and their concentrations

The tensile stress in cobalt deposited in solutions containing chlorides ranged from 17 to 56 kg/sq mm, (24,000 to 80,000 psi), (12) (13) (14) which stresses are about twice the limits reported for nickel deposited in similar solutions. A low current density (1.6 amp/sq dm) (12) or a high solution pH (13) favored the lower values in this range. Another report indicated that stress decreased as the temperature of the bath was increased. (5)

Stress in cobalt deposited in chloride-free sulfate solutions was usually in the range of 10 to 24 kg/sq mm (14,000 to 31,000 psi). (15) (16) These limits are only one-half those reported for cobalt from chloride or chloride-sulfate baths.

A stress of 14 kg/sq mm (20,000 psi) was reported for cobalt deposited in a sulfamate bath. (17) Cobalt electroplated in fluoborate baths was low or high in stress 3.2 kg/sq mm and 12.6 kg/sq mm, respectively) depending on the concentration of the cobalt fluoborate (18)

Stress ranges for four types of cobalt plating baths are compared with hardness data in Table 5.3. The harder deposits exhibited high stress values, as a rule.

A saccharin addition (1.6 g/l) to a cobalt chloride bath at 25 C reduced the stress in cobalt deposited at 1.6 amp/sq dm from 19 to only 9.5 kg/sq mm (from 27,000 to 13,500 psi). (12) Sodium benzene sulfonate had little effect on stress. (19) Aromatic carboxylic acids, esters, and amides were used to eliminate stress. (20) Thiourea reduced stress in concentrations above 25 mg/l, and at 0.5 g/l a slight

TABLE 5.3

Hardness and Stress Ranges for Cobalt Electrodeposits

Type of Bath	Hardness, kg/sq mm	Stress, kg/sq mm	Stress, psi	Reference
Chloride or sulfate-chloride	370	17–56	24,000–80,000	5, 8, 12–14
Fluoborate at 25 or 30 C	330–443	12.6–37.2	18,000–53,000	6, 9, 18
Fluoborate at 45 or 50 C	180–350	—	—	6
Sulfate (chloride-free)	290–310	10–24	14,000–31,000	15, 16
Sulfamate	250–330	14	20,000	3, 17
Chloride with a saccharin addition	—	9.5	13,500	12

compressive stress was induced in the deposit, (21) whereas stress in the deposit produced without thiourea was about 30 kg/sq mm.

Magnetic Properties

Magnetic coatings for computer systems are important for both static and dynamic switching. Hard magnetic coatings with a high coercive force above 200 oersteds are needed for high-density permanent storage (typically on drums or disks), whereas soft magnetic coatings with a coercivity of only about 2 oersteds are desired for fast switching memory devices. Thus coercivity is an important property of cobalt and cobalt alloy coatings intended for computer systems. Some electrodeposited cobalt films exhibit high coercivity, but low coercivity has been reported for others.

Coercive force is the demagnetizing force that must be applied to reduce the induction to zero. Coercivity is the maximum value of coercive force for a specific material.

In determining the characteristics of a hysteresis loop, magnetic induction in gauss (expressed on the vertical axis) is determined for a range of coercive forces (expressed on a horizontal axis). The complete cycle includes measurements of induction at equal and opposite values of coercive force. Remanent induction (B_r) is defined as the flux remaining in the material when the magnetizing field is removed. The ratio of remanent induction to the maximum induction (B_m) obtained by applying large fields defines the squareness of the hysteresis loop. A square loop (or B_r/B_m ratio approaching 1.0) is desired for both high-density permanent storage and fast switching.

Coercive force, remanent induction and maximum induction are properties frequently reported for cobalt and cobalt alloy coatings. Spontaneous magnetization, which occurs without the application of a magnetic field (the inherent magnetization of a material), is sometimes reported, as are the anisotropy constants for coatings of a specific thickness.

Property Data

Coercivity of a coating is dependent on the crystalline structure, thickness and composition of the material. For cobalt, increasing thickness usually reduces coercivity. A hexagonal-close-packed structure appears to favor high coercivities in comparison with a mixture of hexagonal and face-centered-cubic structures. Buffering agents, stress reducers, and other solution constitutents have profound effects on the crystalline structure and magnetic properties of electrodeposited cobalt. Temperature, pH, and current density also affect the magnetic properties.

Low coercive forces of 9 to 12 oersteds were reported for 7-μm-thick deposits obtained at a current density of 1.6 amp/sq dm in a 25 C, dilute cobalt chloride solution containing 0.5 to 1.6 g/l saccharin added to reduce stress. (12) The deposits were oriented with (1010) planes parallel with the substrate, whereas an orientation of (1020) planes was detected for deposits obtained with no addition of saccharin. With no saccharin, coercive force increased to 35 oersteds. Higher concentrations of cobalt chloride increased coercive force to 70 oersteds.

An ultrasonically agitated cobalt chloride bath that did not contain saccharin was used at a low pH of 1.0 to obtain a coercive force of about 10 oersteds. (22) Supplemental information on solution composition and conditions is given in Table 5.4. Coercivity was appreciably increased when the pH was raised to 2.0 or 5.0. Eliminating ultrasonic agitation also increased coercivity.

Table 5.4 also lists cobalt sulfate baths used for depositing low coercivity deposits. All solutions were operated at a low temperature. With one exception, all chloride and sulfate baths were maintained at a low pH (1.6 or 2.0), which tends to favor the hcp crystal form oriented magnetically in a direction perpendicular to the substrate surface. (25) An increase in the pH of the sulfate baths, an increase in temperature, or a decrease in current density below 1.0 amp/sq dm increased coercivity. (15)

High coercivities were reported for cobalt deposits in some solutions with a relatively high pH or a high temperature. For example, coercivity of 1000 to 1200 oersteds was reported for cobalt deposited in a 0.2 N sulfate bath operated at 40 C. (24) The other conditions are summarized in Table 5.4. Reducing the temperature or the concentration of cobalt sulfate reduced coercivity. A coercivity of 600 oersteds was obtained for a thin, 0.05-μm film deposited in a cobalt chloride bath at 90 C. (25) A decrease in the temperature reduced coercivity to 350 oersteds. Films with a thickness greater than 1.0 μm exhibited coercivities of 100 oersteds, or less.

Intermediate values of coercivity ranging from 65 to 230 oersteds were associated with a pH above 2.0 and/or a temperature in the range of 30 to 70 C. (15, 26–28) These values are similar to coercivities reported for wrought cobalt, which tend to increase as the ratio of hcp:fcc crystal forms increases. (The theoretical coercivity value of hcp cobalt is 5,000 oersteds.)

TABLE 5.4

Conditions for Depositing Low and High Coercivity Cobalt

Coercivity, oersteds	Squareness of Hysteresis Loop (Br/Bm)	Plating Bath	pH	Temperature, C	Current Density, amp/sq dm	Reference
8	0.87–0.88	20 g/l $CoCl_2 \cdot 6H_2O$ with saccharin	2.0	25	1.6	12
10	0.57	150 g/l $CoCl_2 \cdot 6H_2O$ and 30 g/l H_3BO_3 [a]	1.0	—	10.0	22
9–17	—	195 g/l $CoSO_4 \cdot 7H_2O$ and 92 g/l $(NH_4)_2SO_4$	—	—	4.0	23
15	0.4	500 g/l $CoSO_4 \cdot 7H_2O$ and 31 g/l H_3BO_3	6.5	20	1.0	15
25	0.75	280 g/l $CoSO_4 \cdot 7H_2O$ and 3 g/l H_3BO_3	1.6	20	1.0–4.0	15
1,000–1,200	—	560 g/l $CoSO_4 \cdot 7H_2O$, 10 g/l H_3BO_3 and 65 g/l $Na_2SO_4 \cdot 10H_2O$	4.8	40	1.5	24
600	—	350 g/l $CoCl_2 \cdot 6H_2O$ and 20 g/l H_3BO_4	—	90	0.6	25

[a] Ultrasonic agitation (1 watt/sq cm) was applied in this solution.

Data satisfying a relationship between coercivity and stress were reported for cobalt deposited in unbuffed solutions. (12) The relationship was as follows:

$$H_c = \sigma\lambda/I_s$$

where H_c represents coercivity, σ is the stress, λ is magnetostriction, and I_s is the saturation induction.

Heat treatment at 440 C increased the squareness of the hysteresis loop and decreased coercive force of coatings with intermediate values of coercivity and squareness. (12)

Saturation magnetization (B_m) depends on directional aspects (orientation). Values as high as 15,000 (29) (30) and 16,600 gausses (26) (29) have been reported by different investigators. Remanence (b_r) values as high as 9,000 to 10,000 gausses (26, 28) have also been reported. The highest B_r/B_m ratio reported for cobalt deposits was 0.98. (31) A ratio of 0.7 to 0.88 has been reported by several investigators. Examples are given in Table 5.4.

The anisotropy constant of electrodeposited cobalt shifted with thickness from about 4 to 9 $\times$ 10^6 erg/cu cm as film thickness decreased. (32) A value of 8.2 $\times$ 10^6 erg/cu cm was reported for a 0.017 μm film deposited by another investigator. (33) Annealing at 450 to 700 C increased the value of the constant. (32) These cobalt deposits (up to 1.0 μm) were deposited at 1.0 amp/sq dm in a bath containing 1.1 M cobalt sulfate, 0.5 M boric acid, and 0.35 M sodium chloride.

Cobalt Alloys

Magnetic Properties

Cobalt alloy electrodeposits of chromium, iron, nickel plus iron, 13 percent phosphorus or 50 percent tungsten exhibited low coercivities, < 50 oersteds, as indicated in Table 5.5. Coercivities of < 7 oersteds were measured for the alloy containing 64 percent nickel and 14 percent iron, the 13 percent phosphorus alloy, and the 50 percent tungsten alloy. Cobalt alloy containing only 5 percent phosphorus or 30 percent tungsten was appreciably higher in coercivity.

Higher coercivities of 1400 to 1750 oersteds were obtained for cobalt alloys containing nickel and/or phosphorus. The phosphorus content of these coatings ranged from 1 to 3 percent. Sodium hypophosphite was the agent used for supplying the phosphorus content of these alloys. Supplementary information is detailed in Table 5.5.

Films about 0.1 μm in thickness with a coercivity of about 1,000 oersteds sustained bit-of-information densities up to 15,700/cm (40,000/inch), which was useful for reducing the circuitry of magnetic, flexible tape. (52)

TABLE 5.5

Magnetic Properties of Cobalt Alloy Electrodeposits

Codeposited Metal	Coercivity, oersteds	Remanence (Br), kilogauss	Squareness of Hysteresis Loop (Br/Bm)	Plating Bath
Chromium, 10%	20–50	10–16	0.2–0.75	Sulfate[a]
Iron	15–50	8–12	No data	Sulfamate (37)
Manganese, 1% and phosphorus, 5%	600–850	6–9	0.65–0.75	Chloride[b]
Nickel, 3.6%	140	8.6	0.65	Sulfate or chloride (26,29)
Nickel, 11.6%	95	4.2	0.32	Sulfate or chloride (26,29)
Nickel, 18%	120	11.0	0.73	Sulfate-chloride[c]
Nickel, 18%	50	11.0	0.65	Sulfamate[d]
Nickel, 21%	200	8.4	0.66	Sulfate or chloride (26,29)
Nickel, 25%	100–180	6.0	Near 1.0	Sulfate (27)
Nickel, 34%	500	7.4	0.70	Sulfamate[d]
Nickel, 35%	200–260	4.0–6.0	No data	Sulfate[e]
Nickel, 35%	160	3.6	No data	Chloride (48)
Nickel, 64% and iron, 14%	6	No data	No data	Sulfate-chloride (49)
Nickel, 14% and phosphorus, 1%	300–500	No data	No data	Sulfate[f]
Nickel, 19% and phosphorus, 2%	1750	No data	No data	Sulfate[f]
Nickel and phosphorus, 3%	300–900	No data	No data	Chloride[g]
Nickel and phosphorus	600–1750	No data	0.75	Sulfate[h]
Nickel, 1% and phosphorus, 7%	800–1400	6.0	0.65–0.70	Chloride[i]
Nickel and phosphorus, 2%	800–1000	No data	No data	Sulfate-chloride[j]
Phosphorus, 1%	300–1400	No data	No data	Sulfate-chloride[k]
Phosphorus, 2-½%	20–800	5.5	0.5–0.8	Chloride[l]

Phosphorus, 3%	1200–1400	No data	No data	Alkaline[m]
Phosphorus, 5%	700–750	5.3	0.65	Chloride[n]
Phosphorus, 13%	2	3.0	0.4	Chloride[n]
Tungsten, 30%	130–530	0.18–0.32	No data	Citrate[o]
Tungsten, 50%	6–10	0.002–0.003	No data	Citrate[p]

[a] 5 μm coatings deposited at 2 to 10 amp/sq dm in a bath containing 340 g/l $Cr_2K_2(SO_4)_4$, 140 g/l $CoSO_4 \cdot 7H_2O$, 150 g/l aminoacetic acid and 0.05 g/l sodium lauryl sulfate, 20 to 40 C, with a pH of 2.0 to 2.5. (34–36)

[b] Deposited at 1.5 amp/sq dm in a bath at 20 to 60 C containing 200 g/l $CoCl_2 \cdot 6H_2O$, 25 g/l $MnCl_2 \cdot 4H_2O$, 25 g/l NaH_2PO_2 and 10 g/l of the sodium salt of ethylenediamine tetraacetic acid. (38–42) Reducing the pH from the range of 2.5 to 3.5 to 1.0 reduced coercivity to the range of 20 to 80 oersteds.

[c] Deposited at 10 amp/sq dm in a 65 C bath containing 1 M Ni, 1 M Co, 0.4 M $HOCH_2COOH$ and 1.2×10^{-4} M KCNS with a pH of 3.4. (43,44) Superimposing alternating current increased coercivity to 250 oersteds.

[d] Deposited at 0.45 amp/sq dm in a room-temperature bath containing 59.4 g/l sulfamate ions, 4.96 g/l $^{2+}$, 14.85 g/l Ni^{2+}, and 6.65 g/l Cl^-. (45,46) With a pH of 2.0, an 18% Ni film was deposited with a coercivity of 50 oersteds. Adjusting the pH to between 4.0 and 6.0 increased the nickel content to 34% and the coercivity to 500 oersteds. Similar results were obtained in a Watts-type bath, but the maximum coercivity of these deposits was only 300 oersteds, obtained at a nickel content of 30%.

[e] Deposited at 1.0 to 2.0 amp/sq dm in a 50 to 60 C bath containing about 135 g/l $NiSO_4 \cdot 7H_2O$, 115 g/l $CoSO_4 \cdot 6H_2O$, 25 g/l H_3BO_3 and 12 g/l KCl with a pH of 4.0 to 5.0. (47)

[f] Deposited at 3.2 amp/sq dm in a bath containing 40 g/l $NiSO_4 \cdot 7H_2O$, 60 g/l $CoSO_4 \cdot 7H_2O$, 70 g/l NH_4Cl, 2 g/l NaH_2PO_2, and 5 ml/l of a wetting agent with the pH adjusted to 3.0. (44,50)

[g] A coercivity of 900 oersteds was obtained at 1.0 amp/sq dm in a 25 C bath containing 110 g/l $NiCl_2 \cdot 6H_2O$, 50 g/l $CoCl_2 \cdot 6H_2O$, 50 g/l Na_2SO_4, 25 g/l NH_4Cl, and 10 g/l NaH_2PO_2 with a pH of 1.0. (51) A higher current density or a higher temperature reduced coercivity. Treatment with acid increased coercivity to about 1200 oersteds.

[h] A coercivity of 1750 oersteds was obtained for deposits made at 2 to 12 amp/sq dm in a 50 C bath containing 0.15 M $NiSO_4$, 0.214 M $CoSO_4$, 1.28 M NH_4Cl, 20 g/l $NaH_2PO_2 \cdot H_2O$ and 4 ml/l of a wetting agent. (43) With a smaller concentration of NaH_2PO_2 (3 g/l), coercivity was reduced to only 825 oersteds.

[i] Deposited at 5 to 10 amp/sq dm in 40 C bath containing 140 g/l $NiCl_2 \cdot 6H_2O$, 100 to 200 g/l $CoCl_2 \cdot 6H_2O$, and 25 to 100 g/l NaH PO with the pH adjusted to between 2.5 and 3.0. (38)

[j] Deposited at current densities up to 12 amp/sq dm in a 53 C bath containing 0.214 M $CoSO_4$, 0.15 M $NiSO_4$, 1.28 M NH_4Cl, 0.02 to 0.19 M NaH_2PO_2, and a wetting agent with the pH adjusted to 3.0. (52)

[k] A coercivity of 1400 oersteds was measured for cobalt deposited at 5 amp/sq dm in a 55 C bath containing 100 g/l

$CoCl_2 \cdot 6H_2O$, 100 g/l $CoSO_4 \cdot 7H_2O$, 100 g/l NH_4Cl, and 15 g/l $NaH_2PO_2 \cdot H_2O$ with the pH adjusted to 5.5. (53) Lower coercivities were obtained with smaller concentrations of sodium hypophosphite.

(l) A coercivity of 20 oersteds was obtained for cobalt deposited at 1.0 amp/sq dm in a 0.5 M $CoCl_2$ bath containing 100 g/l NH_4Cl and 15 g/l $NaH_2PO_2 \cdot H_2O$ with the pH adjusted to 4.7. (54) Increasing the current density to 6.0 amp/sq dm increased coercivity to 800 oersteds.

(m) 1 μm deposits produced with alternating current superimposed on direct current in a 75 C bath containing 35 g/l H_3PO_4, 20 g/l $Na_2HPO_2 \cdot H_2O$, and 150 g/l Rochelle salts. (55) Reducing the concentration of Al_2SO_4 from 50 to 20 g/l or omitting the superimposed alternating current reduced coercivity to <400 oersteds.

(n) Deposited at 2.5 amp/sq dm in a 20 C bath containing 200 g/l $CoCl_2$. $6H_2O$ and 25 g/l $NH_4H_2PO_2$. (56) The high coercivity values of 700 to 750 oersteds were obtained when the pH was adjusted to 1.7 to 2.0, whereas the low coercivity value of 2 oersteds was measured for cobalt deposited in a bath with a pH of 1.0 to 1.5.

(o) Deposited at 2 to 20 amp/sq dm in a 70 C bath containing 110 g/l $CoSO_4 \cdot 7H_2O$, 25 g/l $Na_2WO_4 \cdot 2H_2O$ and 200 g/l citric acid. (48)

(p) Deposited at 10 or 20 amp/sq dm in a 70 C bath containing 12.5 g/l $CoSO_4 \cdot 7H_2O$, 39 g/l $Na_2WO_4 \cdot 2H_2O$ and 66 g/l citric acid with a pH of 4.0 to 8.0. (48)

Several cobalt-nickel-iron alloys with zero magnetostriction were identified, including the following: (57)

Nickel, percent	Iron, percent
64.0	16.0
68.5	17.5
72.0	18.0
74.0	19.0

Mechanical and Physical Properties

Tensile strength and ductility have been measured for cobalt alloys of manganese, nickel, phosphorus, and tungsten, with the results summarized in Table 5.6. Cobalt alloy containing 1/2 percent nickel exhibited a tensile strength of 120 to 130 kg/sq mm (170,000 to 190,000 psi), a density of 8.86 g/cu cm, and endurance limit of 39 kg/sq mm (55,000 psi) at room temperature and a dynamic modulus of 20,000 kg/sq mm (29 × 10^6 psi). (57)

The ultimate tensile strength of all cobalt alloys reached a maximum of 19.2 kg/sq mm (275,000 psi) for alloy containing 50 percent nickel. (58) A cobalt alloy containing 60 percent nickel retained much of its tensile strength and hardness after heat treatment at 340 C, as shown in Figure 5.2. (3) However, unalloyed cobalt

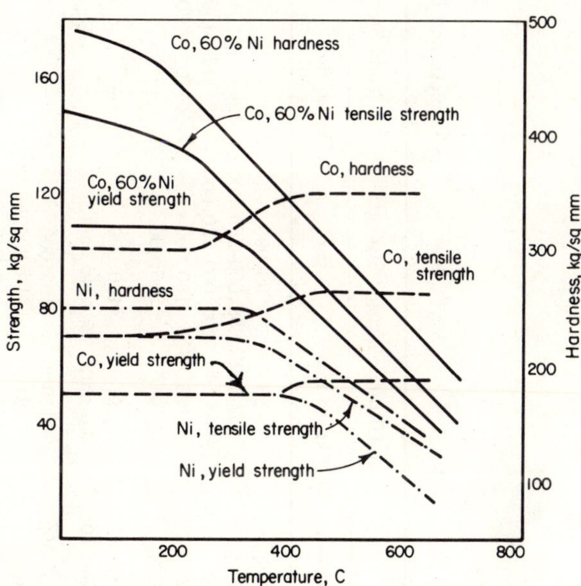

Figure 5.2. Strength and Hardness of Cobalt, Cobalt-Nickel Alloy, and Nickel after Heat Treatment. (3)

TABLE 5.6

Strength and Ductility of Cobalt Alloys

Codeposited Metal	Ultimate Tensile Strength		Yield Strength		Elongation, percent	Modulus of Elasticity		Hardness, kg/sq mm	Plating Bath
	kg/sq mm	psi	kg/sq mm	psi		kg/sq mm	psi × 10^6		
Nickel, 20%	56	80,000	31.5	45,000	1	No data	No data	250	Sulfamate[a]
Nickel, 50%	180–192	255,000–275,000	126	180,000	—	No data	No data	—	Sulfamate[b]
Nickel, 60%	140–154	200,000–220,000	98–112	140–160,000	2–4	20,300	29.0	450	Sulfamate[a]
Nickel, 80%	126	180,000	56–70	80–100,000	3–6	No data	No data	300	Sulfamate[a]
Nickel, 99%	144	206,000	—	—	4	18,000	25.8	510	Sulfate-chloride[c]
Phosphorus, 1%	91	130,000	—	—	7	No data	No data	380	Chloride[d]
Phosphorus, 10%	No data	No data	—	—	Brittle	No data	No data	700	Chloride[e]
Tungsten, 20%	110	155,000	—	—	Brittle	17,000	25.0	500	Alkaline[f]
Tungsten, 28%	84	119,800[i]	—	—	—	No data	No data	546	Alkaline[j]
Tungsten, 35%[g]	82	116,500	53	74,700	1	No data	No data	450–500	Alkaline[h]
Tungsten, 35%[k]	75	106,600	—	—	—	No data	No data	532	Alkaline[l]
Tungsten, 30% and Nickel, 5%	35	49,600[m]	—	—	—	No data	No data	550	Alkaline[n]
Tungsten, 35%, iron 4% and nickel, 1%	71	102,000[o]	—	—	—	No data	No data	470	Alkaline[p]

[a] Deposited at 3 amp/sq dm in a 49 C sulfamate solution containing about 90 g/l Ni^{2+} + Co^{2+}, 14 g/l $NiBr_2$ and 30 g/l H_3BO_3 with the pH adjusted to about 3.0. (3)

[b] Deposited at 5.4 amp/sq dm in a 20 C sulfamate bath containing 75 g/l Ni^{2+}, 75 g/l Co^{2+}, 30 g/l H_3BO_3, and a wetting agent with a pH of 4.0. (58)

[c] Bright plate deposited at 5 amp/sq dm in a 65 C bath containing 240 g/l $NiSO_4 \cdot 7H_2O$, 20 g/l $CoSO_4 \cdot 7H_2O$, 30 g/l $NiCl_2 \cdot 6H_2O$, 30 g/l H_3BO_3, 0.75 g/l NH_4Cl, 45 g/l $Ni(HCO_2)_3$, and 2.5 g/l CH_2O with a pH of 3.7. Stress was 10.5 kg/sq mm (15,000 psi) in tension. (59)

[d] Deposited at 5 to 40 amp/sq dm in a 75 to 95 C bath containing 210 g/l $CoCl_2 \cdot 6H_2O$, 44 g/l $Co_3(PO_4)_2$, 50 g/l H_3PO_4, and 2.0 g/l H_3PO_3 with a pH of 0.5 to 1.0. (13,60)

[e] Deposited at 5 to 40 amp/sq dm in a 75 to 95 C bath containing 180 g/l $CoCl_2 \cdot 6H_2O$, 88 g/l $Co_3(PO_4)_2$, 50 g/l H_3PO_4, and 40 g/l H_3PO_3 with a pH of 0.5 to 1.0. (13,60)

[f] Deposited at 5 amp/sq dm in a 90 C bath containing 160 or 200 g/l $CoCl_2 \cdot 6H_2O$, 27 or 36 g/l $Na_2WO_4 \cdot 2H_2O$, 400 g/l Rochelle salts and 50 g/l NH_4Cl with the pH adjusted to 8.5(2). (61)

[g] Property data for specimens heat treated 16 hours at 1093 C.

[h] Deposited at 5 amp/sq dm in a 85 C bath containing 120 g/l $CoSO_4 \cdot 7H_2O$, 30 g/l $Na_2WO_4 \cdot 2H_2O$, 380 g/l Rochelle salts, 50 g/l NH_4Cl and 0.5 g/l of a wetting agent with the pH adjusted to between 8.0 and 9.0. (61)

[i] Property values for specimens heat treated at 1350 C and aged for 16 hours at 850 C.

[j] Deposited at 5 amp/sq dm in a hot solution containing 100 g/l $CoSO_4 \cdot 7H_2O$, 42 g/l $Na_2WO_4 \cdot 2H_2O$, 365 g/l Rochelle salts, 55 g/l NH_4Cl and 0.5 g/l of a wetting agent with the pH adjusted to between 8.5 and 9.0. (62)

[k] Property values for specimens heat treated at 1122 C and aged at 1300 C.

[l] Deposited at 5 amp/sq dm in an 85 C bath containing 115 g/l $CoSO_4 \cdot 6H_2O$, 43 g/l $Na_2WO_4 \cdot 2H_2O$, 380 g/l Rochelle salts, 50 g/l NH_4Cl and 0.5 g/l of a wetting agent with the pH adjusted to between 8.0 and 9.0. (61)

[m] Strength of specimens heat treated 4 hours at 1140 C and aged 4 hours at 700 C.

[n] Deposited at 5 amp/sq dm in a 21 C bath containing 115 g/l $CoSO_4 \cdot 7H_2O$, 5 g/l $NiSO_4 \cdot 7H_2O$, 43 g/l $Na_2WO_4 \cdot 2H_2O$, 380 g/l Rochelle salts, 50 g/l NH_4Cl and 0.5 g/l of a wetting agent with the pH adjusted to between 8.0 and 9.0. (61)

[o] Strength of specimens heated treated at 1340 C and aged 16 hours at 750 C.

[p] Deposited at 4 amp/sq dm in a 90 C bath containing 86 g/l $CoSO_4 \cdot 7H_2O$, 7 g/l $NiSO_4 \cdot 7H_2O$, 9 g/l $FeSO_4 \cdot 7H_2O$, 48 g/l $Na_2WO_2 \cdot 2H_2O$, 40 g/l NH_4Cl, 110 g/l $(NH_4)HC_6H_5O_7$ with the pH adjusted to between 8.5 and 9.0. (63)

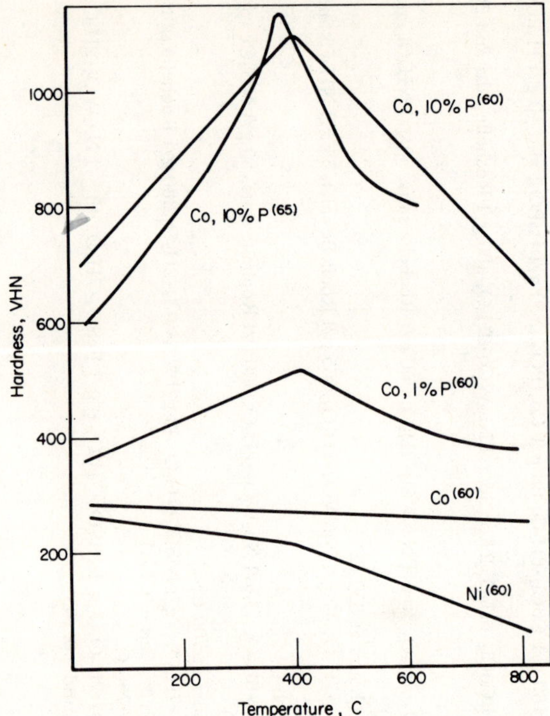

Figure 5.3. Hardness of Cobalt-Phosphorus Alloy Electrodeposits after Heat Treating.

retained a higher strength and hardness after heating to 620 C. Supplementary property data for a 60 percent nickel alloy are as follows: (64)

Density, g/cu cm	9.0
Linear coefficient of expansion (20 to 260 C), microinch/inch/C	13.1
Specific resistivity, microhm-cm	7.1
Modulus of elasticity, kg/sq mm	20,300
Stress (compressive), kg/sq mm	5.0

The agent or procedure used to induce a compressive stress was not disclosed.

Cobalt alloys containing 1 or 10 percent phosphorus were hardened after heat treatment at 400 C and retained their original hardness after heating at 800 C, as indicated in Figure 5.3. Figure 5.4 shows hot-hardness data for the cobalt-phosphorus alloys.

Cobalt alloys containing 25 to 35 percent tungsten remained harder at high temperatures than cobalt-phosphorus alloys. The hot-hardness data for these alloy electrodeposits are shown in Figure 5.4. Figure 5.5 shows tensile strength data at high temperatures for cobalt alloys of nickel and tungsten. An alloy containing 35 percent tungsten, 3 percent iron, and 1 percent nickel retained high tensile strength (50 kg/sq mm) at 900 C. (63)

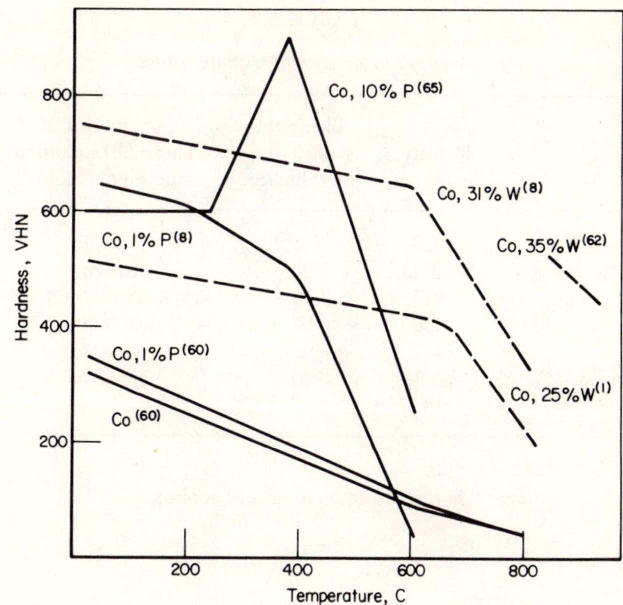

Figure 5.4. Hot Hardness of Cobalt-Phosphorus and Cobalt-Tungsten Alloys.

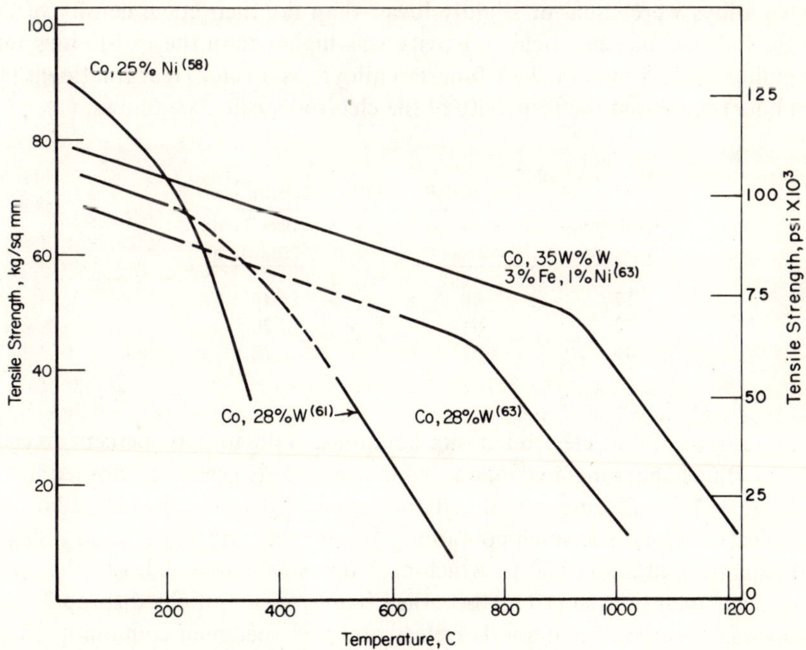

Figure 5.5. Tensile Strength at High Temperatures of Cobalt Alloys of Nickel and Tungsten.

TABLE 5.7

Physical Properties of Cobalt Alloys

Codeposited Metal	Density, g/cu cm	Electrical Resistivity, microhm-cm	Coefficient of Thermal Expansion, microinch/inch/C	Reference
Nickel, 60%	9.0	7.1	13.1	64
Phosphorus, 1.3%	No data	28	No data	60
Tungsten, 17%	9.7	58[a]	13.8 (20–950 C)	1
Tungsten, 30%	10.6	84	16.2 (20–1050 C)[b]	1, 66
Tungsten, 34%	10.7	93	No data	1
Tungsten, 34%, iron, 3%, and nickel, 1%	12.3	100	No data	63

[a] Contained 0.165% O_2.
[b] Contracted 0.8% in length as a result of heating and cooling.

Density, electical resistivity, and thermal expansion data for cobalt alloys of nickel, phosphorus, and tungsten are given in Table 5.7. Density values for the tungsten alloys were equal or slightly lower than the theoretical density of the metallurgical alloys, but electrical resistivity was higher than the resistivities reported for metallurgical forms of cobalt-tungsten alloys, as a rule. Heat treatment (1200 C for an hour) decreased the resistivity of the electrodeposits, as follows: (1)

Tungsten, percent	Electrical Resistivity, microhm-cm: Before Heat Treatment	After Heat Treatment
17	58	46
30	84	70
34	93	76

Similar tungsten alloy electrodeposits contained 0.06 to 1.14 percent oxygen and 0.008 to 0.011 percent hydrogen, which probably were combined as water associated with small amounts of salts occluded in the deposit. One dilation curve for an electroformed specimen containing 10 percent tungsten showed a disproportional expansion at about 850 C, which probably was associated with the volatilization of such inclusions. (1) On the other hand, only a slight disproportional expansion was observed when another electroformed specimen containing 30 percent tungsten was dilated to 1050 C. (66) The oxygen content of this specimen was not determined.

Stress is compared in Table 5.8 with hardness and electrical resistivity data for

TABLE 5.8

Electrical Resistivity, Stress, and Hardness of Electrodeposited Cobalt Alloys

Electrodeposited Metal	Electrical Resistivity, microhm-cm	Hardness, kg/sq mm	Stress[a] kg/sq mm	Stress[a] psi	Reference
Copper, 10%	<2	205	No data	—	67
Copper, 30%	10	285	No data	—	67
Chromium, 10%	No data	400–500	12–26	17,000–37,000	34
Nickel, 11%	11.6	350	6	9,000	2, 68[b]
Nickel, 18%	No data	—	27.5[c]	39,000	50
Nickel, 18%	No data	400	9.0	13,000	68[b]
Nickel, 24%	11.6	360	—	—	2
Nickel, 31%	—	500	11	16,000	68[b]
Nickel, 39%	11.6	440	—	—	2
Nickel, 44%	—	500	13	19,000	68[b]
Nickel, 49%	—	450	15	21,000	68[b]
Nickel, 61%	10.8	390	—	—	2
Nickel, 98%	11.5	510	10.5	15,000	59
Phosphorus, 1.3%	28	380	25.3[d]	36,000	60
Palladium, 40%	6.5	300, 450	10	14,200	69[e], 70[f]
Palladium, 50%	7.5	500	—	—	70[f]
Palladium, 65%	7.0	340,400	85[g]	120,000	69[h], 70[f]
Silver, 14%	20.0	290(370)	—	—	71[i]
Silver, 50%	19.0	275(360)	—	—	71[i]
Silver, 68%	7.0	150(220)	—	—	71[i]
Silver, 80%	4.0	120(130)	—	—	71[i]
Vanadium, 4%	5.0	450	—	—	72[j]
Vanadium, 6%	5.0	500	—	—	72[j]
Vanadium, 10%	3.0	425	—	—	72[j]
Tungsten, 34%, iron, 2% and nickel, 1%	100	—	53.5[k]	76,400	63

[a] A negative sign indicates compressive stress.

[b] Stress and hardness data are given for deposits obtained at 5.4 amp/sq dm in a 60 C bath containing 600 g/l nickel sulfamate, 2.0 to 10.5 g/l cobalt sulfamate and 40 g/l H_3BO_3 with a pH of 4.0. (68)

[c] Superimposing alternating current reduced stress from 27 to 6.4 kg/sq mm. (50)

[d] The addition of saccharin reduced stress to 18.3 kg/sq mm. (60)

[e] Stress data for alloy deposited at 5 amp/sq dm in a 20 C bath containing 40 g/l $[Pd(NH_3)_4]Cl_2$, 24 g/l $[Co(NH_3)_6]Cl_2$, and 50 g/l NH_4Cl with the pH adjusted to between 9.0 and 10.0. (69)

[f] Resistivity data for alloy deposited at 0.3 amp/sq dm in a 20 C bath containing 7 g/l $Pd^{2+} + Co^{2+}$, 200 to 500 g/l $K_4P_2O_7$, and 20 g/l Rochelle salts with a pH of 7.9. (70)

[g] Adding 0.5 g/l saccharin reduced stress to 30 kg/sq mm. (69)

[h] Stress data for alloy deposited at 1.5 amp/sq dm in the solution described in Footnote (e). (69)

[i] Data for alloy deposited in cyanide-pyrophosphate solutions. The microhardness figures in parentheses refer to data after aging 1.0 year. (71)

[j] Deposited at about 6 amp/sq dm in a 20 C bath containing cobalt chloride and sodium vanadate with a pH < 1.0. (72)

[k] Reduced to 24.3 kg/sq mm with an addition of octylphenoxyethanol to the alkaline bath. (63)

several cobalt alloys. Most of the alloys are highly stressed, but stress in some nickel alloys was reduced with one of three special procedures, as follows:

(a) Increasing the concentration of cobalt sulfamate to 600 g/l reduced the stress of nickel alloys to between 9 and 15 kg/sq mm. (68)
(b) Superimposing alternating current reduced stress of an 18 percent nickel alloy from 27 to 6.4 kg/sq mm. (50)
(c) Adding saccharin to the alloy plating solution reduced the stress of cobalt alloy deposits containing nickel, phosphorus, or palladium. A 1.5 g/l addition reduced the stress of a nickel alloy from 40 to 5.2 kg/sq mm. (72) Saccharin reduced the stress of a 1.3 percent phosphorus alloy from 25 to 18 kg/sq mm. (60) Only 0.5 g/l reduced the stress of a 65 percent palladium alloy from 85 to 30 kg/sq mm. (69)

Structure

The microstructure of cobalt electrodeposits is usually columnar or fibrous. Figure 5.6 shows the fibrous structure of cobalt deposited in a sulfamate bath at 49 C. A fibrous structure also was obtained for cobalt alloy containing 17 or 22 percent nickel. A banded structure like that in Figure 5.7 was obtained for cobalt alloy containing 53, 73, or 75 percent nickel, however. (3,58) A banded structure or a mixed banded and fibrous structure was observed for cobalt alloys containing 2 or 3 percent phosphorus (37,60) or 23 percent tungsten. (1,61) The structure of alloy containing about 3 percent phosphorus is shown in Figure 5.8.

After heat treating at 600 C, initially banded nickel alloys were finegrained and equiaxed. (3) Heating at this temperature also recrystallized the phosphorus alloys. (37,60) Figure 5.9 shows the structure of a recrystallized alloy containing about 3 percent phosphorus. Cobalt-tungsten alloy deposits were recrystallized (and hardened) at 730 C (61) or 800 C. (1)

A mixture of hexagonal-close-packed and face-centered-cubic lattice structures has been observed by several investigators in cobalt and cobalt-rich alloy electrodeposits, whereas the face-centered-cubic structure is characteristic of cobalt alloys containing more than 30 percent nickel (73,74) or 5 percent iron. (75) A low pH (< 2.5) favored the cubic form for cobalt, whereas a high pH or a high temperature increased the proportion of hexagonal cobalt in deposits produced in a sulfate bath with organic addition agents. (76) With no organic additive, the hexagonal form was predominant. The hexagonal structure was also approached for cobalt deposited in sulfate baths containing no addition agents when the pH was > 2.9. (33,-77,78) Applying a magnetic field during deposition favored hexagonal cobalt. (33) Agitation of a sulfate bath also favored hexagonal cobalt. (79)

A high deposition temperature also induced the hexagonal structure for cobalt deposited in a chloride bath. (80) High current densities favored the cubic form for cobalt deposited in sulfamate solutions. (81)

For cobalt, transformation from the hexagonal to the cubic form was observed upon heating to 425 C. (11)

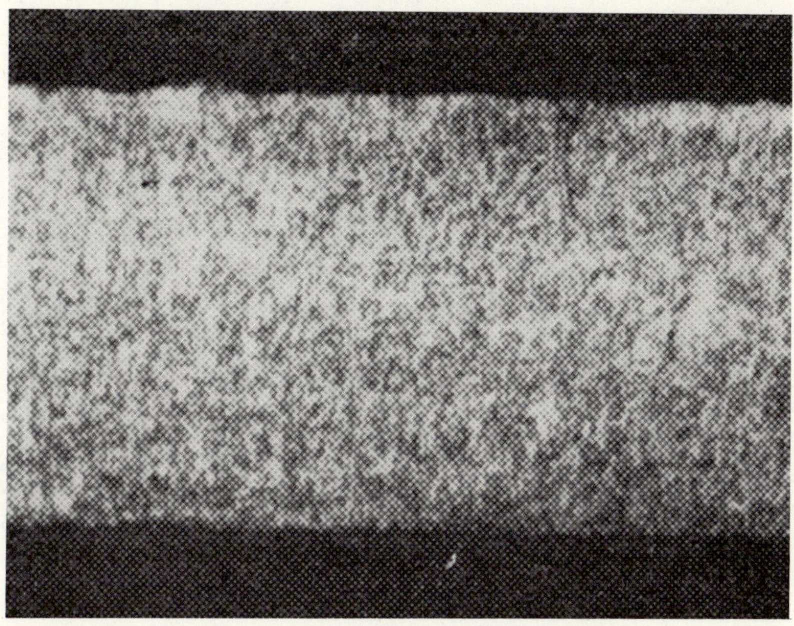

Figure 5.6. Fibrous Structure of Cobalt Electrodeposited in a Sulfamate Solution. (300×) (3)

Figure 5.7. Banded Structure of Cobalt Alloy Containing 53 Percent Nickel. (300×) (3)

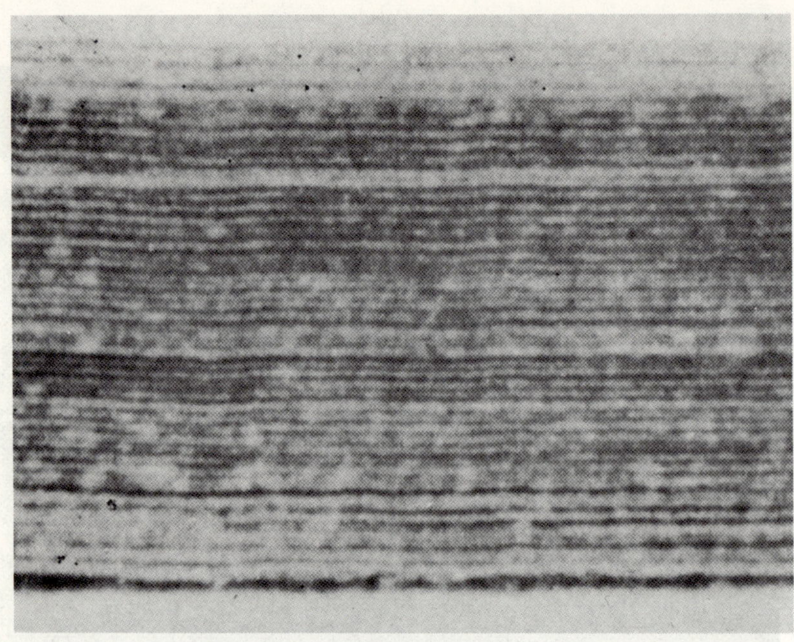

Figure 5.8. Microstructure of a 770-Oersted Cobalt-Phosphorus Deposit Showing Laminations Oriented Parallel to Basis Metal. (1000×) Etchant = H_3PO_4 + H_2O_2 (37)

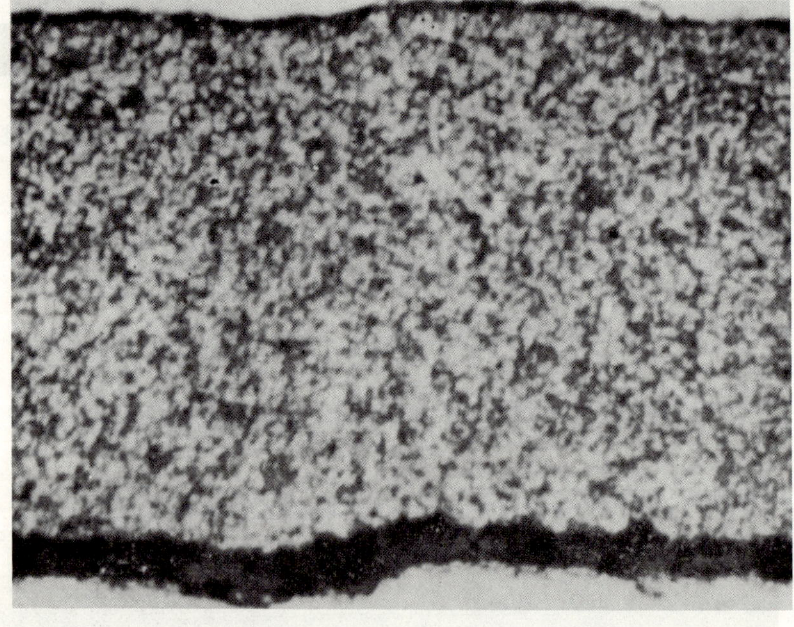

Figure 5.9. Photomicrograph of Cobalt-Phosphorus Alloy Heat-Treated for 4 Hours at 600 C. The columnar and laminated structure has disappeared and the precipitate has coarsened and agglomerated. (1000×) Etchant = H_3PO_4 + H_2O_2 (37)

Cobalt-phosphorus alloy electrodeposits are regarded as supersaturated solid solutions, (1,65,81) which transform to two phases, including Co_2P, upon heat treating to 320 C. The crystallite size is less than 100 angstroms. (51,82) A mixture of hexagonal and cubic cobalt was identified in deposits containing 1 to 7 percent phosphorus. (83) The hexagonal form was dominant for cobalt deposited in a solution with the pH > 3.0, which was associated with an increase in coercivity and a reduction in stress. (84)

The cobalt alloys containing 10 to 33 percent tungsten also are supersaturated solutions, (1,48,85,86) but Co_3W patterns appeared during X-ray examination of some deposits produced at high current densities. (48, 86) A hexagonal structure was observed in cobalt alloy containing up to 35 percent tungsten, deposited in a tartrate bath. (87,88) Crystals were oriented with the (1010) and (1011) planes parallel with the substrate surface.

Cobalt alloys containing up to 20 percent copper deposited in pyrophosphate solutions also were identified as supersaturated solid solutions. (67)

A high degree of preferred orientation has been observed for cobalt deposited in sulfate baths, but cobalt deposited in chloride solutions was randomly oriented. (80) Current density affects the orientation, with high current densities favoring a random distribution. (74) Current density also influenced the orientation of a cobalt-nickel-phosphorus alloy (3% phosphorus). (51) With a low current density, the (0002) plane paralleled the substrate surface, whereas the (1010) plane was parallel to the surface when cobalt was deposited at a high current density.

References

(1) Brenner, A., Burkhead, P., and Seegmiller, E., "Electrodeposition of Tungsten Alloys Containing Iron, Nickel, and Cobalt," Research Paper No. 1834, *Journal Research National Bureau of Standards, 39* (4), 351–383 (1947).

(2) Fedot'ev, N. P., Vyacheslavov, P. M., and Kashcheeva, E. A., "An Investigation of Some 'Structure-Sensitive' Properties of Ni-Co Galvanic Alloy," *Journal Applied Chemistry (USSR), 36* (11), 2396–2398 (1963); *Zh. Prikl. Khimii, 36* (11), 2474–2477 (1963).

(3) Endicott, D. W., and Knapp, Jr., J. R., "Electrodeposition of Nickel-Cobalt Alloy: Operating Variables and Physical Properties of the Deposits," *Plating, 53*, 43–60 (1966).

(4) Read, H. J., and Whalen, T. J., "The Ductility of Plated Coatings," *Tech. Proceedings Am. Electroplaters' Soc., 46*, 318–325 (1959).

(5) Lindsay, J. H., and Read, H. J., "Some Properties of Electrodeposited Cobalt," *Plating, 57* (5), 497–503 (1970).

(6) Fegredo, D. M., and Balachandra, J., "Studies in the Electrodeposition of Cobalt from Fluoborate Bath," *Journal of the Indiana Institute of Science, 35B*, 191–201 (1953).

(7) Fedot'ev, N. P., Aleskovskii, V. B., Vyacheslavov, P. M., Volkhonskii, N. V., and Romanova, D. L., "Influence of Electrolysis Conditions on the Roughness and Hardness of Electrolytic Cobalt Deposits," *Journal Applied Chemistry* (*USSR*), *32* (7), 1575–1578 (1959); *Zh. Prikl. Khimii, 32* (7) 1542–1546 (1959).

(8) Brenner, A., "A Microhardness Tester for Metals at Elevated Temperatures," *Plating, 38* (4), 363–366 (1951).

(9) Marikar, Y. M. F., and Vasu, K. I., "Iron, Cobalt, and Nickel Deposition—A Study of Fluoborate Baths," *Metal Finishing, 67* (8), 59–62, (1969).

(10) Unpublished data in the files of the Cobalt Information Center, Columbus, Ohio.

(11) Chubb, W., "Contribution of Crystal Structure to the Hardness of Metals," *Journal of Metals, AIME Transactions, 203*, 189–192 (1955).

(12) Fisher, R. D., "The Influence of Residual Stress on the Magnetic Characteristics of Electrodeposited Nickel and Cobalt," *Journal Electrochem. Soc., 109* (6), 479–485 (1962).

(13) Brenner, A., Couch, D. E., Williams, E. K., "Electrodeposition of Alloys of Phosphorus and Nickel or Cobalt," Research Paper No. 2061, *Journal Research National Bureau of Standards, 44* (1) 109–122 (1950).

(14) Hothersall, A. W., "Stress in Electrodeposited Metals," *J. Inst. Metals, Symposium on Internal Stresses in Metals and Alloys, Reprint No. 1082*, (1947), 12 pp.

(15) Polukarov, Yu. M., "The Structure and Magnetic Characteristics of Electrolytic Deposits of Ferromagnetic Metals and Alloys Prepared Under Different Conditions," *Russian Journal of Physical Chemistry, 34* (1), 68–71 (1960). II. "Dependence of the Structure and Coercive Force of Cobalt Deposits on the Electrochemical Conditions of Their Preparation," *Zh. Fiz. Khimii, 34* (1), 150–156 (1960).

(16) Bodnevas, A., and Matulis, J., "Internal Stresses of Electrolytic Cobalt Deposits. I. The Influence of the Electrolyte pH, the Current Density, and the Thickness of the Deposit on the Internal Stress," *Lietuvos TSR Mokslu Akad. Darbai Ser. B, 1*, 49–60 (1963); and "II. Influence of Additions on Internal Stress," *Lietuvos TSR Mokslu Akad. Darbai, Ser. B, 2*, 9–21 (1963).

(17) Barrett, R. C., "Plating of Nickel, Cobalt, Iron, and Cadmium from Sulfamate Solutions," *Tech. Proceedings Am. Electroplaters' Soc., 47*, 170–175 (1960).

(18) Fegredo, D. M., and Balachandra, J., "Measurement of Internal Stress of Cobalt Deposited Electrolytically from Cobalt Fluoborate baths," *Journal of Scientific and Industrial Research (India), 13B*, 753–755 (1954).

(19) Gaigalas, K., Bodnevas, A., Matulis, J., "Effect of Some Organic Sulfur-Containing Additives on Internal Stresses of Electrolytic Cobalt Deposits," *Issled. Obl. Elektroosazhdeniya Metal., Mater. Respub. Kong, Elektrokhim, Litov. SSR., 10*, 50–54 (1968).

(20) Fisher, A., "Electrodeposition of Stress-Free Metal Deposits," U. S. Patent 3,338,804 (August 29, 1967). Assigned to Kewanee Oil Company.

(21) Vyagis, Yu. K., Bodnevas, A. I., and Matulis, Yu., "Some Properties of Electrolytic Deposits of Nickel and Cobalt Obtained in the Presence of Thiourea," *Protection of Metals, 1* (5), 468–471 (1965); *Zashchita Metallov, 1* (5), 525–529 (1965).

(22) Abramova, N. I., Balashova, N. N., Meshcherskaya, E. I., "Porosity, Brightness, and Hardness of Cobalt-Nickel-Phosphorus Electrolytic Alloy Prepared in an Ultrasonic Field," *Electrochemistry, 5* (5), 528–529, (1969); *Elektrokhimiya, 5* (5), 574–575 (1969).

(23) Bursuc, I., Ojog, A., and Tutovan, V., "Contributions to the study of Magnetism of Electrodeposited Cobalt," *Physica Status Solidi, 4*, 161–165 (1964).

(24) Gofman, N. T., Nikolaishvili, T. N., "Electrolytic Production of Cobalt Deposits with Given Characteristics," *Izvestiya Vysshikh Uchebnykh Zavedeniy SSSR, Khimiya i Khimicheskaya Tekhnologiya, 8* (1), 104–110 (1965).

(25) Reimer, L., "Magnetic Investigations of Electrolytically Deposited Thin Cobalt Layers," *Zeitschrift für Naturforschung, 12a*, 1014–1015 (1957).

(26) Quinn, H. F., "Further Studies of the Transition Region of a Cobalt Sulfate Electrolyte," *Electrochemical Society, Extended Abstracts*, Fall Meeting, Detroit, Michigan, 109 (1961).

(27) Kaznachey, B. Ya., and Zhogina, V. M., "Electrodeposition of Magnetohard Alloys," *Trudy Vsesoyuznogo Nauchno-issledo-Vatelsuogo HN–TA Zvukozapisi, 6*, 119–135 (1959).

(28) Zappone, P. P., U.S. Patent 2,619,454 (November 25, 1952). Assigned to Brush Development Company.

(29) Quinn, H. F., and Croll, I. M., "Electrocrystallography of Cobalt and Cobalt-Nickel Alloys," *Advances in X-ray Analysis, 4*, 151–174 (1961).

(30) Procopiu, S., Viscrian, I., and Naum, C., "The Magnetization of Thin Cobalt Electrodeposits on Platinum," *Comptes Rendus, 261*, 341–344 (1965).

(31) Tyndall, E. P. T., and Wertz-Baucher, W. W., *Phys. Rev., 35*, 292 (1930).

(32) Andra, W., and Spinarke, C., "Magnetocrystalline Anisotropy of Electrolytically Deposited Single-Crystal Nickel-Cobalt Films," *Physica Status Solidi, 33* (2), 1155–1157 (1969).

(33) Goddard, J., and Wright, J. G., "The Effect of Solution pH and Applied Magnetic Field on the Electrodeposition of Thin Single-Crystal Films of Cobalt," *British Journal of Applied Physics, 15* (7), 807–814 (1964).

(34) Potapov, I. I., Kudryavtsev, N. T., and Kudakova, T. I., "Electrodeposition of a Cobalt-Chromium Alloy From a Solution of Chromium-Potassium Alum," *Protection of Metals, 4* (3), 296–297 (1968); *Zashchita Metallov, 4* (3), 323–324 (1968).

(35) Kudryavtsev, N. T., Pshiluski, Ya. B., Potapov, I. I., Smirnova, T. G., and Cupak, T. E., "Cr and Cr Alloy Electrodeposition from Solutions of Trivalent Cr Salts," *Korose Ochrana Mater., 10* (4), 7375 (1966).

(36) Kudryavtsev, N. T., Potapov, I. I., and Mel'nikova, M. M., "Study of the Process of Electrodeposition of Cobalt-Chromium Alloy," *Protection of Metals, 2* (2), 181–184 (1966); *Zashchita Metallov, 2* (2), 216–220 (1966).

(37) Zentner, V., "Electrodeposited and Electroless Magnetic Alloys for Computers," *Plating, 52* (9), 368–872 (1965).

(38) Bondar, V. V., Mel'nikova, M. M., and Polukarov, Yu. M., "Electrodeposition of Magnetically Hard Alloys. II. Electrodeposition of Cobalt-Nickel-Phosphorus and Cobalt-Manganese-Phosphorus Alloys," *Inform. Sistemy, Moscow*, 124–127 (1964).

(39) Bondař, V. V., Mel'nikova, M. M., and Polukarov, Yu. M., "Electrodeposition of Magnetically Hard Alloys I. Electrodeposition of Cobalt-Phosphorus Alloys," *Inform. Sistemy, Moscow*, 117–123 (1964).

(40) Bondař, V. V., Mel'nikova, M. M., and Polukarov, Yu. M., "Electrodeposition of Hard Magnetic Alloys," *Nekotorye Vopr. Teorii i Praktiki Ispol'z. v Galvanotekhn. Neyadovit. Elektrolitov, Kazan*, 68–71, (1964).

(41) Bondař, V. V., Mel'nikova, M. M., and Polukarov, Yu. M., "Electrodeposition of Hard Magnetic Co-Mn-P Alloys," *Protection of Metals, 1* (5), 476–479 (1965); *Zashchita Metallov, 1* (5), 534–538 (1965).

(42) Bondař, V. V., and Mel'nikova, M. M., "Mechanism of Electrodeposition of Cobalt-Phosphorus Alloys," *Protection of Metals, 1* (5), 472–475 (1965); *Zashchita Metallov, 1* (5), 530–533 (1965).

(43) Koretzky, H., "Electrodeposited Magnetic Films: A Critical Survey," *Proceedings of 1st Australian Conf. on Electrochemistry*, New York City (1965).

(44) Bate, G., Speliotis, D. E., and Morrison, J. R., "Static Magnetization Reversal in Thin Films of Co-Ni-P," *Journal Applied Physics, 35* (3), 972–973 (1964).

(45) Tsu, I., "A New Cobalt-Nickel Sulfamate Bath," *Plating 48*, 1207–1210 (1961).

(46) Tsu, I., and Fritsch, M. C., "Electrodeposition of Magnetic Cobalt-Nickel Alloys," U.S. Patent 3,111,463 (November 19, 1963).

(47) Zhogina, V. M., and Kaznachei, B. Ya., "Electrodeposition of Magnetic Hard Alloys," *Trudy Soveschaniya po Elektrokhimii, 4*, 506–511 (1956); published in 1959.

(48) Polukarov, Yu. M., "The Production of Coatings of Special Magnetic Properties," *Electrodeposition of Alloys*, Edited by V. A. Averkin. Translated by the Israel Program for Scientific Translations, Jerusalem, 52–69 (1964). *OTS* 64-11015.

(49) Wolf, I. W., "Electrodeposition of Magnetic Materials," *Journal Applied Physics, Supplement, 33* (3), 1152–1159 (1962).

(50) Bate, G., "Thin Metallic Films for High-Density Digital Recording," *IEEE Transactions on Magnetics, MAG-1* (3), 193–205 (1965).

(51) Elmore, G. V., and Bakos, P., "Cobalt/Nickel Electrolytes Containing Phosphite or Benzoate Additive Developed for Magnetic Plating," *Journal Electrochem. Soc., 111* (11), 1244–1248 (1964).

(52) Judge, J. S., Morrison, J. R., and Speliotis, D. E., "Flexible Recordings Surfaces of Electrodeposited Cobalt-Nickel-Phosphorus," *Plating, 53* (4), 441–450 (1966).

(53) Tsu, I., and Cappell, R. J., "The Influence of Plating-Bath Variables on the Composition and Coercivity of Cobalt-Phosphorus Electrodeposits," *Plating, 52*, 1123–1126 (1965).

(54) Sallo, J. S., and Carr, J. M., "Studies of High Coercivity Cobalt-Phosphorus Electrodeposits," *Journal Applied Physics, Supplement, 33* (3), 1316–1317 (1962).

(55) Wilding, B. R., "Plated Magnetic Coatings," *Trans. Inst. Metal Finish., 46* (3), 101–106 (1968).

(56) Bondar̆, V. V., Mel'nikova, M. M., and Polukarov, Yu. M., "Magnetic Properties and Structure of Electrolytic Cobalt-Phosphorus Alloys as a Function of Electrolyte pH," *Protection of Metals, 4* (3), 317–318 (1968); *Zashchita Metallov, 4* (3), 341–342 (1968).

(57) Toledo, E., and Mo, R., "Effects of Cobalt on Properties of Plated Wire," *Plating, 57* (1), 43–45 (1970).

(58) McFarlen, W. T., "An Evaluation of Electroformed Nickel/Cobalt Alloy Deposits," *Plating, 57* (1), 46–50 (1970).

(59) Brenner, A., Zentner, V., and Jennings, C. W., "Physical Properties of Electrodeposited Metals," *Plating, 39*, 865–927, 933 (1952).

(60) Brenner, A., Couch, D. E., and Williams, E. K., "Electrodeposition of Alloys of Phosphorus and Nickel or Cobalt," *Plating, 37*, 36–42; 161–163 (1950).

(61) Turns, E. W., and Hildebrand, J. F., "Electroforming Refractory Metal Alloys," *Proceedings Am. Electroplaters' Soc., 51*, 150–162 (1964).

(62) Browning, M. E., and Turns, E. W., "Alloy Electroforming for High-Temperature Aerospace Applications," Symposium on Electroforming—Applications, Uses and Properties of Electroformed Metals, *ASTM Special Technical Publication No. 318*, 107–121 (1962).

(63) Browning, M. E., Leavenworth, Jr., H. W., Webster, Jr., W. H., and Dunkerley, F. J., "Deposition Forming Processes for Aerospace Structures," *Technical Documentary Report No. ML-TDR-64-26, Final Report under Contract AF 33(657)-7016*, Am. Machine & Foundry Co. (January, 1964), 133 pp.

(64) Communication from Chemical Casting Company, to Cobalt Information Center, Columbus, Ohio (February, 1970).

(65) Kanbe, T., Nemoto, K., Mitani, H., and Shoji, K., "Phase Changes in Electrolytic Co-P Deposits During Heat Treatment," *Journal of the Metal Finishing Society of Japan, 18* (4), 142–145 (1967).

(66) Safranek, W. H., and Schaer, G. R., "Properties of Electrodeposits at Elevated Temperatures," *Tech. Proceedings Am. Electroplaters' Soc., 43*, 105–117 (1956).

(67) Fedot'ev, N. P., Vyacheslavov, P. M., Burkat, G. K., and Mokrova, L. N., "Electrodeposition of Cobalt-Copper Alloys," *Journal Applied Chemistry (USSR), 41* (2), 401–404 (1968); *Zh. Prikl. Khimii, 41* (2), 424–427 (1968).

(68) *Report No. 4125*, International Nickel Limited (1968).

(69) Vinogradov, S. N., Kudryavtsev, N. T., "Internal Stresses in a Palladium-Cobalt-Electrolytic Alloy," *Protection of Metals, 5* (6), 601–602 (1969); *Zashchita Metallov, 5* (6), 686–687 (1969).

(70) Vyacheslavov, P. M., Fedot'ev, N. P., Burkat, G. K., and Semenova, N. L., "Electrodeposition of a Palladium-Cobalt Alloy," *Protection of Metals, 5* (3), 311–312 (1969); *Zashchita Metallov, 5* (3), 348–349 (1969).

(71) Fedot'ev, N. P., and Vyacheslavov, P. M., "The Phase Structure of Binary Alloys Produced by Electrodeposition," *Plating, 57* (7), 700–706 (1970).

(72) Vyacheslavov, P. M., Fedot'ev, N. P., Burkat, G. K., and Nikol'skaya, M. A., "Electrodeposition of A Cobalt-Vanadium Alloy," *Protection of Metals, 5* (3), 313–314, (1969); *Zashchita Metallov, 5* (3), 350–351 (1969).

(73) Weil, R., and Read, H. J., "The Structure of Electrodeposited Metals," *Metal Finishing, 53* (11), 60; (12), 60 (1955).

(74) Newman, R. C., "The Lattice Spacings of Thin Electrodeposits of Cobalt and Nickel on a Copper Single Crystal," *Proc. Phy. Soc., 69B*, 432–440 (1956).

(75) Gofman, N. T., and Nikolaishvili, T. N., "Electrolytic Deposition of Cobalt Deposits with Given Characteristics," *Izvestiya Vysshikh Uchebnykh Zavedeniy SSSR, Khimiya i Khimicheskaya Tekhnologiya, 8* (1), 104–110 (1965).

(76) Gaigalas, K. I., Bodnevas, A. I., Matulis, Yu. Yu., "Phase Transformations in Electrodeposits of Cobalt in Relation to Conditions of Electrolysis Composition," *Trudy Akademii Nauk Litovskoi SSR, Seriia B, 4* (55), 59–68 (1968).

(77) Calusaru, A., Ripeanu, S., and Benes, L., "Structure and Magnetic Properties of Electrolytic Cobalt Layers Used for Neutron Polarization," *Acad. Rep. Popular Romine, Studii Cercetari Fiz (Romania), 14* (3), 239–247 (1963).

(78) Okund, G., "Crystal Structure of Electrodeposited Cobalt. I. II," *Bulletin of the University of the Osaka Prefecture, Series A, 4*, 89–100; 101–110 (1956).

(79) Pangarov, N. A., and Rashkov, St., "Electrolytic Deposition of Alpha and Beta Modifications of Cobalt," *Comptes rendus de l'academi bulgare des Sciences, 13*, 439–442 (1960).

(80) Sard, R., Schwartz, C. D., and Weil, R., "Deposition Modes of Cobalt from Sulfate-Chloride Solutions," *Journal Electrochem. Soc., 113* (5), 424–428 (1966).

(81) Cadorna, L., and Cavallotti, P., "Cobalt Plating from Sulfamate Baths," *Electrochim. Metal. (Politec, Milan), 1* (3), 364–374 (1966).

(82) Bate, G., and Speliotis, D. E., "Hard Magnetic Films of Co-Ni-P," *Journal Applied Physics, 34* (4), Part 2, 1073–1074 (1963).

(83) Kanbe, T., Mitani, H., and Shoji, K., "Phase Change in Cobalt-Phosphorus Electrodeposits During Heat Treatment. II. Properties of Co-P Electrodeposits for Several Kinds of Anodes," *Kinzoku Hyomen Gijutsu, 19* (2), 50–54 (1968).

(84) Gorbunova, K. M., Bondar, V. V., Moiseev, V. P., Popova, O. S., Polukarov, Yu. M., and Sutyagina, A. A., "The Structure of Electrolytically Deposited Metals from the Data of Structural Investigations," *Physics of Metals and Metallography, 9* (1), 62–67 (1960); *Fizika Metallov i Metallovedenie, 9* (1), 73–80 (1960).

(85) Offermans, H., and Stackelberg, M. V., "Electrodeposited Tungsten-Cobalt, Tungsten-Nickel, and Tungsten-Iron Alloys," *Metalloberflaeche, 1*, 142–144 (1947).

(86) Polukarov, Yu. M., Rastorguev, L. N., and Shevkun, I. G., "Magnetic Properties and Structure of Electrodeposited Cobalt-Tungsten Alloys," *Russian Journal of Physical Chemistry, 36* (6) 685–689 (1962); *Zh. Fiz. Khimii, 36* (6), 1299–1305 (1962).

(87) Yoshioka, S., and Yamamoto, H., "The Structure of Tungsten-Cobalt Electrodeposits Obtained from a Tartrate Bath," *Trans. Japan Inst. Metals, 3* (4), 210–214 (1962).

(88) Yamamoto, H., "Structure of W-Co Electrodeposits Obtained from a Tartrate Bath," *Nippon Kinzoku Gakkaishi, 25*, 292–296 (1961).

Chapter 6

Copper

Copper is electrodeposited for numerous engineering and decorative applications requiring a wide range of mechanical and physical properties. This range extends from properties superior to full-hard, wrought copper to properties equivalent to annealed, pure copper. Resistivity, elastic modulus, and coefficient of thermal expansion are important physical properties which can be controlled. Strength, hardness, ductility, and resistance to fatigue are critical mechanical properties which the engineer may wish to specify. Dimensional stability and internal stress level must be known if, for example, the electrodeposit is to be heated in service or subsequently fabricated. The desired properties will be obtained only when the plating baths and operating conditions are properly selected. Undesired properties result from uncontrolled variations in electroplating conditions or solution composition. Examples of the ranges of properties which can be obtained with electrodeposited copper are presented in Table 6.1.

Mechanical Properties

High Strength Electrodeposits

PROPERTIES. Coppers with tensile strengths ranging from 42 to 66 kg/sq mm (60 to 93 kpsi) and yield strengths from 27 to 40 kg/sq mm (38 to 57 kpsi) have been electrodeposited in copper cyanide baths and in acid and alkaline copper sulfate solutions containing organic amines. The properties of representative deposits are given in Table 6.2. Elongations of 3 to 18 percent, ample ductility for many electroforming applications, were measured for high-strength copper. The densities of these deposits were virtually the same as those of rolled copper sheet. (1) However, the tensile strength of the electrodeposited copper can exceed 35 kg/sq mm (50 kpsi), the tensile strength of hard, cold-rolled sheet. The higher strength of the electrodeposited copper is partially due to its extremely fine grain size. Microstructures of four representative high-strength copper deposits are shown in Figures 6.1 through 6.4.

High-strength copper deposited in copper cyanide solutions had higher ductility than copper of comparable strength deposited in acid or alkaline copper sulfate

TABLE 6.1

Ranges of Properties of Copper Electrodeposits

Outstanding Characteristic	Tensile Strength		Elongation, Percent (2 in.)	Internal Stress,[a]		Hardness, (VHN_{200}), kg/sq mm	Electrical Resistivity, microhm-cm	Plating Bath
	kg/sq mm	kpsi		kg/sq mm	psi			
High Strength	45–63	64–90	4–18	−4.2 or 3.6 to 5.5	−6,000 or 5,000 to 7,800	131–159	1.75–2.02	See Table 6.2
Hardness	3.5–55	5–79	0–10	−4.2 or 3.5	−6,000 or 5,000	193–350	1.96–4.60	See Tables 6.3A and 6.3B
Low Electrical Resistivity	18–27	26–38	15–41	−0.5 to 1.6	−700 to 2,200	48–64	1.70–1.73	Copper sulfate or fluoborate solutions containing no addition agents. See Table 6.8
Near Zero Stress	14–23	20–33	8–24	−0.08 to 0.06	−110 to 80	56–57	1.71–1.72	Copper sulfate solutions with low current densities. See Table 6.4
Good Leveling	36	51	14–19	2.0	2,900	128–137	1.82	Copper sulfate solutions with proprietary addition agent. See Table 6.13
Thermal Stability[b]	22.5–30	32–43	26–39	0.5 to 2.9	700 to 4,100	55–106	1.73–1.76	See Table 6.12

[a] Negative values indicate a compressive stress.
[b] Deposits that changed $< 0.02\%$ in length after heating to 400 C. See Table 6.12.

TABLE 6.2

Properties of High-Strength Electrodeposited Copper[a]

Tensile Strength,		Yield Strength,		Elongation,	Internal Stress[b]		Density	Electrical Resistivity,	Plating	Bath Temp.	Current Density,
kg/sq mm	kpsi	kg/sq mm	kpsi	percent (in. 2 in.)	kg/sq mm	psi	g/cu cm	microhm-cm	Bath	C	amp/sq dm
50	71[c]	30	43	7	5.0	7,100	8.913	1.89	d	30	5
65.5	93[e]	—	—	3	—	—	—	—	e	—	—
57	81[c]	—	—	4[f]	5.5	7,800	8.913	1.75	g	60	2
46.5	66[c]	—	—	18	5.2	7,400	8.908	1.76	g	60	1
55	78[c]	—	—	6[f]	4.7	6,700	8.912	1.79	h	80	4
60.5	86[c]	40	57	10	3.6	5,100	8.917	2.02	i	80	6
45	64	27	38	12		—	—	1.82	j	80	6
42	60	31	44	4	−4.2	−6,000	8.906	2.10	k	55	4
48.5	69[l]	29	41	4	−4.0	−5,700	—	2.08	k	55	2

[a] Data for 0.4 to 0.5 mm-thick copper deposited in copper sulfate solutions or for 0.04 to 0.05 mm-thick copper deposited in cyanide solutions; 0.05 mm copper deposited in copper sulfate baths was usually stronger than thicker deposits, and 0.01 mm copper deposited in cyanide solutions usually exhibited a lower tensile strength than the 0.05 mm deposits.

[b] A minus sign designates a compressive stress.

[c] Knoop hardness values (100 to 200-g load) ranged from 131 to 159 kg/sq mm.

[d] The solution contained 0.75 M/1 (187 g/l) $CuSO_4 \cdot 5H_2O$, 0.75 M/1 (74 g/l) H_2SO_4 and 3.5 g/l triisopropanolamine. (1) A tensile strength of 46 to 63 kg/sq mm (65 to 90 psi) and an elongation of 8 to 12 percent were reported in a patent disclosing this solution. (2)

[e] Vickers hardness was 170. $CuSO_4$ with molasses and thiourea. Exact composition was not reported. (3)

[f] Significant strain determined for thin deposits by means of a bulge test. (1)

[g] The solution contained 25 g/l CuCN, 6 g/l free NaCN and 15 g/l Na_2CO_3. (1)

[h] The solution contained 40 g/l CuCN, 10 g/l free NaCN and 30 g/l Na_2CO_3; the pH was in the range of 11.8 to 12.8. (1)

[i] The solution contained 75 g/l CuCN, 10 g/l free KCN, 40 g/l KOH, and 2 g/l KSCN; and pH was 13.6. The current was reversed for 5 seconds after each 15 seconds of direct plating. (1)

[j] The solution contained 75 g/l CuCN, 10 g/l free KOH and 40 g/l KOH; the pH was 13.6. The current was reversed for 5 seconds after each 15 seconds of direct plating. (1)

[k] The solution contained 0.4 M/1 (100 g/l) $CuSO_4 \cdot 5H_2O$, 20 g/l $(NH_4)_2SO_4$, 80 ml/l ethylene diamine and 4 ml/l conc. NH_3. (1)

[l] Knoop hardness values (200-g load) ranged from 169 to >202 kg/sq mm.

Figure 6.1. Cross Section of Copper Deposited at 5 amp/sq dm in a Copper Sulfate Solution Containing 2 g/l Triisopropanolamine. (1)

baths. Elongations from 10 to 18 percent and 3 to 8 percent were reported for deposits from the cyanide and sulfate solutions, respectively. The cyanide copper deposits produced without addition agents had resistivities of from 1.75 to 1.79 microhm-cm (98 to 96% IACS), while the resistivities of copper deposited in copper sulfate solutions containing an amine were 1.89 to 2.10 microhm-cm (91 to 82%

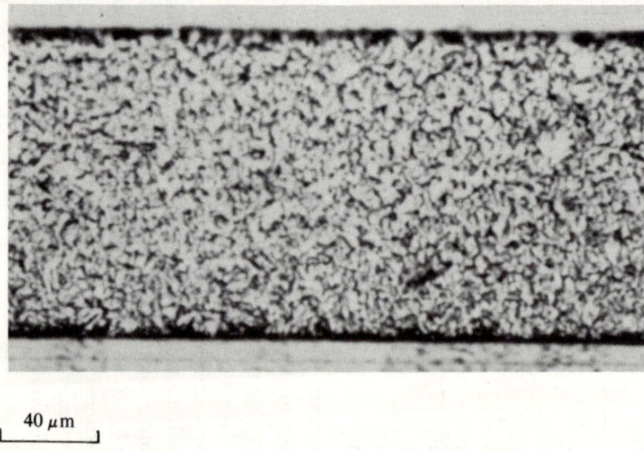

Figure 6.2. Cross Section of Copper Deposited at 6 amp/sq dm in a Copper Cyanide Bath at 80 C. (1)

IACS). Higher values of resistivity from the sulfate solution probably were due to small amounts of organic additives included in the metal and concentrated at grain boundaries.

Effect of Thickness. Data in Table 6.2 are for deposits of 0.4- to 0.5-mm (0.016- to 0.020-inch)-thick copper plated in copper sulfate solutions and for 0.04- to 0.05-mm (0.0016- to 0.002-inch)-thick copper plated in cyanide baths. Thin deposits from copper sulfate solutions are usually stronger than thicker ones. For example, strength increased 10 to 20 percent for copper deposited in acid sulfate solutions when thickness was decreased from 0.45 to 0.05 mm (0.018- to 0.002 inch). (1) In contrast, tensile strengths for copper plated in cyanide solutions decreased slightly as thickness decreased from 0.45 to 0.05 mm (0.018 to 0.002 inch). The tensile strength of bright acid copper electroplate of the type used for undercoats for nickel and chromium increases with decreasing thickness. Tensile strengths of 59 kg/sq mm (84,000 psi) have been reported for thicknesses of only 0.01 mm, while the strength was 36 kg/sq mm (51,000 psi) for 0.4 to 0.5 mm (0.016 to 0.020 inch) bright acid copper deposited under similar conditions.

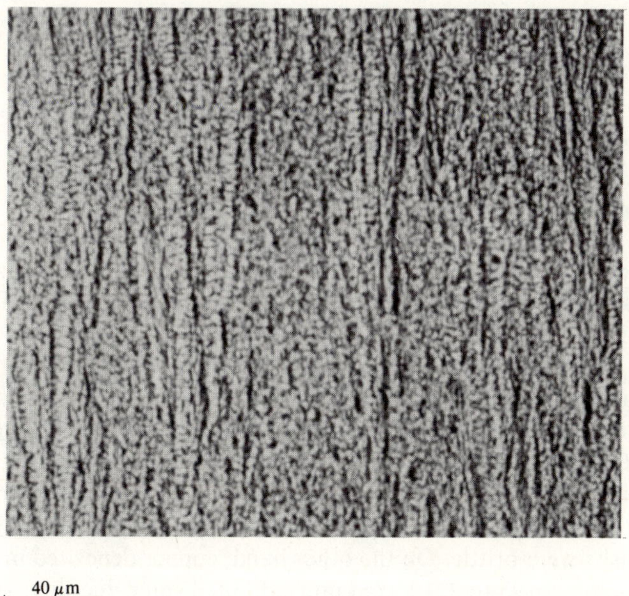

40 μm

Figure 6.3. Cross Section of Copper Deposited at 6 amp/sq dm with Periodically Reversed Current in a Copper Cyanide Bath at 80 C. (1)

Figure 6.4. Cross Section of Copper Deposited at 4 amp/sq dm in a Copper Sulfate-Ethylene Diamine Bath at 55 C. (1)

Hard Electrodeposits

Hard copper electrodeposits are desired when wear resistance is important. Data on the hardest deposits reported in the literature are summarized in Table 6.3A for cyanide solutions and in Table 6.3B for sulfate solutions. Some, but not all of the high-strength copper deposits listed in Table 6.2 were very hard. Some of the hard copper electrodeposits listed in Table 6.3A exhibited high strength and fine grain structures like those in Figures 6.2 and 6.3. However, hardness cannot be directly correlated with tensile strength for electrodeposits of copper.

As shown in Table 6.3B, several organic addition agents in copper sulfate baths increased the hardness of electrodeposited copper to greater than 200 kg/sq mm, but such deposits were brittle. On the other hand, copper deposited in cyanide baths and with hardness of 193 to 224 kg/sq mm exhibited some ductility. In comparison, deposits of soft copper have hardness of 50 to 60 kg/sq mm and may have elongations exceeding 25 percent.

No change in hardness or grain size was observed for copper deposited in copper sulfate solutions as a result of using ultrasonic agitation. (4) On the other hand, ultrasound increased the hardness of copper deposited in a cyanide bath (5) and

TABLE 6.3A

Hard Copper Electrodeposits—Copper Cyanide Solutions

Hardness (KHN_{200}) kg/sq mm	Tensile Strength kg/sq mm	Tensile Strength kpsi	Elongation, percent	Electrical Resistivity, microhm-cm	Addition Agent	Bath Temp, C	Current Density, amp/sq dm
224	46.5	65	45[a]	1.96	Proprietary brightener and PR[b]	80	6
193	55.5	79	10	1.98	KCNS with PR[c]	80	6
200[d]	—	—	—	—	NaCNS and betaine[e]	80	1.5
220–250	—	—	—	—	KCNS with interrupted current[f]	72	3.5
248	—	—	—	—	Rochelle salts with ultrasonic agitation[g]	60	8
237	—	—	—	—	$(NH_4)_2PO_4$[h]	60	2 to 7

[a] Significant strain determined for thin deposits by means of a bulge test. (1)

[b] The solution contained 75 g/l CuCN, 10 g/l free KCN, 22.5 g/l KOH and a proprietary brightener containing a selenium compound. The current was reversed for 5 seconds after every 15 seconds of direct plating. (1)

[c] The solution contained 75 g/l CuCN, 40 g/l KOH and 2 g/l KCNS; pH was 13.6. The current was reversed for 5 seconds after every 15 seconds of direct plating. (1)

[d] Vickers diamond pyramid values, which are approximately 5% lower than Knoop values at a 200 g load.

[e] The solution contained 90 g/l NaCN, 20 g/l Na_2CO_3, 1 g/l betaine and 10 g/l NaCNS. (8)

[f] The solution contained 85 g/l CuCN, 20 g/l free KCN, 15 g/l KOH, 4 g/l KCNS and a wetting agent. The current was interrupted for 5 seconds after every 46 seconds of direct plating. (9)

[g] The solution contained 45 g/l CuCN, 53 g/l KCN, 60 g/l Na_2CO_3 and 60 g/l Rochelle salts. (5) Hardness without ultrasonic agitation was 167 kg/sq mm.

[h] The solution contained 90 g/l CuCN, a total of 132 g/l KCN, 7 g/l $(NH_4)_2HPO_4$ and 3 g/l furfuryl alcohol. (10)

TABLE 6.3B

Hard Copper Electrodeposits—Copper Sulfate Solutions

Hardness (KHN_{200}) kg/sq mm	Tensile Strength		Elongation, percent	Electrical Resistivity, microhm-cm	Addition Agent	Temp, C	Current Density, amp/sq dm
	kg/sq mm	kpsi					
265[a]	—	—	—	46	15 g/l asparagine[b]	20	4
249	—	—	—	95.5	6.7 g/l citric acid[c]	—	—
305[a]	—	—	—	—	0.2 g/l gelatin[d]	25	2
305[a]	—	—	—	—	0.1 g/l gelatin[e]	25	10
200–300[a]	—	—	—	—	0.075 g/l gelatin, casein, etc.[f]	—	3
230[a]	—	—	—	—	Naphthalene disulfonic acid and thiourea[g]	20	3
225[a]	—	—	—	—	1.8 g/l β-naphthoquinoline[h]	25	2
216–344	—	—	—	—	Polyacrylic acid amide[i]	—	—
263[a]	3.5	5	—	3.7	0.2 g/l Rochelle salts[j]	18	1
240[a]	—	—	—	—	1 g/l Rochelle salts[j]	18	1
200[a]	25.8	37	—	—	0.0025 g/l Rochelle salts[j]	18	1
249[a]	—	—	—	—	12 g/l tartaric acid[k]	20	—
350[a]	—	—	—	—	0.2 g/l thiourea[l]	20	3
212[a]	—	—	—	—	Triethanolamine[m]	25	1.1
202[a]	35	50	1	2.1	(n)	55	2

[a] Vickers diamond pyramid values, which are approximately 5 percent lower than Knoop hardness numbers at a 200-g load.

[b] The solution contained 250 g/l $CuSO_4 \cdot 5H_2O$, 98 g/l H_2SO_4 and 15 g/l asparagine. (11,12)

[c] The solution contained 200 g/l $CuSO_4 \cdot 5H_2O$, 98 g/l H_2SO_4 and 6.7 g/l citric acid. (11)

[d] The solution continued 125 g/l $CuSO_4 \cdot 5H_2O$, 49 g/l H_2SO_4 and 0.2 g/l gelatin. (13)

[e] The solution contained 188 g/l $CuSO_4 \cdot 5H_2O$, 200 to 300 g/l H_2SO_4, 0.075 g/l gelatin, 0.15 g/l casein, 13 ml/l ethyl alcohol and 6.5 g/l acetic acid. (14)

[f] The solution contained 220 g/l $CuSO_4 \cdot 5H_2O$, 200 to 300 g/l H_2SO_4, 0.075 g/l gelatin, 0.15 g/l casein, 13 ml/l ethyl alcohol and 6.5 g/l acetic acid. (15)

[g] The solution contained 250 g/l $CuSO_4 \cdot 5H_2O$, 50 g/l H_2SO_4, 0.5 g/l naphthalene disulfonic acid, 0.005 g/l thiourea. (16) Stress in copper deposited in solutions containing thiourea was as high as 9,300 psi. (17)

[h] The solution contained 125 g/l $CuSO_4 \cdot 5H_2O$, 49 g/l H_2SO_4 and 0.9 or 1.8 g/l β-naphthoquinoline. (18,19)

[i] The solution contained 25 g/l $CuSO_4 \cdot 5H_2O$, 98 g/l H_2SO_4 and polyacrylic acid amide. (20)

[j] The solution contained 200 g/l $CuSO_4 \cdot 5H_2O$, 5 g/l H_2SO_4 and Rochelle salts. (21,22)

[k] The solution contained 250 g/l $CuSO_4 \cdot 5H_2O$, 98 g/l H_2SO_4 and 12 g/l tartaric acid. (11)

[l] The solution contained 250 g/l $CuSO_4 \cdot 5H_2O$, 50 g/l H_2SO_4 and 0.2 g/l thiourea. (23)

[m] The solution contained 125 g/l $CuSO_4 \cdot 5H_2O$, 49 g/l H_2SO_4, 0.006 g/l chloride ions and 15 g/l triethanolamine. (24)

[n] An alkaline copper sulfate-ethylene diamine solution containing 100 g/l $CuSO_4 \cdot 5H_2O$, 20 g/l $(NH_4)_2SO_4$, 80 ml/l ethylene diamine, and 4 ml/l NH_4OH. Deposit was 0.25 mm (0.01 inch) thick. (1)

reduced the hardness of copper deposited in a pyrophosphate solution. (6) Electroless copper deposited with a proprietary process exhibited a hardness of 215 kg/sq mm and a tensile stress of 18 kg/sq mm (25,000 psi). (7)

The electrical resistivity of representative hard deposits was at least 1.96 microhm-cm (88% IACS) and was as high as 95.5 microhm-cm (1.8% IACS), which is indicative of occluded organic impurities.

Some of the deposits obtained in copper sulfate solutions containing organic additives were compressively stressed 4.2 kg/sq mm (6,000 psi). On the other hand, internal tensile stresses of about 3.5 kg/sq mm (5,000 psi) were reported for representative, hard deposits produced in copper cyanide baths. These levels of internal stress are unlikely to be detrimental in applications where wear resistance is an important reason for using hard electrodeposited copper.

Internal Stresses in Copper Electrodeposits

STRESS LEVELS. Unlike many metals, electrodeposited copper rarely exhibits high internal stresses. However, some applications for copper require almost complete freedom from internal stresses. Near zero internal stress is important for preventing changes in dimensions when an electroformed shape is separated from its mandrel. For example, electroformed reflectors must be dimensionally accurate to within 0.01 percent.

Internal stresses are higher in the high-strength copper deposits in Table 6.2 in comparison to low-strength copper deposited in copper sulfate solutions containing no addition agents. Hence low-strength copper is preferred for electroforming applications requiring very high precision, i.e., dimensional accuracy of better than 10 microns (0.0004 inch). On the other hand, the stress level of the high-strength copper is low enough to be safe for less rigid standards of dimensional accuracy, such as 50 microns (0.002 inch). The compressive (negative) internal stresses characteristic of copper deposited in the copper sulfate-ethylene diamine bath may be advantageous for certain electroformed parts because it facilitates separation from the mandrel.

Internal stresses in copper electrodeposited from a number of plating baths are presented in Table 6.4. Copper deposited in fluoborate solutions at current densities from 2 to 8 amp/sq dm had low internal stresses, that is, less than 0.55 kg/sq mm or 800 psi. (1) The highest current density of 20 amp/sq dm modestly increased stress to 1.1 kg/sq mm or 1600 psi.

The temperature of the pyrophosphate baths influenced the internal stress level. Internal stress decreased from 1.2 kg/sq mm (1700 psi) to −1.3 kg/sq mm (−1900 psi) when the temperature was raised from 50 to 60 C. (1) Reducing the current density from 4 to 2 amp/sq dm for copper deposited in the 50 C bath also induced compressive stresses (−1.1 kg/sq mm or −1600 psi). High tensile stress of 6.6 kg/sq mm (9,400 psi) as well as high compressive stresses of −7.0 kg/sq mm (−10, 000 psi) have been reported for copper deposited in proprietary solutions. (25,26)

Internal stresses were less than 0.55 kg/sq mm (800 psi) (1) in copper deposited at low current densities (2 to 4 amp/sq dm) in copper sulfate baths containing 0.25 or 0.4 N sulfuric acid and operated at 30 C. Higher current densities or lower temperatures increased stress, as shown in Figure 6.5. A high concentration of sulfuric acid (1.1 N) increased stress slightly to 1.3 kg/sq mm (1,850 psi).

By increasing the temperature of copper sulfate solutions to 40 to 80 C, internal stresses can be maintained at nearly zero while the current density is increased. (31) Thus, stress was zero for copper deposited at (a) 2 amp/sq dm in a 40 C solution, (b) 4 amp/sq dm in a 60 C solution, and (c) 8 amp/sq dm in an 80 C bath.

EFFECTS OF ADDITION AGENTS. Some addition agents such as phenol sulfonic acid reduce stress slightly or induce a slight internal compressive stress instead of a tensile stress. (1) Ethylene diamine, amidopyrinemethionine, arginine, and glycine have the same effect. (32,33) Grain refining agents like naphthalene disulfonic acid and lactic acid had little or no effect on internal stress, while some such as β-naphthoquinoline and selenic acid caused an increase in internal stress. Internal stresses in copper deposited from copper sulfate solutions containing these addition agents are given in Table 6.4.

Gelatin in small concentrations (0.01 g/l) had very little effect on stress, but a higher concentration of 0.1 g/l induced an appreciable compressive stress (−3.0 kg/sq mm or −4,200 psi). (1) A high internal stress of 10.5 kg/sq mm (15,000 psi) was reported in another investigation of gelatin additions of 0.1. (13) Either the purity of the gelatin or its hydration and decomposition products evidently affects its influence on stress.

Rochelle salt additions to copper sulfate baths had significant effects. Internal stress increased to 4.0 kg/sq mm (5,700 psi) with a concentration of 0.1 g/l and decreased to −4.0 kg/sq mm (−5,700 psi) with 0.17 g/l. (30,34) Zero stress was reported for copper deposited in a bath containing 0.134 g/l. At a concentration of 0.2 or 1 g/l, Rochelle salts also induced a compressive stress in copper deposited from a copper sulfate solution containing only 5 g/l sulfuric acid. (21,22) A compressive stress of −10 kg/sq mm (−14,000 psi) was obtained with 0.2 g/l Rochelle salts in a solution at 18 C. (34)

Results with thiourea are conflicting. One investigator reported that internal stresses varied from 6.5 kg/sq mm (9,300 psi) with 0.1 g/l to −3.4 kg/sq mm (−4,800 psi) with 0.3 g/l. (16) Another reported 2.0 kg/sq mm (2,800 psi) for copper deposited in a bath containing 0.1 g/l thiourea. (35) A third report cited stress values of 8.0, 0, and −4.8 kg/sq mm (11,200, 0, −6,700 psi) for copper deposited in solutions containing 0.025, 0.1 and 0.3 g/l thiourea, respectively. (21)

Internal stresses in copper deposited in cyanide solutions are higher than in copper from other solutions. Data in Table 6.4 show values from 4.9 to 10 kg/sq mm, (7,000 to 14,000 psi) for copper deposited in cyanide baths at 40 or 60 C. A reduction in stress to a minimum of 2.4 or 2.7 kg/sq mm (3,500 or 3,900 psi) was obtained by adding Rochelle salts, or increasing the temperature of the solutions to 80 C, respectively. There is no record of compressive stress in copper deposited from cyanide solutions.

TABLE 6.4

Internal Stresses in Electrodeposited Copper

Plating Bath	Addition Agent	Temp, C	Current Density, amp/sq dm	Stress[a] Reference (1) kg/sq mm	Stress[a] Reference (1) kpsi	Stress[a]. Other References kg/sq mm	Stress[a]. Other References kpsi	Reference
Copper fluoborate[b]	None	30–60	2–8	0–0.6	0–0.8	1.0	1.5	25
Ditto	Ditto	50	20	1.1	1.6	—	—	—
″	″	60	4	−1.3	−1.9	—	—	—
Copper pyrophosphate[d]	None	50	2	−1.1	−1.6	6.6	9.4	25
Ditto	Ditto	50	4	1.2	1.7	−7.0	−10.0	26
Copper sulfate[e]	None	20	2–4	1.0–3.0	1.4–4.3	0.8, 1.0	1.13, 1.4 (25 C)	27, 13, 28
Ditto	Ditto	30	2–4	<0.6	<0.8	0.35, 1.9	0.5, 2.7	22, 17
″	″	30	8	3.0	4.3	—		—
″	Aminobenzoic acid, 1 g/l	20	4	—		−0.1	−0.150	29
″	β-naphthoquinoline, 0.1 g/l	30	2	3.7	5.3	10.5 (25 C)	15.0 (25 C)	13
″	Gelatin, 0.1 g/l	30	2	−3.0	−4.2	7.6, 10.5 (25 C)	11.0, 15.0 (25 C)	22, 13
″	Selenic acid, 10^{-3} g/l	30	2	2.0	2.8	—	—	—
″	Triisopropanolamine, 3.5 g/l	30	2	1.5	2.1	—	—	—

Copper cyanide[f]	None	40	1	10.0	14.0	—	—	—
Ditto[g]	Ditto	40–60	1–4	4.9–6.6	7.0–9.5	7.0	9.9	28
"	"	80	1–4	4.2–4.7	6.0–6.7	4.5 (70 C)	6.4 (70 C)	30
"	Rochelle salts, 45 g/l	40–60	1–4	2.4–4.0	3.5–5.8	6.0 (70 C)[h]	8.7 (70 C)[h]	30
"[i]	None	80	4–8	2.7–3.2	3.9–4.5	—	—	—
" with PR[i]	Ditto	80	6	3.2	4.6	—	—	—

(a) A negative sign indicates compressive stress.

(b) Solutions containing 177 or 366 g/l $Cu(BF_4)_2$, 12 or 22 g/l HBF_4 and 12 or 22 g/l H_3BO_3.

(c) A compressive stress of 0.14 kg/sq mm (−200 psi) was measured for one deposit.

(d) A solution containing 90 g/l $Cu_2P_2O_7 \cdot 3H_2O$, 350 g/l $K_4P_2O_7$, 80 g/l K_3PO_4, 15 g/l KNO_3 and 2 ml/l NH_3 with a pH of 8.5.

(e) Solutions containing 0.75 M (187 g/l) $CuSO_4 \cdot 5H_2O$ and 0.25 or 0.4 M (25 or 39 g/l) H_2SO_4.

(f) A solution containing 15 g/l CuCN, 6 g/l free NaCN and 15 g/l Na_2CO_3 with a pH of 11.4 to 12.8.

(g) Solutions containing 25 to 40 g/l CuCN, 6 to 10 g/l free NaCN, (51) to 30 g/l Na_2CO_3 with a pH of 11.7 or 11.8.

(h) Plating with periodically reversed current increased stress to 8.1 kg/sq mm (11,600 psi).

(i) A solution containing 75 g/l CuCN, 8 g/l KCN and 40 g/l KOH with a pH of 13.6.

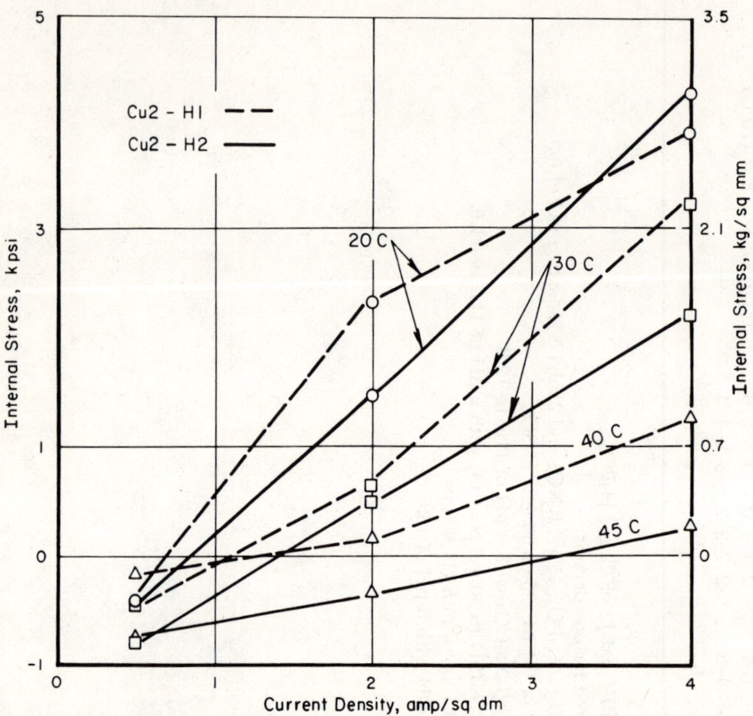

Figure 6.5. Effects of Increasing Temperature and Current Density on the Stress in Copper Deposited in Copper Sulfate Solutions

Cu2-H1 contained 157 g/l $CuSo_4$ and 25 g/l H_2SO_4.
Cu2-H2 Contained 187 g/l $CuSo_4$ and 39 g/l H_2SO_4.

STRESS MEASUREMENTS. The stress values in Table 6.4 were measured by plating on helices by the Brenner-Senderoff method. (36) In comparative tests conducted by an ASTM Committee B8 group, lower values were obtained when stress was measured by plating on one side of a rigid strip. (37) For example, stress in copper deposited at 4 amp/sq dm in a copper sulfate solution at 27 C containing 75 g/l of sulfuric acid was 1.5 kg/sq mm (2,200 psi) on a helix, but only 0.15 kg/sq mm (200 psi) on a rigid strip.

Microstresses determined by X-ray diffraction analysis of copper deposits are higher than macrostresses measured by plating on one side of a rigid strip. (38) For example, a microstress of 0.6 kg/sq mm (900 psi) was determined for copper with a compressive macrostress of 0.05 kg/sq mm (75 psi). These were typical values for copper deposited at 1.0 amp/sq dm in a purified solution of 1 N copper sulfate and 1 N sulfuric acid.

Fatigue Strength

The fatigue strenths of electrodeposited coppers are given in Table 6.5. In copper sulfate solutions, additives increased fatigue strength 50 percent or more. However, copper deposited in cyanide baths exhibited the highest fatigue limit, about twice the limit for annealed copper and copper electroplated in copper sulfate solution containing no addition agents. (1)

Effects of Temperature

As shown in Table 6.6, in comparison with annealed, wrought copper, tensile strength at 325 C was slightly lower for the copper deposited in sulfate and fluoborate baths. (1) The pyrophosphate solution yielded bright plate similar in strength at 325 C to annealed, wrought copper, whereas copper electroplated in cyanide baths was about 15 percent stronger. Only the copper deposited in the pyrophosphate bath exhibited good ductility at 325 C.

At −78 C, tensile strengths of copper electrodeposits were usually 10 to 30 percent greater than at 22 C. (1)

Effects of Processing

Cold rolling to 50 percent reduction increased the strength of electrodeposits from 22-26 kg/sq mm (31,000 to 36,000 psi) to about 40 kg/sq mm (56,000 psi). (1) Yield strengths of these and other deposits were about the same as the yield strength of wrought copper and ranged from 32 to 35 kg/sq mm (45,000 to 50,000 psi). Elongation of the electrodeposits after cold rolling was 2 to 7 percent. Hardness after cold working was 30 to 50 percent higher than for wrought copper cold-worked about the same amount. Cold rolling had little or no effect on the modulus of elasticity except for the modulus of copper deposited in the fluoborate bath which was increased from 9,000 to 12,000 kg/sq mm (from 13,000,000 to 17,000,000 psi).

Annealing electrodeposited copper at 300 C after cold working to wire restored good ductility. (39) The tensile strength and ductility of the annealed wire was usually in the following ranges:

Tensile Strength	21.5 to 22 kg/sq mm or 30,600 to 31,400 psi
Elongation	29 to 47 percent

The influence of annealing on the properties of electrodeposits is given in Table 6.7. Copper deposited in a copper sulfate bath containing triisopropanolamine retained a high strength (44 kg/sq mm or 62,000 psi) after annealing at 325 C. This and other representative copper deposits exhibited about the same strength as wrought copper after annealing at 500 C (20 to 25 kg/sq mm or 28,000 to 35,000 psi).

TABLE 6.5

Fatigue Strengths of Electrodeposited Copper (1)

Property	Copper Sulfate Solutions: No Addition Agents	Copper Sulfate Solutions: With Addition Agents	Copper Fluoborate Solutions	Copper Pyrophosphate Solutions	Copper Cyanide Solutions	Annealed Wrought Copper
Fatigue Limit						
10^6 cycles, kg/sq mm	9–10	13.5[a,b], 17.5	11	15.5	20[d]	10.5
10^6 cycles, kpsi	13–14	19[a,b], 25[c]	16	22	28[d]	15[e]
10^7 cycles, kg/sq mm	8–8.5	10[b], 13[c]	9	12	14[d]	8.5
10^7 cycles, kpsi	11–12	14[b], 18[c]	13	17	20[d]	12[e]
Ratio of Fatigue Limit (10^6 cycles) to Tensile Strength	0.5	0.58[a], 0.44[b], 0.36[c]	0.5	0.4	0.35	0.5

(a) Phenolsulfonic acid, 1.0 g/l.

(b) Selenic acid, 10^{-3} g/l.

(c) Triisopropanolamine, 3.5 g/l.

(d) Deposited in a solution containing 40 g/l KOH and 2 g/l KCNS, while the current was periodically reversed. The fatigue limit for 10^6 cycles for a similar deposit obtained in a KCNS-free solution was 25,000 psi.

(e) The fatigue limits at 10^6 and 10^7 cycles for hard, cold-worked, wrought copper were 24,000 and 18,000 psi, respectively.

TABLE 6.6

Influence of Temperature on the Strength and Ductility of Electrodeposited Copper (1)

Property	Copper Sulfate Solutions		Copper Fluoborate Solutions	Copper Pyrophosphate Solutions	Copper Cyanide Solutions[b]	Annealed Wrought Copper
	No Addition Agents	With Addition Agents[a]				
Tensile Strength at −78 C						
kg/sq mm	27	54.5	31	34	56	28 to 52.5
kpsi	38	78	44	49	80	40 to 75
Elongation at −78 C, percent	41	14	33	39	11	35 to 60
Tensile Strength at 325 C						
kg/sq mm	8	8.5	10	11	14	12
kpsi	11	12	14	16	20	17
Elongation at 325 C, percent	7	4	8	25	7	18

(a) Triisopropanolamine, 3.5 g/l.
(b) Deposited in a solution containing 40 g/l KOH and 2 g/l KCNS, while the current was periodically reversed.

TABLE 6.7

Influence of Annealing Temperature on the Strength and Ductility of Electrodeposited Copper (1)

Property	Copper Sulfate Solutions		Copper Fluoborate Solutions	Copper Pyrophosphate Solutions	Copper Cyanide Solutions[b]	Annealed Wrought Copper
	No Addition Agents	With Addition Agents[a]				
Tensile Strength after Annealing at 325 C						
kg/sq mm	21.5	44	23	25	27	—
kpsi	31	62	33	36	38	—
Elongation after Annealing at 325 C	40	11	29	46	33	—
Tensile Strength after Annealing at 500 C						
kg/sq mm	21	22.5	19.5	22	24.5	22
kpsi	30	32	28	31	35	31
Elongation after Annealing at 500 C, percent	44	32	41	56	39	45

[a] Triisopropanolamine 3.5 g/l.

[b] Deposited in a solution containing 40 g/l KOH and 2 g/l KCNS, while the current was periodically reversed.

Effects on Substrates

Like many other electrodeposited metals, copper electrodeposits reduced the fatigue strength of a high strength steel, but the amount of reduction was less than 20 percent for 25 micron-thick copper on a hard steel with a fatigue limit of 76 kg/sq mm (109,000 psi). (40)

The hydrogen embrittlement of high strength steel frequently observed after copper plating in cyanide solutions was avoided by adjusting the total cyanide and copper concentrations in the solution to maintain a cyanide to copper ratio of 2.6 to 2.9, and equalizing the free cyanide and hydroxide concentrations. (41) A typical solution operated at 60 C contained 60 g/l CuCN, 90 g/l sodium cyanide, and 16 g/l NaOH.

Physical Properties

Electrical Resistivity

Copper with an electrical resistivity of 1.70 to 1.73 microhms-cm (101 to 99.5% IACS) was obtained at relatively low current densities in either copper fluoborate or sulfate solutions containing no addition agents. (1) These values are between those of cold-rolled and annealed sheet, 1.77 and 1.69 microhms-cm respectively.

The plating conditions and the composition of the copper fluoborate bath yielding low resistivity copper was as follows: (1)

Copper fluoborate, $Cu(BF_4)_2$	177 g/l (0.75 Moles/l)
Fluoboric acid, HBF_4	12 g/l
Boric acid, H_3BO_3	12 g/l
pH	0.8 – 1.0
Temperature	30 or 60 C
Cathode current density	2 or 8 amp/sq mm

The copper deposited in a higher concentration copper fluoborate bath had slightly higher resistivity (1.74 microhms-cm).

Conditions for depositing low resistance copper in copper sulfate solutions were as follows: (1)

Copper sulfate, $CuSO_4 \cdot 5H_2O$	87–275 g/l (0.35–1.1 M/l)
Sulfuric acid, H_2SO_4	25–74 g/l (0.25–0.75 M/l
Temperature, C	30–45 C
Cathode current density	0.5 to 4 amp/sq dm

Copper deposited at a higher cathode current density (8 amp/sq dm) in the copper sulfate bath, or a lower temperature (20 C), exhibited higher resistivity (1.74 to 1.80 microhms-cm). An addition of gelatin, β-naphthoquinoline, thiourea or triisopropanolamine increased resistivity to 1.77 to 2.10 microhms-cm. Much higher resistivities have been reported for copper deposited in solutions containing citric, aspartic or metaphosphoric acid. (42) However, annealing reduced these high resistivities as shown in Figure 6.6.

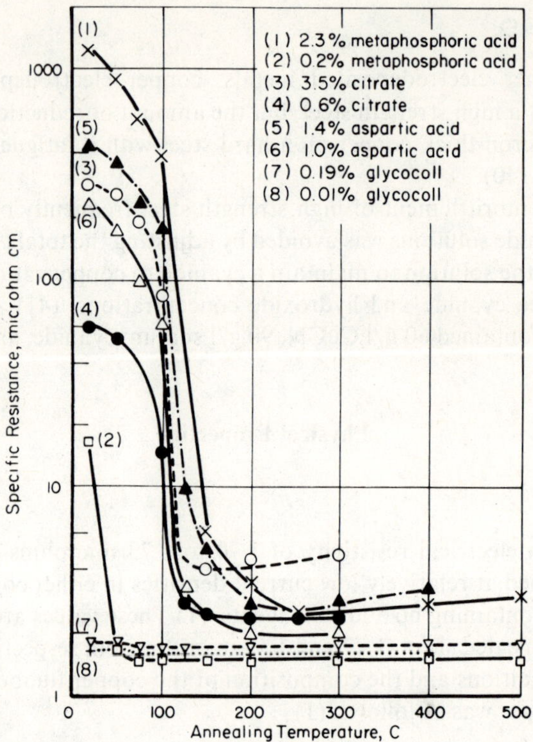

Figure 6.6. Changes in Electrical Resistivity of Copper as a Function of Annealing Temperature. The copper was deposited in copper sulfate solutions containing addition agents. (42)

The low electrical resistivity data cited for copper deposited in copper sulfate and fluoborate solutions was for 0.4 to 0.5 mm-thick copper. Values about 4 percent higher were reported thinner, 0.08 mm copper electrodeposited under similar conditions. (1)

The electrical resistivity of 0.04 to 0.05 mm-thick copper deposited in cyanide solutions generally ranged from 1.78 to 1.87 microhm-cm, although a few deposits obtained at low current densities of 1 or 2 amp/sq dm had resistivities of 1.74 to 1.76 microhms-cm. The resistivity of copper deposited in cyanide solutions containing potassium thiocyanate ranged from 1.98 to 2.02. Low-resistivity copper contained fewer impurities than high resistivity copper. The correlation between impurities and resistivity may be obserfed in Table 6.8 for copper deposited in cyanide, fluoborate, pyrophosphate, and sulfate solutions. The same data are presented graphically in Figure 6.7. Note that the oxygen content was usually roughly 20 percent of the total impurity level.

TABLE 6.8

Impurities in Electrodeposited Copper (1)

Electrical Resistivity, microhm-cm	Oxygen Content, weight %	Hydrogen Content, weight %	Nitrogen Content, weight %	Carbon Content, percent	Sulfur Content, percent	Total Impurities, percent	Type of Plating Bath
1.73	0.0005	<0.0001	0.0010	0.001	0.0011	0.0035	Copper sulfate[a]
1.73	0.0005-0.0008	<0.0001	0.0017	(b)	—	0.0041	Copper fluoborate[c]
1.75	0.0007	<0.0001	0.0015	—	—	0.003–0.004	Copper pyrophosphate[d]
1.76	0.0010	<0.0001	0.0021	—	—	0.0053	Copper cyanide (80 c)[e]
1.79	0.0033	<0.0001	0.0055	—	—	0.01	Copper cyanide (40 c)[f]
1.89	0.0050	0.0006	0.0094	0.0062	—	0.022	Copper sulfate with 3.5 g/l triisopropanolamine[g]
2.10	0.0350	0.0040	0.0535	0.0530	—	0.15	Copper sulfate with 0.1 g/l gelatin[h]

[a] The bath contained 0.75 M/l $CuSO_4 \cdot 5H_2O$ and 0.4 M/l H_2SO_4 and was operated at 30 C. Cathode current density was 2 amp/sq dm. The oxygen, hydrogen and nitrogen contents reported elsewhere (44) for similarly electrodeposited copper were 0.0001, 0.0002 and 0.00004 weight percent, respectively.

[b] The fluorine and boron contents were 0.0017 and <0.0001% respectively.

[c] The bath contained 1.77 g/l $Cu(BF_4)_2$ 12 g/l HBF_4 and 12 g/l H_3BO_3 and was operated at 30 C. Cathode current density was 8 amp/sq dm.

[d] The bath contained 90 g/l $Cu_2P_2O_7 \cdot 3H_2O$, 350 g/l $K_4P_2O_7$, 80 g/l K_3PO_4, 15 g/l KNO_3, 2 ml/l NH_3 and was operated at 50 C; the pH and current density were 8.5 and 2 amp/sq dm.

[e] The bath contained 75 g/l CuCN, 10 g/l free KCN and 40 g/l KOH (pH of 13.6). Current density was 6 amp/sq dm.

[f] The bath contained 15 g/l CuCN, 6 g/l free NaCN and 15 g/l Na_2CO_3 (pH of 12.8). Current density was 1 amp/sq dm.

[g] The bath contained 0.75 M/l $CuSO_4 \cdot 5H_2O$, 0.75 M/l H_2SO_4 and 3.5 g/l triisopropanolamine and was operated at 30 C. Current density was 5 amp/sq dm.

[h] The bath contained 0.75 M/l $CuSO_4 \cdot 5H_2O$, 0.75 M/l H_2SO_4 and 0.1 g/l gelatin and was operated at 30 C. Cathode current density was 2 amp/sq dm. The sulfur content in copper deposited in a copper sulfate solution containing 50 g/l of sulfuric acid was increased from about 0.02 to 0.2 percent when cysteine or methionine was added to the plating bath. (43)

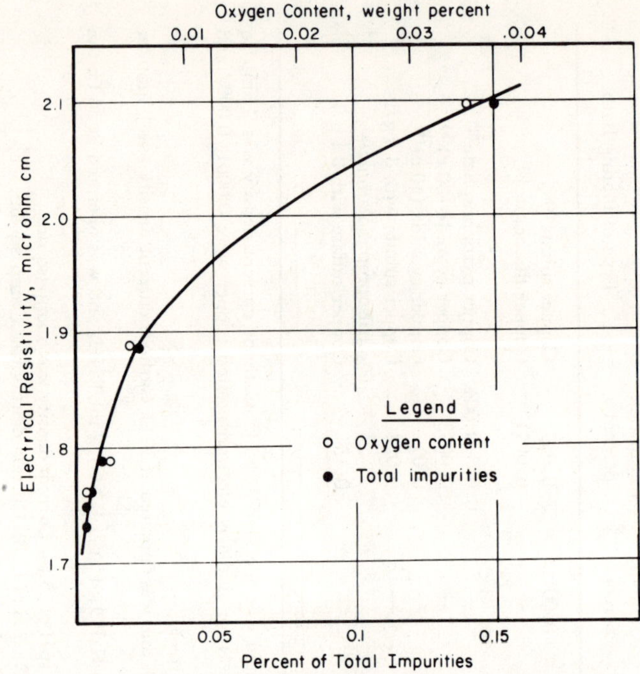

Figure 6.7. Electrical Resistivity of Copper Electrodeposits as a Function of Impurity Content. (1)

Modulus of Elasticity

Values for modulus of elasticity were nearly identical for electrodeposited and annealed wrought copper as shown in Table 6.9. Organic addition agents in copper sulfate solutions had no influence on the modulus. (1) The explanation for the low modulus from the fluoborate baths is now known. It may be related to the large grain size and preferred orientation of copper deposited from fluoborate baths.

Thermal Expansion

The values of the average coefficients of the thermal expansion over the range of 20 to 400 C are shown in Table 6.10. Coefficients for copper deposited in fluoborate, pyrophosphate, and sulfate solutions containing no addition agents were slightly lower than the coefficient for annealed, wrought copper. The use of an organic addition agent in the copper sulfate bath increased the coefficient about 30 percent because porosity developed as the coating was heated. The coefficient for copper deposited in cyanide solutions was reported to be about 12 percent lower than the coefficient for annealed, wrought copper.

TABLE 6.9

Modulus of Elasticity of Electrodeposited Copper (1)

Plating Bath	As Deposited		Annealed (500 C, 15 min)		Cold-Rolled (50% Reduction)	
	10^3 kg/sq mm	-10^6 psi	10^3 kg/sq mm	10^6 psi	10^3 kg/sq mm	10^6 psi
Sulfate[a]	11	16[b][c]	8.5	12[d]	10.5	15[d]
Sulfate[e]	11	16[f]	7	10[d]	10	14[d]
Fluoborate	8.5	12	4	6[d]	11	16
Pyrophosphate	12	17	10	14[d]	12	17[d]
Cyanide[g]	11–12	16–17[h]	10.5	15[d]	10	14
Wrought copper	—	—	12	17	12	17

[a] Solution containing 6.75 Moles/l $CuSO_4$ and 0.4 Moles/l H_2SO_4, 20 C, 2 amp/sq dm. The modulus was reduced to 14×10^6 psi when the temperature was increased to 30 C.

[b] Values from 5.6 to 8.1×10^6 psi and a value of 11×10^6 psi were reported for electrodeposited and annealed, wrought copper, respectively, when modulus was determined by a dynamic method involving the vibration of a reed cut from sheet stock, which may have reflected the effects of plastic deformation.

[c] In comparison with this value, moduli of (1) 14,000,000 and (2) 16,100,000 psi were reported for copper deposited in a similar purified copper sulfate bath. Value (1) was measured for copper that was not machined. Value (2) was obtained after the outer surface was machined. (45)

[d] Values from stress-strain curves. The other values are from resistance strain gage measurements.

[e] Solution containing 0.75 Moles/l $CuSO_4$, 0.75 Moles/l H_2SO_4 and 3.5 g/l triisopropylamine, 30 C, 5 amp/sq dm.

[f] Moduli of (1) 15.9×10^6 and (2) 16.9×10^6 psi were measured for copper deposited in a copper sulfate solution containing 50 g/l ethyl alcohol.

[g] Deposited in a solution containing 40 g/l KOH and 2 g/l KCNS, while the current was periodically reversed.

[h] Moduli of (1) 15.9×10^6 and (2) 16.9×10^6 psi were measured for copper deposited with PR current in a cyanide bath containing proprietary brighteners. Value (1) was obtained for unmachined deposits and Value (2) for deposits with a machined surface. (45)

TABLE 6.10

Coefficient of Thermal Expansion for Electrodeposited Copper (1)

Plating Bath	Coefficient of Thermal Expansion, μm/m/C	
	20–200 C	20–400 C
Sulfate[a]	17.1	17.8
Sulfate[b]	18.9	25.8
Fluoborate	16.7	17.6
Pyrophosphate	16.7	17.5
Cyanide[c]	16.7	16.9
Cold-rolled copper	17.3	18.1

[a] Solution containing 6.75 Moles/l $CuSO_4$ and 0.4 Moles/l H_2SO_4, 2 amp/sq dm.

[b] Solution containing 0.75 Moles/l $CuSO_4$, 0.75 Moles/l H_2SO_4 and 3.5 g/l triisopropylamine, 30 C, 5 amp/sq dm.

[c] Deposited in a solution containing 40 g/l KOH and 2 g/l KCNS, while the current was periodically reversed.

Structural Features

Grain Size and Shape

The grain structure of a copper electrodeposit generally fits one of four types; columnar, fibrous, equiaxed, or banded. (1,46) Figure 6.8 illustrates the *columnar* structure of copper deposited in copper sulfate solutions containing no addition agents. Some addition agents induce the *fibrous* structure in Figure 6.9, which is typical of copper deposited in fluoborate baths. The *fine-grained* structure in Figure 6.10 is representative of copper deposited in copper cyanide solutions at high temperatures (70–80 C), copper pyrophosphate baths or copper sulfate solutions containing naphthoquinoline, triisopropanolamine, and other amines. Figure 6.11 shows the *banded* structure obtained with periodically reversed current in copper cyanide baths containing proprietary brighteners or copper sulfate solutions containing thiourea.

Structure is correlated with properties in Table 6.11. Fine-grained copper usually is stronger and harder than columnar or fibrous structures. Some, but not all of the copper exhibiting a fine-grained structure also had a high electrical resistivity.

Fine grained deposits may show grain growth as a result of annealing at 500 C. (1) Figures 6.12 and 6.13 illustrate the grain growth (and voids) induced by annealing copper deposited in a sulfate bath containing 2 g/l triisopropanolamine. In contrast, the fine-grained deposits obtained in the pyrophosphate bath showed no significant change in grain size as a result of annealing, as shown by Figures 6.14 and 6.15. Columnar and fibrous deposits showed grain growth after annealing. Figure 6.16

illustrates the structure of the fibrous copper deposited in a fluoborate bath after annealing at 500 C.

Orientation

Crystal planes are preferentially oriented in many copper deposits. Copper deposited in sulfate solutions containing no addition agents has a predominance of crystals oriented with their (110) planes parallel to the surface of the substrate. (1) A similar orientation was observed in copper deposited in the fluoborate bath, but for copper deposited in pyrophosphate and high-temperature cyanide baths, (111) and (200) planes, respectively, were parallel to the substrates.

Addition agents in the copper sulfate bath sometimes affect orientation. A random orientation was observed for copper deposited in a solution containing triisopropanolamine. (1) Thiourea resulted in a (100) orientation. (48,49,50) With operating temperatures above 30 C, low copper concentration in the copper sulfate bath, and no organic agents, orientation shifted from (110) to (100). (51) High current densities of

Figure 6.8. Coarse Columnar Copper Deposited in a Copper Sulfate Bath Containing no Addition Agents.

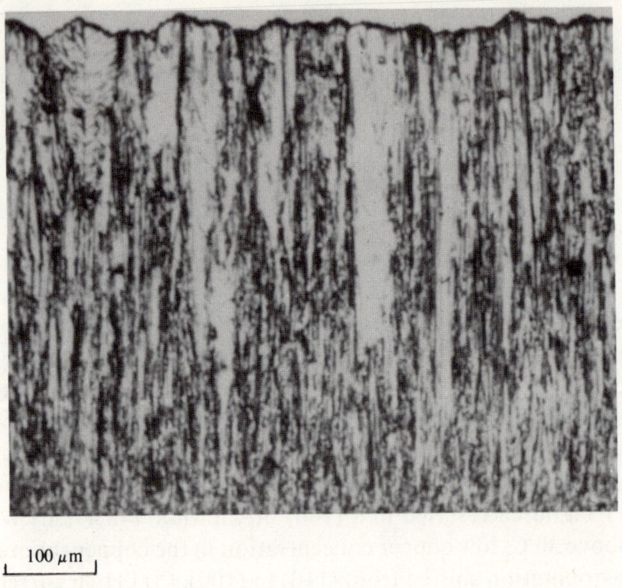

Figure 6.9. Fibrous Structure of Copper Deposited in Fluoborate Baths or in Copper Sulfate Solutions Containing Phenol Sulfonic Acid and Gelatin.

Figure 6.10. Fine-Grained Equiaxed Copper Deposited in High-Temperature Cyanide or Pyrophosphate Baths or in Copper Sulfate Solutions containing Amines.

125 to 175 amp/sq dm resulted in a (113) orientation. (52) Low temperatures (5 to 20 C) favored a (111) orientation. (53)

A (111) orientation for copper deposited in a cyanide bath was changed to a random orientation as overvoltage was slightly increased, to a (110) orientation as overvoltage was further increased, back to a (111) orientation at a high overvoltage level. (54) Additions of sulfur compounds with brightening characteristics such as thiocyanide and thiosulfate induced a strong (111) orientation in deposits from cyanide baths. (55)

Thermal Stability

Copper deposited in copper fluoborate, pyrophosphate, and sulfate solutions containing no addition agents showed little or no permanent change in dimensions after heating to 400 to 500 C. For example, the change in length for a typical deposit produced in a copper sulfate bath without additives was 15 microinches/inch. Significant expansion, indicative of the formation of porous structures, was found for copper deposited in the following solutions:

1. Copper sulfate baths containing dextrin, gelatin glycerol, naphthalene disulfonic acid, phenol sulfonic acid, selenic acid, thiourea or triisopropanolamine
2. Copper sulfate-ethylene diamine solutions.
3. Copper cyanide solutions containing potassium thiocyanate.

Changes in dimensions and densities for typical deposits are given in Table 6.12. The coefficient of thermal expansion from 20 to 400 C for stable copper deposits was 17.5 or 17.8 microinch/inch/C, essentially the same as for wrought copper.

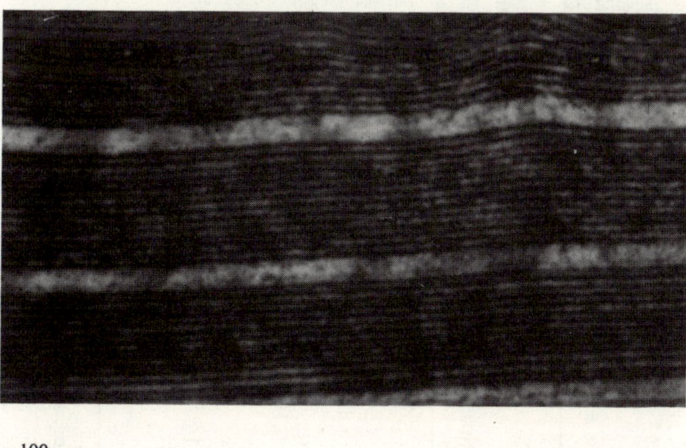

100 μm

Figure 6.11. Banded Copper Deposited with PR Current in Cyanide Baths Containing Proprietary Brighteners or Copper Sulfate Solutions Containing Thiourea.

TABLE 6.11

Correlation of Structure and Properties (1)

Type of Structure	Tensile Strength		Elongation, percent	Hardness, KNH_{200} kg/sq mm	Density, g/cu cm	Electrical Resistivity, microhm-cm	Plating Solutions
	kg/sq mm	kpsi					
Columnar—Fig. 6.8	18–24	25–35	15–35	60–75	8.92–8.93	1.71–1.75	Copper sulfate—no addition agents
Fibrous—Fig. 6.9	20–27	28–38	15–32	60–75	8.925	1.73–1.75	Copper fluoborate and copper sulfate with some addition agents
Fine-grained, equiaxed—Fig. 6.10	28–>56	40–>80	25–>40	80–>150	8.91–8.925	1.75–>2.0	Copper cyanide at 80 C, copper pyrophosphate and copper sulfate with some addition agents such as triisopropanolamine[a]
Banded—Fig. 6.11	42–49	60–70	to 45	>180	—	>2.0	Copper cyanide with proprietary brighteners and PR current and copper sulfate with thiourea

[a] Electron microscopic studies also showed a very fine grain size for copper deposited in cyanide and pyrophosphate solutions. (47)

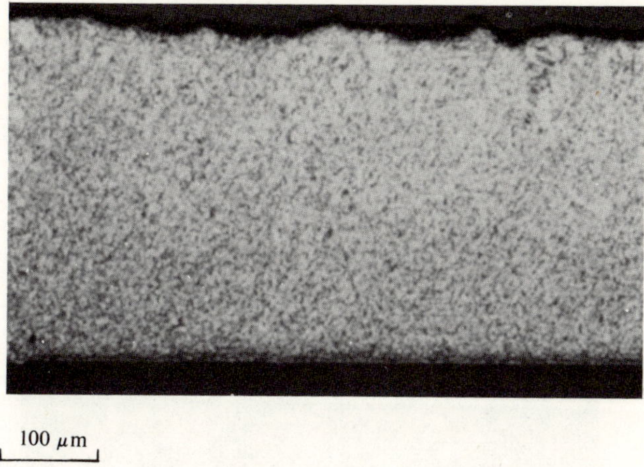

Figure 6.12. Structure of Copper Deposited in a Sulphate Bath Containing 2 g/l Triisopropanolamine.

During heating in a vacuum to 400 or 600 C, large amounts of gas evolved from copper that had been deposited in copper sulfate solutions containing citrate, tartrate or borate ions. (56) The gas probably was water vapor released from the water of crystallization associated with entrapped molecules of the addition agents. Electrical resistivity of such copper deposits declined markedly as a result of heating to only

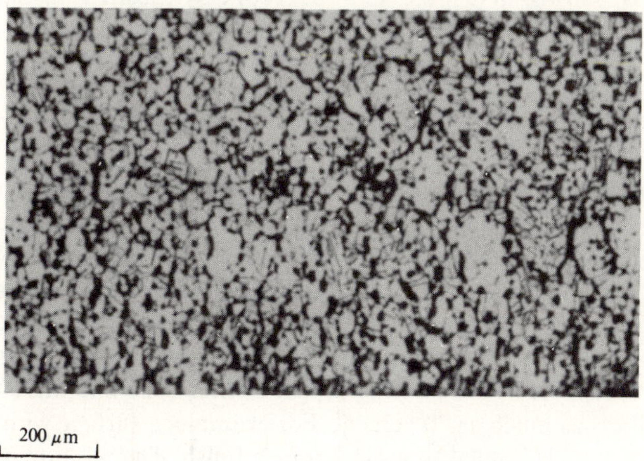

Figure 6.13. Structure after annealing 15 Minutes at 500 C of Copper Deposited in a Sulfate Bath Containing 2 g/l Triisopropanolamine.

Figure 6.14. Structure of Copper Deposited in a Pyrophosphate Bath.

about 150 C, which is the temperature for the release of water from hydrated sodium citrate or tartrate. The decrease in resistivity with increasing heat treating temperature is shown in Figure 6.6 for copper deposited in sulfate solutions containing organic addition agents or metaphosphoric acid. (42)

Bright Leveling Deposits

Bright, fine-grained copper with good leveling characteristics is used for many cladding and forming purposes and nearly all decorative applications. Proprietary brighteners developed for cyanide baths promote leveling, particularly when the current is periodically reversed. Bright copper electroplated in pyrophosphate baths also exhibits leveling power. However, proprietary brighteners for copper sulfate solutions are more effective for plating mirror-like, very smooth copper over rough substrates. (58) A layer of this leveling copper as thin as 20 microns (0.8 mil) reduces surface roughness as much as 70 percent. For example, a surface with a roughness of 14 microinches rms, is leveled to a 4-microinch finish. The relation between leveling and plating thickness is shown in Figure 6.17.

Bright leveling copper electroplated in copper sulfate baths is more effective than other electroplates for smoothing rough substrates. Combining it with leveling semibright and bright nickel smooths very rough surfaces, i.e., 20 microinches, rms to mirror-like finishes of less than 4 microinches, rms.

The properties of four kinds of copper and their leveling characteristics are given in Table 6.13. The leveling deposits exhibited high strength and good ductility, as a general rule, although some were stronger and harder than others. All were appreciably more ductile than bright copper deposited in copper sulfate solutions containing thiourea. Typical deposits exhibit a fine-grained structure like that in Figure 6.10.

Influence of Plating Variables

As may be seen from the preceding data, the properties and structure of copper depend on the plating solution and the addition agents employed to refine grain structure or brighten the deposits. Also, impurities frequently have significant effects. For example, a cellulose product leached from filter paper or cotton by a copper sulfate solution improved the strength and ductility of the deposits. (1) Copper with a tensile

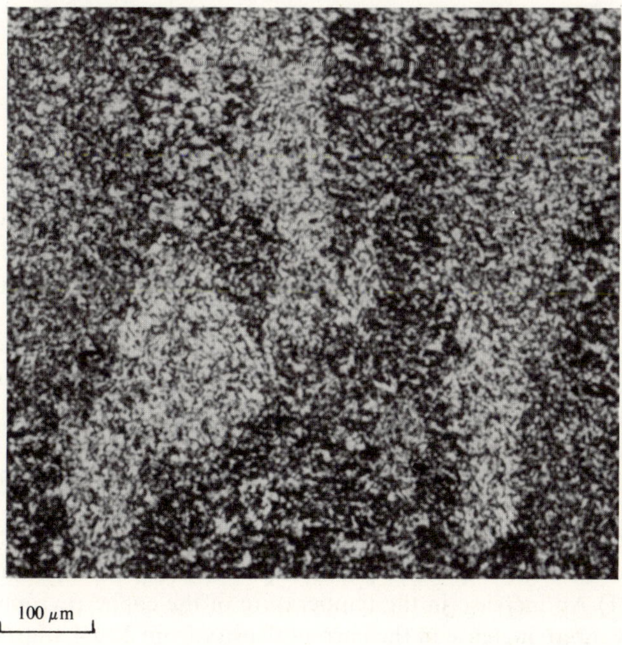

Figure 6.15. Structure after Annealing 15 Minutes at 500 C of Copper Deposited in a Pyrophosphate Bath.

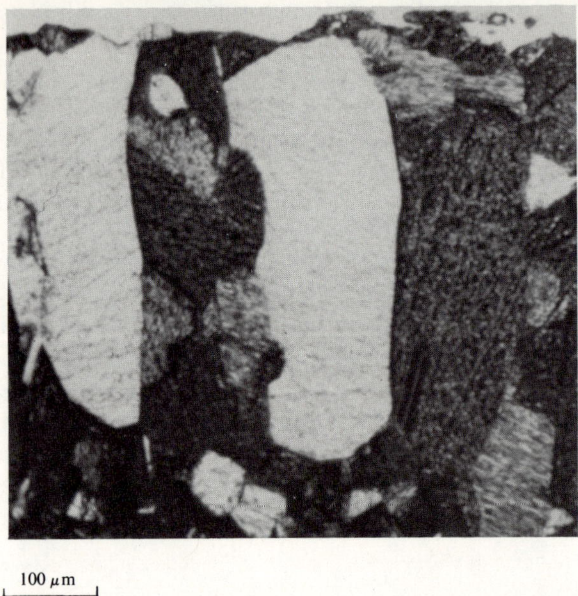

Figure 6.16. Structure after Annealing 15 Minutes at 500 C of Copper Deposited in a Fluoborate Bath. (Compare with Figure 6.9)

strength of 22.5 kg/sq mm (32,000 psi) and an elongation of 34 percent was deposited in a solution containing this impurity, whereas a nonuniform, coarse-grained deposit with a tensile strength of only 10.5 kg/sq mm (15,000 psi) and an elongation of only 8 percent was obtained after the solution was purified by electrolysis.

Pyrophosphate Baths

The pyrophosphate bath is noteworthy because its deposits have high purity, good ductility (31 to 39 percent elongation), fairly high tensile strength (27 to 30 kg/sq mm or 38,000 to 42,000 psi), low electrical resistivity (1.73 or 1.74 microhm cm), and low internal stresses (1.2 or −1.1 kg/sq mm, equivalent to 1,700 psi or −1,600 psi, respectively). (1) An increase in the temperature of the copper pyrophosphate bath from 50 to 60 C or an increase in the current density from 2 to 4 amp/sq dm did not affect the strength or ductility of the deposits. (1) Internal stress was increased from −1.1 kg/sq mm (−1600 psi) to 1.2 kg/sq mm (1700 psi) when the current density was increased. Air agitation or a combination of air and mechanical agitation resulted in the same strength and ductility. A proprietary grain-refining agent has been used to increase the hardness of deposits from pyrophosphate baths from about 84 to 185 kg/sq mm. (63)

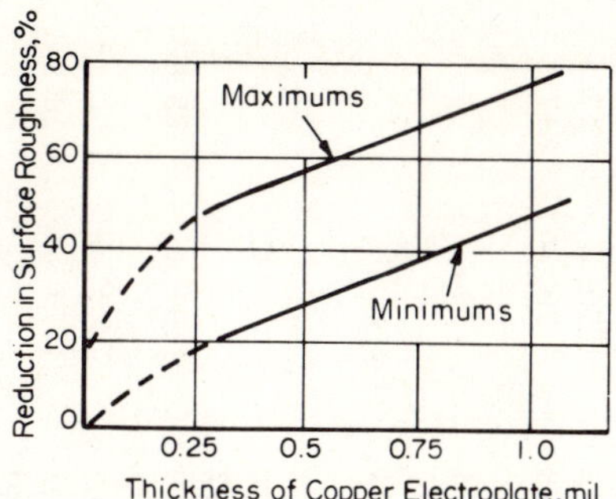

Figure 6.17. The Leveling Power of Bright Copper Deposited in Copper Sulfate Solutions Containing a Proprietary Brightener. (56)

Cyanide Baths

Changes in the concentrations of the major constituents of the plating solutions also affect properties and structure. The effects of such changes in copper cyanide baths are summarized in Table 6.14. Hard and strong copper with a tensile strength of at least 43 kg/sq mm (61,000 psi) was deposited with periodically reversed current in solutions at 80 C containing 75 g/l copper cyanide. Effects on structure have not been examined individually for all variables. However, copper deposited in low concentration copper cyanide baths at low temperatures (40 C) showed coarse-grained columnar structures. The fibrous structure typical of copper deposited at 60 C was changed to a fine-grained, equiaxed structure by adding Rochelle salts or increasing the copper cyanide concentration to 75 g/l and the temperature to 80 C.

The impurities in representative deposits from cyanide baths were in the range of 50 to 100 ppm, which was higher than the impurity contents of copper deposited in fluoborate, pyrophosphate, or sulfate baths containing no organic addition agents. (1) The oxygen contents in the cyanide copper plates ranged from 5 to 33 ppm. A bright deposit obtained with a selenium salt addition agent contained 11 ppm selenium.

Fluoborate Baths

The fluoborate bath, like the sulfate solution, is capable of producing high-purity, thermally stable copper with a low electrical resistivity. Impurities totaling 41 ppm in a representative deposit included 5 ppm oxygen, 17 ppm percent nitrogen, 17 ppm fluorine, and 1 ppm boron. (1) The effects of modifying solution composition, temperature, and current density on the properties of the copper deposited from fluoborate baths are summarized in Table 6.15.

TABLE 6.12

Thermally Stable and Unstable Copper Electrodeposits (1)

Plating Bath	Addition Agent	Temp, C	Current Density, amp/sq dm	Change in Length After Heating, percent[a]	Density, g/cu cm		Electrical Resistivity, microhm-cm		Coefficient of Thermal Expansion
					Before Heating	After Heating	Before Heating	After Heating[a]	microinch/inch/C[a]
$Cu(BF_4)_2$[b]	None	30	8	0.000	8.926	—	1.73	1.70	17.6
$Cu_2P_2O_7$[c]	KNO_3, 15 g/l	50	2	−0.017	8.926	8.926	1.75	1.70	17.5
$CuSO_4$[d]	None	30	2	−0.0015	8.926	8.924	1.73	1.70	17.8
$CuSO_4$[e]	Dextrin, 0.02 g/l	30	2	0.06	—	—	—	—	—
$CuSO_4$[e]	Dextrose, 0.05 g/l	30	2	0.04[f]	8.922	—	1.73	—	—
$CuSO_4$[e]	Gelatin, 0.1 g/l	30	2	1.2[f]	8.86	—	2.10	—	—
$CuSO_4$	Naphthalene disulfonic acid, 1.8 g/l	30	2	0.04[f]	8.923	—	1.71	—	—
$CuSO_4$[e]	β-naphthoquinoline			0.14	8.923	—	1.78	—	—
$CuSO_4$	SeO_2, 10^{-4} g/l	30	2	0.06	8.929	—	1.76	—	—
$CuSO_4$	Thiourea, 0.015 g/l	30	2	0.4[f]	8.918		1.77		
$CuSO_4$[e]	Triisopropanolamine, 3.5 g/l	30	5	0.33	8.913	7.35 (porous)	1.89	1.94	25.8
$CuSO_4$	Proprietary agent[g]	22	4	3.6	8.917	—	1.82	—	—
CuCN[h]	KCNS, 2 g/l	80	6	−0.03–0.05	8.917	8.917	2.02	1.79	16.9
$CuSO_4$-Ethylene diamine[i]	NH_3, 4 ml/l	55	2	0.16[f]	8.891	—	2.10	—	—

[a] Change in length for a 6-inch specimen after heating to either 400, 500 or 550 C. The thermal coefficient is given for the range of 20 to 400 C.

[b] The solution contained 177 g/l $Cu(BF_4)_2$, 12 g/l HBF_4 and 12 g/l H_3BO_3; the pH was 0.8–1.0.

[c] The solution contained 90 g/l $Cu_2P_2O_7$, 350 g/l $K_4P_2O_7$, 80 g/l K_3PO_4, 15 g/l KNO_3 and 2 ml/l concentrated NH_3; the pH was 8.5.

[d] The solution contained 187 g/l $CuSO_4 \cdot 5H_2O$ and 39 g/l H_2SO_4.

[e] The solution contained 187 g/l $CuSO_4 \cdot 5H_2O$ and 74 g/l H_2SO_4.

[f] Ductility determined by bending was reduced after heating and cooling.

[g] A sulfonated organic compound and an azo dye. (57)

[h] The solution contained 75 g/l CuCN, 10 g/l free KCN, 40 g/l KOH and 2 g/l KCNS; pH was 13.6. The current was reversed for 5 seconds after every 15 seconds of direct plating.

[i] The solution contained 100 g/l $CuSO_4 \cdot 5H_2O$, 20 g/l $(NH_4)_2SO_4$, 80 ml/l ethylene diamine and 4 ml/l concentrated NH_3.

TABLE 6.13

Properties of Bright, Leveling Electroplated Copper

Plating Bath	Tensile Strength		Elongation, percent	Hardness VHN_{200}, kg/sq mm	Density, g/cu cm	Internal Stress		Electrical Resistivity, microhm-cm	
	kg/sq mm	kpsi				kg/sq mm	psi	Before Annealing	After Annealing at 500 C
Cyanide with KSCN and PR current[a]	60[b]	86[b]	10[b]	144	8.917	3.6	5,100	2.02	1.79
Cyanide with proprietary brightener and PR current[c]	46[b]	66[b]	45[b]	224	—	—	—	1.96	—
Pyrophosphate[d]	28.5[e]	41[e]	35[e]	92	8.926	1.2	1,700	1.74	1.70
Copper sulfate with proprietary brightener[f]	36.5,[e] 59[g]	52,[e] 84[g]	19[e]	128	8.918	2.0	2,900	1.82	—
Copper sulfate with no addition agent[h]	21	30	24	53	8.922	1.5	2,200	1.72	1.70

[a] The solution contained 75 g/l CuCN, 10 g/l free KCN, 40 g/l KOH and 2 g/l KSCN, maintained at a pH of 13.6 and a temperature of 80 C. Cathode current density was 6 amp/sq dm. (1)

[b] Tensile strength and elongation for 0.05 to 0.1 mm-thick deposits. (1)

[c] The solution contained 75 g/l CuCN, 22.5 g/l free KCN, 22.5 g/l KOH and a selenium-type proprietary brightener, maintained at a temperature of 80 C. Cathode current density was 6 amp/sq dm. (1)

[d] The solution contained 90 g/l $Cu_2P_2O_7 \cdot 3H_2O$, 350 g/l $K_4P_2O_7$, 80 g/l K_3PO_4, 15 g/l KNO_3 and 2 ml/l NH_4OH to adjust the pH to 8.5. Temperature was 50 C and the cathode current density was 4 amp/sq dm. (1)

[e] Tensile strength and elongation for 0.4 to 0.5 mm-thick-deposits. (1)

[f] The solution contained 225 g/l $CuSO_4 \cdot 5H_2O$, 50 g/l H_2SO_4 and a proprietary brightener. (57) It was operated at 27 or 28 C with a current density of about 4 amp/sq dm. (1)(59)

[g] Tensile strength for 0.01-mm-thick deposits. (59)

[h] Data for nonleveling copper included for comparison purposes with leveling copper deposited in a similar solution (187 g/l $CuSO_4 \cdot 5H_2O$ and 39 g/l H_2SO_4) with similar conditions (30 C, 4 amp/sq dm). (1)

TABLE 6.14

Effects of Modifying Solution Composition and Operating Conditions on the Properties of Copper Deposited in Cyanide Baths (1)

Variable[a]	Effect On Tensile Strength	Effect On Ductility	Effect On Hardness	Effect On Stress	Effect On Electrical Resistivity
Increase in copper cyanide concentration	Reduced appreciably	Increased appreciably	Reduced	Increased	None
Increase in free cyanide concentration	Increased appreciably	Reduced appreciably	None[b]	None	None
Increase in sodium carbonate concentration	None	None	None	None	None
Replaced sodium with potassium salts	Reduced appreciably	Reduced	Reduced	Increased	Decreased
Change in pH	Minor	Minor	Minor	Minor	—
Addition of Rochelle salts	Minor	Minor	Minor	Decreased[c]	Increased slightly
Addition of KCNS	Increased	Reduced	Increased	Minor[d]	Increased appreciably
Use of PR current	Increased appreciably	Reduced[e]	Increased[e]	Increased slightly[f]	None
Increase in temperature	Minor or slightly increased	Usually increased	Minor	Minor from 40 to 60 C, decreased from 60 to 80 C	Minor
Increase in current density	Increased with some baths; no change with others	Usually decreased	Increased with some baths; no change with others	Minor	Increased slightly

[a] Changes in these variables had no effect on the modulus of elasticity or density. However, density was consistently high (8.918 to 8.925) when high temperatures (60 or 80 C) were used for high-copper cyanide baths. Density ranged from 8.900 to 8.918 for copper deposited in low-copper cyanide solutions at 40 C.

[b] A slight increase in hardness was observed for copper deposited in 65 C solutions containing 20 or 80 g/l CuCN, whereas a slight decrease was noted for copper deposited in a bath containing 120 g/l CuCN. (60)

[c] A slight increase in stress from 4.5 to 6.1 kg/sq mm (6400 to 8700 psi) was reported previously. (61)

[d] A KSCN addition induced a compressive stress of −2.8 or −3.3 kg/sq mm (−4,000 or −4,700 psi), according to a previous report. (61)

[e] A large increase in hardness and a large decrease in ductility was noted previously. (62)

[f] An increase in stress from 6.1 to 8.2 kg/sq mm (8,700 to 11,600 psi) was reported previously. (61)

TABLE 6.15

Effects of Modifying Solution Composition and Operating Conditions on the Properties of Copper Deposited in Fluoborate Solutions (1)

Variable	Effect On Tensile Strength	Effect On Ductility	Effect On Hardness	Effect On Stress	Effect On Electrical Resistivity
Increasing concentration of all solution constituents[a]	None	None	Reduced slightly	None	Increased from 1.73 to 1.74
Increasing temperature (from 30 to 60 C)	Reduced slightly from 26 to 22.5 kg/sq mm	Reduced from 31 to 16%	Reduced from 81 to 56	Reduced from 0.5 to −0.15 kg/sq mm	None
Decreasing current density from 8 to 2 amp/sq dm	Reduced from 26 to 14 kg/sq mm	Reduced from 31 to 7%	Reduced from 81 to 53	Reduced from 0.5 kg/sq mm to 0	None
Addition of 1.2 ml/l molasses[b]	Increased from 12 to 21 kg/sq mm	Increased from 7.3 to 11 percent	Increased from 40 to about 60	—	—
Addition of 2 g/l Dacolyte[b]	Increased from 12 to 21 kg/sq mm	Increased from 7.3 to 10.5 percent	Increased from 40 to 48[c]	—	—

[a] Refers to increases in the concentrations of $Cu(BF_4)_2$ from 177 to 366 g/l, HBF_4 from 12 to 22 g/l and H_3BO_4 from 12 to 22 g/l for deposits obtained at 8 amp/sq dm, 30 C. (1)

[b] Refers to a solution containing 224 g/l $Cu(BF_4)_2$, 4 g/l HBF_4 and 15 g/l H_3BO_3 at 49 C and a current density of 32 amp/sq dm. (64)

TABLE 6.16

Effects of Modifying Solution Composition and Operating Conditions on the Properties of Copper Deposited in Sulfate Solutions With No Addition Agents (1)

Variable	Effect on Tensile Strength	Effect on Ductility	Effect on Hardness	Effect on Stress	Effect on Density	Effect on Electrical Resistivity	Effect on Structure
Increasing copper sulfate concentration	None[a]	Usually increased[a]	None	Variable but slight	Increased slightly	None	Little or none
Increasing sulfuric acid concentration from 0.25 to 0.5 M	Decreased	Decreased	Increased	None	None	None	None
Increasing sulfuric acid concentration from 0.5 to 0.75 M	Increased	Increased	Variable, but slight	Usually increased	Increased slightly	None	None
Increasing temperature	Decreased	Decreased	Usually decreased slightly	Decreased	None	None	Increased grain size
Increasing current density	Increased[b]	Increased slightly[c]	Variable, but slight	Increased	Variable, but slight[d]	Increased slightly[d]	Refined grain size
Periodic current reversal	Increased	Decreased	Increased	Increased slightly[e]	None	Increased slightly	Refined grain size
Additions of gelatin, 0.01 and 0.1 g/l	Increased appreciably	Decreased	Increased	Variable[f]	Reduced	Increased	—
Additions of phenol sulfuric acid, 1 g/l	Increased slightly	Increased slightly	Increased	Reduced	None	None	—

Additions of selenic acid, 10^{-3} g/l	Increased	Increased slightly	Increased	Increased	Increased	Increased slightly	Refined grain size
Additions of thiourea, 0.015 g/l	Reduced slightly	Reduced	Increased slightly	No significant effect	Reduced slightly	Increased slightly	Refined grain size
Additions of thiourea and naphthalene disulfonic acid, 0.5 g/l[g]	Increased	None	Increased	Ditto	None	Increased slightly	Refined grain size
Additions of triisopropanol-amine, 3.5 g/l	Increased[h]	Decreased	Increased	Increased	Decreased	Increased	Refined grain size appreciably
Other addition agents	(i)	(i)	(j)	(j)	—	—	—

[a] A low tensile strength of only 20 kg/sq mm (20,000 psi) and a low elongation of only 8 percent were measured for copper deposited at a low current density (0.5 amp/sq ft) in a solution containing a low copper sulfate concentration (87 g/l $CuSO_4 \cdot 5H_2O$) and a high sulfuric acid concentration (74 g/l).

[b] An increase was observed for increases in current density from 0.5 to 2 amp/sq dm and from 2 to 4 amp/sq dm. No further increase was noted when current density was increased to 8 amp/sq dm.

[c] An increase was observed for increases in current density from 0.5 to 2 amp/sq dm. With a supplemental increase in current density, elongation was sometimes increased slightly and sometimes reduced slightly.

[d] Density was reduced significantly to only 8.86 g/l cu cm at a very high current density of 20 amp/sq dm (200 amp/sq ft). Electrical resistivity was increased from 1.72 to 1.90 microhm-cm.

[e] The increase in stress caused by periodic current reversal was noted previously. (65)

[f] A relatively high concentration (0.1 g/l) induced a compressive stress.

[g] An addition of naphthalene sulfonic acid with no thiourea had little effect on the properties, but ductility was improved.

[h] The increase in tensile strength was especially significant at a relatively high current density of 5 amp/sq dm to 50 kg/sq mm or 71,000 psi.

[i] Additions of dextrin reduced tensile strength and ductility. Dextrose severely embrittled the copper. Additions of β-naphthoquinoline, glycerol, glycine, lactic acid and a mixture of ethyl alcohol and molasses had little influence on tensile strength, but reduced ductility. (1) Casein or molasses increased tensile strength to the range of 28 to 35 kg/sq mm (40,000 to 50,000 psi). (66) Triethanolamine (5–7.5 g/l) increased tensile strength to as much as 51 kg/sq mm (73,000 psi) and reduced elongation to 6 or 9 percent. (67)

[j] Other addition agents which influence hardness and stress in electrodeposited copper have been noted. (68, 69)

TABLE 6.17

Comparison of Structure and Properties of Copper Deposited at 4 amp/sq dm in Several Different Copper Solutions (1)

Plating Solution	Structure	Modulus of Elasticity, kg/sq mm	Modulus of Elasticity, 10^6 psi	Tensile Strength, kg/sq mm	Tensile Strength, kpsi	Yield Strength, kg/sq mm	Yield Strength, kpsi	Elongation, percent (2 in.)	Density, g/cu cm	Electrical Resistivity, microhm-cm	Hardness, KHN_{200}
Amine[a]	Fine-grained (Fig. 6.4)	—	—	42	60	31	44	4	8.906	2.10	169[b]
Cyanide Low Cu conc. and low temperature[c]	Columnar	11,000	16	31	44	—	—	7	8.900	1.80	144
Low Cu conc. and High temperature[d]	Fine-grained (Fig. 6.10)	11,000	16	55	78	—	—	6	8.912	1.79	131
High Cu conc. and high temperature[e]	Ditto		—	26	37	—	—	22	8.919	1.80	114
Ditto with 2 gl KSCN[f]	″	12,000	17	60	86	40	57	10	8.917	2.02	144
Fluoborate[g]	Fibrous (Fig. 6.9)	8,500	12	26	37	11	16	31	8.926	1.73	56

Pyrophosphate[h]	Fine-grained (Fig. 6.10)	12,000	17	28	40	15	22	33	8.926	1.74	92
Sulfate[i]	Coarse, columnar (Fig. 6.8)	10,000	14	21	30	6	9	24	8.922	1.72	53
Sulfate with 10^{-3} g/l Se[j]	Fibrous	11,000	16	35	50	20	28	20	8.928	1.75	108
Sulfate with 3.5 g/l triisopropanol-amine[k]	Fine-grained	10,500	15	50	71	30	43	7	8.913	1.89	144

(a) Solution at 55 C containing 100 g/l $CuSO_4 \cdot 5H_2O$, 20 g/l $(NH_4)_2SO_4$, 4 ml/l NH_4OH and 80 ml/l ethylene diamine.

(b) Hardness of copper deposited in a similar bath at a slightly lower temperature was 140 to 150 kg/sq mm. (70)

(c) Solution at 40 C containing 40 g/l CuCN, 6 g/l free NaCN and 30 g/l Na_2CO_3 with a pH of 11.7 to 12.8.

(d) Solution at 80 C containing 40 g/l CuCN, 10 g/l free NaCN and 30 g/l Na_2CO_3 with a pH 11.8 to 12.8.

(e) Solution at 80 C containing 75 g/l CuCN, 10 g/l free KCN and 40 g/l KOH with a pH of 13.6.

(f) Copper deposited at 6 amp/sq dm.

(g) Solution at 30 C containing 177 g/l $Cu(BF_4)_2$, 12 g/l HBF_4, and 12 g/l H_3BO_3.

(h) Solution at 50 C containing 90 g/l $Cu_2P_2O_7 \cdot 3H_2O$, 350 g/l $K_4P_2O_7$, 80 g/l K_3PO_4, 15 ml/l KNO_3 and 2 ml/l NH_3 with a pH of 8.5.

(i) Solution at 30 C containing 187 g/l $CuSO_4 \cdot 5H_2O$, and 39 g/l H_2SO_4. In comparison with data for these deposits, commercial electrolytic copper (produced with addition agents) exhibited a tensile strength of 28 to 34 kg/sq mm (40,000 to 48,000 psi). (71)

(j) Solution at 30 C containing 187 g/l $CuSO_4 \cdot 5H_2O$, 74 g/l H_2SO_4 and 10^{-3} g/l Se (2 amp/sq dm).

(k) Solution at 30 C containing 187 g/l $CuSO_4 \cdot 5H_2O$, 74 g/l H_2SO_4 and 3.5 g/l triisopropanolamine (5 amp/sq dm).

Copper Sulfate Baths

The copper sulfate bath, with or without addition agents, has been adopted for more diverse applications than any other type. The properties of the copper deposited in sulfate baths are appreciably influenced by addition agents, bath composition and operating conditions. A summary of these effects is given in Table 6.16.

An alkaline, copper sulfate bath prepared with 80 ml/l of ethylene diamine produced strong, hard deposits, as shown in Table 6.17 but no data have been developed on the influence of varying operating conditions.

The structures and properties of copper deposited from sulfate, cyanide, fluoborate, and pyrophosphate baths are compared in Table 6.17.

References

(1) Lamb, V. A., Johnson, C. E., and Valentine, D. R., "Physical and Mechanical Properties of Copper Deposits from Sulfate, Fluoborate, Pyrophosphate, Cyanide and Amine Baths," *Final Report on Project 21 to the American Electroplaters' Society*, (February, 1969), 316 pp.; *Journal Electrochem. Soc. 117,* 291c–318c, 341c,–352c, 381c–404c. (1970).

(2) Max, A. M., "Copper Plating Solution," U.S. Patent 2,482,354 (September 20, 1949). Assigned to Radio Corporation of America.

(3) Heussner, C. E., Balden, A. R., and Morse, L. M., "Some Metallurgical Aspects of Electrodeposits," *Plating, 35,* 554–561, 577–578, 719–723, 768 (1948).

(4) Levi, F. A., "Electrodeposition of Copper in Thin Layers on a Rotating Cathode in Presence of Ultrasonic Waves," *Nuovo Cimento, 12,* 493 (1959).

(5) Lanyi, R. J., Lane, D. H., Forbes, C. A., and Ricks, H. E., "The Application of Ultrasonic Energy to Metal Processing," *SAE Trans., 71,* 520 (1963).

(6) Trofimov, A. N., "Electrodeposition of Metals in Ultrasound," *Primenenie Ul'traakust. k Issled. Veshchestva, No. 12,* 113 (1960).

(7) Riedel, W., "X-Ray Investigations of Copper Deposits," *Galvanotechnik, 58,* 704 (1967).

(8) Hofer, E., and Javet, Ph., "Modern Methods of Structural Studies of Electrolytic Deposits," *Microtecnic, 17* (3), 104 (1963).

(9) Chocianowicz-Biestek, T., and Szmidt, K., "Methods of Obtaining Bright Copper Coatings from a Cyanide Bath," *Prace Inst. Mech. Precyz., 8* (30), 45 (1960).

(10) Steponavicius, A., and Visomirskis, R., "Cyanide Electrolyte for Bright Copper Plating," *Protessy Uprochneniya Detalei Mashin, Akad. Nauk SSSR*, 200 (1964).

(11) Raub, E., "Properties and Characteristics of Galvanically Deposited Metals with High Non-Metallic Contents," *Z. Metall., 39,* 33 (1948).

(12) Solokhina, V. G., Kudryavtsev, N. T., and Lapatukhin, V. S., "Effect of Organic Additives on the Structure and Hardness of Electrodeposited Copper on Gravure Printing Rolls," *Sb. Nauch. Rabot. Vsesoyuz. Nauch.-Issled. Inst. Poligraf. Prom.*, (8), 48 (1958).

(13) Fischer, H., Huhse, P., and Pawlek, F., "Internal Stress in Galvanic Copper Deposits," *Z. Metall., 47,* 43 (1956).

(14) Sammour, H. M., "Copper Electroplating. II. Microhardness and Structure of Electrolytic Copper Plate," *Metallurgia ital., 50,* 397 (1958).

(15) Svatek, L., and Vitek, J., "Additive for Acid Copper-Plating Bath," Czech. Patent 117,058 (December 15, 1965); Belg. Patent 660,618 (July 1, 1965). Lucek, J., and Svatek, L., "Hard Electrolytic Copper and Its Applications," *Tech. Dig. (Prague), 9* (6), 383 (1967).

(16) Vagramyan, A. T., and Petrova, Yu. S., *The Mechanical Properties of Electrolytic Deposits*, USSR Acad. Sci. Press, Institute of Physical Chemistry, Moscow (1960). English translation, Consultants Bureau, Inc., N.Y., (1962).

(17) Lyzlov, Yu. V., and Samartsev, A. G., "Internal Stress in Electrolytic Copper Deposits," *Zh. Fiz. Khimii., 33,* 1345 (1959); *Plating,* 46, 266 (1959).

(18) Elze, J., "Deposition Conditions and Structure of Compact Electrolytic Metal Deposits," *Metall, 8,* 519 (1954).

(19) Binder, H., and Fischer, H., "Relations Between Internal Stresses, Texture, and Hardness in Electrolytically Deposited Copper Layers," *Z. Metall., 53,* 161 (1962).

(20) Haensel, G., and Resch, H., "Investigations of the Effects of Amines, Amino Acids, Sulfonic Acids and Polyacrylic Acid Amide on the Electrocrystallization of Copper from Sulfuric Acid Electrolyte," *Metalloberflaeche, 21,* 268 (1967).

(21) Fedot'ev, N. P., "Physical and Mechanical Properties of Electrodeposited Metals," *Plating, 53,* 309 (1966).

(22) Fedot'ev, N. P., and Kruglova, E. G., "Reinforcement of Silver Mirrors by Copper Plating," *Journal Applied Chemistry (USSR), 28,* 251 (1955).

(23) Vagramyan, A. T., and Solov'eva, Z. A., "Hardness of Electrodeposits," *Electroplating and Metal Finishing, 15,* 45 (1962).

(24) Haensel, G., and Resch, H., "The Influence of Organic Nitrogen Compounds on the Deposition Potential and Deposit Structure of Copper by Electrodeposition from Sulfuric Acid Electrolytes," *Metalloberflaeche, 17,* 325 (1963).

(25) Dini, J. W., "Plating Through Holes in Printed Circuit Boards: Evaluation of Some Copper Baths," *Plating, 51,* 119 (1964).

(26) Diggin, M. B., "Modern Electroforming Solutions and Their Applications," *Symposium on Electroforming—Applications, Uses and Properties of Electroformed Metals, ASTM Spec. Tech. Pub. No. 318,* 10–31 (1962).

(27) Sadek, H., "Density Variation During Copper Deposition," *Z. Physik. Chem., 217,* 207 (1961).

(28) Phillips, W. M., and Clifton, F. L., "Stress in Electrodeposited Nickel," *Proceedings Am. Electroplaters' Soc., 34,* 97 (1947).

(29) Sadek, H., and Rizk, A. S., "Stress During the Electrodeposition and Dissolution of Copper on Copper Substrate. The Effect of Addition Agents," *Egypt. J. Chem., 1,* 105 (1958).

(30) Fedot'ev, N. P., and Pozin, Yu. M., "Effect of Surface Active Substances on the Mechanical Properties of Electrolytic Deposits," *Journal Applied Chemistry (USSR), 31,* 406 (1958).

(31) Shibasaki, Y., "Studies on the Internal Stress of the Electrodeposited Layer (Part I). The Relation Between the Internal Stress of Deposited Copper Layer from Acidic Copper Sulfate Solution and Current Density or Temperature," *J. Electrochem. Soc., Japan, 19,* 82 (1951).

(32) Lebedeva, V. N., and Popereka, M. Ya., "Electrodeposition of Copper from a Sulfate Solution in the Presence of Nitrogen-Containing Organic Compounds," *Protection of Metals, 3,* 267 (1967).

(33) Lebedeva, V. N., and Popereka, M. Ya., "Effect of Organic Additives on Initial Crystallization Stresses in Electrodeposited Copper." *Protection of Metals, 3,* 519 (1967).

(34) Fedot'ev, N. P., and Pozin, Yu. M., "Internal Stresses in Copper and Nickel Electrodeposits," *Protective Metallic and Oxide Coatings, Metal Corrosion and Electrochemistry,* Edited by N. P. Fedot'ev, Akad. Nauk SSSR. Translated by the Israel Program for Scientific Translations, Jerusalem, 55–59 (1968).

(35) Polukarov, Yu, M., and Kuznetsov, V. A., "Aging of Electrodeposited Copper," *Zh. Fiz. Khimii, 36,* 2382 (1962); *Chem. Abstr., 58,* 5261a.

(36) Brenner, A., and Senderoff, S., "A Spiral Contractometer for Measuring Stress in Electrodeposits," *Journal Research National Bureau of Standards, 42,* 89 (1949).

(37) Borchert, L. C., "Investigation of Methods for the Measurement of Stress in Electrodeposits," *Proceedings Am. Electroplaters' Soc., 50,* 44 (1963).

(38) Schneider, R., and Weil, R., "Internal Stresses in Copper Single-Crystal Electrodeposits," *Plating 55,* 1063 (1968).

(39) Edwards, J., and Wall, A. J., "Energy Requirements in the Electrodeposition of Copper, Including a Summary of Work on the Deposition of Copper for Direct Working to Wire," *Final Report to International Copper Research Association.* British Non-Ferrous Metals Research Association, (December, 1964).

(40) Stareck, J. E., Seyb, E. J., and Tulumello, A. C., "The Effect of Different Chromium Deposits on the Fatigue Strength of Hardened Steel," *Plating, 42,* 1395 (1955); *Proceedings Am. Electroplaters' Soc., 42,* 129 (1955).

(41) Rogers, R. R., "Ways to Electroplate High-Strength Steels without Embrittlement," *Metal Progress, 93,* (6), 91 (1968).

(42) Raub, E., and Mueller, K., *Fundamentals of Metal Deposition,* Elsevier Publishing Co., Barking, Essex (1967), 268 pp.

(43) Lebedeva, V. N., and Popereka, M. Ya., "Effect of Organic Substances on Copper Electrodeposition. I. Amino Acids," *Elektrokhimiya, 3* (12), 1454 (1967).

(44) Nishihara, K., Tsuda, S., Matsumura, Y., and Koga, H., "The Gas in Electrodeposited Copper," *Suiyokwai-Shi, 13* (10), 789 (1959).

(45) Read, H. J., and Graham, A. H., "The Elastic Modulus and Internal Friction of Electrodeposited Copper," *Journal Electrochem. Soc., 108,* 73 (1961).

(46) Safranek, W. H., "Physical and Mechanical Properties of Electroformed Copper," *Symposium on Electroforming-Applications. Uses and Properties of Electroformed Metals, ASTM Spec. Tech. Pub. No. 318,* 44–55 (1962).

(47) Stoebe, T. G., Hammad, F.H., and Rudee, M. L., "Transmission Electron-Microscope Observations of the Structure of Electrolytically Deposited Copper and Its Annealing Behavior," *Electrochimica Acta, 9,* 925 (1964).

(48) Clifton, F. L., and Phillips, W. M., "Bright Copper Plating in Acid Baths," *Proceedings Am. Electroplaters' Soc., 30,* 92 (1942).

(49) Barnes, S. C., "The Effect of Thio Compounds on the Structure of Copper Electrodeposits," *Electrochem. Soc. Extended Abstracts,* 43, (October, 1961).

(50) Barnes, S. C., "The Effect of Thio Compounds on the Structure of Copper Electrodeposits," *Journal Electrochem. Soc., 111,* 296 (1964).

(51) Breckpot, R., "The Crystallization Front in Electrodeposition of Metals," *Anales Real Soc., Espan. Fis. Quim. (Madrid), Ser. B, 61* (1), 51 (1965).

(52) Krushev, L., Pangarova, V., and Pangarov, N., "The Effect of Crystal Orientation on the Strength of Copper Deposited at High Current Densities," *Plating, 55,* 841 (1968).

(53) Leont'ev, A. V., "Dependence of the Grain Structure of Copper Deposited at Various Current Densities on Electrolyte Temperature," *Izvestiya Vysshikh Uchebnykh Zavedeniy, SSR, Kimiya, Khimicheskayo Tekhnologiya, 9* (1), 98 (1966).

(54) Pangarov, N., Nenov, I., and Khristova, I., "Predominant Orientation of Electrodeposited Copper and Nickel," *Izv. Inst. Fizikokhim, Bulgar. Akad. Nauk, 3,* 133 (1963).

(55) Fujino, T., "Effects of Plating Conditions and Additives on Crystalline Structures of Copper Deposits from Cyanide Baths. Studies on Bright and Leveled Copper Deposits from Cyanide Baths. (Part 4)," *Journal of the Metal Finishing Society of Japan, 18,* 382 (1967).

(56) Raub, E., and Wolff, K., "Thermal Expansion Measurements of Binary Alloys Having Reversed Solubilities and of Metals Having Strong Lattice Interferences," *Z. Metall., 40,* 126 (1949).

(57) Creutz, H., Stevenson, R. M., and Romanowski, E. A., "Electrodeposition of Copper from Acidic Baths," U.S. Patent 3,267,010 (August 16, 1966). Assigned to Udylite Corporation.

(58) Safranek, W. H., and Beach, J. G., "Electroplating Bright Leveling and Ductile Copper," *CDA Report 122/8 to Copper Development Association,* Battelle Memorial Institute (March, 1968), 11 pp.

(59) Chart, J. E., and Read, H. J., "Effects of Mechanical Properties of Copper Electrodeposits on Corrosion Resistance of Copper-Nickel-Chromium Coatings," *Report to International Copper Research Association,* Pennsylvania State University, Project N-138, August, 1969.

(60) Fukutsuka, T., and Kawasaki, M., "Hardness and Adhesion of Electrodeposited Copper from Cyanide Solutions," *Repts. Osaka Pref. Ind. Research Inst., 4* (1), 18 (1952).

(61) Graham, A. K., and Lloyd, R., "Stress Data on Copper Deposits from Alkaline Baths," *Plating, 35,* 449, 506 (1948).

(62) Such, T. E., "The Physical Properties of Electrodeposited Metals," *Metallurgia, 56,* 61 (1957).

(63) Couch, R. W., and Bikales, R. G., "Plating of Printed Circuits with Pyrophosphate Copper and Tin-Nickel," *Proceedings Am. Electroplaters' Soc., 48,* 176 (1961).

(64) Struyk, C., and Carlson, A. E., "Copper Plating from Fluoborate Solutions," *Monthly Rev. Am. Electroplaters' Soc., 33,* 923 (1946).

(65) Fedot'ev, N. P., and Khonikevich, A. A., "The Influence of Current Density and Sulfuric Acid Concentration on the Magnitude of the Internal Strains in Electrolytic Copper Deposits," *Journal Applied Chemistry (USSR), 32,* 2566 (1959).

(66) Max, A. M., "Application of Electroforming to the Manufacture of Disk Records," *Symposium on Electroforming-Applications, Uses and Properties of Electroformed Metals, ASTM Spec. Tech. Pub. No. 318,* 71–85 (1962).

(67) Max, A. M., "Electrodeposition of Copper," U.S. Patent 2,475,974 (1949).

(68) Lamb, V. A., and Valentine, D. R., "Physical and Mechanical Properties of Electrodeposited Copper. II. The Sulfate Bath," *Plating, 53,* 86 (1966).

(69) Subramanian, R., Balasingh, C., and Shenoi, B. A., "Internal Stress in Electrodeposits. Parts 1–4," *Metal Finishing, 65* (3), 67; (4), 54; (5), 58; (6), 101 (1967).

(70) Belyakova, L. A., and Gudin, N. V., "Deposition of Bright Coatings of Copper and Its Alloys from Ethylenediamine Electrolytes," *Theory and Practice of Bright Electroplating* (1963). Translated by the Israel Program for Scientific Translations, Jerusalem, 161–169 (1965).

(71) Prater, T. A., and Read, H. J., "The Strength and Ductility of Electrodeposited Metals. II. Some Data on Acid Copper Deposits," *Plating, 37,* 830 (1950).

(64) [illegible], C. and [illegible], "Copper Plating from [illegible] Solutions," *Metal Finishing*, [illegible] ... *Electrochem. Soc.*, 31, 922 (1950).

(65) [illegible], N. P., and [illegible], "The Influence of Current Density and [illegible] Concentration on the Magnitude of the Internal Strain in Electrolytic Copper Deposits," *Journal of Chemistry* (USSR), 32, 2586 (1959).

(66) Max, A. M., "Application of Electroforming to the Manufacture of [illegible] ... *Electroforming* ... *Properties of Electrodeposited Metals*, [illegible], Tech. Pub. No. 318, [illegible] (1962).

(67) Max, A. M., "Electrodeposition of Copper," U.S. Patent 2,[illegible] (1949).

(68) Lamb, V. A., and Valentine, D. R., "Physical and Mechanical Properties of Electrodeposited Copper, II, The Sulfate Bath," *Plating*, 53, 86 (1966).

(69) Subramanian, R., [illegible], and [illegible], "Internal Stress in Electrodeposits," [illegible] (1963).

(70) [illegible] ... *Electrodeposition: Theory and Practice of Electroplating*, 1961 (Translated by The Israel Program for Scientific Translations, Jerusalem, 1965).

(71) [illegible], J. A., and Read, H. J., "The Strength and Ductility of Electrodeposited Metals, II, Some Data on Acid Copper Deposits," *Plating*, 53, [illegible] (1966).

Chapter 7

Copper Alloys

Bronze and brass are the most important commercial copper alloys applied by electrodeposition. Brass was the first alloy produced by electrodeposition and is still used extensively for decorative purposes and for promoting rubber adhesion to steel and other metals. In each case, deposits with a thickness < 10 μm are customary. To provide protection against corrosion, nickel is frequently deposited first on steel before it is plated with brass. When brass (or bronze) is plated for decoration, a lacquer is usually applied over the alloy electroplate to protect it from tarnishing.

Decorative yellow brass plate ordinarily contains 70 percent copper and 30 percent zinc. The alloy applied for improving rubber adhesion contains 70 to 74 percent copper and 26 to 30 percent zinc. A white brass alloy containing 30 to 40 percent copper and 60 to 70 percent zinc was used extensively during World War II to conserve nickel. A similar alloy electroplate containing 45 to 56 percent copper, balance zinc, is still being electrodeposited under nickel and chromium electroplates on approximately 5,000,000 bumpers per year. Alloy deposits containing 56 percent copper, combined with nickel and chromium provides better corrosion protection than deposits containing less copper. (1)

Alloy containing 90 percent copper and 10 percent zinc is deposited to simulate bronze, but true bronze alloy contains 10 to 20 percent tin and no zinc. As an undercoat for 6 μm of nickel plus chromium, bronze alloy deposits were better than copper. (2) A combination of 25 μm of bronze, 2.5 μm of nickel and chromium provided better corrosion protection than 25 to 30 μm of nickel plus chromium on steel. (3,4) Bronze containing 6 to 9 percent tin has been adopted for bearings. (5) White alloy containing 40 to 45 percent tin, balance copper, is called speculum, which is deposited (over copper) on some hardware items in place of nickel and chromium. Although speculum tends to oxidize, its tarnish film ordinarily has the same color as the alloy.

The customary solutions for plating brass, bronze, and other copper alloys are based on cyanide complexes, but some baths for copper-tin alloy plating also contain pyrophosphate ions useful for complexing stannous tin. (2) Brighteners have been developed for both bronze (6,7) and brass. (8-12) Several noncyanide baths for depositing brass and bronze, which are of little or no commercial import, have been reviewed comprehensively by Brenner. (13)

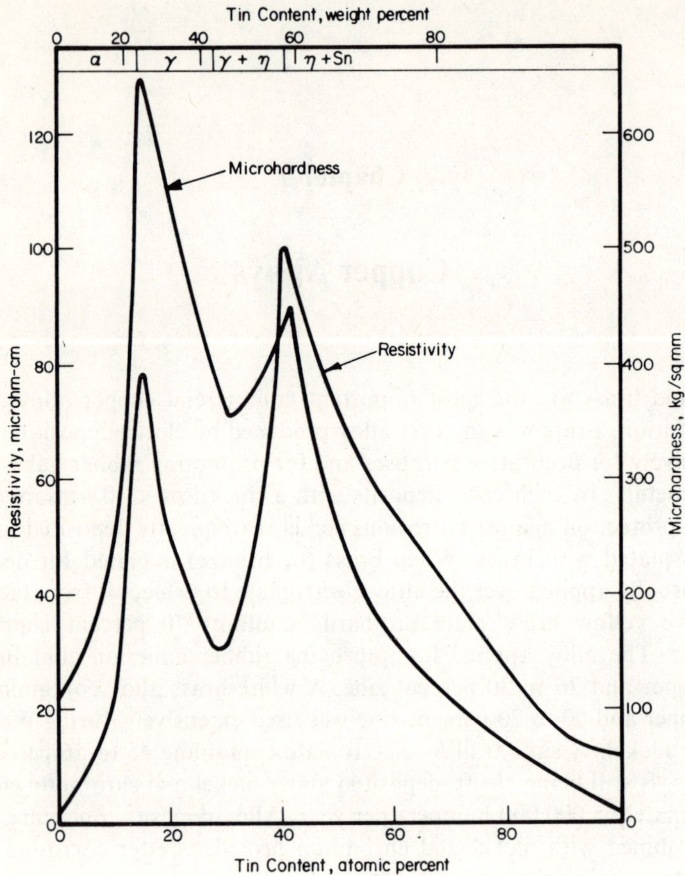

Figure 7.1. Electrical Resistivity and Microhardness of Copper-Tin Alloy Electrodeposits. (15, 16)

Until 1972, published data on the mechanical and physical properties of copper-zinc alloys were limited in scope to information on microhardness and electrical resistivity. AES Research Project 33 data (14) have partly filled some important gaps in the list of the mechanical properties of brass. However, similar data for copper alloys containing tin and other metals have not yet been developed in like degree.

Copper-Tin Alloys

Electrical Resistivity

The resistivities of copper-tin alloy deposits are shown in Figure 7.1 as a function of the tin content. These alloys were deposited at 2 amp/sq dm in a bath at 65 C that contained 0.3 N potassium cyanide, 0.2 N sodium hydroxide, 0.02 to 0.4 N stannic tin ions, and 0.01 to 0.4 N cuprous ions. (15) High (maxima) resistivities of 78 and 100 microhm-cm were observed at tin contents of 15 and 41 percent. The microhardness of the alloys also reached maxima at the same compositions.

The resistance of copper alloy deposits containing 49 percent tin was reduced about 50 percent by heat treating at 220 C for an hour. (17)

Microhardness

Supplementing the microhardness data in Figure 7.1, Table 7.1 lists measurements from six different reports, which show remarkable correlation with the tin

TABLE 7.1

Microhardness Data for Copper-Tin Alloys

Tin Content, weight percent	Microhardness, kg/sq mm	Reference
3.7	183	18
4.5	166	18
7.3	195	18
7.7	273	18
8.0	283	19
9.0	201	18
9.8	253	18
10	200	16
10	260–280	3
11.8	216	18
12.5	300[a]	20
16	314	19
18	370	17
20	300–320	3
21	350[a]	20
25	680	16
32	540	17
40	560	17
42	520[a]	20
46	530	17
48	400[a]	20
58	460[a]	20
58	440	16
63	250	17
64	129	20

[a] Diamond pyramid hardness values with a load of 50 or 100 grams. Values with a 400-gram load were 25 to 45 percent lower.

content of the deposits. In all cases, alloy containing 7.3 to 10 percent tin ranged in hardness from 195 to 283 kg/sq mm. Alloy deposits containing 12.5 to 21 percent tin exhibited hardness values of 300 to 370 kg/sq mm and deposits containing 25 to 46 percent tin ranged from 520 to 680 kg/sq mm in hardness. A microhardness of 400 to 460 kg/sq mm was reported for alloy containing 48 to 58 percent tin.

Composition is not the only factor affecting the hardness of copper alloys containing 4 to 11 percent copper. (18) Deposits produced at 40 or 50 C were slightly harder than deposits obtained at 70 C, for example.

One hour of heat treatment at 220 C reduced the hardness of speculum alloy containing 49 percent tin from 530 to 330 kg/sq mm. (17) On the other hand, heat treatment at 370 C had only a slight effect on the hardness of alloy containing about 10 percent tin, which was reduced from 270 to 240 kg/sq mm. (5)

Stress

A tensile stress range from 4.5 to 8 kg/sq mm (6,500 to 11,500 psi) was reported for copper alloy containing about 10 percent tin, which was deposited in a cyanide-pyrophosphate bath at 62 C. (3) From a similar solution containing a higher concentration of stannous tin, alloy containing 20 percent tin exhibited a stress of only 1.4 kg/sq mm (2,000 psi). No stress data have been reported for bronze or speculum deposited in cyanide-sodium stannate solutions. Alloy containing 25 or 26 percent tin deposited in copper sulfate-stannous sulfate-sulfuric acid baths was highly stressed. (21)

Structure

Metallographic examination of bronze with a tin content of 10 percent shows a columnar or fibrous grain structure (3,18) unless brighteners are added to induce further grain refinement. However, some deposits from a cyanide bath at 60 C exhibited a mixture of fibers and bands. (18) Alloy containing 18 to 20 percent tin deposited in a cyanide-pyrophosphate bath exhibited small, equiaxed grains, whereas alloy deposits containing 40 to 45 percent tin were banded or lamellar. (3)

Copper alloy containing 9 to 22.8 percent tin, which was deposited in solutions prepared with sodium cyanide, sodium hydroxide, copper cyanide, sodium stannate, and Rochelle salts, exhibited a face-centered-cubic structure. (22) Deposits containing 39 to 45 percent tin had a hexagonal-close-packed structure.

A supersaturated solid solution containing up to 17.8 percent tin in copper was revealed by X-ray examination of alloy deposits from cyanide-sodium stannate baths, whereas two phases were identified in deposits containing >17.8 percent tin. (23) Two phases (alpha and gamma) were observed by other investigators in deposits containing about 16 weight percent tin. (15,16) The metastable gamma phase was the only phase identified in deposits containing 23 to 42 weight percent tin by these investigators. Another investigator also reported the metastable gamma phase for bright speculum containing about 40 percent tin and a mixture of gamma and eta phases for dull deposits. (24) Crystal size was estimated at 0.02 and 0.04 μm (200 and 400 angstroms) for the bright and dull deposits, respectively.

A metastable phase that was mixed with alpha phase in deposits containing 17.5 to 33.3 percent tin and mixed with eta phase (Cu_6Sn_5) in deposits containing 35 to 57.5 percent tin was also identified by still another team of investigators. (17) However, this phase was identified as delta, which decomposed on heating at 220 C to the epsilon (Cu_3Sn) form. In alloy deposits containing 18 to 48 percent tin, which approximates the compositions of the metastable alloy deposits described by several authors in another paper, (25), the eta phase (Cu_6Sn_5) was reported. Eta phase also was identified in all alloy deposits containing >58 percent tin.

A relatively early study revealed preferred orientation of bright speculum (about 40 percent tin) with (110) and (100) planes perpendicular to the substrate surface. (24) Crystal faces were randomly oriented in a dull deposit. In a subsequent report, the orientation of crystal faces was correlated with the electrode overpotential, as follows: (26)

Overpotential, mv	Preferred Orientation Parallel with Cathode Surface
300	(111)
400	(100)
500	(100) and (211)
600	(100)
700	(110)

Copper-Zinc Alloys

Electrical Resistivity

Resistivity values for copper-zinc alloy deposits in Figure 7.2 are, as a rule approximately 25 to 45 percent higher than the resistivities of the corresponding metallurgical alloys. The investigators attributed the higher resistivity of the electrodeposits to their finer-grain structure. (27) The alloys were deposited at 0.4 amp/sq dm in a solution at 50 C containing 0.03 to 0.5 N copper cyanide, 0.015 to 0.6 N zinc cyanide, 0.15 N free sodium cyanide, 0.63 N sodium hydroxide, and 0.014 N ammonium hydroxide, which is a typical bath for plating decorative brass and other copper-zinc alloys.

Tensile Strength and Elongation

Tensile strengths from 64 to 77 kg/sq mm (91,000 to 110,000 psi) were determined for 56 to 82-μm-thick deposits containing 68 to 73 percent copper and 27 to 32 percent zinc. (14) Alloy foils deposited at 2.0 amp/sq dm were slightly

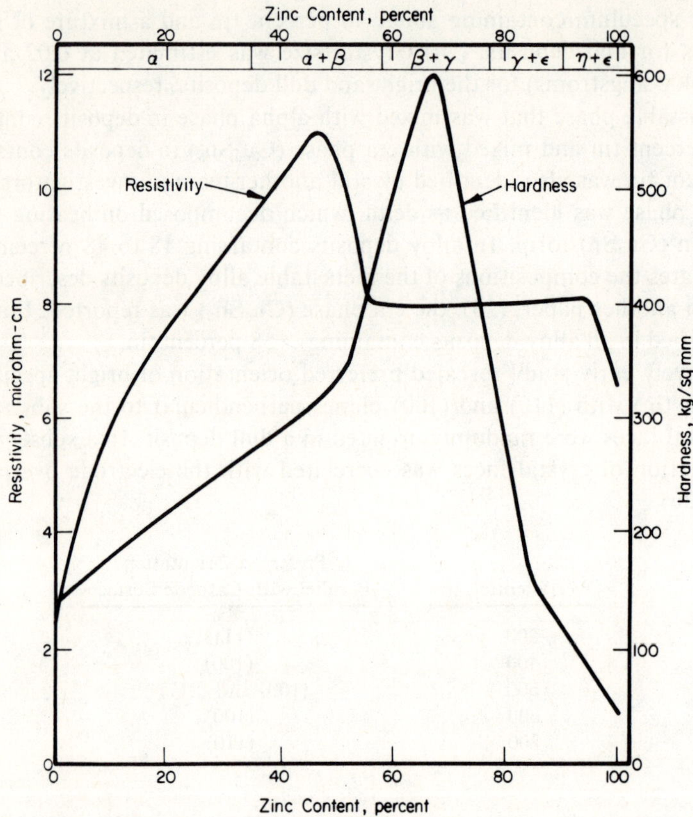

Figure 7.2. Electrical Resistivity and Microhardness of Copper-Zinc Alloy Electrodeposits. (27)

stronger than foils deposited at 1.33 amp/sq dm. Elongation was in the range of 1.0 to 1.5 percent. All deposits were obtained in a solution at 65 C containing 2.5 N sodium cyanide, 0.67 N copper cyanide, 0.065 N zinc cyanide, and 1.1 N sodium hydroxide.

Thinner (14 μm) foils deposited in less concentrated solutions fractured during bulge testing at loads corresponding to 48–56 kg/sq mm (69,000–80,000 psi) when alloy composition was adjusted to 70 percent copper and 30 percent zinc. (28) Tensile strengths for alloy foils containing 90 percent copper and 10 percent zinc were in about the same range, but foils containing 80 percent copper and 20 percent zinc were slightly stronger, ranging in tensile strength from 62 to 70 kg/sq mm (88,500 to 100,000 psi). Lower strength values were obtained for all three alloys when thickness was limited to 7.5 or 11 μm, as shown in Table 7.2. The ductility of these foils measured by strain during bulge testing was about the same as the ductility of thicker (56 to 82 μm) foils deposited in a more concentrated cyanide bath.

The tensile strengths of the thin (14 μm, maximum) foils listed in Table 7.2 are comparable with the ultimate strength of half-hard, cold-rolled brass sheet of the same composition. However, the values in Table 7.2 for the thicker (56 or 82 μm) foils are higher than the ultimate tensile strength of full-hard, cold-rolled alloy also containing about 30 percent zinc. The ductility of all electrodeposits was slightly lower than the elongation of spring sheet.

Microhardness and Wear

The microhardness of alloy deposits containing 10 to 80 percent zinc ranged from 180 to 602 kg/sq mm, as shown in Figure 7.2. The hardest alloy contained 67 percent zinc and exhibited a mixture of the body-centered-cubic beta phase and the hexagonal gamma phase. (27) Most of the alloys containing 10 to 67 percent zinc were two to four times harder than their annealed, wrought metal counterparts. The deposits containing 85 or 95 percent zinc were 35 to 40 percent harder than the corresponding metallurgical alloys.

TABLE 7.2

Tensile Strength and Ductility of Copper-Zinc Alloy Electrodeposits

Zinc Content, percent	Foil Thickness, μm	Tensile Strength[a]		Strain at Fracture[a]
		kg/sq mm	psi	
10	7.4	45	63,600	0.017
10	10.7	48	68,200	0.019
10	13.7	51	72,500	0.021
20	7.7	54	76,000	0.015
20	11.0	63	89,700	0.018
20	14.0	66	94,500	0.018
30	7.5	41	58,500	0.017
30	10.8	49	69,100	0.015
30	13.9	52	73,900	0.016
30	56	75, 77[b]	107,000; 110,000[b]	0.016–0.017[b]
30	82	64, 68[b]	91,000; 97,000[b]	0.021–0.021[b]

[a] Average fracture strength or strain at fracture for 11 to 16 replicates subjected to bulge tests (except for the values in the bottom two lines). The alloy foils were deposited at 1.9 amp/sq dm on stainless steel cathodes in a solution at 55 or 60 C containing 1.0 to 1.45 N sodium cyanide, 0.34 to 0.42 N copper cyanide, 0.2 to 0.3 N zinc cyanide, and 0.41 to 0.72 N sodium carbonate (pH 10.6 to 11.0). (28)

[b] Tensile and strain data for foils deposited at 1.3 or 2.0 amp/sq dm in a solution at 65 C containing 2.5 N sodium cyanide, 0.67 N copper cyanide, 0.065 N zinc cyanide, and 1.1 N sodium hydroxide. (14)

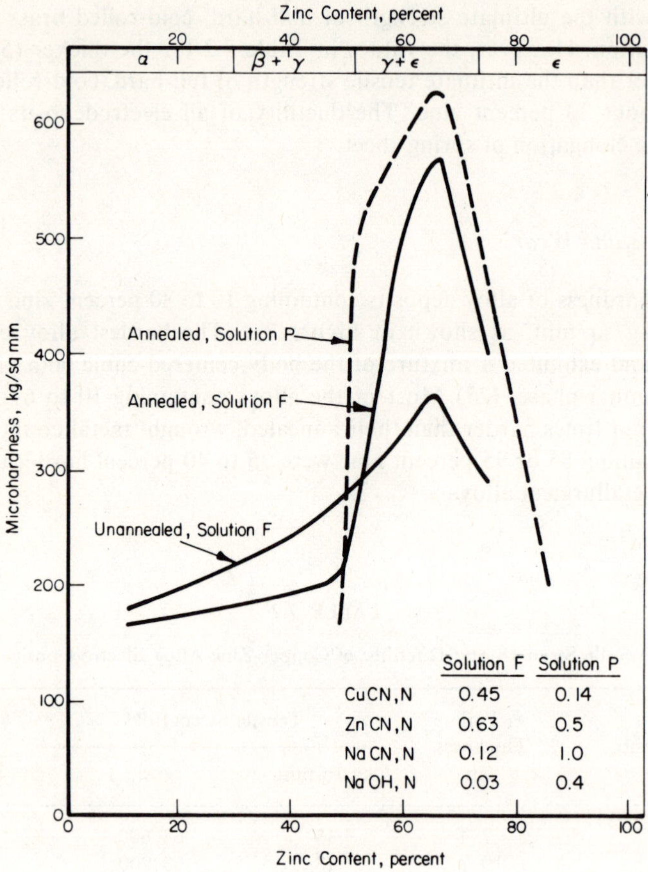

Figure 7.3. Microhardness of Copper-Zinc Alloy Deposits before and after Heat Treating 15 Hours at 200 C. (32)

Figure 7.2 shows a microhardness of about 250 kg/sq mm for alloy deposits containing 70 percent copper and 30 percent zinc. Harder deposits from 319 to 330 kg/sq mm were reported for 70/30 brass deposited at a lower current density in a more concentrated cyanide solution at 65 C. (14) A microhardness of 200 to 300 kg/sq mm was reported for alloy containing about 30 percent zinc deposited in 20 to 30 C cyanide baths. (29) Harder deposits from 300 to 500 kg/sq mm were reported by the same authors for alloy containing 66 percent zinc. The 70/30 alloy deposited in an ethylene diamine solution exhibited a hardness of 230 to 235 kg/sq mm. (30) The hardness of a similar alloy deposited in a pyrophosphate bath at 60 C was 214 kg/sq mm. (31)

Alloy deposits containing 52 to 79 percent zinc were increased in hardness by heat treating 15 hours at 200 C, as shown in Figure 7.3. Aging for 10 days also increased their hardness, to about the same degree. (32) For example, alloy deposits containing 29 or 30 percent copper and 70 or 71 percent zinc increased in hardness from the range of 340–390 kg/sq mm to about 550 kg/sq mm. Deposits containing <49 percent zinc were slightly softened by heat treating at 200 C.

Yellow, 70/30 brass deposited at 1.33 amp/sq dm in a 65 C cyanide bath exhibited about the same resistance to wear as 1/2-hard, rolled-brass sheet containing 63 percent copper, 35 percent zinc, and 2 percent lead. (14) Slightly softer deposits obtained at a current density of 2 amp/sq dm were slightly less resistant to wear.

Stress

Stress in 56 and 84-μm-thick deposits containing 66 to 70 percent copper and 30 to 34 percent zinc ranged from 8.3 to 9.3 kg/sq mm (11,800 to 13,300 psi). (14) The alloy was deposited at 1.33 or 2.0 amp/sq dm in a 65 C solution containing 2.5 N sodium cyanide, 0.67 N copper cyanide, 0.065 N zinc cyanide, and 1.1 N sodium hydroxide.

Structure

The fine-grained deposit in Figure 7.4 was typical of brass containing about 30 percent zinc, which was deposited at 1.33 or 2.0 amp/sq dm in a 65 C cyanide solution containing 1.1 N sodium hydroxide. (14) X-ray diffraction indicated a crystal size of only 250 angstroms. Only the face-centered-cubic alpha phase was detected in this investigation, which is in agreement with the results of other investigators. (27,32) Similar alloy deposits from 60 C pyrophosphate bath also consisted of the alpha phase. (31) However, beta phase was identified in the 70/30 brass deposited by another team of investigators in a 55 C cyanide solution containing sodium carbonate in place of sodium hydroxide. (28) The alpha phase is characteristic of annealed wrought alloys containing up to about 38 percent zinc.

A mixture of the alpha and beta phases was reported for brass alloy deposits containing 46 to 56 percent zinc by one team of investigators, (27) but either beta or a mixture of beta and gamma phases was detected in another study. (32) Only the beta phase was reported for alloy containing 43 percent zinc deposited in a pyrophosphate solution. (31) The phase compositions of these and other electrodeposited alloys are compared in Table 7.3 with the phase compositions of the wrought, metallurgical alloys.

Brass deposits were refined in grain size by using ultrasonic agitation in the plating bath, which also induced a leveling effect. (33) Alloys containing 20 to 45 percent zinc, balance copper, deposited in copper iodide-zinc iodide-formamide solutions at 40 C were coarsely columnar in structure. (34) On the other hand, deposits obtained in a similar solution at 25 C were lamellar. Nodules were observed on the deposits in both cases.

Figure 7.4. Cross Section of a Brass Alloy Electrodeposit Containing 30 Percent Zinc, Etched with Ferric Chloride Solution. (500×) (14)

Other Copper Alloys

Table 7.4 lists microhardness data for copper alloy deposits containing antimony, cobalt, nickel, lead, or silver. Copper alloy containing 35 percent antimony or about 50 percent nickel was the hardest of these alloys, but the copper-nickel alloy was not as hard as bronze containing 25 to 40 percent tin (Table 7.1).

Figure 7.5 shows resistivity data for copper alloys of cobalt, nickel, and lead. The resistivities of the copper-nickel alloy deposits were 4 to 12 percent higher than the resistivities of the corresponding wrought alloys.

The resistivity of copper alloy containing 6.5 percent lead was reduced from about 10 to 2.5 microhm-cm by annealing at 200 C. (43) This change can be attributed to the precipitation of lead from the supersaturated solid solution characteristic of copper alloy deposits containing up to 10 percent lead. These deposits were obtained in alkaline citrate-tartrate solutions.

Alloy containing up to about 20 percent cadmium also was deposited in the form of a supersaturated solid solution from a cyanide-citrate bath (44) or acid sulfate solutions. (45) Up to about 20 percent cobalt was retained in a supersaturated solution form in copper alloys deposited in pyrophosphate solutions. (37) Copper alloy deposits containing up to 35 percent nickel from cyanide baths consisted of a solid solution, (46) like their metallurgical counterparts. Alloys from a pyrophosphate bath which contained 31 to 75 percent nickel also corresponded in structure to the alpha solid solution characteristic of the cast alloys. (39) In contrast, alloys deposited in oxalate solutions consisted of a mixture of the crystals of copper and nickel. (47) Deposits obtained at low current densities in sulfate baths were solid solutions, but others at higher current densities consisted of a mixture of the solid solution and nickel. (48)

Alloy deposits from cyanide-tartrate baths containing up to about 14 percent antimony exhibited a face-centered-cubic structure. (36) A hexagonal structure corresponding to epsilon phase was identified in alloy deposits containing >14 and up

TABLE 7.3

Phases Identified in Copper-Zinc Electrodeposits

Zinc Content, percent	Type of Plating Bath	Phases Identified by X-Ray Examination		Reference
		Electrodeposits	Wrought Alloy	
10 or 20	Cyanide with Na_2CO_3	Alpha	Alpha	28
30	Ditto	Beta	Alpha	28
30	Cyanide with NaOH	Alpha	Alpha	14
30	Ditto	Alpha	Alpha	27
30	"	Alpha	Alpha	32
30	"	Alpha	Alpha	29
30	Pyrophosphate	Alpha	Alpha	31
43	Pyrophosphate	Beta	Alpha and beta	32
46 to 56	Cyanide with NaOH	Alpha and beta	Beta	27
48	Ditto	Beta	Beta	29
49	"	Beta	Beta	32
52	"	Beta and gamma	Beta and gamma	32
66	"	Gamma	Gamma	29
66 to 70	"	Gamma and epsilon	Gamma	32
67	"	Beta, gamma, and epsilon	Gamma	27
73 to 79	"	Epsilon	Gamma and epsilon	32
82	"	Epsilon	Epsilon	29
85	"	Epsilon	Epsilon	32
85	"	Gamma and epsilon	Epsilon	27
95	"	Epsilon and eta	Epsilon and eta	27

TABLE 7.4
Microhardness Data for Copper Alloys

Codeposited Metal	Type of Plating Bath	Microhardness, kg/sq mm	Reference
Antimony, 5%	Sulfate-tartrate	214–230 (50 g)	35
Antimony, 35%	Cyanide-tartrate	550	36
Cobalt, 1%	Pyrophosphate (20 C)	125	37
Cobalt, 10%	Ditto	150	37
Cobalt, 20%	"	180	37
Cobalt, 35%	"	285	37
Cobalt, 70%	"	285	37
Nickel, 10%	Pyrophosphate (60 C)	190[a]	38
Nickel, 18 to 50%	Pyrophosphate	250–300	39
Nickel, 20%	Pyrophosphate (60 C)	200[a]	38
Nickel, 30 to 50%	Sulfate-citrate (55 C)	200–260[b]	40
Nickel, 32%	Pyrophosphate (60 C)	260[a]	38
Nickel, 46%	Ditto	360[a]	38
Nickel, 56%	"	400[a]	38
Lead, 10 to 12%	No data	200–250[c]	41
Silver, 10%	No data	140 (25 g)	42
Silver, 30%	No data	160 (25 g)	42

[a] Hardness was about two times the hardness of the corresponding wrought alloy.
[b] Stressed in compression (21 kg/sq mm or 30,000 psi).
[c] Hardness increased to 300–350 kg/sq mm after annealing at 100 C.

to 43 percent antimony. At 43 percent antimony the tetragonal structure of zeta phase appeared, and above 58 percent antimony a mixture of the zeta and antimony lattice structures was observed.

As shown by X-ray diffraction studies, copper-bismuth alloy deposits consisted of a mixture of face-centered-cubic copper and hexagonal bismuth lattice structures. (36) Deposits consisting of $CuSe_2$ with semiconducting properties were obtained in a solution of copper sulfate and selenic acid. (49)

Composites

Copper containing 6.5 volume percent of 0.1 to 0.2-μm particles of alumina exhibited a microhardness of 148 kg/sq mm, in comparison with 110 kg/sq mm for the hardness of alumina-free copper deposited in a similar copper sulfate-sulfuric acid bath at 30 C. (50) Magnetic stirring was used to suspend 67 g/l of the alumina powder. The amount of codeposited alumina decreased from 6.5 to 5.5 volume percent when the current density was reduced from 10 to 5 amp/sq dm. Heat treating the copper-alumina composites at 600 or 800 C softened the deposits appreciably (50 to 56 kg/sq mm), but very little grain growth was observed after such a heat treatment. During another investigation of incorporating alumina particles in copper, no particles were obtained with the copper sulfate-sulfuric acid electrolyte, but an ethylene diamine solution was satisfactory for producing composites containing a large volume of alumina. (51)

Russian investigators were successful in codepositing up to 2.8 volume percent of corundum, 12 volume percent of nickel particles, and 14 volume percent of silver particles in a copper sulfate bath, using current densities of 12, 1.0, and 1.0 amp/sq dm, respectively. (52) Microhardness values ranged from 120 to 134 kg/sq mm for the corundum composites and 115 to 150 kg/sq mm for the copper-nickel composites. The microhardness of the silver composites was approximately 100 kg/sq mm.

A composite of copper containing 40 volume percent of tungsten filaments, produced by continuously winding tungsten wire on the cathode during copper deposition, exhibited a tensile strength of 135 kg/sq mm (192,000 psi). (53) The electrolyte was a copper sulfate-sulfuric acid bath.

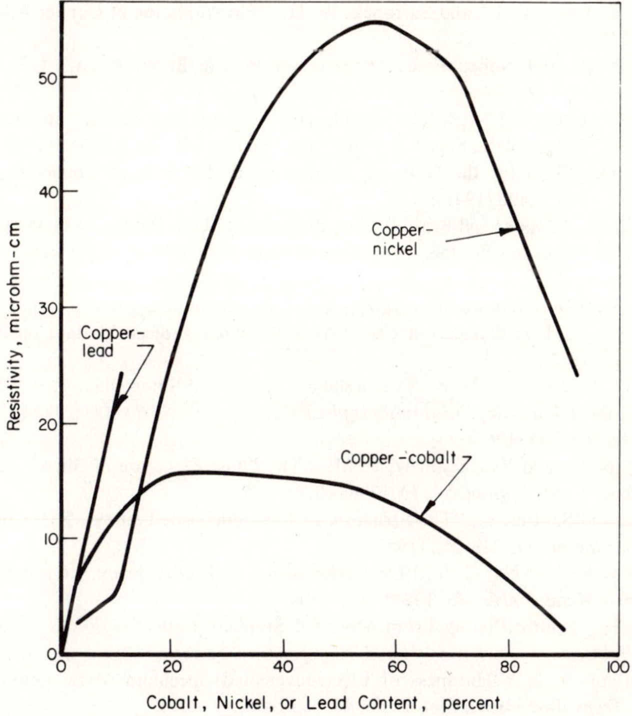

Figure 7.5. Electrical Resistivity Data for Copper Alloy Deposits of Cobalt, (37) Nickel, (38) and Lead. (43)

Graphite flakes incorporated in copper deposited in a copper sulfate solution reduced friction for specimens held against a rotating disk. (54) A deposit containing 10 volume percent of the carbon flakes had a dynamic coefficient of 0.05, in comparison with a coefficient of 0.08 for commercial bronze.

References

(1) Donakowski, W. A., and Springer, W. S., "Brass Plating for Improved Bumper Durability," *Plating, 58*, 1094–1098 (1971).

(2) Lowenheim, F. A., "The Corrosion Resistance of Alloy Electrodeposits," *Proceedings Am. Electroplaters' Soc., 44*, 42–46 (1957).

(3) Safranek, W. H., and Faust, C. L., "Copper-Tin Alloy Plating," *Proceedings Am. Electroplaters' Soc., 41*, 201–208 (1954); and *Plating, 41*, 1159–1170 (1954).

(4) Safranek, W. H.; Hespenheide, W. G., and Faust, C. L., "Bronze and Speculum Plates Provide Good Corrosion Protection for Steel," *Metal Finishing, 52*, 70–73, 78 (1954).

(5) Safranek, W. H., Neill, W. J., and Seelbach, D. E., "Bronze Plating Solves Design and Corrosion Problems," *Steel, 133* (35), 102–109 (1953).

(6) Schmerling, G., and Heymann, E., "Electrodeposition of Alloys Containing Copper and Tin," U.S. Patents 2,722,508 (1955) and 2,793,990 (1957), and British Patents 732,538 and 741,864 (1955).

(7) Bride, J. E., Faust, C. L., and Safranek, W. H., "Electroplating of Copper Alloys," U.S. Patent 2,886,500 (1959).

(8) Ostrow, B. D., and Nobel, F. I., "Process for Plating Bright Brass," U.S. Patent 2,817,627 (1957).

(9) Weiner, R. (Deut. Gold-u. Silber-Scheideanstalt—formerly Roessler—in Frankfurt-am-Main), "Baths for the Electrolytic Separation of Brass," German Patent 688,155 (1940).

(10) Hull, R. O., "Bath for the Galvanic Production of Deposits of Copper or Copper Alloys," German Patent 708,490 (1941).

(11) France, D. R., "Copper and Brass Plating Brightener," U.S. Patent 2,838,448 (1958).

(12) Tomono, R., "Glossy Plating," Japanese Patent 5402 (1959); see *Chem. Abstr., 54*, 9563h (1960).

(13) Brenner, A., *Electrodeposition of Alloys, Vol. I*, Academic Press, 411–543 (1963).

(14) Johnson, C. E., Communication citing AES Research Project 33 data on copper-zinc alloys (January 10, 1972).

(15) Burkat, G. K., Fedot'ev, N. P., Vyacheslavov, P. M., and Smorodina, T. P., "Structure of Electrolytic Copper-Tin Alloy," *Journal Applied Chemistry (USSR), 39*, 658–661 (1966); *Zh. Prikl. Khimii, 39*, 702–705 (1966).

(16) Fedot'ev. N. P., and Vyacheslavov, P. M., "The Phase Structure of Binary Alloys Produced by Electrodeposition," *Plating, 57*, 700–706 (1970).

(17) Raub, E., and Sautter, F., "The Structure of Electrodeposited Alloys. XII. Copper-Tin Alloys," *Metalloberflaeche, 11*, 249–252 (1957).

(18) Menzies, I. A., and Ng, C. S., "Electrodeposition of Bronze From a Cyanide-Stannate Bath," *Trans. Inst. Metal Finish., 46*, 137–143 (1968).

(19) Lee, W. T., "Bronze Plating From Modified Stannate Baths," *Trans. Inst. Metal Finish., 36*, 51–57 (1958).

(20) Ramanathan, V. R., "Hardness of Electrodeposited Speculum Metal and Other Tin-Copper Alloys," *Trans. Inst. Metal Finish., 36*, 48–50 (1958).

(21) Gerenrot, Yu. E., Dymarskaya, P. I., and Éichis, A. P., "The Structure of Copper-Tin Electrodeposited Coatings," *Protection of Metals, 3*, 395–400 (1967); *Zashchita Metallov, 3*, 465–471 (1967).

(22) Inagaki, H., "A Study of Copper-Tin Alloy Deposition," *Journal of the Metal Finishing Society of Japan, 9*, 338–341 (1958).

(23) Kryzhanovskii, A. V., "Structure of Electrodeposited Alloys of the Cu-Sn System," *Nauch. Zapiski Fiz.-Mat. Fak. Odessk. Gosudarst. Pedagog. Inst., 22* (1), 81–86 (1958).

(24) Rooksby, H. P., "The X-Ray Structure of Speculum Electrodeposits," *Journal Electrodepositors' Technical Soc., 26*, 119–124 (1950).

(25) Smart, R. F., Angles, R. M., and Robins, D. A., "The Structure of Some Electrodeposited Tin Alloy Coatings," *J. Inst. Metals, 89*, 349–353 (1961).

(26) Menzies, I. A., and Ng, C. S., "The Structure, Texture and Growth of Bronze Electrodeposits on Polycrystalline Substrates," *Trans. Inst. Metal Finish., 47*, 156 (1969).

(27) Andreeva, G. P., Fedot'ev, N. P., Vyacheslavov, P. M., and Pal'mskaya, I. Ya., "A Study of the Structure and Physicochemical Properties of Electrolytic Brass," *Journal Applied Chemistry (USSR), 36*, 1229–1235 (1963); *Zh. Prikl. Khimii 36*, 1283–1290 (1963).

(28) Nystrom, W. A., "The Mechanical Properties of Electrodeposited Brass," M.S. Thesis in Metallurgy, Penn. State Univ. (1961).

(29) Raub, E., and Krause, D., "The Structure of Electrodeposited Alloys. II. Cu-Zn Alloys," *Z. Elektrochem., 50* (4/5), 91–96 (1944).

(30) Belyakova, L. A., "Bright Brass Deposits From Ethylenediamine Electrolyte," *Blestyashchie Komb. Metal. Pokrytiya, No. 2*, 48–53 (1967).

(31) Sree, V., and Rama Char, T. L., "Electrodeposition of Copper-Zinc Alloys From the Pyrophosphate Bath," *Electroplating and Metal Finishing, 12* (9), 326–330; (10), 385–388 (1959).

(32) Cole, J., and O'Neill, H., "Hardness and Structural Transformations in Electrodeposited Copper-Zinc Alloys," *J. Inst. Metals, 88* (4), 159–164 (1959).

(33) Kenahan, C. B., Schlain, D., and Chin, E., "Effects of Ultrasonics on Electrodeposition of Copper Alloys From Cyanide Electrolytes," *Report of Investigation 6938, U.S. Bureau of Mines* (April, 1967), 32 pp.

(34) Menzies, I. A., Badelek, P. S. C., and Moreland, P. J., "Codeposition of Zinc and Copper From Organic Solutions and Electron-Probe Microanalysis of Alloy Deposits," *Electrochim. Acta, 13* (3), 429–442 (1968).

(35) Yurinskaya, L. V., and Popov, S. Ya., "Deposition of Bright Copper-Antimony Alloys From Sulfate-Tartrate Electrolytes," *Theory and Practice of Bright Electroplating*. Translated Israel Program for Scientific Translations, Jerusalem, by Ch. Nisenbaum, 176–180 (1963).

(36) Raub, E., "The Structure of Electrodeposited Alloys. VII. The Copper-Bismuth and the Copper-Antimony Alloys," *Z. Erzberg. und Metallhüttenwesen, 5* (4), 153–160 (1952).

(37) Fedot'ev, N. P., Vyacheslavov, P. M., Burkat, G. K., and Mokrova, L. N., "Electrodeposition of Cobalt-Copper Alloys," *Journal Applied Chemistry (USSR), 41*, 401–404 (1968).

(38) Andreeva, G. P., Fedot'ev, N. P., and Vyacheslavov, P. M., "Study of the Structure and Physicochemical Properties of the Electrolytic Copper-Nickel Alloy," *Journal Applied Chemistry (USSR), 36*, 1875–1878 (1963); *Zh. Prikl. Khimii, 36*, 1932–1936 (1963).

(39) Rama Char, T. L., "Electrodeposition of Nickel Alloys From the Pyrophosphate Bath," *Proceedings Am. Electroplaters' Soc., 46*, 76–80 (1959).

(40) Priscott, B. H., "Electrodeposition of Copper-Nickel Alloys From Citrate Solutions," *Trans. Inst. Metal Finish., 36*, 93–96 (1959).

(41) Bollenrath, F., "Some Physical Properties of Electrolytically Produced Pb-Cu Alloys," *Luftfahrt-Forschung, 20*, 284–287 (1943).

(42) Kimata, H., and Nishi, S., "Properties of Electrodeposited Silver-Copper Alloys. Metallurgical Studies on Electrical Contact Materials (Part 4)," *Journal of the Metal Finishing Society of Japan, 18* (8), 293–298 (1967).

(43) Raub, E., and Engel, A., "The Structure of Electrodeposited Alloys. III. Copper-Lead and Silver-Bismuth Alloys," *Z. Metall., 41* (12), 485–491 (1950).

(44) Polukarov, Yu. M., and Grinina, V. V., "Theory of Electrodeposition of Alloys. XIII. Phase Structure of Electrolytic Cu-Cd Alloys Deposited From Complex Electrolytes," *Elektrokhimiya, 1* (4), 433–438 (1965).

(45) "Electrodeposition of Copper Alloys," *Report No. 88*, International Copper Development Council, Brussels, Belgium (November 1965), 13 pp.

(46) Palatnik, L. S., and Lukov, I., "X-Ray Studies of Electrolytic Deposits. III. Copper-Nickel Alloys," *Zh. Tekh. Fiz., 7*, 2111–2114 (1937).

(47) Silvestroni, P., and Sartori, G. A., "On the Nature of Products Obtained by Electrolytic Codeposition of Copper and Nickel From Solutions of Complex Oxalates," *Ricerca Sci., 17*, 1156–1160 (1957).

(48) Polukarov, Yu. M., and Gorbunova, K. M., "The Problems in the Theory of Electrodeposition of Alloys," *Zh. Fiz. Khimii, 32*, 762–767 (1958).

(49) Pacauskas, E., Janickis, J., and Pilkauskiene, V., "Simultaneous Electrodeposition of Selenium and Copper," *Lietuvos TSR Mokslu Akad. Darb. Ser. B, No. 4*, 11–21 (1967).

(50) Hoashi, J., "Electrodeposited Copper-Aluminum Oxide and Silver-Aluminum Oxide Coatings," *Journal of the Metal Finishing Society of Japan, 15* (7), 258–262 (1964).

(51) Val'be, R. S., "Codeposition of Dispersed Particles in Nickel-Plating and Copper-Plating Electrolytes," *Kult. Izdeliya*, 30–37 (1967).

(52) Saifullin, R. S., "Copper Coatings With Included Solid Particles," *Journal Applied Chemistry (USSR), 40*, 2185–2188 (1967); *Zh. Prikl. Khimii, 40*, 2273–2276 (1967).

(53) Withers, J. C., and Abrams, E. F., "The Electroforming of Composites," *Plating, 55*, 605–611 (1968).

(54) Martin, P. W., "Electrodeposition of Composite Materials. 2," *Metal Finishing Journal, 11* (131), 447–451 (1965).

Chapter 8

Gold and Gold Alloys

Escalating applications for gold and gold alloys since about 1960, chiefly for electrical contacts, have encouraged the development of property data. Physical and mechanical properties of gold deposited in three different types of solutions have been reported. Acid cyanide baths prepared with organic acids such as citric acid and neutral solutions containing phosphates now supplant the alkaline cyanide baths for many applications, partly because the new solutions are easier to control and partly because they produce better deposits that contain fewer pores.

Porosity in gold electrodeposits has been the subject of many investigations because the continuity of the coating is an important characteristic of gold used for electrical contacts. Although porosity decreases with increasing thickness, users want to minimize thickness to reduce costs. However, a minimum of about 2.5 μm (0.1 mil) is the thinnest gold that approaches freedom from porosity, even with the smoothest of substrates and the best of alternative plating conditions. Hence, 2.5 μm frequently appears as a minimum thickness requirement in purchaser specifications for electrical contacts.

Although the electrical resistivities of electrodeposited and wrought gold are nearly equal, the contact resistance of electrodeposits frequently increases after thinly plated contacts are exposed to corrosive atmospheres because oxidation products of the substrate material exposed through pores diffuse over the contact surfaces. Furthermore, the resistance of thinly plated contacts operated at slightly elevated temperatures increases considerably as a result of the diffusion of the substrate material with gold, unless a thermal barrier such as nickel is employed. Thus contacts frequently are plated with nickel, prior to the deposition of at least 2.5 μm of gold to eliminate porosity and diffusion and improve reliability.

Several gold alloy deposition processes have been developed to inhibit wear. For example, data are available on alloys of antimony, cobalt, copper, nickel, and silver. Of these, the alloys containing about 0.1 percent cobalt or nickel, which are deposited in acid cyanide baths, have become the most popular. Such alloy deposits are relatively free from pores and have only a slight effect on contact resistance, relative to pure gold. Most importantly, these alloys are more resistant to wear. Their frictional resistance is appreciably lower than that of pure electrodeposited or wrought gold or electrodeposited alloys produced in alkaline solutions, which accounts for the superior wear resistance of the gold alloys of cobalt and nickel.

TABLE 8.1

Density and Electrical Resistivity of Electrodeposited Gold

Type of Plating Bath	Density, g/cu cm	Resistivity, microhm-cm	Reference
Acid[a]	19.26 to 19.27	—	1
	19.1 to 19.2	2.2 to 3.5[b]	2
	19.2	2.34	3, 4
	—	2.29	5
	—	2.2	6
Alkaline[c]	19.2	7.49	3, 4
	19.13 to 19.22	—	1
	18.9	2.2	2
Neutral[d]	19.1	2.3	2
	—	2.38	7
Proprietary	17.3 to 18.2	2.13 to 2.78	8

[a] Cyanide baths containing an organic acid such as citric acid, operated at a pH <5 and a temperature >60 C.

[b] Relatively soft deposits (52 to 78 kg/sq mm) exhibited a resistivity of 2.2 microhm-cm, whereas harder deposits (78 to 129 kg/sq mm) had a higher resistivity of 3.5 microhm-cm.

[c] Cyanide baths with a pH >10.

[d] Cyanide baths containing phosphates.

Surprisingly high tensile strengths > 100 kg/sq mm (> 140,000 psi) have been reported for gold alloy deposits containing copper or copper plus silver. The copper alloys are hardened and relieved of stress by heat treatment.

Physical Properties of Gold

Density

Density values reported for electrodeposited gold with a thickness of at least 5 μm range from 17.3 to 19.27 g/cu cm, as noted in Table 8.1. The relatively high density range of 19.1 to 19.27 g/cu cm for high-purity gold from acid cyanide and neutral solutions compares with 19.3 g/cu cm for high-purity cast, worked and annealed gold. A high density > 19 g/cu cm also was cited for some deposits from alkaline solutions. However, densities of < 19 g/cu cm were reported for other gold deposits from alkaline baths and some proprietary solutions. The grain size of the relatively high-density gold deposits was much smaller than the grain size of lower density gold deposited in an alkaline cyanide-phosphate solution. (1)

Densities of only 7.7 to 16.2 g/cu cm were calculated for thin deposits (2 to 3 μm) from an acid gold bath, based on beta back scatter methods for measuring thickness. (9) These data presumably reflect porosity in thin coatings, which varies appreciably, depending on substrate surface smoothness and other factors.

Gold electroplated in acid solutions contained fewer pores than gold from alkaline cyanide baths, according to pore counts after electrolytic or sulfur dioxide tests. (10–15) Porosity decreases appreciably with decreasing substrate surface roughness, based on data for gold over 1.5 μm of nickel on polished copper surface ranging in roughness from 0.04 to 0.7 μm. (16,17) Smoothing the substrate by electropolishing or chemical polishing reduced porosity. (11, 18) Interposing nickel between a copper substrate and the gold electroplate reduces pore counts in the gold electroplate. (14,18–21) An intermediate layer of tin-nickel alloy also reduced porosity in electrodeposited gold overlays. (10) Increasing gold thickness also reduces porosity, (11,12,14,16–18,21,22) but a few pores were still detected in 10 μm-thick gold deposited in an alkaline cyanide bath. (23) This amount of porosity was approximately duplicated by deposits from an acid or neutral bath with a thickness of only 4 μm. (18)

A study of the effects of varying solution temperature and other conditions showed decreasing porosity in gold deposits as the temperature of acid, alkaline, and nearly phosphate solutions was increased from 20 to 60 or 80 C. (24) Porosity was minimized with a current density of 0.3 to 0.5 amp/sq dm, in comparison with a lower or higher current density. Increasing the free potassium cyanide concentration from 20 to 140 g/l in the alkaline bath increased porosity appreciably, but an increase in the potassium carbonate concentration from 20 to 160 g/l reduced porosity. In the acid citrate or phosphate baths, a pH of 6.0 was best for minimizing porosity. Increasing the concentration of citrate ions from 10 to 400 g/l reduced porosity.

Electrical Resistivity and Contact Resistance

The lowest resistivity reported for gold electrodeposits was 2.13 microhm-cm, as noted in Table 8.1, which compares with 2.04 microhm-cm for cast, worked, and annealed gold. Most of the resistivity values reported for gold deposited in acid and neutral solutions were fairly low, ranging from 2.2 to 2.34 microhm-cm, whereas some values reported for gold from alkaline cyanide baths ranged up to 7.49 microhm-cm, indicative of impurities, inclusions, and/or discontinuities. Heating at 300 C reduced resistivity from 7.49 to 6.17 microhm-cm in the case of a deposit from an alkaline bath, or from 2.34 to 2.20 microhm-cm, for gold deposited in an acid solution. (3)

The contact resistance of electroplated gold is about equal to that of wrought gold, (25,26) but the nature of the substrate caused some differences in the contact resistance of thin, high-purity gold electrodeposits. (27) Gold deposits on nickel resulted in a larger increase in resistance than gold on copper or silver, in comparison with gold electroplate on wrought gold. Exposure to corrosive environments or heating in air increases the contact resistance of thin gold deposits on copper or silver substrates, (25,27,28) presumably because of the resistance effects of substrate metal corrosion or oxidation products. Exposure to corrosive atmospheres increases the contact resistance of thin gold deposits over nickel, but the degree of increase with the nickel substrate was relatively small in comparison with the increase noted for gold deposited directly on copper. (27) The corrosion of thin gold electroplate on microwave surfaces also affected the performance of microwave energy transistors. (29)

Diffusion during heating substrate materials with gold also increases contact resistance. For example, the rapid diffusion of copper through gold and the subsequent formation of copper oxide demonstrates the advisability of a diffusion barrier. (30)

To prevent sticking of miniature reed switches fabricated from nickel iron alloy, sequential nickel and gold deposits were diffused by heating at 800 C in a reducing atmosphere. (31) Subsequent oxidation of iron and nickel which diffused to the surface increased contact resistance, but nickel oxide at the surface prevented sticking during 10^6 operations at 100 C.

Thermal Properties

The thermal conductivity of electrodeposited gold was reported to be 0.707 cal/cm/C/sec. (32) The coefficient of thermal expansion at 100 C was 14.3 $\times 10^{-6}$/C for a deposit from a conventional bath (17) and varied from 14.18 to 14.60 $\times 10^{-6}$/C for tubes electroformed in proprietary solutions. (8)

Mechanical Properties of Gold

Tensile Strength and Elongation

The tensile strength of electroformed foils with a thickness of about 25 μm ranged from about 11.2 to 21.7 kg/sq mm (16,000 to 31,000 psi), depending on the type of plating bath and the current density, as shown in Table 8.2. Gold from an acid bath and a neutral phosphate bath was stronger, on the average, than gold deposited in an alkaline cyanide solution. (33) An increase in current density from 0.1 or 0.3 to 0.5 amp/sq dm increased tensile and yield strengths appreciably for gold deposited in all three baths. The tensile strengths of the thin gold deposits obtained in the acid bath were about the same as tensile strength values for thicker (100 μm) tubes electroformed in a similar acid bath, which ranged from 15.5 to 21.2 kg/sq mm (22,000 to 30,000 psi). (34) Thin foils electroformed in an alkaline cyanide bath were similar in tensile strength to previously reported values for gold deposited in a similar solution, which ranged from 12.7 to 14.2 kg/sq mm (18,000 to 20,000 psi). (35)

Each of the tensile and yield values given in Table 8.2 for thin (25 μm) deposits is an average of 4 to 12 values that spanned broad ranges. (33) In at least one group

TABLE 8.2

Tensile Properties of Electroformed gold[a]

Plating Bath	Current Density, amp/sq dm	Tensile Strength, kg/sq mm	Tensile Strength, psi	Yield Strength, kg/sq mm	Yield Strength, psi	Elongation, percent
Acid[b]	0.1	15.5	22,000	9.6	13,600	7.2
	0.3	17.4[c]	24,800[c]	10.3	14,700	8.3
	0.5	21.7[c]	30,900[c]	13.0	18,500	6.9[c]
Alkaline[d]	0.1	11.2	15,900	12.4	10,700	3.7
	0.3	12.8[e]	18,100[e]	8.3	11,800	4.1[e]
	0.5	18.0	25,500	12.2	17,400	3.9[e]
Neutral[f]	0.1	16.0	22,700	9.0	12,900	3.5
	0.3	14.4	20,500	9.8	14,000	3.0
	0.5	21.2	30,100	13.0	18,600	6.3

[a] Each value is an average of 4 to 12, 25-μm-thick specimens plated in two or three laboratories. (33)

[b] A cyanide bath at 65 C prepared with $KAu(CN)_2$ containing 10 g/l Au, 40 g/l $H_3C_6H_5O_7 \cdot H_2O$ and 120 g/l $K_3C_6H_5O_7 \cdot H_2O$ with a pH of 4.8.

[c] A tensile strength of 15.5 to 21.2 kg/sq mm (22,000 to 30,000 kg/sq mm) was reported for 100-μm-thick gold deposited at 0.3 to 0.5 amp/sq dm in a similar bath containing 7.5 g/l of gold. (34)

[d] A cyanide bath at 65 C prepared with $KAu(CN)_2$ containing 10 g/l Au, 30 g/l KCN, 30 g/l K_2CO_3, 30 g/l $K_2HPO_4 \cdot 3H_2O$ and 4.6 g/l KOH with a pH of 11.5.

[e] A tensile strength of 12.7 to 14.2 kg/sq mm (18,000 to 20,000 psi) was reported for gold deposited in another cyanide bath. (35) Elongation ranged from 30 to 45 percent.

[f] A cyanide bath at 65 C prepared with $KAu(CN)_2$ containing 10 g/l Au, 60 g/l KH_2PO_4 and 60 g/l K_2HPO_4 with a pH of 7.0.

(the foils electroformed at 0.5 amp/sq dm in a neutral bath), the weaker specimens with an average tensile strength of 12.8 kg/sq mm (18,200 psi) exhibited a coarser grain size during metallographic examination than stronger foils produced by another electrodepositor, which had an average tensile strength of 27.4 kg/sq mm (39,000 psi). This difference in structure and strength may have been due to a difference in the purity of the plating bath.

Elongation of the thin (25 μm) foils ranged from only 3.0 to 8.3 percent. The foils from the acid bath were more ductile than those from the alkaline or neutral solutions. Thicker deposits from an acid bath exhibited better ductility, because elongations ranged from 22 to 29 percent. (34) Other reports on the ductility of gold deposits cited elongations of 30 to 45 percent (35) and 24 to 55 percent. (13).

Annealing at 200 or 300 C reduced the tensile strength of 25-μm-thick electroformed gold foils and increased their ductility, as shown in Table 8.3. Tensile strength was reduced 30 to 40 percent. Elongation was increased about 400 percent. (33)

TABLE 8.3

Effect of Annealing on the Tensile Properties of Electrodeposited Gold[a]

Annealing Temp, C	Tensile Strength of Gold Electroformed at 0.1 amp/sq dm		Tensile Strength of Gold Electroformed at 0.5 amp/sq dm		Elongation of Gold Electroformed at 0.1 amp/sq dm	Elongation of Gold Electroformed at 0.5 amp/sq dm
	kg/sq mm	psi	kg/sq mm	psi		
No annealing	14.4	20,500	22.2	31,500	5.0	3.8
200	10.2	14,600	13.5	19,200	10.6	9.2
300	10.7	15,300	13.2	18,700	17.0	16.1

[a] Tensile properties for 25-μm gold foils electroformed in an acid bath at 65 C containing 10 g/l Au, 40 g/l $H_3C_6H_5O_7 \cdot H_2O$ and 100 g/l $K_3C_6H_5O_7 \cdot H_2O$ with a pH of 4.8. (33)

Hardness and Wear Resistance

High-purity gold as soft as 30 kg/sq mm was reported for deposits obtained in a neutral solution. (7) A range of about 44 to 82 kg/sq mm was cited in other publications, however, as Table 8.4 shows. A similar range for gold deposited in acid and alkaline baths has been reported by several investigators. A hardness of 100 kg/sq mm was cited in a few cases for gold deposited in all three solutions,

TABLE 8.4

Hardness of Electrodeposited Gold

Type of Plating Bath	Current Density, amp/sq dm	Bath Temp, C	Hardness, kg/sq mm[a]	Reference
Acid[b]	No data	No data	43 to 82 (25g)	36
	0.3 to 1.0	20 to 60	50	37
	0.3 to 0.5	65	50 to 60 (100g)	34
	0.3	65	52 to 78 (25g)	2
	No data[c]	65 to 70	65	38
	0.1	65	71 to 76	33
	0.5	65	84 to 93	33
	1 to 1.5[d]	18 to 24	90 to 100	39
	No data	No data	85 to 90	40
	0.5[e]	50	115 (25g)	41
Alkaline[f]	0.3	65	47 to 86 (25g)	2
	0.3 to 1.0	20 to 60	50	37
	0.4 to 1.0	60 to 65	60	42
	0.2	66	61 to 65 (25g)	40
	0.2	25	65 to 70 (25g)	40
	No data	No data	65	4
	0.25	60	96	43
	3.0	80	100	44
	0.4 to 1.0	60 to 65	130[g]	42
	0.1	65	68 to 76	33
	0.5	65	70 to 100	33
Neutral	No data	No data	30 (25g)	7
	0.3	65	44 to 82 (25g)	2
	0.1	65	71 to 76	33
	0.5	65	80 to 100	33

[a] Figures in parentheses refer to the applied load in grams.

[b] Unless otherwise noted, a cyanide bath prepared with citric acid to reduce the pH to the range of 4.0 to 4.8.

[c] A bath with a pH of 6.0 prepared with succinic acid and dihydroxylethyl glycine.

[d] Contained phosphate and citrate ions.

[e] An acid bath prepared with potassium gold cyanide, ammonium citrate and $NH_4Cr(SO_4)_2$.

[f] Solutions with a pH $>$ 10.0 containing free cyanide.

[g] 0.5 g/l of turkey red oil was added to the bath to reduce porosity and increase hardness.

when the current density was raised to at least 0.5 amp/sq dm or when solution temperature was reduced to < 30 C.

Hardness was consistently increased (usually about 20 percent) by an increase in current density from 0.1 to 0.5 amp/sq dm. (33) This effect was observed for 25-μm-thick deposits from acid, alkaline, and neutral baths.

The addition of turkey red oil to an alkaline bath greatly increased the hardness of deposits at 0.4 to 1.0 amp/sq dm in a 60 to 65 C alkaline cyanide bath. (42) Deposits averaged 60 and 130 kg/sq mm in hardness before and after this addition. A hardness of 115 kg/sq mm was reported for gold deposited in an acid bath containing ammonium chromium sulfate. (41)

The load applied during hardness measurements was not always cited by the authors of the reports on hardness data. Low loads below 25 grams give low readings and higher loads result in high hardness numbers, according to data for bright gold, as follows: (45)

Load, grams	Hardness, kg/sq mm	Load, grams	Hardness, kg/sq mm
2	61	35	106
4	70	60	103
5	74	100	105
10	82		

Another report cited a hardness of 150 kg/sq mm for a gold alloy containing 40 percent copper when the applied load was 2 grams and a hardness of 375 kg/sq mm when the load was increased to 50 grams. (46)

Acid gold deposits exhibit better wear resistance than deposits from alkaline or neutral baths. In one case, the improved wear resistance for the acid gold was observed for sockets used with palladium-plated plugs. (47) In another, wear was measured on plated rotating spindles. (2) Wear rates for high-purity (soft) deposits from acid, alkaline and neutral solutions were 0.7, 1.5, and 5.0 mg/hr/sq cm, respectively. With harder deposits from acid and alkaline baths, the wear rates were 0.6 and 4.2 mg/hr/sq cm.

Octadecylamine surface films reduced the friction coefficient from about 1.0 to 0.2 for sliding, pin, and socket contacts. (48) Heating to 100 C or exposure to corrosive atmospheres did not reduce the effectiveness of the films.

Stress

Little or no stress has been detected in the high-purity deposits obtained in typical acid, alkaline, or neutral baths. A slight tensile stress (0.5 to 1.1 kg/sq mm or 700 to 1500 psi) was observed in 100-μm-thick acid gold deposits. (34) A slight compressive stress (−0.65 to −1.4 kg/sq mm or −940 to −2000 psi) has been reported for typical deposits from alkaline cyanide baths, (2,13) but high tensile stress to 7.0 kg/sq mm (10,000 psi) was reported in other cases, as Table 8.5 shows. No stress was detected in deposits from neutral solutions.

TABLE 8.5

Stress Data for Unalloyed Gold Deposits

Type of Plating Bath	Temp, C	Current Density, amp/sq dm	Stress[a] kg/sq mm	psi	Reference
Acid[b]	65	0.3	0	0	2
	No data	No data	0	0	3
	65	0.3	0.5 to 1.1	700 to 1,500	34
Alkaline[c]	65	0.3	−0.7 to −1.4	−1,000 to −2,000[d]	2
	No data	No data	−0.65 to −1.1	−940 to −1,600	13
	50[e]	0.3	5.6 to 7.0	8,000 to 10,000	2
	No data	No data	7.0	10,000	49
Neutral	65	0.3	0	0	2

[a] Data for deposits with a thickness of at least 5 μm. A negative sign denotes compressive stress.

[b] Cyanide solutions prepared with citric acid or another organic acid to reduce the pH below 5.0.

[c] Solutions with a pH > 10.0 prepared with gold potassium cyanide if not otherwise noted.

[d] Compressive stress increased to 6 kg/sq mm (8,500 psi) when deposit thickness was only 1 μm

[e] A noncyanide, proprietary alkaline bath.

Hydrazine sulfate additions to acid gold baths reduce stress in thick deposits. (50) Periodic current reversal supposedly reduced stress in deposits obtained at 0.5 to 1.5 amp/sq dm in alkaline cyanide bath operated at 60 to 70 C. (51)

Physical Properties of Gold Alloys

Density

The densities of some gold alloys are listed in Table 8.6. The density of alloys containing only about 0.1 percent cobalt and nickel was only 17.8 or 17.9 g/cu cm, which is 6 or 7 percent lower than the density of high-purity gold. This difference indicates considerable porosity and/or codeposition of nonmetallic material. On the other hand, thin deposits of these alloys appear to be about as free from pores as high-purity deposits from acid baths. (2, 14) Some porosity tests indicate equivalent freedom from porosity for high-purity gold and the gold alloys containing about 0.1 percent of cobalt or nickel, whereas other test data showed that porosity was slightly greater for the alloy deposits. (2) However, the alloy deposits containing nickel or cobalt contained fewer pores than unalloyed gold deposited in alkaline cyanide solutions. (2,12) Limiting the current density to 0.5 amp/sq. dm, agitating the plating bath or increasing the deposit thickness from about 1 to 4 or 5 μm favored a reduction in the porosity of alloy deposits containing cobalt or nickel. (18)

TABLE 8.6

Density and Electrical Resistivity Data for Gold Alloys

Codeposited Material	Type of Plating Bath	Density, g/cu cm	Resistivity, microhm-cm	Reference
Aluminum oxide, 8%[a]	Alkaline (60 C)	18.2[b]	—	52
Antimony, 1.9%	Alkaline (60 C)[c]	—	8.8	43
Antimony, 4%	Ditto	—	10.0	53
Antimony, 12%	"	—	22.0	53
Antimony, 45%	"	—	55.0	53
Cadmium, 1.4%	Alkaline (50 C)	18.9	4.2	2
Cadmium, 24%[d]	Alkaline (55 C)	15.7	14.1	2
Cobalt, 0.1%	Acid	—	2.85	7
Cobalt, 0.1%	Acid (35 C)	17.8	4.8	2
Cobalt, 0.9%	Ditto	17.3	9.7	2
Cobalt, 1%	Acid		3.15	7
Cobalt, [e]	No data	17.3 to 18.2	6.1 to 17.9	8
Copper, 15%	Alkaline	—	17.0	54
Copper, 26%	Alkaline	—	17.5[f]	55
Nickel, 0.1%	Acid	—	2.96	7
Nickel, 0.1%	Acid (35 C)	17.9	3.8	2
Nickel, 0.8%	Acid (35 C)	17.4	4.1	2
Nickel, 1%	Acid	—	3.38	7
Nickel, 1%	Neutral (50 to 60)[g]	—	2.0	56
Nickel, 18%	Ditto	—	3.0	56
Nickel, 63%	"	—	6.0	56
Silver, 20%	Alkaline (50 C)	—	12.0	57

[a] Eight volume percent of 0.02 μm Al_2O_3 particles.

[b] Density wss reduced to only 15 g/cu cm after heating for an hour at 1000 C because of the expansion of occluded electrolyte (estimated as 18 volume percent).

[c] Deposits at 0.25 amp/sq dm from a cyanide-citrate bath containing 8 to 10 g/l of free KCN. Alloy containing 1.9% antimony was a solid solution, whereas alloys containing 5 to 40% antimony consisted of a mixture of $AuSb_2$ and gold.

[d] The alloy also contained copper. The total cadmium and copper content was about 24%.

[e] Electroformed 5 to 10-μm-thick tubes. Cobalt content was not reported.

[f] Maximum resistivity for gold-copper alloy electrodeposits.

[g] A cyanide-pyrophosphate-citrate solution with a pH of 7 to 8.

Another report on the influence of codepositing a second metal with gold indicated that alloy deposits containing nickel or silver contained fewer pores than unalloyed gold. (24) The gold-silver alloy was deposited in an alkaline cyanide bath and the gold-nickel alloy was obtained in the acid citrate bath with a pH of 4.0. Porosity decreased appreciably with increasing nickel content from 0.5 to 4 percent.

Electrical Resistivity and Contact Resistance

The codeposition of nickel has less influence on increasing the resistivity of gold deposits than the codeposition of antimony, cadmium, or copper, according to the data in Table 8.6. Even so, the resistivities of gold-nickel alloy deposits were about 50 percent higher than those for the metallurgical alloys. (5) One report cited about equal resistivities for gold alloy deposits containing the same amount of cobalt or nickel, (7) but another listed higher resistivities for the alloy deposits containing cobalt. (2) The resistivity measured for gold-antimony alloy as a function of the antimony content is shown in Figure 8.1. A maximum resistivity of 70 microhm-cm appears at an antimony content of 65 atomic percent. (53)

Some reports on gold-copper alloys containing 15 or 26 percent copper show a resistivity of 17 or 17.5 microhm-cm, (54,55) but another cites higher resistivities from 150 to 250 microhm-cm. (58) However, heat treatment for about 2 hours at 150 C reduced the resistivity of these alloys to about 35 microhm-cm. A 1-minute heat treatment at 200 C reduced resistivity to about 14 microhm-cm.

The contact resistance of gold alloy deposits is usually higher than that of high-

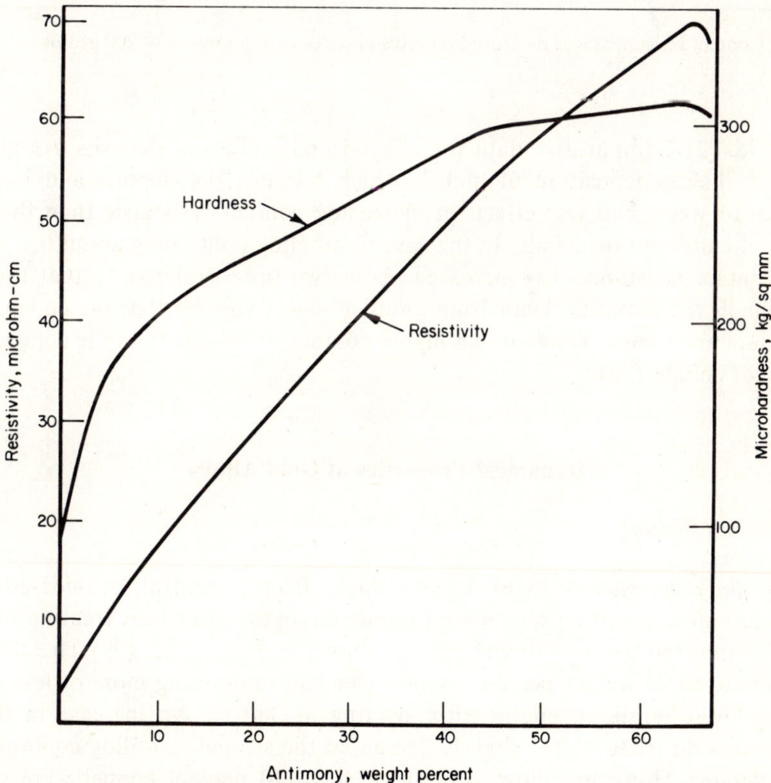

Figure 8.1. Specific Resistivity and Microhardness of Gold-Antimony Electrodeposits. (43, 53)

TABLE 8.7

Contact Resistance Data for Gold Alloys (2)

Codeposited Metal	Type of Plating Bath	Contact Resistance, milliohm[a] 1.25 μm-thick Deposits	5-μm-thick Deposits	25-μm-thick Deposits
None	Acid (65 C)	0.23	0.24	0.20
	Alkaline (65 C)	0.22	0.25	0.20
	Neutral (65 C)	0.23	0.24	0.20
Cadmium, 0.4%	Alkaline noncyanide (50 C)	0.24	0.30	0.45
Cobalt, 0.1%	Acid (35 C)	0.34	0.63	0.80
Cobalt, 0.9%	Acid (35 C)	0.41	0.75	1.05
Nickel, 0.1%	Acid (35 C)	0.27	0.39	0.41
Nickel, 0.8%	Acid (35 C)	0.32	0.48	0.55
Nickel, 18%	Acid (35 C)	1.74	4.90	—

[a] Contact resistance was measured on crossed wires with a tension of 200 grams.

purity gold. (2) Comparative data for alloyed and unalloyed deposits are given in Table 8.7. The codeposition of nickel, which hardens the deposit and improves resistance to wear, had less effect on increasing contact resistance than the codeposition of cadmium or cobalt. In the case of an alloy containing about 0.1 percent nickel, contact resistance was increased about two times, relative to that for high-purity, unalloyed deposits. Data from another report showed that the codeposition of cadmium and copper resulted in a higher contact resistance than the codeposition of nickel or cobalt. (37)

Mechanical Properties of Gold Alloys

Tensile Strength Data

Gold-copper deposits (0.25 to 0.5 μm thick) from a neutral, cyanide-ethylenediamine tetraacetic acid bath exhibited tensile strengths after heat treating of 35 to 120 kg/sq mm (50,000 to 170,000 psi), as shown in Figure 8.2. (59) The strongest alloys contained 20 weight percent copper. Deposits containing more or less copper exhibited lower tensile strengths after heating at 350 C. An increase in the annealing temperature to 450 C slightly decreased the strength of alloy containing 20 percent copper. However, alloys containing 30 to 40 percent copper were slightly stronger after heating at 450 C, by comparison with tensile strength data for similar alloys heated at 350 C. Annealing was required to eliminate stress and brittleness.

As a rule, the yield strength of the gold-copper alloys containing 10 to 30 percent copper was 5 to 10 percent lower than tensile strength, in the case of foils annealed at 300 or 450 C. However, a few foils exhibited a yield strength only 2 or 3 percent below their tensile strength. Most of these specimens were still brittle, with elongation values below 3 percent. In general, the elongation of foils containing 12 to 15 percent copper and annealed at 300 C ranged from 5 to 7 percent, whereas elongation values for similarly heated foils containing about 20 percent copper ranged from 10 to 13 percent.

According to another report, the tensile strength of gold alloy reached 175 to 185 kg/sq mm (245,000 to 265,000 psi) when copper and silver contents were adjusted to about 1.0 and 2.0 percent, respectively. (60) With gold alloy deposits containing 0.2 percent copper and 0.5 percent silver, the investigators claimed a tensile strength of 140 to 145 kg/sq mm (200,000 to 206,000 psi).

The yield strength of gold deposits containing a dispersion of 0.02 μm particles of aluminum oxide was 400 to 500 percent higher than the yield strength of pure elec-

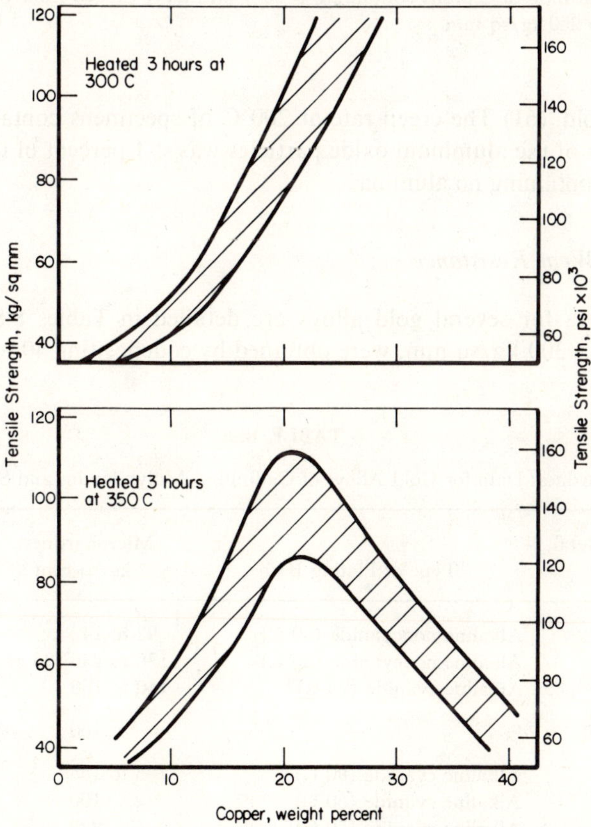

Figure 8.2. Tensile Strength of Gold-Copper Electrodeposits after Heat Treating. (59)

TABLE 8.8

Hardness Data for Gold-Antimony Alloys[a]

Antimony Content, percent	Microhardness, kg/sq mm	Reference
1.5	204	62
1.9	203	43
5	220	43
10	240	43
20	260	43
40	320	43
60	330[b]	43

[a] All solutions were alkaline baths prepared with gold potassium cyanide and operated at 60 C.

[b] Maximum hardness for alloys containing antimony. The hardness of deposits containing >60% antimony ranged from 150 to 260 kg/sq mm.

trodeposited gold. (61) The creep rate at 200 C of specimens containing 1.0 or 1.2 volume percent of the aluminum oxide particles was < 1 percent of the creep rate of gold deposits containing no alumina.

Hardness and Wear Resistance

Hardness data for several gold alloys are detailed in Tables 8.8 to 8.11. Very hard deposits, > 300 kg/sq mm, were obtained by codepositing 40 to 60 percent an-

TABLE 8.9

Hardness Data for Gold Alloys of Cadmium, Molybdenum and Silver

Codeposited Metal, percent	Type of Plating Bath	Microhardness, kg/sq mm	Reference
Cadmium, 1.0%	Alkaline noncyanide (50 C)	92 to 142	36
Cadmium, 1.4%	Alkaline noncyanide (50 C)	176 to 236 (25 g)	2
Cadmium, 12%[a]	Alkaline cyanide (25 C)	140 to 160	63
Molybdenum, 2%	—	160	62
Silver, 8%	Alkaline cyanide (80 C)	145 to 155	44
Silver, 20%	Alkaline cyanide (50 C)	190	57
Silver, 50%	Alkaline cyanide (50 C)	220	64

[a] The alloy deposits also contained 2 to 15% Ag.

timony (Table 8.8), 20 to 25 percent copper (Table 8.10) or 18 to 20 percent nickel (Table 8.11). Gold alloys with a hardness > 200 kg/sq mm contained smaller amounts of these metals. A gold alloy containing 50 percent silver also exhibited a hardness of > 200 kg/sq mm. (64) The hardness of another alloy containing less silver was increased to 220 kg/sq mm by the adoption of ultrasonic agitation. (71)

The codeposition of only 0.1 percent of cobalt or nickel increased the hardness of gold deposited in acid baths about two times, to a range of 135 to 196 kg/sq mm. Cobalt and nickel were about equally effective. An increase in the cobalt or nickel content to 0.9 or 1.0 percent further increased hardness to as much as 238 kg/sq mm. Gold alloy deposits containing 18 or 20 percent nickel reached microhardness values as high as 312 to 500 kg/sq mm. However, codeposition of 18 percent nickel in gold deposits from a neutral cyanide-pyrophosphate bath exhibited a hardness of only 200 kg/sq mm.

An electroless gold alloy containing about 1 percent phosphorus exhibited a hardness of approximately 100 kg/sq mm. (72) The deposit was obtained in a gold cyanide bath at 92 to 96 C, which contained citric acid, ammonium chloride, and sodium hypophosphite.

Gold-copper alloy deposits with a hardness of 250 kg/sq mm were hardened by heat treating and reached values of 400 kg/sq mm. (73) A 3-hour heat treatment at 200 C increased the hardness of alloy containing 25 percent copper from about 380 to 475 kg/sq mm. (59) A 1-hour heat treatment at 300 C increased the hardness of this alloy to 490 kg/sq mm. After longer annealing periods (8 hours or more) at 200 C, the hardness of the copper alloy was 450 kg/sq mm.

Some reports cite data showing decreasing wear with increasing hardness. For example, alloys containing 30 to 40 percent copper, which exhibited microhardness

TABLE 8.10

Hardness Data for Gold Copper Alloys

Copper Content, percent	Type of Plating Bath	Microhardness, kg/sq mm	Reference
10	Alkaline cyanide (60 C)	194	49
15	Alkaline cyanide (60 C)	242	49, 54
16	Alkaline cyanide (60 C)	180	62
19[a]	Acid (50 C)	240 to 250	67
20 to 40	Neutral	450	65
25	Neutral	380	59
25	Neutral	325	66
40	—	250	62
40	Alkaline cyanide (60 C)	260	49
60	Alkaline cyanide	220	49

[a] The alloys also contained 0.5 to 1.0 percent Sb.

TABLE 8.11

Hardness Data for Gold Alloys of Cobalt and Nickel[(a)]

Codeposited Metal, percent	Microhardness, percent	Reference
Cobalt, 0.1%	100	66
Cobalt, 0.1%	137 to 196 (25 g)	2
Cobalt, 0.1%	172	7
Cobalt, 0.9%	194 to 238 (25 g)	2
Cobalt, 1.0%	195	7
Cobalt, 2.0%	285	66
Nickel, 0.1%	135 to 167 (25 g)	2
Nickel, 0.1%	180	7
Nickel, 0.1%	300	68
Nickel, 1%	190	7
Nickel, 1%	167 to 206 (25 g)	2
Nickel, 1%[(b)]	98	56
Nickel, 1.8%[(c)]	150 to 190	23
Nickel, 2%	160 to 180	39
Nickel, 2 to 8%[(d)]	200	69
Nickel, 5%[(c)]	180	70
Nickel, 18%	312 to 392 (25 g)	2
Nickel, 18%[(b)]	200	56
Nickel, 20%	~500	66
Nickel, 63 to 72%[(b)]	300	56
Nickel, 78%[(b)]	450	56

(a) Acid cyanide baths, if not otherwise noted.
(b) A neutral, cyanide-pyrophosphate bath at 50 to 60 C.
(c) An alkaline cyanide bath at 20 to 60 C.
(d) An alkaline cyanide bath at 25 C.

values of 240 to 270 kg/sq mm, were three to four times as resistant to wear as unalloyed gold. (49) In another case, gold alloy with a hardness of 86 kg/sq mm resisted abrasion in Taber Abraser tests for 3300 cycles in comparison with only 2400 cycles for gold with a hardness of 47 kg/sq mm. (74) Alloy deposits containing 5 percent silver with a hardness of 190 kg/sq mm were five times as resistant to wear as unalloyed gold with a hardness of about 100 kg/sq mm. (56)

Wear data for several alloyed and unalloyed deposits, based on the weight loss of the deposits on spindles rotating in sand, showed a greater weight loss for the harder alloys containing cobalt and nickel than the unalloyed gold deposited in acid or alkaline cyanide baths. (2) On the other hand, alloy containing 0.1 or 0.9 percent cobalt was removed at about the same rate as unalloyed gold deposited in a neutral bath. Hardness and wear data are listed in Table 8.12.

TABLE 8.12

Hardness and Wear of Gold Alloys (2)

Codeposited Metal, percent	Type of Plating Bath	Hardness, kg/sq mm[(a)]	Wear, mg/hr/sq cm[(b)]
None	Acid (65 C)	52 to 129	0.07
None	Alkaline cyanide (65 C)	47 to 86	0.15
None	Neutral (65 C)	44 to 82	0.50
Cadmium, 1.4%	Alkaline noncyanide (50 C)	176 to 236	0.15
Cobalt, 0.1%	Acid (35 C)	137 to 196	0.68
Cobalt, 0.9%	Acid (35 C)	194 to 238	0.70
Nickel, 0.1%	Acid (35 C)	135 to 167	1.01
Nickel, 1.0%	Acid (35 C)	167 to 206	0.23
Nickel, 18%	Acid (35 C)	194 to 238	0.51

(a) 25-gram load.

(b) Wear on electroplated spindles rotating in a sand bed for 96 hours.

Another report also showed no correlation between hardness and wear. (75) Gold deposits with a hardness of 200 to 250 kg/sq mm exhibited more wear than deposits with a hardness of 100 to 150 kg/sq mm. According to another investigator, alloy deposits containing 1.5 percent antimony, which had a hardness of 204 kg/sq mm, were five times as resistant to wear as a gold alloy containing 40 percent copper, which had a hardness of 250 kg/sq mm. (62)

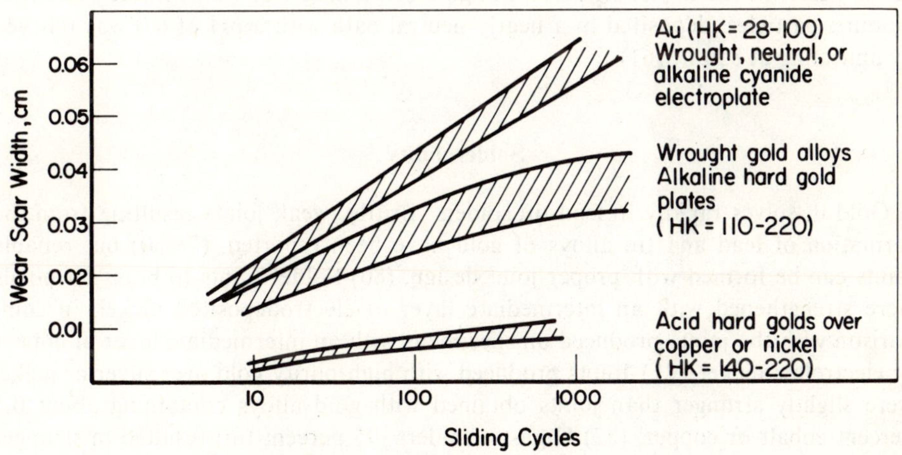

Figure 8.3. Wear of Wrought and Electroplated Gold Alloys. (7)

During a study of wear from sliding contacts, gold deposits from acid baths that contained at least 0.1 percent cobalt showed mild wear in contrast to severe wear exhibited by wrought gold alloys or electrodeposited alloys from alkaline baths, even though some of the wrought alloys were almost as hard as the acid gold-cobalt alloys. (7) The relationship of wear and hardness for the electrodeposited and wrought materials is shown in Figure 8.3. The superior wear resistance of the gold-cobalt alloy electrodeposits was attributed to their lower coefficient of friction, in comparison with the coefficients for other materials, which are listed in Table 8.13. (7) The low friction of the acid gold alloys was, in turn, attributed to randomly dispersed, codeposited organic polymer detected by scanning electron microscopy. Electron probe analysis showed these deposits to be rich in carbon, probably combined with nitrogen and oxygen.

Stress

In contrast with the low tensile or compressive stress of unalloyed gold deposited in acid or alkaline cyanide baths (Table 8.5), gold alloys from similar solutions exhibit relatively high tensile stress. For example, stress ranging from 21 to 42 kg/sq mm (30,000 to 60,000 psi) was reported for gold-cobalt alloys from acid solutions. (3) A stress as high as 12 kg/sq mm (17,000 psi) was reported for gold-copper alloys deposited in alkaline cyanide baths. (49) Codeposition of copper with gold in a neutral bath also resulted in a high stress. (59) Quantitative data for gold alloys are given in Table 8.14.

High stress for gold alloy containing cobalt or nickel, deposited in an acid bath with a pH of 4.2 to 4.5, was reduced to low values by increasing the temperature of the solution from 20 to 50 C. (76) The deposits contained about 0.3 percent nickel or 0.4 percent cobalt. The high stress of gold containing 1 or 1.25 volume percent of alumina particles, deposited in a nearly neutral bath with a pH of 6.0 was relieved by annealing at 850 C. (61)

Solderability

Gold dissolves rapidly in lead-tin solders. Brittle, weak joints resulting from the formation of lead and tin alloys of gold have been reported, (78-80) but reliable joints can be formed with proper joint design. (80) Solder joints to brass terminals were strengthened with an intermediate layer of electrodeposited nickel, in comparison with the joints produced on specimens with an intermediate layer of copper or electroless nickel. (81) Joints produced with high-purity gold over silver or nickel were slightly stronger than joints obtained with gold alloys containing about 0.5 percent cobalt or copper. (82) High-tin solders (95 percent tin) resulted in stronger joints than the common solder containing 60 percent tin and 40 percent lead, especially when the soldering temperature was raised to 274 C, or higher. A lower

TABLE 8.13

Correlation of Hardness, Friction, and Wear (7)

Gold Alloy	Initial Microhardness, kg/sq mm	Surface Hardness After 1000 Cycles, kg/sq mm	Coefficient of Friction		Width of Scar Wear at 1000 Cycles, cm
			In N_2 – 20% O_2	In Vacuum	
Pure wrought gold	77	135	1.7 to 1.9	1.9 to 2.3	0.058
Wrought, 19% silver	107	150 to 175	1.1 to 1.3	1.5 to 1.8	0.040
Wrought, 35% silver	155	150 to 175	1.1 to 1.3	1.5 to 1.8	0.039
Electrodeposit, 0.1% cobalt	172	180 to 190	0.8 to 1.0	0.8 to 1.1	0.010
Wrought, 0.1% cobalt	85	140 to 150	1.2 to 1.4	1.8 to 2.1	0.040
Palladium	—	400 to 500	—	—	0.030
Platinum	—	1000 to 1200	—	—	0.028

TABLE 8.14

Stress Data for Gold Alloys

Codeposited Metal, percent	Types of Plating Bath	Stress[a] kg/sq mm	psi	Reference
Cadmium, 1.4%	Alkaline noncyanide (50 C)	13 to 15.5	18,500 to 22,000[b]	2
Cobalt, 0.1%	Acid (35 C)	27 to 28	38,000 to 40,000[b]	2
Cobalt, 0.9%	Acid (35 C)	22 to 29	31,000 to 41,000[b]	2
Copper, 8%	Alkaline cyanide	10.5	15,000	49
Copper, 12%	Ditto	12.0	17,000	49
Copper, 20%	"	8.5	12,000	49
Copper, 35%	"	7.0	10,000	49
Nickel, 0.1%	Acid (35 C)	16 to 21	23,000 to 30,000[b]	2
Nickel, 0.8%	Acid (35 C)	14 to 15.5	20,000 to 22,000[b]	2
Nickel, 18%	Acid (35 C)	−1.4	−2,000[c]	2
Silver, 1%	Alkaline cyanide (20 C)	3.5	5,000[d]	2

[a] A negative sign indicates compressive stress.

[b] Stress range for deposits ranging from 1 to 12 μm in thickness.

[c] Stress of a 12-μm deposit. Thinner deposits (1 to 5 μm) showed no stress.

[d] Stress of a 12 μm deposit. Thinner deposits were stressed in compression (−1.4 to −5 kg/sq mm).

temperature of 230 or 245 C was better for making strong joints in the case of the common, 60–40, tin-lead solder.

In another paper that discussed the influence of soldering temperature, gold-plate thickness and purity, a temperature of 230 C was better than a higher temperature of 285 C for soldering to 1 to 3 μm-thick gold deposits. (83) To achieve strong joints to gold-nickel and gold-cobalt alloy deposits, a reduction in temperature to 190 C and a limit in thickness to 1.5 μm were recommended. Gold plate thickness in the range of 1.2 to 2.5 μm was recommended in another report. (84)

Structure

Electrodeposited gold develops a continuous structure at a lower thickness level than vacuum deposited gold films. (85) Continuity was observed for electrodeposits with a thickness < 50 angstroms. Thinner, 15-angstrom-thick deposits were continuous on (100) copper crystal faces but discontinuous on (111) copper planes.

High-purity gold deposited in acid, alkaline, and neutral baths exhibit columnar grains, but deposits from the acid solutions are finer in grain size. A grain size of 0.02 to 0.1 μm was reported for the acid gold deposits in comparison with 1.5 to 3

μm for the grain size of gold deposited in alkaline cyanide solution. (1) Less hydrogen is absorbed in the finer-grained gold deposited in the acid baths. (87)

Increasing the current density from 0.3 to 0.6 or 0.7 amp/sq dm enlarged the grains of gold deposited in acid or nearly neutral phosphate baths. (86) The grain size of very thin deposits is influenced by the grain size of the substrate.

Some proprietary addition agents for alkaline and neutral solutions refine grain size. The codeposition of about 0.1 percent cobalt in gold from an acid bath produces a mixed structure of columns and bands. (2) Only bands could be seen in the cross sections of alloy deposits containing about 0.9 percent cobalt or 0.1, 0.8, or 18 percent nickel. The alloy with 18 percent nickel contained cracks.

Banding also was observed in the cross sections of gold alloys containing about 25 percent copper which were deposited in neutral or slightly alkaline solutions prepared with potassium gold cyanide, the copper salt of diethylene triamine pentaacetic acid and phosphate ions. (59) Annealing at 350 C (8 hours) had little effect on the structure, but 800 C for 0.5 hour completely annealed the alloy. No cracking was reported in these alloys at thicknesses up to 0.5 mm. On the other hand, similar alloys deposited in alkaline cyanide baths contained cracks. (88)

The crystal size of some gold alloy deposits is much smaller than the crystal size of unalloyed gold. Data on crystal size and orientation are summarized in Table 8.15. Alloys of gold and cobalt, copper, or nickel had crystal sizes of 100 to 1300 angstroms, in comparison with 10,000 or > 10,000 angstroms for unalloyed gold deposited in acid or alkaline cyanide solutions.

X-ray data for gold-copper alloys deposited in alkaline cyanide baths disclosed a mixture of gold and copper. (54,91) After reducing the free cyanide concentration, a

TABLE 8.15

Crystal Size and Orientation of Gold and Gold Alloy Deposits

Codeposited Metal	Type of Plating Bath	Crystal Size, angstroms	Preferred Orientation[a]	Reference
None	Alkaline cyanide	200 to 20,000	(211)	89
None	Alkaline cyanide	10,000	—	90
None	Acid cyanide	>10,000	(311)	36
Cadmium, 1%	Alkaline noncyanide	>10,000	(100)	36
Cobalt, <0.1%	Alkaline cyanide	1,000	—	90
Cobalt, 0.1%	Acid cyanide	600	Random	36
Cobalt, 2.5%[b]	Acid cyanide	340	—	36
Copper	Alkaline cyanide	100	—	58
Nickel, 0.15%	Acid cyanide	1,300	(100)(111)	36
Silver, 1%	Alkaline cyanide	>10,000	Random	36

[a] Parallel to substrate surface.

[b] The alloy also contained 1% indium.

solid solution of copper in gold mixed with separate crystals of copper and gold was observed. (91) The solution contained 4 percent copper. Solid solutions characteristic of the metallurgical alloys also were noted by another investigator. (58) Heat treating at 300 C reduced crystal size from 10^{-6} to 10^{-3} cm. In another paper, it was noted that heat treating or aging transformed a mixture of gold and copper to a single, solid solution. (54)

Solid solutions were reported for gold alloy deposits containing 1.9 percent antimony, but alloy containing 5 to 40 percent antimony consisted of gold and $AuSb_2$. (43) Gold-silver alloys were solid solutions, like their metallurgical counterparts. (92, 93) Gold alloy containing up to 30 atomic percent iron was deposited as a supersaturated solid solution. (94) Two phases were observed in gold alloys containing 1.8 to 28 percent zinc or 49 to 65 percent cadmium, but only a single phase with an enlarged lattice was noted in alloys deposits containing 1.8 to 28.8 percent cadmium. (95) The solid solubilities of lead and bismuth in electrodeposited gold is very low, like their metallurgical counterparts. (96)

The alloys of nickel and cobalt deposited in acid baths are metastable, supersaturated solid solutions, (6,93) which are transformed to two phases by heating. (93) Like gold, these alloys are face-centered-cubic. (40)

References

(1) Craig, S. E., Jr., Harr, R. E., Henry, J., and Turner, P., "A Comparison of Various 24K Gold Electrodeposits," *Journal Electrochem. Soc., 117,* 1450–1456 (1970).

(2) Duva, R., and Foulke, D. G., "Properties of Some Gold Electrodeposits," *Plating, 55,* 1056–1062 (1968).

(3) Foulke, D. G., "Quality Control in Precious Metal Plating," *Metal Finishing, 63,* 42 (July 1965).

(4) Foulke, D. G., "Addition Agents for Gold Plating Baths," *Metal Finishing Journal, 9,* 487 (1963).

(5) Grossman, H., and Merl, W., "Specific Resistance of Electrodeposited Gold Films," *Metalloberflaeche, 23,* 100–103 (1969).

(6) Danemark, M. A., "Physical Properties of Electrodeposited Pure Gold and Gold-Cobalt Alloys," *Galvanotechnica, 15,* 129 (1964).

(7) Abbot, W. H., "Precious Metal Electrodeposits for Electrical Contact Applications," *MCIC-72 05, Symposium on Electrodeposited Metals as Materials for Selected Applications.* Available from NTIS, Springfield, Virginia, 22151.

(8) Danemark, M. A., "Properties of Electrodeposited Gold and Gold-Cobalt Alloys," *Galvano, 35* (348), 31–37; (349), 101–104 (1966).

(9) Cooley, R. L., and Lemons, K. E., "Densities of Electroplated Gold," *Plating, 56,* 511–515 (1969).

(10) Clarke, M., and Leeds, J. M., "Porosity in Single and Multi-Layer Electrodeposits and Its Relation to the Basis Metal State," *Trans. Inst. Metal Finish., 13,* 50 (1965).

(11) Ehrhardt, R. A., "Acid Gold Plating," *Tech. Proceedings Am. Electroplaters' Soc., 47,* 78 (1960).

(12) Frant, M. S., "Porosity Measurements on Gold-Plated Copper," *Journal Electrochem. Soc., 108,* 774 (1961).

(13) Lee, W. T., "Gold Plating: Which Gold to Use," *Electroplating and Metal Finishing, 18,* 40 (1965).

(14) Tweed, R. E., "Manufacturing Methods for Electroplating Silver, Gold, and Rhodium on

Electrical Connector Contacts," *Interim Engineering Progress Reports III, V, and VI*, Air Force Contract AF33 (657)-9752, Nu-Line Industries, Incorporated, (September 1963 - December 1962; July 1964 - January 1965; and January 1965 - February 1965). AD425933; AD457143; and AD458539.

(15) Nozaki, C., Owada, K., and Aotani, K., "Gold Plating. I. Electrolytic Characteristics of Gold Plating and Its Corrosion Resistance," *Kinzoku Hyomen Gijutsu, 21*, 180–188 (1970).

(16) Antler, M., "Current Research in the Precious Metal Plating of Electric Contacts," *Plating, 54*, 915 (1967).

(17) Garte, S. M., "Effect of Substrate Roughness on the Porosity of Gold Electrodeposits," *Plating, 53*, 1335 (1966).

(18) Ashurst, K. G., and Neale, R. W., "The Porosity of Gold Electrodeposits," *Trans. Inst. Metal Finish., 45*, 75–82 (1967).

(19) Modjeska, R. S., and Kann, S. N., "Diffusion Phenomena in Plated Precious Metals," *Tech. Proceedings Am. Electroplaters' Soc., 50*, 117 (1963).

(20) Garte, S. M., "Determination of the Porosity of Gold Electrodeposits by Quantitative Electrography," *Plating, 53*, 1331 (1966).

(21) Garte, S. M., "Porosity of Gold Electrodeposits: Effect of Substrate Surface Structure," *Plating, 55*, 946–951 (1968).

(22) Lee, W. T., "Latest Techniques in Technical Gold Plating," *Corrosion Technology, 10*, 59 (1963).

(23) Atanasyants, A. G., Kudryavtsev, N. T., and Karataev, V. M., "Hard Gold Plating," *Zh. Prikl. Khimii, 30*, 876 (1957).

(24) Leeds, J. M., and Clarke, M., "The Effects of Plating Conditions on Porosity in Gold Electrodeposits," *Trans. Inst. Metal Finish., 46*, 1 (1968).

(25) Angus, H. C., "Properties and Behavior of Precious-Metal Electrodeposits For Electrical Contacts," *Trans. Inst. Metal Finish., 39*, 20 (1962).

(26) Kleinle, A., and Loebich, O., "Electroplated Precious Metal Coatings for the Electrical Industry," *Galvanotechnik, 53*, 62 (1962).

(27) Walton, R. F., "Effect of Type and Porosity of Gold Deposits on the Contact Resistance of Electroplated Materials," *Plating, 53*, 209–216 (1966).

(28) Baker, R. G., "Coatings for Contact Surfaces," paper presented at Reliable Electrical Connections EIA Conference, Dallas, Texas, 219 (1958).

(29) Klima, S. J., Smith, W. F., and Sikolowski, T., "Corrosion: Its Effects on Precious Metal Plated Microwave Surfaces," *Extended Abstracts of Electrodeposition Division, 2, Electrochem. Soc.*, 11–15. (October 1964).

(30) Frant, M. S., "Copper Oxides on the Surface of Gold Plate," *Plating, 48*, 1305 (1961).

(31) Ohki, Y., and Maruyama, H., paper presented at the 13th National Relay Conference (April 26–29, 1965). Summarized by Frank J. Oliver, *Electro-Technol. (New York), 76*, 69 (1965).

(32) Anderson, J. S., "Physical and Engineering Properties of Electro-Deposited Metals," *Electroplating, 3*, 84 (1949).

(33) Cady, J. R., Private Communication citing AES Project 25 data (November, 1971).

(34) Dini, J. W., Sandia Corporation, Private Communications to Battelle Memorial Institute, Columbus Laboratories (May 13 and September 13, 1965).

(35) Dytrt, J. F., "Electroplating and Metallizing Printed Wiring," *Industrial and Engineering Chemistry, 51*, 286 (1959).

(36) Antler, M., "Current Research in the Precious Metal Plating of Electric Contacts," *Plating, 54*, 915–922 (1967).

(37) Loebich, O., and Zilske, W., "Comparative Evaluation of Electroplated Precious Metal Coatings in the Electrical Industry," *Metalloberflaeche, 22*, 34–38 (1968).

(38) Jones, D., and Dalton, I. M., "Electrodeposition of Gold," British Patent, 1,193,615 (June 3, 1970).

(39) Firoiu, C., and Neata-Balescu, M., "Electrolytic Production of Au-Ni Alloys," *Metalurgia (Bucharest), 18*, 508–510 (1966).

(40) Foulke, D. G., "The Effect of Addition Agents on the Structure and Physical Properties of Gold Electrodeposits," *Plating, 50,* 39 (1963).

(41) Schumpelt, K., "Acid Gold-Electroplating Bath," U.S. Patent 3,367,853, (February 6, 1968).

(42) Seegmiller, R., and Gore, J. K., "A Cyanide Bath for Heavy Gold Plating," *Tech. Proceedings Am. Electroplaters' Soc., 47,* 74 (1960).

(43) Fedot'ev, N. P., Vyacheslavov, P. M., and Volyanuk, G. A., "Deposition Technology and Physicochemical Properties of Electrolytic Gold-Antimony Alloy," *Journal Applied Chemistry (USSR), 40,* 1693–1696 (1967); *Zh. Prikl. Khimii, 40,* 1759–1753 (1967).

(44) Poptsova, Z. P., "Galvanic Coatings by Gold-Silver Alloys," *Tr. Kazansk. Khim.-Tekhnol. Inst., 33,* 151–154 (1964).

(45) Wilson, G. A., "Thickness and Hardness of Gold Deposits," *Metal Finishing, 58,* 50 (1960).

(46) Wilson, G. A., "Quality Control of Gold Deposits. Assessment of Thickness and Hardness," *Industrial Finishing (London), 12,* 50 (1960).

(47) Reid, F. H., "Palladium Plating-Processes and Applications in the United Kingdom," *Plating, 52,* 531 (1965).

(48) Anderson, J. R., and Saunders, J. B., "Stable Sliding Connector Contacts," *Final Report for December 1, 1961 to September 30, 1963 Contract DA36-039SC-89150,* Stanford Research Institute (September 30, 1963), 64 pp. AD426113. PB167424.

(49) Fedot'ev, N. P., Kruglova, E. G., and Vyacheslavov, P. M., "Electrodeposition of Gold-Copper Alloys," *Zh. Prikl. Khimii, 32,* 2014 (1959).

(50) "Gold Plating," Belgian Patent 641,175 (December 12, 1963). Assigned to Sel-Rex Corporation.

(51) Bondarev, V. V., Krupnikova, E. I., and Rodionova, L. L., "Electrolytic Application of Thick Gold Coatings," *Tr., Gos. Nauch.-Issled. Proekt. Inst. Splavov Obrab. Tsvet. Metal, 31,* 93–100 (1970).

(52) Chen, E. S., and Sautter, F. K., "Porosity in Electrodeposited Gold-Alumina Alloys," *Journal Electrochem. Soc., 117,* 726–728 (1970).

(53) Fedot'ev, N. P., and Vyacheslavov, P. M., "The Phase Structure of Binary Alloys Produced by Electrodeposition," *Plating, 57,* 700–706 (1970).

(54) Fedot'ev, N. P., "Electrodeposited Silver and Gold Alloys in Radio and Electrical Engineering," *Theory and Practice of Bright Electroplating,* Akad. Nauk Lit. SSR, Inst. Khim. i Khim, Tekhnol., Proc., of All-Union Conference, Vilnyus (1962). Translated by the Israel Program for Scientific Translations, Jerusalem, 254–258 (1965).

(55) Andreeva, G. P., Fedot'ev, N. P., and Vyacheslavov, P. M., "Physico-Chemical Properties and Structure of Gold-Copper Electrodeposits," *Zh. Prikl. Khimii, 37,* 146 (1964).

(56) Fedot'ev, N. P., Vyacheslavov, P. M., Lokshtanova, O. G., and Kruglova, E. G., "Electrodeposition Technology and Physicochemical Properties of Gold-Nickel Alloy," *Journal Applied Chemistry (USSR), 40,* 2167–2171 (1967); *Zh. Prikl. Khimii, 40,* 2253–2258 (1967).

(57) Fedot'ev, N. P., Vyacheslavov, P. M., Andreeva, G. P., Kruglova, E. G., and Volyanyuk, G. A., "Electrodeposition of Gold-Silver Alloys and Their Properties," *Protective Metallic and Oxide Coatings, Metal Corrosion and Electrochemistry,* Edited by N. P. Fedot'ev, Akademiya Nauk SSSR. Translated by the Israel Program for Scientific Translations, Jerusalem, 209–217 (1968).

(58) Rochat, R., "Effect of Heat-Treatment on Some Electroplated Alloys," *Bull. Ann. Soc. Suisse Chronomet. Lab. Suisse Recherches Horl., 4,* 45 (1957).

(59) Wiesner, H. J., and Distler, W. B., "Physical and Mechanical Properties of Electroformed Gold-Copper Alloys," *Plating, 56,* 799–804 (1969).

(60) Mel'nikov, P. S., "Protective Galvanic Coatings of Gold Alloys of Increased Hardness and Wear Resistance," *Zashchita Metallov, 6,* 365-7 (1970).

(61) Sautter, F. K., "Electrodeposition of Dispersion-Strengthened Gold-Al_2O_3 Alloys," *Technical Report under U.S. Army Contract DA Project 1-F-5-23801-A-289-01,* Benet R and E Laboratories, Watervliet Arsenal (December 1963), 11 pp. AD 435 622.

(62) Korovin, N. V., "Alloy Deposition-Theory, Practice and Current Applications," *Electroplating and Metal Finishing, 17,* 151; 249; 269 (1964).

(63) Marakhtanova, Z. N., Balashova, N. N., Smagunova, N. A., ad Sokolova, L. S., "Electrodeposition of a Gold-Silver-Cadmium Alloy," *Zashchita Metallov 5,* 437–40 (1969).

(64) Vrobel, L., "Bath for Electroplating Bright Gold Coatings with High Content of Silver," Czech. Patent 128,666 (August 15, 1968).

(65) Korbelak, A., and Duva, R., "Precious Metal Plating and Solderability," *Tech. Proceedings Am. Electroplaters' Soc., 48,* 142 (1961).

(66) Korbelak, A., "Precious Metal Electroplating," *Encyclopedia of Electrochemistry,* Reinhold Publishing Company 987–990 (1964).

(67) Cathrein, R., and Danemark, M., "Electrolyte and Method for Coating Articles with a Gold-Copper-Antimony Alloy," U.S. Patent 3,380,814 (April 30, 1968).

(68) Ostrow, B. D., and Nobel, F. I., "Recent Developments in Gold Electroplating," *Tech. Proceedings Am. Electroplaters' Soc., 47,* 68 (1960).

(69) Matsumoto, S., "Electroplating of Au-Cu-Ni Alloy. Study of Electroplating of Gold Alloys, Pt I," *Journal of the Metal Finishing Society of Japan, 11* 59 (1960).

(70) Kawai, S., "Studies of Structures of Au-Ni Electrodeposited Alloys," *Journal of the Metal Finishing Society of Japan, 19,* 487–91 (1968).

(71) Vrobel, L., "The Influence of Ultrasonic Vibrations of the Electrodeposition of Gold," *Trans. Inst. Metal Finish., 44,* 161–164 (1966).

(72) Socha, Jan, and Zak, Tadeusz, "Chemical Gold Plating," *Prace Inst. Mech. Precyz, 16,* 2–7 (1968).

(73) Foulke, D. G., "Engineering Applications for Precious Metal Plating," *Metal Progress, 84,* 107 (1963).

(74) Rinker, E. C., "Bright Gold Plating," *Plating, 40,* 861 (1953).

(75) Fischer, J., and Weimer, D. E., *Precious Metal Plating,* Robert Draper Ltd., Teddington, England (1964), 271 pp.

(76) Socha, J., and Zak, T., "Properties of Gold Coatings Deposited from Acid Baths," *Prace Inst. Mech. Precyz, 14,* 1–5 (1966).

(77) Foster, F. G., "Embrittlement of Solder by Gold From Plated Surfaces," *Special Tech. Pub. No. 319,* American Society for Testing and Materials, 13–19 (1962).

(78) Keller, J. D., "Printed Wiring Surface Preparation Methods--Elimination of Gold Plating as a Surface Preparation for Printed Circuits and Development of a Contamination-Free Surface," *Special Tech. Pub. No. 319,* American Society of Testing and Materials, 3–12 (1962).

(79) Thwaites, C. J., "Solderability of Coatings for Printer Circuits, Pt 4," *Trans. Inst. Metal Finish., 43,* 143 (1965).

(80) Whitfield, J., and Cubbin, A. J., "Experimental Observations on the Effect of Gold and Palladium on Soldered Joints," *ATE Journal* (January 1965).

(81) Thompson, J. M., and Bjelland, L. K., "Evaluation of Solderability of Electroplated Coatings," *Tech. Proceedings Am. Electroplaters' Soc., 48,* 182 (1961).

(82) Weil, R., and Diehl, R. P., "Solderability of Some Gold Electrodeposits," *Plating, 52,* 1142 (1965).

(83) Harding, W. B., and Pressly, H. B., "Soldering to Gold Plating," *Tech. Proceedings Am. Electroplaters' Soc., 50,* 90 (1963).

(84) Day, B. H. S., "Electroplating Applied to Printed Circuit Production," *Plating, 52,* 228 (1965).

(85) Lawless, K. R., "Growth and Structure of Electrodeposited Thin Metal Films," *Journal of Vacuum Science and Technology, 2,* 24 (1965).

(86) Clark, M., and Chakrabarty, A. M., "A Technique for the Metallographic Examination of Gold Coatings Using an Electrodeposited Iron Support," *Trans. Inst. Metal Finish., 48,* 124–130 (1970).

(87) Taran, L. A., Berezina, S. I., and Vozdvizhenskii, G. S., "Effect of Hydrogen on the Structure and Properties of Gold Coatings Deposited from Cyanide-Citrate Electrolytes," *Navodorozhivanie Metal. Bor'ba Vodorodn, Khrupkost'yu,* 184–7 (1968).

(88) Wullhorst, B., "Basic Principles of Gold Electroplating From Cyanide Solutions and Its Practical Application," *Metalloberflaeche, 7,* A49 (1953).

(89) Schloetterer, H., "Production and Surface Properties of Oriented Electrolytic Deposits," *Metalloberflaeche, 18,* 33 (1964).

(90) Fedot'ev, N. P., Ostroumova, N. M., and Vyacheslavov, P. M., "Hard Electroplated Gold," *Zh. Prikl. Khimii, 29,* 489–492 (1956).

(91) Raub, E., and Sautter, F., "The Growth of Electrodeposited Alloys. X. Gold-Copper Alloys," *Metalloberflaeche 10,* 65–72 (1956).

(92) Raub, E., "Alloy Electrodeposits," *Metalloberflaeche, 7A,* 12–17 (1953).

(93) Danemark, M. A., "A Review of the Principles of Electroplating Gold Alloys," *Metal Finishing Journal (London), 10,* 483 (1964).

(94) Lokshtanova, O. G., Fedot'ev, N. P., and Vyacheslavov, P. M., "Technology of the Electrodeposition and Some Physical-Chemical Properties of Gold-Iron Alloys," *Zashchita Metallov, 4,* 548–552 (1968).

(95) Raub, E., and Disam, A., "Electrolytic Zinc-Gold and Cadmium-Gold Alloys," *Metalloberflaeche, 17,* 17 (1963).

(96) Raub, E., and Disam, A., "Electrolytic Bismuth-Gold and Lead-Gold Alloys," *Metalloberflaeche, 16,* 317 (1962).

Chapter 9

Iron and Iron Alloys

Electrodeposited iron has been used for electroforming electrotypes, glass molds, record stampers, and printing plates and for building up worn or mismachined parts. Use as a wear-resistant coating has been reported in Russia for hard iron deposits. Soldering-iron tips are plated with iron to decrease copper attrition and improve life. Interest in iron foil has spurred the development of physical property data.

Data have been reported for iron deposited in fluoborate, sulfamate, and sulfate solutions, but deposits from chloride baths have received more attention. Electrodeposited iron is surprisingly resistant to corrosion in mildly corrosive environments, evidently because of its high purity, but a protective film such as a cathodically precipitated chromic oxide film improves corrosion resistance in humid atmospheres.

Mechanical Properties

Tensile Data

Iron with a coarse-grained structure electrodeposited in chloride solutions exhibited tensile strengths ranging from 26 to 60 kg/sq mm (37,000 to 86,000 psi), but 90 kg/sq mm (128,000 psi) was reported for fine-grained iron deposited in a chloride solution containing glycerol. The tensile strength of iron deposited in sulfamate and chloride-fluoborate solutions ranged from about 70 to 109 kg/sq mm (100,000 to 155,000 psi). Table 9.1 details mechanical property data for iron deposited in four plating baths.

No systematic study of the influence of thickness on tensile properties has been reported. However, increasing thickness from 0.2 to 1.0 mm appeared to have little effect on the strength or ductility of iron deposited in a ferrous chloride-manganous chloride solution. (8) Some slight effect may have been masked by the large influence of increasing the temperature of the bath. For a temperature of 71 to 78 C, tensile strength ranged from 60 to 77 kg/sq mm (85,000 to 110,000 psi). Much lower strength values from 33 to 51 kg/sq mm (47,500 to 73,000 psi) were reported for iron deposited in solutions heated to 90 or 98 C.

TABLE 9.1

Mechanical Properties of Electrodeposited Iron

Solution	Ultimate Tensile Strength		Elongation, percent	Hardness, kg/sq mm	Reference
	kg/sq mm	psi			
Chloride[a]	43 to 44	61,000 to 63,000	5 to 18	—	1
Chloride[b]	48 to 80	68,000 to 113,000	0 to 16	—	1
Chloride ($CaCl_2$)[c]	38 to 39	55,000 to 56,000	6 to 27	—	2
Chloride ($CaCl_2$)[d]	35	50,000	40	155 to 165	3
Chloride ($CaCl_2$)[e]	37 to 44	52,000 to 62,000	2 to 13	145 to 147	4
Chloride ($CaCl_2$)[f]	26 to 55	37,000 to 78,000	0 to 13	120 to 148	5
Chloride (NaCl)[g]	45 to 60	65,000 to 86,000	4 to 18	150 to 530	6
Chloride (glycerol)[h]	80 to 90	114,000 to 128,000	—	360 to 380	7
Chloride ($MnCl_2$)[i]	60 to 77	85,000 to 110,000	6 to 23	—	8
Chloride ($MnCl_2$)[j]	33 to 51	47,500 to 73,000	17 to 50	—	8
Chloride-fluoborate[k]	54 to 71	77,000 to 101,000	1.1 to 3.5	—	9
Chloride-fluoborate[l]	71 to 105	101,000 to 149,000	1.1 to 1.6	—	9
Sulfamate[m]	69 to 109	98,000 to 155,000	—	—	10
Sulfate[n]	33 to 38	47,000 to 54,000	5 to 7	180	11
Sulfate[o]	58 to 62	83,000 to 89,000	3.0 to 3.5	250	11

Footnote	Fe^{2+} Conc., N	pH	Other Compounds, g/l	Temp, C	Current Density, amp/sq dm	Stress Relief
(a)	5.0	<1.0	None	97 to 102	10	None
(b)	5.0	—	None	95 to 106	20	Ditto
(c)	No data	(0.01 N)	$CaCl_2$	90	5 to 7	"
(d)	2.3	0.15 to 1.5	150 $CaCl_2$	90	8.6	"
(e)	1.8	0.9	100 $CaCl_2$	90	18	"
(f)	1.5	1.2 to 1.6	150 $CaCl_2$	95	12	"
(g)	2.5	<1.0	50 NaCl	90 to 95	20	"
(h)	2.0	<1.0	90 glycerol + 150 NaCl	90	20	2 hr, 200 to 300 C
(i)	1.5	1.8 to 2.1	5 MnCl	71 to 78	5	None
(j)	1.5	1.5 to 2.1	5 MnCl	90 to 98	10	None
(k)	2.1	3.1 to 3.2	130 $Fe(BF_4)_2$	70	5	¼ hr, 400 C
(l)	2.1	3.2	130 $Fe(BF_4)_2$	80	20	Ditto
(m)	1.8	2.0 to 2.5	30 $NH_4SO_3NH_2$	43 to 49	5.4 to 16	24 hr, 190 C
(n)	2.0	2.8 to 3.5	None	75	10	None
(o)	2.0	2.5	None	75	10	None

The tensile strength of iron deposited in 90 C chloride baths was reduced when the pH of the bath was increased from 0.9 to 1.4 or 2.0. (4) Although strength increased slightly when the ferric iron concentration was raised from < 0.01 to 0.25 g/l, elongation decreased from 11 to 3 percent. Increasing the temperature from 90 or 95 to 105 C reduced tensile strength and ductility. (5) The addition of 0.2 g/l copper ions or 1 g/l nickel or manganese ions slightly increased tensile strength and ductility, as a rule. An increase in ferric iron concentration from 0.1 to 0.5 g/l in these solutions improved strength and ductility.

At a current density of 20 amp/sq dm, iron deposited in a bath at 80 C containing chloride and fluoborate ions exhibited a tensile strength of 71 to 105 kg/sq mm (101,000 to 149,000 psi). (9) Elongation ranged from 1.1 to 1.6 percent. These data were measured for foils with a thickness of about 25 μm.

A tensile strength as high as 109 kg/sq mm (155,000 psi) was reported for 25-μm-thick iron deposited in a sulfamate solution. (10) The deposits were slightly reduced in tensile strength, but improved in ductility when the temperature of the solution was increased from 43 to 54 C. Increasing the current density from 5.4 to 16 amp/sq dm had no effect on tensile strength, but ductility was reduced.

An increase in the pH of a sulfate bath from 2.5 to 3.5 or a reduction in current density from 10 to 5 amp/sq dm reduced tensile strength appreciably. (11) Ductility was slightly improved. By reducing temperature from 75 to 50 C, tensile strength was increased appreciably from 38 to 65 kg/sq mm (50,000 to 93,000 psi). Elongation for iron deposited in the 50 C bath was 4 percent, which was only slightly lower than the elongation for iron deposited in the 75 C solution.

Iron deposited in fluoborate or sulfate baths was slightly ductile and exhibited elongations of 1 to 7 percent. On the other hand, lower-strength iron deposited in chloride solutions elongated up to 50 percent before breaking.

Microhardness

Hardness data in Table 1 are approximately proportional to tensile strength values. The hardness of deposits with a tensile strength below 56 kg/sq mm (80,000 psi) ranged from 120 to 180 kg/sq mm. Iron with a tensile strength greater than 80 kg/sq mm (114,000 psi) exhibited hardness values from 350 to 380 kg/sq mm. Harder deposits with very little or no ductility were obtained by operating at low bath temperatures below 65 C or adding grain-refining agents such as glycerol, sugar, formic acid, acetic acid, citric acid, or ascorbic acid. The conditions for producing these hard deposits are given in Table 9.2.

A high current density in the range of 10 to 40 amp/sq dm and a low bath temperature in the range of 20 to 50 C favored high hardness. Typical deposits with a hardness of 400 to 1100 kg/sq mm contained inclusions of iron oxide or hydroxide, which were responsible for the hardening effect. (13, 18, 24, 27) The weight loss of two such deposits during heating in nitrogen at 950 C for 1.0 hour ranged from 1.0 to 1.55 percent. Density before heating was 6.3 and 7.25 g/cu cm, 20 and 8 percent below the density of high-purity, wrought iron.

Iron with a hardness of 1000 to 1100 kg/sq mm was deposited in a sulfate bath at 25 to 35 C containing citric acid. (30) These deposits contained 0.64 to 0.7 percent carbon. The carbon in iron deposits from solutions containing citric acid is believed to be combined as iron carbide. (31) A carbon content of 0.7 percent also was reported for iron deposited at 20 amp/sq dm in a 90 C, 2 N ferrous chloride bath containing glycerol and sugar. (7) A carbon content of 0.05 to 0.08 percent was found in softer iron deposited in a 90 C chloride bath containing phenol. (32)

Additions of ammonium, magnesium, potassium, or sodium sulfate increased the hardness of iron deposited in fluoborate baths, but citric acid additions had a much larger effect. (31) Superimposing alternating current reduced hardness. (33) Periodically reversing the current for 1 second after 20 seconds of direct plating increased hardness slightly in the case of iron deposited in a 35 C sulfate bath. (34) Hardness and resistance to wear were increased by periodically reversing the current for plating iron in a chloride bath. (35)

Iron with a hardness in the range of 350 to 650 kg/sq mm and deposited in a sulfate bath containing 1 g/l oxalic acid was more resistant to wear than chromium electrodeposits. (36) Resistance to wear was better for iron with a hardness of 450 to 600 kg/sq mm than for hardened steel. (14) Wear resistance was improved four- to five-fold, with iron exhibiting a hardness of 1000 to 1100 kg/sq mm, by comparison with conventional, soft iron deposits. (30) Ultrasonics improved the wear resistance of iron deposited in a chloride bath at 45 C, but induced cracks. (37)

Iron with a hardness of 390 kg/sq mm deposited at 6.4 amp/sq dm in a 43 C, 1.5 N ferrous sulfate bath containing 42 g/l ferrous chloride and 10 g/l ammonium chloride was reduced in hardness to about 180 kg/sq mm by heat treating at 650 C. (38) Iron deposited in a fluoborate bath was reduced in hardness to the range of 50 to 90 kg/sq mm by annealing at 910 C. (39) Tensile strength was reduced to the range of 25 to 33 kg/sq mm (35,000 to 47,000 psi) and elongation was increased to between 25 and 40 percent. The hydrogen content of iron deposited in a sulfate solution was reduced from 0.019 to 0.00001 percent by heating at 700 to 750 C, which also improved ductility. (40)

Stress

Iron deposited in chloride and sulfate baths are usually stressed in tension. The lowest stress value reported for iron deposited in chloride baths was 2.1 kg/sq mm (3,000 psi). (41) An addition of a reaction product of sodium naphthalene sulfonate, a high temperature > 80 C and a low current density of 2.7 amp/sq dm inhibited stress in this iron deposit. Higher current densities increased stress appreciably. For example, the tensile stress reached 31.5 kg/sq mm (45,000 psi) for iron deposited at a current density of 22 amp/sq dm in this 90 C, 3.7 N ferrous chloride bath containing the sulfonated naphthalene product.

In the absence of a stress-reducing compound, stress ranged from at least 12 to >50 kg/sq mm (17,000 to $>$ 71,000 psi), as indicated in Table 9.3. The explanation for this broad range is not obvious. However, the deposits exhibiting a rela-

TABLE 9.2

Conditions for Depositing Hard Iron Deposits

Hardness, kg/sq mm	Solution	pH	Fe^{2+} Conc., N	Supplementary Compounds	Temp., C	Current Density, amp/sq dm	Reference
400	Chloride	<1.0	1.5	200 g/l $MnCl_2$, 70 g/l NH_4Cl	65	8 to 10	12
400 to 520[(a)]	Ditto	3.0 to 5.5	1.2 to 1.5	None	25	10 to 40	13
450 to 600	"	<1.0	1.5 to 3.0	150 g/l NaCl, 100 g/l glycerin, and 40 g/l sugar	80 to 95	20 to 30	14
450 to 800	Phenol sulfonate	3.5 to 4.5	0.7	None	20 to 60	8 to 12	15
480 to 720	Sulfate	2.2	0.5	25 g/l H_3BO_3, 180 g/l urea	45	4 to 10	16
522 to 534	Chloride	<1.0	2.5	100 g/l NaCl, 80 g/l glycerol, and 40 g/l sugar	90	20	17
500 to 800[(b)]	Phenol sulfonate	—	1.2	None	20	5	18
545 to 610	Chloride	<1.0	1.0	None	50	10 to 20	19
500 to 565	Ditto	<1.0	1.0	None	70 to 90	40	19
575	Fluoborate	2.5	1.3	18 g/l H_3BO_3, 2 g/l HBF_4	20	15	20

600 to 615[c]	Sulfate	2.2 to 2.5	1.5	150 g/l K_2SO_4	20	2 to 3	21
615 to 640[c]	Ditto	2.2 to 2.5	1.5	150 g/l K_2SO_4, 4 g/l $H_2C_2O_4$	20	3 to 4	21
600 to 700	Chloride	2.0 to 4.0	3.0	50 g/l NaCl, 50 g/l $MnCl_2$, and 2 g/l $C_6H_8O_6$	20 to 40	20 to 40	22
600 to 850	Sulfosalicylate	1.8 to 2.1	2.3	None	60	—	23
700[d]	Chloride	3.0 to 6.5	0.075	100 g/l NH_4Cl	20, 50	10 to 100	24
725	Ditto	<1.0	3.0	2 g/l $C_6H_6O_6$	60	5	25
800	"	—	1.0	270 g/l KCl, 20 g/l HCO_2H	55 to 60	10	26
800 to 825[e]	"	6.5	0.3	100 g/l NH_4Cl, 5 g/l $HC_2O_2H_3$	50	—	27
800 to 1000	"	<1.0	2.5	100 g/l NH_4Cl, 100 g/l $MnCl_2$[g]	—	20 to 30	28
900	"	0.5 to 0.7	3.1 to 3.8	50 g/l $MnCl_2$	40	30 to 40	29
1000 to 1100[f]	Sulfate	5.5	1.0	50 g/l NaCl, 100 g/l $H_3C_6O_7H_5$	25 to 35	1	30

(a) The FeO + FeOH content was 3 to 5 percent.

(b) Nonmetallic inclusions were dispersed in the deposit.

(c) Increasing the bath temperature to 40 C or reducing the current density to 1 or 2 amp/sq dm reduced hardness appreciably.

(d) FeO inclusions ranged from 0.2 to 2.0 percent.

(e) The addition of acetic acid induced FeOH inclusions which hardened the deposits, which contained <94 percent iron.

(f) The deposits contained 0.64 to 0.7 percent carbon.

(g) The solution also contained 80 g/l glycerol.

TABLE 9.3

Stress Data for Iron Deposited in Chloride Baths

Stress		Fe^{2+}			Current	
kg/sq mm	psi	Conc., N[a]	Supplementary Compounds	Temp., C	Density, amp/sq dm	Reference
2.1 to 5.6	3,000 to 8,000	3.7	Sulfonated reaction product[b]	82 to 93	2.7 to 5.5	41
4.2 to 7.0	6,000 to 10,000	3.7	Ditto	88 to 93	8.0 to 11.0	41
12	17,000	1.0	100 g/l KCl, 20 g/l MnCl	20 to 60	35	42
12.5	18,000	3.5	None	40	20	43
13 to 15	18,500 to 21,500	2.5	100 g/l NaCl, 80 g/l glycerol, and 35 g/l sugar	90	20	17
20	28,500	1.0	100 g/l KCl, 20 g/l MnCl	20 to 40	5	42
20	28,500	2.0	Ditto	60	15	44
25	35,500	2.0	"	20	15	44
28	40,000	2.25	150 g/l $CaCl_2$	90	8.6	3
30	42,500	3.0	100 g/l KCl, 20 g/l MnCl	20	15	44
30	42,500	3.0	Ditto	20 to 60	5	42
31.5	45,000	3.7	Sulfonated reaction product[b]	88 to 93	22	41
40 to 50	57,000 to 71,000	1.0	None	90	10 to 20	19
59	84,000	1.0	None	70	20	19
68.4 to 69.8	98,000 to 100,000	1.0	None	50	10 to 20	19
79	112,000	2.25	100 g/l NaCl, 20 g/l $MnCl_2$	—	30	45

(a) All solutions contained at least 1.0 g/l hydrochloric acid.
(b) A condensation product of sodium naphthalene sulfonate. (41)

tively low stress (below 20 kg/sq mm) may have been relieved by cracking. A reduction in the bath temperature usually increased stress. (19, 44) An increase in the ferrous iron concentration also increased stress. (44) Stress and hardness increased proportionally as the current density was increased or the temperature reduced for depositing iron in a 1.0 N ferrous chloride bath containing 0.8 g/l hydrochloric acid. (19) Figure 9.1 shows this relationship.

Additions of glycerol, gelatin, or citric acid to 2.5 N ferrous chloride solutions containing 100 g/l sodium chloride and 1.2 to 1.4 g/l hydrochloric acid increased stress in iron deposited at 20 amp/sq ft. (46) A 65 g/l addition of dextrin shifted stress from the tensile to a compressive condition. Stressed and cracked deposits resulted from adding phenol to a ferrous chloride bath. (32)

Stress in iron deposited in sulfate baths containing 45 or 50 g/l sodium chloride ranged from 8.4 to 42 kg/sq mm (12,000 to 60,000 psi), depending on the temperature, current density, and pH. Data in Table 9.4 show that a decrease in

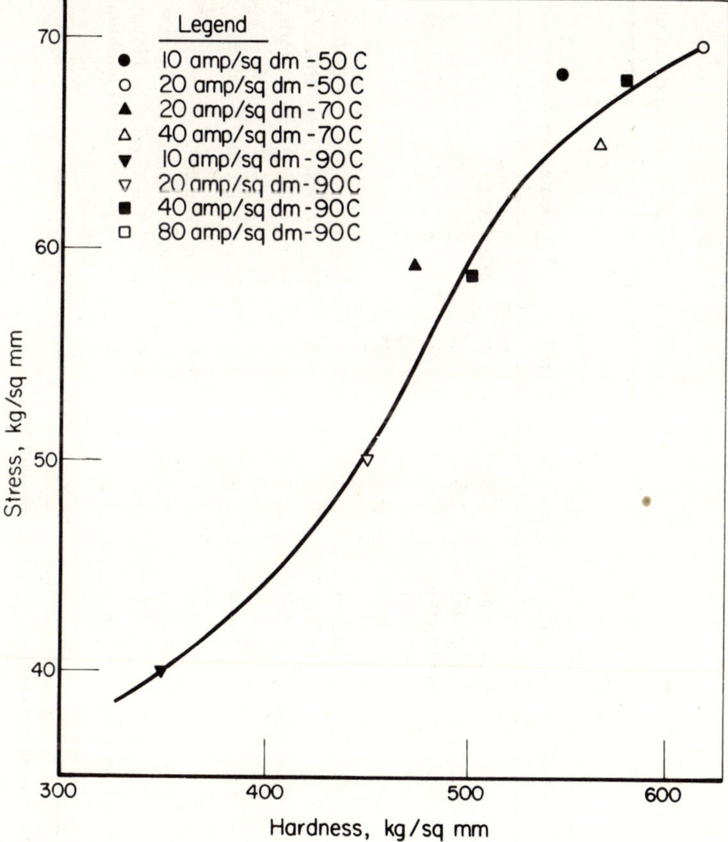

Figure 9.1. Relationship of Hardness and Stress in Iron Deposited in a 1.0 N Ferrous Chloride Bath. (19)

TABLE 9.4

Stress Data for Iron Deposited in Sulfate Solutions

Stress[a]		Fe^{2+}				Current	
kg/sq mm	psi	Conc., N	pH	Supplementary Compounds	Temp., C	Density, amp/sq dm	Reference
8.4	12,000	2.5	3.3	50 g/l NaCl	80	10	47
21.0	30,000	2.5	2.25	Ditto	80	10	
28.0	40,000	2.5	1.9	"	80	10	
42.0	60,000	2.5	2.25	"	60	10	
30.0	43,000	2.5	2.25	"	80	15	
28.0	40,000	1.2	3.0	45 g/l NaCl	20	4	48
−10.0	−14,200	1.2	3.0	45 g/l NaCl, 1.0 g/l thiourea	20	4	
3.0	4,300	1.2	3.0	45 g/l NaCl, 1.0 g/l tartaric acid	20	4	
4.0	5,700	1.2	3.0	45 g/l NaCl, 1.0 g/l citric acid	20	4	
5.0	7,100	1.2	3.0	45 g/l NaCl, 0.1 g/l glycine	20	4	
38	54,000	1.2	2.9	45 g/l NaCl	75	6	49
−1.0	−1,400	1.2	2.9	45 g/l NaCl, 1.0 g/l arginine	75	6	
8.0	11,400	1.2	2.9	45 g/l NaCl, 1.0 g/l amidopyrine	75	6	
1.0	1,400	1.2	2.9	45 g/l NaCl, 1.0 g/l naphthalene sulfonic acid	75	6	

[a] A negative sign indicates compressive stress.

temperature from 80 to 60 C or a reduction in the pH from 3.3 to 1.9 increased stress appreciably. (47) An increase in current density also increased stress. Additions of tartaric or citric acid reduced stress. (48) Thiourea (48) and arginine (49) additions shifted stress from tensile to compressive values. Deposits produced with thiourea and other organic additives contained sulfides and carbides and were not ferromagnetic. (48)

Deflections due to high tensile stresses were reported for iron deposited in a fluoborate bath with a pH of 2.0. (50) This stress was similar to that for iron deposited in a chloride bath. However, 8-mil-thick iron deposited in the fluoborate bath with a pH of 2.8 exhibited a compressive stress. The 0.63 to 0.85 N ferrous iron bath contained 1.1 g/l ferric iron and 3.8 g/l sodium chloride. It was operated at a temperature of 60 C. The current density was 3 amp/sq dm.

The procedure for measuring stress was not reported by some investigators, but others measured the deflection of a flexible strip or helix to calculate stress. In at least one investigation, the deflection was directly proportional to electroplate thickness, up to a thickness of about 60 μm. (47) Hence stress was independent of thickness.

A reduction in the impact strength of a 0.3 percent carbon steel from 20 to 11 kg-m/sq cm was reported as a result of depositing 1.0 mm of iron on the steel specimen. (51) A fatigue strength of only 18 kg/sq mm (25,500 psi) was reported for a normalized steel plated with about 0.2 mm iron at 20 amp/sq dm in a 3.5 N ferrous chloride bath operated at 40 C. (52) An addition of molybdenum disulfide to the plating bath purportedly reduced stress but had only a minor effect on the fatigue strength of the iron-plated steel.

Magnetic Properties of Iron and Iron Alloys

Coercivity values for iron electrodeposits, which range from 15 to 40 oersteds, were reduced to < 10 oersteds by codepositing 3 to 5 percent nickel and to 5 oersteds by codepositing 8 percent nickel. Coercive force is the demagnetizing force that must be applied to reduce induction to zero. Low coercivity is desired for soft, fast-switching memory devices. Data for iron and iron-nickel alloy films are given in Table 9.5. Heat treatment at temperatures < 300 C decreased the coercivity of thin, iron-nickel films deposited in a sulfate bath, which was attributed to the evolution of hydrogen gas and relief of stress. (64) Heat treatment at higher temperatures increased coercivity due to the formation of the Ni_3Fe phase.

The ratio of remanent induction (B_r) and maximum induction (B_m) for the iron-nickel alloys ranged from 0.77 to >0.95, which also is desirable for fast switching. The saturation induction of 10,000-angstrom-thick iron deposited in a ferrous ammonium sulfate bath, measured by the deflection of an electron beam, was 30,000 gausses, (65) which corresponded to values reported previously.

Iron alloy containing 13.5 percent phosphorus, described as a metastable solid solution of an amorphous character, exhibited lower coercivity than iron containing less phosphorus. (66)

TABLE 9.5

Magnetic Property Data for Iron and Iron-Nickel Alloys

Codeposited Metal	Coercivity, oersteds	Squareness of Hysteresis Loop (B_r/B_m)	Fe^{2+} Conc., N[(a)]	pH	Supplementary Compounds	Temp., C	Cathode Current Density, amp/sq dm	Reference
None	20	No data	1.5	1.0	238 g/l $CaCl_2$	80	—	53
None	30	Ditto	1.5	0.9 to 1.5	180 g/l $CaCl_2$	65 to 90	3.5	54
None	15 to 40	"	—	2.8	15–18 g/l H_3BO_3 and a wetting agent	20	0.6	55
Nickel, 1–2%	16.5 to 18	"	1.5	1.0	238 g/l $CaCl_2$	80	—	53
Nickel, 2%	15	"	1.5	0.9 to 1.5	180 g/l $CaCl_2$	65 to 90	3.5	54
Nickel, 3%	1 to 10[(b)]	0.8 to 0.9	1.5	0.3	228 g/l $CaCl_2$	—	2.2 to 9.0	56
Nickel, 5%	8 to 10	>0.95	1.85	1.0	220 g/l $AlCl_3 \cdot 6H_2O$	65	2.1	57
Nickel, 5%	8 to 9	No data	—	—	—	—	—	58
Nickel, 8%	5	Ditto	1.5	0.9 to 1.5	180 g/l $CaCl_2$	80	3.5	54
Nickel, 25%	5[(c)]	"	0.02[(d)]	3.1	218 g/l $NiCl_2 \cdot 6H_2O$, 25 g/l H_3BO_3, 9.7 g/l NaCl	20	5.0	59
Nickel, >50%	4.5[(e)]	0.77	0.04[(d)]	3.7	280 g/l $NiSO_4 \cdot 7H_2O$, 30 g/l H_3BO_3	20	1.0	60
Nickel, >50%	3.6	—	0.1[(d)]	>8.0	384 g/l $K_4P_2O_7 \cdot 3H_2O$, 20 g/l $(NH_4)_3C_6H_5O_7 \cdot H_2O$	—	—	61
Nickel, >50%	1 to 8	—	0.04[(d)]	2.2	250 g/l $NiSO_4 \cdot 7H_2O$, 25 g/l H_3BO_3, 25 g/l $H_3C_6H_5O_7$, 0.8 g/l saccharin	24	2.5	62
Nickel, 5% and copper, 11%	10	0.9	1.6	0.8 to 1.0	180 g/l $CaCl_2$	75	3.2	63

(a) Chloride baths, unless otherwise noted.

(b) Variations of coercive force in this range depended on pH and current density for 1- to 10-μm-thick films. Coercivity decreased as thickness increased from 1 to 10 μm. Thinner films exhibited a higher coercivity, to 50 oersteds.

(c) Coercivity was reduced from 5 to 1 oersteds when 0.6 g/l saccharin was added to the bath.

(d) Ferrous sulfate was the iron compound.

(e) Coercivity was reduced to between 2 and 3 oersteds by adding one of several organic addition agents.

As noted previously, iron deposited in solutions containing thiourea, which contained carbon and sulfur, were not ferromagnetic. (48)

Mechanical Properties of Iron Alloys

Hardness data are listed in Table 9.6 for iron alloys of chromium, nickel, tungsten, and zinc. Codeposition of tungsten produced the hardest alloys, which increased in hardness to > 1300 kg/sq mm as a result of heat treatment at 600 C. (75,76) The heat treatment precipitated a secondary phase from the supersaturated solid solution. The iron-tungsten alloys exhibited no ductility. Young's modulus of elasticity for iron alloy containing 56 percent tungsten was 16,000 kg/sq mm (23,000,000 psi). (76)

Iron alloys containing nickel and/or chromium exhibit good corrosion resistance, but their ability to protect a corrodible metal substrate is limited because of high stress and a tendency to crack. Stress values listed in Table 9.6 range from 19 to 55 kg/sq mm (27,000 to 78,000 psi). Cracks in an iron-chromium-nickel alloy were no longer observed after heat treating at 1000 C, however. (69)

Iron-nickel alloy deposits also are highly stressed, from a minimum of 14 to 38.5 kg/sq mm (20,000 to 48,000 psi). However, a 1.0 g/l saccharin addition to a bath operated with ultrasonic agitation purportedly reduced the tensile stress to a low value. (78)

By incorporating 1.5 to 7.25 percent of corundum particles in 100 μm iron electrodeposits, hardness was increased in proportion to the amount of occluded particles. (79) A temperature of 40 C was better than higher temperatures and a current density of 60 amp/sq dm was better than lower current densities for maximizing the proportion of corundum particles. These dispersion-hardened electrodeposits were produced in a 1.35 N ferrous chloride bath containing 100 g/l manganese chloride and 100 g/l corundum powder.

Stress in iron deposited in a 3.5 N ferrous chloride bath at 40 C was reduced from 12 to 9 kg/sq mm (17,000 to 13,000 psi) by incorporating molybdenum disulfide particles. (43) Either molybdenum disulfide or boron carbide particles decreased hardness, whereas aluminum oxide particles increased hardness. These effects were attributed to the influence of the occluded particles on the structure of the electrodeposited iron.

Structure

Microscopic examinations of iron deposited in chloride, chloride-fluoborate, or sulfate-chloride solutions containing no organic compounds show coarse, columnar or fibrous structures, depending on the current density and the pH and temperature of the plating solution. Current densities above about 10 amp/sq dm favored a fibrous structure for iron deposited in chloride baths operated at 85 C. (41) Although a shift in current density from 5 to 20 amp/sq dm had little influence on the

TABLE 9.6

Hardness and Stress Data for Iron Alloys

Codeposited Metal	Hardness, kg/sq mm	Stress kg/sq mm	Stress psi	Type of Plating Bath	pH	Temp., C	Cathode Current Density, amp/sq dm	Reference
Chromium, 27 to 35%	600	—	—	Sulfate[a]	2.4	20	7 to 12	67
Chromium, 40 to 45%	—	19 to 43[b]	27,000 to 61,000	Sulfate-Sulfamate[c]	1.8	30	11.6	68
Chromium, 18% and nickel, 9%	371[d]	—	—	Sulfate[e]	1.7	25	15	69
Ditto	580 to 600	—	—	Sulfate[f]	0.3	30	25	70
Chromium, 35% and nickel, 12%	—	30 to 55[g]	42,000 to 78,000	Sulfate[h]	1.8	30	18.6	68
Chromium, 35% and nickel, 5%	650	—	—	Sulfate[i]	2.1	20	5 to 12	71
Nickel, 1 to 2%	—	22.5 to 35	32,000 to 50,000	Chloride[j]	1.0	80	—	53
Ditto	—	14.0 to 34	20,000 to 48,000	Ditto	1.0	80	5.0	72
Nickel, 1.6%, etc.[k]	470 to 585	—	—	Chloride[l]	—	85	10 to 35	73
Phosphorus	—	10 to 12	14,000 to 17,000	Chloride[m]	<1.0	80	25	74
Tungsten, 43%	900[n]	—	—	Sulfate[o]	—	—	—	75
Tungsten, 56%	840 to 970[p]	—	—	Sulfate[q]	5.5	90	3 to 6	76
Zinc, 6%	560	—	—	Sulfate[r]	1.7	50	20	77
Zinc, 60%	350	—	—	Sulfate[s]	1.7	50	18	77

(a) The solution contained 160 g/l $Cr_2(SO_4)_3 \cdot 6H_2O$, 30 to 50 g/l $FeSO_4 \cdot 7H_2O$, and 150 g/l NH_2CH_2COOH.

(b) Stress varied with the nature of the substrate. Adding 1.2 g/l saccharin reduced stress to 15 kg/sq mm (21,500 psi).

(c) The solution contained 500 g/l $CrAl(SO_4)_3 \cdot 12H_2O$, 25 g/l $FeSO_4 \cdot 7H_2O$, 40 g/l H_3BO_3, 9 g/l NaF, and 85 g/l $NH_4NH_2SO_3H$.

(d) After heat treating at 1000 C, hardness was reduced to 318 kg/sq mm and cracks were no longer observed.

(e) The solution contained 400 g/l $CrK(SO_4)_2 \cdot 12H_2O$, 56 g/l $NiSO_4 \cdot 6H_2O$, 70 g/l $H_3C_6H_5O_7$, and 8 g/l NaF.

(f) The solution contained 150 to 200 g/l $CrCl_3$, 25 to 50 g/l $NiCl_2$, 12 to 20 g/l $FeCl_2$, 60 g/l $Na_3C_6H_5O_7$, and 130 g/l $AlCl_3$.

(g) Stress varied with the nature of the substrate. The addition of 1.2 g/l saccharin reduced stress in iron alloy on brass from 45 to 13 kg/sq mm (64,000 to 18,500 psi).

(h) The solution was prepared with metal sulfates and contained 45 g/l Cr, 15 to 20 g/l Ni, 5 to 8 g/l Fe, 100 g/l $Na_3C_6H_5O_7$, 60 g/l $(NH_4)_2SO_4$, 30 g/l H_3BO_3, 7 to 10 g/l glycine, and 7 g/l NaF.

(i) The solution contained 2 M $Cr_2K_2(SO_4)_4$, 0.4 M $FeSO_4$, 0.6 M $NiSO_4$, and 2.0 M $NH_2CH_2CO_2H$.

(j) The solution contained 285 g/l $FeCl_2 \cdot 4H_2O$, 6 to 12 g/l $NiCl_2 \cdot 6H_2O$, and 238 g/l $CaCl_2 \cdot 2H_2O$.

(k) The alloy also contained 2.5% Co, 1.5 to 2.0% Mn, 1.4% Cr, and 0.5% C.

(l) Chloride salts of the metals were supplemented with 100 to 150 g/l NH_4Cl, 0.2 to 0.4 g/l gelatin, 40 to 70 g/l dextrin, and 6 to 10 g/l $H_3C_6H_5O_7$.

(m) The solution contained 225 g/l $FeCl_2$, 1.5 g/l $NaH_2PO_2 \cdot H_2O$, and 1.0 to 1.2 g/l HCl.

(n) Heat treating at 600 C for an hour increased hardness to about 1300 kg/sq mm.

(o) The solution contained 70 g/l $Na_2WO_4 \cdot 2H_2O$, 13 g/l $FeSO_4 \cdot 7H_2O$, and 96 g/l $(NH_4)_2C_4H_4O_6$.

(p) Heat treatment at 600 C for an hour increased hardness to 1260–1460 kg/sq mm. Ductility was zero before and after heat treatment.

(q) The solution contained 0.14 M $FeSO_4$, 0.4 M Na_2WO_4, 1.0 M NH_4Cl, and 0.3 to 0.6 M $Na_3C_6H_5O_7$.

(r) The solution contained 248 g/l $FeSO_4 \cdot 7H_2O$, 8.8 g/l $ZnSO_4 \cdot 7H_2O$, 118 g/l $(NH_4)_2SO_4$, 10 g/l KCl, and 0.5 g/l $H_3C_6H_5O_7$.

(s) The solution contained 174 g/l $FeSO_4 \cdot 7H_2O$, 88 g/l $ZnSO_4 \cdot 7H_2O$, 118 g/l NH_4Cl, 10 g/l KCl, 0.5 g/l $H_3C_6H_5O_7$, and a wetting agent.

Figure 9.2. Fibrous Structure of Iron Deposited at 20 amp/sq dm in a 2.5 N Ferrous Sulfate Bath Containing 30 g/l Chloride Ions. (80)

structure of iron deposited in a sulfate-chloride bath, decreasing the temperature from 80 to 60 C or the pH from 3.3 to 1.9 refined grain size. (80) Changing the chloride ion concentration from 10 to 30 g/l had no appreciable effect on the structure. The fibrous deposit obtained with a current density of 20 amp/sq dm in the 80 C, 2.5 N ferrous sulfate bath with a pH of 2.25 is shown in Figure 9.2. In another sulfate bath containing no chloride ions, grain size also decreased with decreasing temperature or pH. (81) Superimposing alternating on direct current increased grain size and reduced stress and hardness. (33)

The pH of a chloride-fluoborate bath influenced structure and mechanical properties appreciably. (82) A fine-grained fibrous structure was observed when the pH of the bath was adjusted to 3.6. Tensile strength and ductility were highest when solutions maintained at a temperature of 68 to 80 C were operated with a pH of 3.6. Increasing the pH to 4.0 enlarged the grains and reduced tensile strength and ductility. The reduction in ductility observed with the increased pH was attributed to inclusions of ferric hydroxide, identified in the structure of typical deposits produced at a pH of 4.0. Although only minor structural changes were observed when the pH was shifted from 3.6 to 3.2, large reductions in strength and ductility were reported. Low values for tensile strength and ductility in this case were attributed to high stress and codeposited hydrogen.

Inclusions of iron oxide or hydroxide have been observed in many iron electrodeposits from chloride and sulfate-chloride baths. Yet few or no such inclusions were noted in iron deposited at 5 amp/sq dm in a 3.75 N ferrous chloride solution containing 5 g/l magnesium chloride and 0.7 g/l hydrochloric acid. (28)

An initial decrease in grain size of iron deposited in chloride or sulfate solutions during 20 hours of storage after plating was followed by grain growth during a subsequent 2100-hour period. (83) This grain growth was associated with a decrease in microhardness and a 50 to 80 percent increase in electrical resistivity. Agglomeration of inclusions at grain boundaries may be the explanation for the increase in resistivity.

The structure of iron deposited in a ferrous chloride bath containing sugar and heat treated at 90 C consisted of martensite and troostite. (7) The deposit contained 0.7 percent carbon. Iron deposited in solutions containing hydroxybenzene sulfonate and heat treated at 700 to 800 C contained Fe_3C, FeO_2, and FeS_2, identified by X-ray examination. (84)

Supersaturated solid solutions were identified for several alloys of iron, including manganese (up to 12.3 percent), (85) tungsten (up to about 50 percent), (86,87) and zinc (up to 25 percent). (88) Gelatin additions to the iron-manganese alloy plating bath introduced laminations in the structure of the alloy deposit. (85)

Iron alloys containing 8 or 18 percent nickel exhibited a body-centered-cubic lattice structure, whereas alloy containing 31 percent nickel contained a mixture of face-centered-cubic and body-centered-cubic structures. (89) With 52 to 80 percent nickel, only the face-centered-cubic structure was detected.

The lattice structure of some iron-cobalt alloys deposited in chloride-pyrophosphate solutions differed somewhat from the structure found for alloys deposited in sulfamate solutions, as follows:

	Lattice Structure	
Cobalt Content, percent	Chloride-Pyrophosphate Solution (90)	Sulfamate Solution (91)
14 to 49	bcc	bcc
72	bcc + hcp	bcc
84	hcp	bcc
100	hcp	hcp

Heating iron-phosphorus alloy for 15 minutes at 400 C precipitated an iron-phosphide phase from the supersaturated solid solution observed before heat treating. (92)

References

(1) Kasper, C., "Notes on the Rapid Electrodeposition of Iron," *Monthly Review, Am. Electroplaters' Soc., 23* (10), 34–36 1936; "Rapid Electrodeposition of Iron From Ferrous Chloride Baths," *Research Paper RP991, Journal of Research, National Bureau of Standards, 18*, 535–541, (1937).

(2) Thomas, C. T., and Blum, W., "The Production of Electrolytic Iron Printing Plates," *Trans. Electrochem. Soc., 57*, 59–77 (1930).

(3) Diggin, M. B., "Modern Electroforming Solutions and Their Applications," Symposium on Electroforming—Applications, Uses and Properties of Electroformed Metals, *ASTM Special Technical Publication No. 318*, 10–31 (1962).

(4) Larson, I., Moulton, R. W., and Putnam, G. L., "Physical Properties of Iron Deposited From Chloride Baths," *Journal Electrochem. Soc. 95* (4), 86C–91C (1949).

(5) Max, A. M., and Van Houten, G. R., "Some Characteristics of Electroformed Iron Deposits," *Proceedings Am. Electroplaters' Soc., 43*, 136–150 (1956).

(6) Melkov, M. P., "Restoration of Machine Parts by Electrodeposition of Steel," *Vestnik Mashinostroeniya, 33* (3), 50–54 (1953).

(7) Petrov, Yu., "Investigation of Quality of Steel Coating Obtained by Plating," *Avtomobil, 29* (2), 31-34 (1951).

(8) Stoddard, W. B., Jr., "Iron Plating," *Trans. Electrochem. Soc., 84*, 305–318 (1943).

(9) Levy, E. M., and Hutton, G. J., "The Electrodeposition and Properties of High Strength Iron Foils Produced From a Fluoborate Bath," *Plating, 55* (2), 138–145 (1968).

(10) Read, H. J., "Prepared Discussion," *Plating, 55* (2), 145-147 (1968).

(11) Kano, T., "Mechanical Properties of Electrodeposited Iron From Sulfate Bath—Conditions of Electrodeposition," *Journal of the Metal Finishing Society of Japan, 18* (3), 81–85 (1967).

(12) Revyakin, V. P., and Myasnikov, P. D., "Iron Plating Machine Parts," *Vestnik Mashinostroeniya, 37* (4), 64-65 (1957).

(13) Trask, H. V., "Electrolytic Process for Producing Iron Products, Process for Producing Brittle Iron Plate, and Electrolytically Deposited Iron Products," U.S. Patents 2,538,990, 2,538,991 and 2,538,992 (1950). Assigned to Buel Metals Company.

(14) Petrov, Yu. N., "Structure and Wear Resistance of Electrodeposited Ferrous Coatings. Effect of Electrolysis Conditions," *Metal Finishing, 57* (10), 72–73 (1959). Translated from *Metallovedenie i Obrabotka Metallov, No. 12*, 53–56 (1958). Zakirov, Sh. Z., and Liedsky, V. B., *Zavodskaia Laboratoria (Plant Laboratory), No. 10* (1955).

(15) Chalaganidze, Sh. I., "Investigation of Iron Plating From a Phenolsulfonate Electrolyte to Restore Auto-Tractor Parts," *Trudy Gruzin. Nauch.-Issledovatel. Inst. Mekhaniz. i Elektrifikatsii Sel'sk. Khoz., No. 4*, 267–277 (1958). Russian Summary.

(16) Kawasaki, M., and Mizumoto, S., "Iron Plating From Sulfamate Baths. Studies on Iron Plating (Part I)," *Journal of the Metal Finishing Society of Japan, 13* (11), 464–467 (1962).

(17) Zakirov, Sh. Z., "Wear Resistance of Iron Plate From a Chloride Bath Containing Organic Additions (Suitable to the Repair of Work)," *Doklady Akad. Nauk Tadzhik. SSR, 20*, 83–86 (1957).

(18) Shishakov, N. A., and Chalaganidze, Sh. I., "Origin of the High Hardness of Electrolytic Iron Deposits," *Russian Journal of Physical Chemistry, 40* (7), 783–785 (1966); *Zh. Fiz. Khimii, 40* (7), 1440–1443 (1966).

(19) Shiryaev, A. M., and Klyushkin, I. E., "Correlation of the Properties of Hard Iron Electrodeposits," *Russian Journal of Physical Chemistry, 37* (12), 1441–1443 (1963); *Zh. Fiz. Khimii, 37* (12), 2663–2667 (1963).

(20) Gindlin, V. K., and Moseev, N. P., "Electrolytic Deposition of Iron From Fluoroborate Electrolytes," *Poligr. i Izd. Delo, Mezhduved. Resp. Nauchn.-Tekhn. Sb., No. 1*, 13–20 (1964).

(21) Vagramyan, A. T., and Solov'eva, Z. A., *Technology of Electrodeposition*. English Translation, Robert Draper Ltd., Teddington (1961), 398 pp.

(22) Petrov, Yu. N., "Chromium Plating and Iron Plating in Maintenance Work," *Mashinostr., Nauchn.-Tekhn. Sb., No. 5*, 58–60 (1964).

(23) Petrov, Yu. N., and Dushevskii, I. V., "Cold Organic Electrolytes for Iron Plating," *Tr. Kishinevsk. Sel'skokhoz. Inst., 33* (2), 31–38 (1964).

(24) Wranglén, G., "Hydroxide Content, Hardness, Internal Stress, and Cathodic Polarization of Electrodeposited Iron," *Svensk Kem. Tid., 61*, 100–113 (1949).

(25) Petrov, Yu. N., "Effect of Electrolysis Conditions on the Resistance of a Chloride Electrolyte to Oxidation and on the Lamination of Iron Coatings," *Tr. Kishinev. Sel'skokhoz. Inst., 40*, 179–184 (1966).

(26) Solov'ev, N. A., "Electrolyte for Electrodeposition of Iron," USSR Patent 133,724 (November 25, 1960).

(27) Wranglén, G., "Effect of Acetic Acid on the Structure and Properties of Electrodeposited Iron," *Svensk Kem. Tid., 61*, 246–252 (1949).

(28) Blasiak, E., Piszczek, L., Korczynski, A., and Lekki, J., "Regenerative Iron Coatings," *Zeszyty Nauk. Politech. Slask., Chem. No. 23*, 21–34 (1964).

(29) Nitsa, N. K., "Restoration of Worn-Out Parts, Such as Shafts and Axles, by Cold Iron Jet Plating," *Tr. Kishinev. Sel'skokhoz. Inst., 40*, 230–235 (1966).

(30) Gogish-Klushin, Yu. V., and Gavrilova, E. F., "Electrolyte for Deposition of Steel," USSR Patent 132,474 (October 5, 1960).

(31) Gogish-Klushin, Yu. V., "Investigation of Iron Electrolytes for Electroplating Articles of Complex Shape," *Sbornik Trudov, Vsesoyuz. Nauch.-Issledovatel. Inst. Goznaka (Upravlenie Proizvodst. Gosudarst. Znakov, Monet i Ordenov), No. 1*, 199–222 (1957).

(32) Jansons, V., Ceske, E., and Timšans, S., "Preparation of Deposits With Increased Hardness by Electrolytic Hardening," *Latvijas Lauksaimniecibas Akad. Raksti, No. 5*, 97–106 (1956).

(33) Popereka, M. Ya., and Avramenko, V. I., "Electrolytic Deposition of Iron With Alternating Current," *Soviet Electrochemistry, 2* (8), 900–903 (1966); *Elektrokhimiya, 2* (8), 971–972 (1966).

(34) Mukai, M., and Ootake, M., "Electrolytic Refining of Iron. XV. Effect of PR Process Applied to Iron Electrodeposition," *Journal of the Metal Finishing Society of Japan, 7*, 54–58 (1956).

(35) Petrov, Yu. N., "Chromium Plating and Iron Plating of Metallic Surfaces for Increasing Their Wear Resistance," *Elektronnaya Obrabotka Materialov, Akad. Nauk Moldavsk. SSR, No. 4*, 21–28 (1966).

(36) Kudryavtsev, N. T., and Yakovleva, L. D., "Electrodeposition of Iron at High Current Densities From Sulfate Electrolytes at Low Temperature," *Zashchitno-Dekorativ. i Spetsial. Pokrytiya Metal., Nauch.-Tekh. Obshchestvo Mashinostroitel. Prom.*, 81–86 (1959).

(37) Petrov, Yu. N., Ginberg, A. I., Granovskii, Yu. V., and Zaidman, G. N., "Wear Resistance and Submicrostructure of Electrolytic Iron Obtained in an Ultrasonic Field," *Elektronnaya Obrabotka Materialov, No. 3*, 45–48 (1968).

(38) Schaffert, R. M., and Gonser, B. W., "A Sulfate-Chloride Solution for Iron Electroplating and Electroforming," *Trans. Electrochem. Soc., 84* (1943), 15 pp. (preprint).

(39) Spencer, Lester F., "Engineering Uses of Plated Coatings," *Metal Finishing, 57* (5), 48–54 (1959).

(40) Kadaner, L. I., and Medyanik, V. N., "Relation Between Plasticity of Electrolytic Deposits and Their Hydrogen Content," *Navodorozhivanie Metal. Bor'ba Vodorodn. Khrupkost'yu*, 182–183 (1968).

(41) Thomas, J. D., Klingenmaier, O. J., and Hardesty, D. W., "Iron Plating Process and Product Characteristics," *Trans. Inst. Metal Finish., 47* (5), 209 (1969).

(42) Domnikov, L., "Iron Coatings From Chloride Solutions. Laminated Structure and Internal Stresses," *Metal Finishing, 63* (5), 53–55 (1965).

(43) Petrov, Yu. N., Mamontov, V. A., Gur'yanov, G. V., and Rybkovskii, V. Ya., "Effect of Nonmetallic Inclusions on Some Physicomechanical Properties of Electrolytic Iron Coatings," *Elektronnaya Obrabotka Materialov, Akad. Nauk Moldavsk. SSR, No. 1*, 41–45 (1967).

(44) Peregudov, F. M., and Kadaner, L. I., "Stratification and Internal Stresses in Deposits Obtained From Chloride Electrolytes for Iron Plating," *Journal of Applied Chemistry (USSR), 35* (12), 2523–2526 (1962); *Zh. Prikl. Khimii, 35* (12), 2624–2628 (1962).

(45) Dawidowicz, J., and Kocanda, S., "Internal Stresses in Hard-Iron Layers," *Biul. Wojskowej Akad. Tech., 14* (1/149), 57–69 (1965).

(46) Zakirov, Sh. Z., "Effect of Internal Stresses in Electrolytic Iron Coatings on Their Hardness and Wear Resistance," *Trudy Tadzhik. Sel'skokhoz. Inst. (Trans. of the Tadzhik Agricultural Institute), 2*, 27–34 (1958).

(47) Levy, E. M., "Iron Electrodeposited From a Sulfate-Chloride Electrolyte: Effect on Internal Stress of Current Density, Temperature, pH, and Chloride Ion Concentration," *Plating, 55* (9), 941–945 (1968).

(48) Popereka, M. Ya., and Koshmanov, V. V., "Effect of Organic Additives in Electrolytes on Internal Stresses in Electrodeposited Iron," *Soviet Materials Science, 2* (6), 453–456 (1966); *Fiziko-Khimicheskaya Mekhanika Materialov, 2* (6), 641–645 (1966).

(49 Popereka, M. Ya., and Avramenko, V. I., "Influence of Organic Additives and Alternating Current on the Cathodic Polarization and Internal Stresses in Electrodeposited Iron," *Protection of Metals, 3* (2), 197–200 (1967); *Zashchita Metallov, 3* (2), 239–242 (1967).

(50) Farmer, M. E., West, D. C., and Darlington, C. G., "Iron Plating From the Fluoroborate Bath Without Organic Additives," *Plating, 56*, 699–704 (1969).

(51) Tashkin, A. E., "Increase in the Surface Hardness and Wear Resistance of Low-Carbon Steel by Electrolytic Iron Plating," *Vestnik Mashinostroeniya, 49* (2), 20–21 (1969).

(52) Andreichuk, V. K., and Gur'yanov, G. V., "Effect of Molybdenum Disulfide Additives on the

Fatigue Strength of Parts With Electrolytic Iron Coatings," *Elektronnaya Obrabotka Materialov, Akad. Nauk Moldavsk. SSR, No. 6*, 68–69 (1969).

(53) Kolk, A., and White, H., "Magnetic Properties of 97 Fe-3 Ni Thin Film Electroplate," *Journal Electrochem. Soc., 110* (2), 98–103 (1963).

(54) Tsu, I., "The Preparation and Properties of Iron-Nickel Films," *Plating, 47* (6), 632–633 (1960).

(55) Zentner, V., "Electrodeposited and Electroless Magnetic Alloys for Computers," *Plating, 52* (9), 868–872 (1965); *J. Intern. Applic. Cobalt, Brussels, Belgium*, 152 (June 1964).

(56) Lems, W., "Magnetic Properties of Electrodeposited FeNi Films. I," *Philips Research Reports, 22* (4), 388–401 (1967).

(57) Underwood, J. D., "An Electroplated 95% Iron 5% Nickel Thin Film for the Destructive Readout Cubic Waffle Iron Store," *Trans. Inst. Metal Finish., 48*, Part 1 (Spring, 1970), 9 pp.

(58) Wolf, I. W., "Electrodeposition of Magnetic Materials," *Journal Applied Physics, Supplement, 33* (3), 1152–1159 (1962).

(59) Smith, R. S., Godycki, L. E., and Lloyd, J. C., "Effects of Saccharin on the Structural and Magnetic Properties of Iron-Nickel Films," *Journal Electrochem. Soc., 108* (10), 996–998 (1961).

(60) Il'yushenko, L. F., and Katsyuk-Kul'gauchuk, L. P., "Decrease in Coercive Force and Increase in Squareness Coefficient of Hysteresis Loops of Electrolytically Deposited Iron-Nickel Films. I," *Vestsi Akad. Navuk Belorus. SSR, Ser. Fiz.-Mat. Navuk, No. 1*, 114–119 (1967).

(61) Kaznachei, B. Ya., Balashova, N. N., Shuvalova, M. A., and Belyaeva, L. A., "Effect of Composition of Pyrophosphate Electrolytes and Conditions of Electrodeposition on the Magnetic Properties of Nickel-Iron Alloys," *Tr. Vses. Nauchn.-Issled. Inst. Magnitn. Zapisi i Tekhnol. Radioveshch. i Televid., No. 1*, 168–179 (1964).

(62) Elmanova, V. A., "Magnetic Properties of Electrochemical Films," *Izv. Akad. Nauk SSSR, Ser. Fiz., 31* (3), 445–447 (1967).

(63) Reekstin, J. F., Jr., "Magnetically Isotropic Iron-Nickel-Copper Alloy Electroplated Films for 'Waffle-Iron' Type Memory Devices," U.S. Patent 3,348,931 (October 24, 1967). Assigned to Bell Telephone Laboratories, Inc.

(64) Lubecka, M., and Schabowska, E., "The Technology and the Magnetic Properties of Electrodeposited Iron-Nickel Thin Films," *Zesz. Nauk. Akad. Gorn.-Hutn. Krakowie, Met. Odlew., 14*, 11–25 (1967).

(65) Yamaguchi, S., "Magnetism of the Electrodeposited Films as Revealed by Electron Diffraction," *Journal Electrochem. Soc., 107* (1), 55–56 (1960).

(66) Bondar, V. V., Gorbunova, K. M., and Polukarov, Yu. M., "Magnetic Properties of Amorphous Layers of Electrodeposited Iron- and Cobalt-Base Alloys," *Fizika Metallov i Metallovedenie, 26* (3), 568 (1968).

(67) Kudryavtsev, N. T., and Smirnova, T. G., "Electrolytic Deposition of Iron-Chromium Alloys," *Tr. Mosk. Khim.-Tekhnol. Inst., No. 44*, 102–107 (1963).

(68) Cleghorn, W. H., Gowri, S., Elsie, P. L., and Shenoi, B. A., "Stress in Electrodeposited Alloys. Iron-Chromium and Iron-Chromium-Nickel," *Metal Finishing, 67* (8), 65–70 (1969).

(69) Domnikov, L., "Iron-Nickel-Chromium Baths. Deposition of Stainless Steel," *Metal Finishing, 62* (3), 61–65 (1964). Kalyuzhnaya, P. F., and Pimenova, K. N., *Zh. Prikl. Khimii, 35* (5), 1057 (1962).

(70) Rotinyan, A. L., Zytner, L. A., and Fedot'ev, N. P., "Iron-Chromium-Nickel Ternary Alloy Electroplating From a Chloride Electrolyte," *Zh. Prikl. Khimii, 41* (10), 2201–2207 (1968).

(71) Kudryavtsev, N. T., Smirnova, T. G., and Volkova, O. P., "Electrodeposition of a Khl8N9-Type Iron-Nickel-Chromium Alloy," *Zh. Prikl. Khimii, 41* (8), 1873–1876 (1968).

(72) Watkins, H., and Kolk, A., "Measurement of Stress in Very Thin Electrodeposits," *Journal Electrochem. Soc., 108* (11), 1018–1023 (1961).

(73) Lekhikoĭnen, M. M., "Investigation of the Conditions of Electrodepositing Iron Plate With Ni, Cr, Co, and Mn Contents," *Trudy Tadzhik. Sel'skokhoz. Inst., 2*, 19–25 (1958).

(74) Lasas, A., "Iron-Phosphorus Coatings," *Mokslas ir Tech., No. 8*, 30 (1965).

(75) Yoshioka, S., Yamamoto, H., and Omi, T., "Electrodeposition of Tungsten-Iron Alloys. I. Improved Process of Electrodeposition of Tungsten-Iron Alloys," *Kinzoku Hyomen Gijutsu, 19* (3), 95–99 (1968).

(76) Brenner, A., Burkhead, P., and Seegmiller, E., "Electrodeposition of Tungsten Alloys Containing Iron, Nickel, and Cobalt," *Research Paper No. 1834, Journal Research National Bureau of Standards, 39* (4), 351–383 (1947).

(77) Jepson, S., Meecham, S., and Salt, F. W., "The Electrodeposition of Iron-Zinc Alloys," *Trans. Inst. Metal Finish., 32*, 160–180 (1955).

(78) Ginberg, A. M., Granovskii, Yu. V., Titov, Yu. E., and Klyachko, Yu. A., "Choosing Optimum Electrodeposition Conditions by the Box-Wilson Method," *Zashchita Metallov, 1* (6), 716–718 (1965).

(79) Konzhina, T. P., Saifullin, R. S., and Skalozub, M. F., "Combined Electrochemical Iron-Corundum Coatings," *Blestyashchie Komb. Metal. Pokrytiya, No. 2*, 88–92 (1967).

(80) Levy, E. M., MacInnis, R. D., and Copps, T. P., "Iron Electrodeposited From a Sulfate-Chloride Electrolyte: Effects on Microstructure and Microhardness of Current Density, Temperature, pH, and Chloride Ion Concentration," *Plating, 56*, 533–542 (1969).

(81) Petrov, Yu. N., Mamontov, E. A., and Rybkovskii, V. Ya., "Effect of Electrolysis Conditions on the Submicrostructure and Microhardness of Electrolytic Iron Obtained From a Sulfuric Acid Electrolyte," *Electronnaya Obrabotka Materialov, Akad. Nauk Moldavsk. SSR, No. 4*, 35–37 (1967).

(82) Levy, E. M., and MacInnis, R. D., "Effect of pH on Structure and Mechanical Properties of Electrodeposited Iron," *Journal Applied Chemistry (London), 18* (9), 281–284 (1968).

(83) Karyakin, V. V., Kozlov, V. M., Mamontov, E. A., and Petrov, Yu. N., "Natural Aging of Electrolytic Iron," *Fizika Metallov i Metallovedenie, 25* (3), 497–500 (1968).

(84) Shishakov, N. A., and Chalaganidze, Sh. I., "The Mechanism of the Great Hardness of Electrodeposited Iron," *Russian Journal of Physical Chemistry, 41* (5), 594–595 (1967); *Zh. Fiz. Khimii, 41* (5), 1129–1130 (1967).

(85) Agladze, R. I., and Gdzelishvili, M. Ya., "Metallographic Study of Manganese Alloys," *Soobshcheniya Akademii Nauk Gruzinskoi SSR, 10*, 615–620 (1949).

(86) Offermans, H., and Stackelberg, M. V., "Electrodeposited Tungsten-Cobalt, Tungsten-Nickel, and Tungsten-Iron Alloys," *Metalloberflaeche, 1*, 142–144 (1947).

(87) Vasu, K. I., and Rama Char, T. L., "Electrodeposition of Tungsten Alloys From the Pyrophosphate Bath. The System Fe-W," *Metalloberflaeche, 16*, 349–354 (1962).

(88) Dalal, H. M., and Gill, D. S., "The Structure of Electrodeposited Iron-Zinc Alloys as Revealed by X-Ray Diffraction," *J. Inst. Metals, 93* (4), 130–131 (1964/65).

(89) Fedorova, N. S., "X-Ray Diffraction Structure Investigation of Electrodeposited Iron-Nickel Alloys," *Zh. Fiz. Khimii, 32*, 1211–1213 (1958).

(90) Subrahmanyam, D. V., and Rama Char, T. L., "Electrodeposition of Iron-Cobalt Alloys From the Pyrophosphate Bath," *Plating, 52* (10), 1035–1039 (1965).

(91) Misra, S. S., and Rama Char, T. L., "Electrodeposition of Iron-Cobalt Alloys From the Sulfamate Bath," *Plating, 51* (5), 423–428 (1964).

(92) Kadaner, L. I., Palatnik, L. S., Isichenko, I. A., and Fomina, L. A., "Effect of Heat Treatment on the Properties and Structure of Iron-Phosphorus Electrodeposits," *Metallovedenie: Obrabotka Metallov, No. 9*, 62–63, (1968).

(76) Brenner, A., Burkhead, P., and Seegmiller, E., "Electrodeposition of Tungsten Alloys Containing Iron, Nickel, and Cobalt," *Research Paper No. 1835, J. Res. Natl. Bur. Standards*, 39 [illegible] (1947).

(77) [illegible] S., [illegible] S., and [illegible], "The Electrodeposition of Iron-Zinc Alloys," [illegible] (19[illegible]).

(78) Gibel', A. M., [illegible] and [illegible], "Choosing Optimum Electrodeposition Conditions by the [illegible]," *Zashchita Metallov*, [illegible] (19[illegible]).

(79) Romanov, [illegible], [illegible], and [illegible], M. I., "Combined Electrochemical Iron-Cadmium Coatings," [illegible] (1967).

(80) Levy, D. M., Martinez, [illegible], "Iron Electrodeposited from a Sulfate-Chloride Electrolyte: Effects on Microstructure and Mechanical Properties of Current Density, [illegible] and [illegible] Ion Concentration," *Plating*, [illegible] (1966).

(81) Petrov, Yu. N., [illegible], and [illegible], V. Ya., "Effect of Electrolysis Conditions on the Substructure and Microhardness of Electrolytic Iron Obtained From Sulfate and Chloride Electrolytes," *[illegible]*, [illegible] (1962).

(82) Levy, D. M., and [illegible], R. D., "Effect of pH on Structure and Mechanical Properties of Iron Electrodeposited from," *Journal Applied Chemistry (London)*, 16 [illegible] (1966).

(83) Karasikov, Y. V., Kozlov, V. M., [illegible], and Petrov, Yu. N., "[illegible] of Electrolytic Iron," *Fizika Metallov i Metallovedenie*, [illegible] (1965).

(84) Shatalov, [illegible] A., and [illegible], "The Mechanism of the Great Hardness of Electrodeposited Iron," *Russian Journal of Physical Chemistry*, [illegible] (19[illegible]).

(85) [illegible], and [illegible], M. Ya., "Metallographic Study of Manganese [illegible]," [illegible] (1959).

(86) [illegible], and [illegible], M. V., "Electrodeposition of Tungsten-Nickel and Tungsten-Iron Alloys," *[illegible]*, [illegible] (1967).

(87) Vasu, K. I., and [illegible], "Electrodeposition of Tungsten Alloys from the Pyrophosphate Bath. The System [illegible]," *Metalloberfläche*, 16, 239 [illegible] (1962).

(88) Dahl, H. M., and [illegible], "The Structure of Electrodeposited Iron-Zinc Alloys as Revealed by X-Ray Diffraction," *[illegible]*, [illegible] (19[illegible]).

(89) Fedorova, [illegible], "X-Ray Diffraction Structure Investigation of Electrodeposited Iron-Nickel Alloys," [illegible] (1958).

(90) [illegible], and [illegible], "Electrodeposition of [illegible] from the Pyrophosphate Bath," *Plating*, 52(10), 1035 [illegible] (1965).

(91) [illegible], and [illegible], "Electrodeposition of Iron-Cobalt Alloys From [illegible] Bath," *Plating*, [illegible] (1968).

(92) [illegible], and [illegible], "[illegible] Properties and Structure of Iron Electrodeposited [illegible]," *[illegible] Metallov*, [illegible] (1968).

Chapter 10

Lead and Lead Alloys

Lead and lead-tin alloys are usually deposited in fluoborate solutions. (1) Alloys with 4 to 10 percent tin are used for corrosion protection of steel. A tin content of 7 to 10 percent is customary for 15-μm-thick overlays on steel or aluminum bearings. Alloys with 10 to 60 percent tin are used as solderable coatings for assembly of electronic equipment.

Other metals such as copper and antimony are codeposited to improve the properties of lead and lead-tin alloys. Also for this purpose, metallic and nonmetallic powders can be dispersed in the plating bath and codeposited as inclusions. Sometimes organic additives in the plating bath or their reduction products are codeposited in sufficient amount to influence the deposit properties. The form of the organic material in the deposit is not known, but the amount can be related to the carbon content, in such cases.

Lead Deposits

Hardness, Strength, and Ductility

The hardness of lead deposits is reported to range from 3 to 20 kg/sq mm (2) which probably includes the 3 kg/sq mm value reported for annealed lead (1) and the 20 kg/sq mm value reported for brush plating with proprietary solutions. (3) A more representative range for lead deposits from fluoborate solutions is from 4 kg/sq mm (4,5) to 7 or 8 kg/sq mm. (6) Values of 6 to 7 kg/sq mm (Vickers hardness with 2- and 5-gram load, respectively) have been reported elsewhere for lead from fluoborate solutions. (7)

Table 10.1 shows the hardness (Knoop 5-gram load), strength, and ductility of electroformed lead from a typical fluoborate bath compared to metallurgical lead. The hardness and purity decreased slightly with solution age: 8 kg/sq mm for 99.993 percent lead at 20 amp-hr/gal and 7 kg/sq mm for 99.96 percent lead at 220 amp-hr/gal. (6)

The same hardness of 4 kg/sq mm (Knoop 25-gram load) was reported for metallurgical lead and 50-μm-thick lead deposited at 10.8 amp/sq dm and 54 C in a solution containing 469 g/l lead fluoborate, 45 g/l free fluoboric acid, 46.5 g/l boric

TABLE 10.1

Properties of Lead Deposits From a Fluoborate Bath Compared to Metallurgical Lead

Property	Electrodeposit[(a)]	Corroding Lead[(b)]	Chemical Lead[(c)]
Purity, percent	99.96 to 99.99	99.73+	99.90+
Hardness, kg/sq mm	7 to 8	3.2 to 4.5	—
Tensile Strength, kg/sq mm	1.39 to 1.57	1.2 to 1.34	1.85 to 2.08
psi	2000 to 2250	1700 to 1900	2650 to 3000
Elongation, percent	50 to 53	30	42 to 50
Density, g/cu cm	11.34	11.34 to 11.36	11.34

(a) Deposits of thickness about 760 μm (0.030 inch) obtained at 2.7 amp/sq dm (25 amp/sq ft) and 32 C from a solution containing 230 g/l lead as lead fluoborate, 74 g/l free fluoboric acid, 20.6 g/l boric acid, 0.23 g/l peptone, and 0.23 g/l resorcinol. (6)

(b) Sand cast. (8)

(c) Rolled lead. (8)

acid, and 10 g/l hydroquinone. (5) A similar hardness value was also reported for alloys containing 5, 7.5, and 10 percent tin. Grain structure was too fine for resolution upon metallographic examination. The procedure was developed for high-speed plating on steel strip.

As shown in Table 10.1, the strength of electrodeposited lead was higher than for metallurgical lead (sand cast), but not as high as for rolled lead. However, the electrodeposit had good ductility compared to both types of metallurgical lead.

Internal Stress

The stress of lead deposits from fluoborate solutions is reported to be nearly zero. (9) A compressive stress of −2.8 kg/sq mm was obtained for lead deposits at 1.08 amp/sq dm from a perchlorate bath with peptone as an addition agent. (10)

Tables 10.2 and 10.3 show that compressive stress in lead deposits from an acetate solution is reduced as thickness is increased. (11) Stress reached −0.02 kg/sq mm at 10-μm thickness and 0.01 kg/sq mm (in tension) at 12.5 μm. Although the thinner deposits are compressively stressed initially, stress changes with time, and all deposits eventually stabilize with a tensile stress. During deposition, the rate of increase of compressive stress with thickness decreased to zero with increasing deposit thickness. An increase in current density resulted in an increase in the crystal size of the deposit and a decrease in the initial compressive stress. At 0.5 amp/sq dm, maximum initial stress and minimum crystal size occurred at 30 C.

Effect on Fatigue Strength of Steel

The fatigue limit (10^6 cycles) of SAE 4140 steel of hardness R_c 43 and tensile strength of 148.5 kg/sq mm (211,000 psi) was reduced only slightly from 76.5 kg/sq

TABLE 10.2

Internal Stress of Lead Deposited From Acetate Solution as a Function of Temperature and Current Density[a]

Temperature, C	Current Density, amp/sq dm	Crystal Size, μm	Stress, kg/sq mm Initial[b]	After Depositing 2.5 μm	50 Minutes After Electrolysis
20	0.5	1.6	−2.5	−0.28	+0.10
20	1.0	5	−1.3	−0.16	+0.06
20	1.5	6	−0.8	−0.10	+0.06
20	2.0	8	−0.3	−0.09	+0.06
20	4.5	14	−0.1	−0.07	+0.05
30	0.5	0.8	−4.0	−0.23	+0.10
40	0.5	1.9	−2.8	−0.21	+0.09
50	0.5	5.2	−2.3	−0.19	+0.08

[a] Acetate solution containing 150 g/l lead acetate, 35 g/l acetic acid, 1 cu cm/l carbon disulfide, 1 cu cm/l orthotoluidine, and 3 g/l gelatin; all deposits on a 6-μm thickness of lead underlayer obtained at 0.5 amp/sq dm and 20 C and stabilized in electrolyte for 70 minutes to minimal stress change. (11)

[b] Initial stress calculated by extrapolating rate of change of stress to zero time. The negative sign indicates compressive stress.

TABLE 10.3

Internal Stress of Lead Deposited from Acetate Solution as a Function of Deposit Thickness[a]

Time, minutes	Deposit Thickness, μm	Mean Internal Stress, kg/sq mm After Deposition	After 65 Minutes
7.5	2.5	−0.28	+0.10
15.0	5.0	−0.11	+0.14
22.5	7.5	−0.04	+0.16
30.0	10.0	−0.02	+0.20
37.5	12.5	+0.01	+0.22

[a] Deposits obtained at 0.5 amp/sq dm and 20 C with the solution and conditions described in Footnote (a), Table 10.2.

TABLE 10.4

Strength and Hardness of Lead-Antimony Alloys[a]

Antimony Content, weight percent	Hardness, kg/sq mm	Tensile Strength, kg/sq mm
0[b]	8	1.5
4.5	15.5	4.0
10	18	5.3
15	21.5	6.5

[a] Alloy deposits of thickness about 760 μm (0.030 inch) obtained at 2.7 amp/sq dm (25 amp/sq ft) and 27 C from a solution containing 230 g/l lead fluoborate, 74 g/l fluoboric acid, 20.6 g/l boric acid, 0.25 g/l peptone, 0.25 g/l resorcinol, and 0 to 11.5 g/l antimony as antimony fluoborate concentrate containing 191 g/l antimony, 36 g/l free fluoboric acid, and 78 g/l free hydrofluoric acid. (13)

[b] Pure lead deposit, Table 10.1, Footnote (a).

mm (109,000 psi) to 74.8 kg/sq mm (105,000 psi) with a 25-μm (0.001-inch) lead deposit. (12) By comparison, the fatigue limit of the steel was reduced to values of 68.5, 61.5, 59.8, 31.6, and 21.1 kg/sq mm for deposits of compressively stressed chromium (42.2 kg/sq mm), copper, compressively stressed bright nickel (2.1 kg/sq mm), Watts nickel, and tensilely stressed chromium (12 kg/sq mm), respectively.

Lead-Antimony Alloys

Table 10.4 shows that the hardness and tensile strength of lead alloy containing 4.5 to 15 percent antimony increased linearly as the antimony content increased. (13) Deposits with a low antimony content had a dull, gray appearance typical of lead deposits from a fluoborate solution. The tensile data agree fairly well with published data for the metallurgical alloys. (8) Elongation of the lead alloy deposits was nearly zero; their structure was fine grained. When the alloy deposits were quickly heated to their melting point, as might occur in joining or soldering operations, vigorous outgassing and bubbling occurred, and an odor similar to that of burnt peptone was evident. In addition, a carbonaceous film would occasionally appear on the surface of the heated part. The phenomenon was attributed to decomposition of organic compounds that had been occluded in the deposit during plating. Of the two organic additives, peptone was the major problem, and both peptone and resorcinol were required to avoid rough, nodular deposits. Subsequent experiments showed that beta-naphthol permitted smooth, thick deposits and caused less bubbling on heating than the peptone-resorcinol system. No physical property data were presented for lead-antimony alloys with beta-naphthol additive.

Lead-Manganese Alloys

Lead alloys with 5 to 23 percent manganese have good antifriction properties and are 2 to 3 times harder than lead. (14) The deposits have satisfactory corrosion resistance to sulfuric acid (10 percent solution) and sodium hydroxide (2 N solution), provided the coatings are nonporous (i.e., thicker than > 10 μm on copper and >25 μm on steel). The electrolyte contained 22 g/l lead oxide dissolved in 40 g/l of sodium ethylenediaminetetraacetate in water plus 22 g/l manganese dissolved in 122 g/l ammonium citrate plus 1 g/l glue, and 0.1 g/l lauryl sulfate. Deposits were obtained at 1 to 4 amp/sq dm and 25 to 35 C with a pH of 5 to 7.

Lead-Tin Alloy Deposits

Hardness and Electrical Resistivity

Table 10.5 shows the electrical resistivity and hardness of lead-tin alloy deposits from fluoborate and fluosilicic acid baths. Such physical property data have been used to estimate the solubility of tin in lead deposits. However, there is no agreement in results or method of interpreting the data. In one report, the maximum in electrical resistivity was used to estimate the solubility as 3.5 percent tin for deposits from fluoborate solutions, (15) which is greater than the limit reported for wrought alloy. Although the maximum in electrical resistivity and hardness was at 3 percent tin, another investigator used the relative minimum in hardness to estimate the solubility as 6 to 7.2 percent tin for deposits from a fluosilicic acid solution. (16) The solubility was estimated to be 8 percent tin for deposits from a fluoborate solution, based on X-ray data, although hardness data showed a relative maximum at 11.7 percent tin and a relative minimum at 13.8 percent tin. (7)

Because the equilibrium content of tin is 2 percent at room temperature, the electrodeposited alloys are metastable and comparable to the metastable solid solutions of tin in lead that can be prepared thermally by quenching. These alloys slowly decompose into separate stable phases at room temperature or at a faster rate at 50 C. Electrodeposited alloys behave similarly. (7) Other studies of tin-lead alloy properties and structure by X-ray, electrical resistivity, and hardness have confirmed that there is no fundamental difference between electrodeposits and alloys that can be produced by thermal crystallization. (17)

The electrical resistivities of the lead-tin alloys shown in Table 10.5 are slightly higher than reported for metallurgical alloys: 20.6, 19.5, 17.5, and 15.6 microhm-cm for 0, 5, 20, and 50 percent tin, respectively. (8)

For high-speed plating of steel strip with lead-tin alloys (terne plate), a solution temperature of 54 C was adopted and hydroquinone was employed as the temperature-stable addition agent. (5) Fine-grained deposits up to 50 μm in thickness were obtained at 10.8 to 24.4 amp/sq dm from a solution containing 211

TABLE 10.5

Electrical Resistance and Hardness of Lead-Tin Alloy Deposits

Tin Content, weight percent	Electrical Resistivity, microhm-cm		Hardness, kg/sq mm		
	(a)	(b)	(b)	(c)	(d)
0	22.9	20	4	7	6
3		21.8[(e)]	9.8[(e)]		
3.5	24.0[(e)]	—	—	—	—
7	23.3	19.6	9.4[(f)]	—	—
9.1	—	—	—	9	7
10	22.7	19.0	9.8	—	—
11.7	—	—	—	12	8
13.8	—	—	—	9[(f)]	7[(f)]
14.1	—	—	—	12	9
20	20.8	17.7	11.0	—	—
40	16.7	15.9	12.0	—	—
60	14.5	14.3	12.0	—	—
80	13.4	13.0	12.0	—	—
100	11.4	12.0	12.0	12	9

(a) Values measured at 25 C for deposits from a fluoborate solution. (15)

(b) Deposits from a fluosilicic acid solution containing 120 g/l lead plus tin fluosilicate, 80 g/l fluosilicic acid, and 1 g/l resorcinol. (16)

(c) VHN (5-gram load); deposits from a fluoborate solution.[(f)]

(d) VHN (2-gram load); same deposits as (c).

(e) Relative maximum.

(f) Relative minimum.

to 443 g/l lead fluoborate, 21.6 to 43.3 g/l tin fluoborate, 20 to 40 g/l free fluoboric acid, 12.5 to 25 g/l boric acid, and 10 g/l hydroquinone. Lead alloys with 0, 5, 7.5, and 10 percent tin all had hardness values near 4 kg/sq mm (Knoop 25-gram load); the 14 percent tin alloy had a hardness of 6.5 kg/sq mm. It was observed that all of the lead-tin alloy deposits resisted casual scratching better than pure lead deposits. The hardness values are lower than those shown in Table 10.5 and the deposits obtained at 54 C are probably closer to stable phases than the supersaturated solutions of tin obtained at lower temperatures.

Ternary and Quaternary Alloys of Lead and Tin

Lead-tin alloys sometimes contain additional elements to produce ternary and quaternary alloys with special properties. For example, a small amount of copper provides greater fatigue resistance for lead-tin alloys used as bearing metals. (1) Other metals such as antimony and arsenic have been codeposited with lead-tin alloys for the purpose of hardening the alloy and improving wear resistance of bearing overlays.

A soft antifriction coating of about 0.5 mm in thickness on bearing liners contained 6 to 9 percent tin, 1 percent copper (maximum), and 0.5 percent antimony, and had a hardness of 28 to 47 kg/sq mm. (18) Deposits were obtained at 2 amp/sq dm and 20 C or 4 amp/sq dm and 30 C from a solution containing 70 to 110 g/l lead fluoborate, 6 to 13 g/l tin fluoborate, 0.5 to 1 g/l copper fluoborate, 0.25 to 0.5 antimony fluoborate, 45 to 60 g/l fluoboric acid, and 0.5 to 2 g/l gelatin. When the solution also contained 5 to 7 g/l resorcinol, the deposit hardness was reduced to 15 to 17 kg/sq mm.

Dispersion-Strengthened Lead and Lead-Tin Alloys

Correlation of Hardness, Strength, and Creep Rate

Table 10.6 shows mechanical property data for electroformed deposits of lead and lead alloy (< 0.2 percent tin) containing dispersions of titanium dioxide, barium sulfate, and/or a carbon compound. (19) Figure 10.1 shows an approximate correlation of strength and steady-state creep rate with hardness. Deposits of 2.5 mm (0.1 inch) were obtained in a fluoborate bath with continuous burnishing during plating to refine grain size and produce a smooth deposit. (20) All of the electrodeposits contained a high percentage of carbon (0.13 to 0.73 percent), which resulted from the addition of lignin sulfonic acid and coumarin to the plating bath to smoothen the deposits. The strengthening effect in the lead and lead-tin alloy electrodeposits was associated with their carbon content. Additional strengthening was achieved with the dispersion of titanium dioxide particles.

The hardness and strength of the lead deposits listed in Table 10.6 are about twice the values normally obtained for lead deposits from a fluoborate bath, as shown previously in Table 10.1, and about three times the values for metallurgical lead (Table 10.6 and Table 10.1). The creep rates of the lead-tin alloy deposit with titanium dioxide was only 20 to 25 percent of the rate of commercial lead with a 1.5 percent dispersion of lead oxide particles prepared by powder-metallurgy techniques, included for comparison in Table 10.6. Creep data obtained at 23 C with an applied load (strain) of 2.1 kg/sq mm (3,000 psi) are compared in Figure 10.2. Both lead-tin alloy and a composite of lead-tin alloy and titanium dioxide were more resistant to creep than the powder-metallurgy composite.

Effect of Carbonaceous Material and Alloying Tin

Hardness increased from 10 to 12.5 kg/sq mm with increasing carbon content from 0.01 to 0.075 percent in the lead and lead alloy deposits listed in Table 10.6 (19)

TABLE 10.6

Mechanical Property Data for Lead Alloys Deposited from a Fluoborate Bath Compared with Commercial Lead and Lead Alloys[a]

Alloy Deposit Panel Number[b]	Alloy Composition weight percent: Carbon[c]	Tin[d]	Titanium[e] Dioxide	Barium[f] Sulfate	Hardness,[g] kg/sq mm	Yield[h] Strength, kg/sq mm	Tensile[i] Strength, kg/sq mm	Elongation at Fracture, percent	Primary[j] Creep Strain, percent	Steady State Creep Rate at 2.1 kg/sq mm, 10^{-4} hr^{-1}
Lead										
1	0.0285	—	—	—	10.4	2.81	3.13	44	7.3	120
2	0.0390	—	—	—	10.8	2.95	3.31	44	7.6	61
3	0.0435	—	—	—	11.3	3.13	3.55	48	6.4	5.5
Lead-Tin										
1	0.0695	0.13	—	—	12.5	3.30	4.01	36	1.4	9.0
2	0.0735	0.11	—	—	12.9	3.59	4.22	26	0.9	5.3
3	0.0720	0.16	—	—	12.5	3.48	4.19	29	0.8	6.6
4	0.0705	0.44	—	—	11.8	3.27	3.66	3	0.5	37
5	0.0690	0.53	—	—	12.0	3.23	3.62	4	0.5	38
6	0.0130	0.11	—	—	10.3	2.50	2.99	35	0.8	350
Lead-Tin-Titanium Dioxide										
1	0.0340	0.12	0.42	—	13.3	3.45	4.12	8	0.6	2.5
2	0.0350	0.17	0.44	—	14.3	3.52	4.12	7	0.4	2.6
3	0.0365	0.10	0.49	—	13.3	3.52	4.05	8	0.4	1.7
4	0.0230	0.05	0.10	—	10.3	2.78	3.16	15	0.7	150

Lead-Barium Sulfate										
1	0.0200	—	—	2.70	10.5	2.74	3.16	12	0.4	220
2	0.0220	—	—	2.65	10.0	2.74	3.16	15	0.5	250
Lead-Tin-Barium Sulfate										
1	0.0435	0.11	—	0.63	12.9	3.20	3.69	16	0.7	17
Metallurgical Lead										
Pb (99.99%)	0.0005	—	—	—	4.5	0.60	1.23	60	—	—
Pb−1.5% PbO	0.0025	—	—	—	9.0	3.03	3.23	18	0.6	10

(a) Electrodeposits about 2.5 mm (0.1 inch) in thickness obtained with burnishing during plating at 5.4 amp/sq dm (50 amp/sq ft) and room temperature from a solution containing 250 g/l lead as lead fluoborate, 20 to 30 g/l fluoboric acid, 30 g/l boric acid, 2 g/l lignin sulfonic acid, and 1 g/l coumarin. (19)

(b) Panel size was 12.7 by 17.8 cm (5 by 7 inches). Each panel provided three tensile and three creep specimens with an overall length of 11.4 cm (4.5 inches) and a gage section 3.3 cm (1.3 inches) long and 6.3 mm (0.25 inch) wide.

(c) Carbon content derived from organic addition agents used. Only half the normal concentration of addition agent was used for preparing Panel 1 Lead deposit. Less than the normal concentration of addition agent was employed for preparing Panel 6 Lead-Tin alloy after filtering to remove titanium dioxide.

(d) Tin alloys were obtained by adding 6 g/l tin fluoborate to the basic solution in Footnote (a) and using lead anodes containing 2 percent tin.

(e) 100 g/l TiO_2 of particle size 0.03 to 0.10 μm was dispersed in the deposition bath by mechanical mixing when TiO_2 was added to the deposit. The titanium dioxide concentration was reduced to 50 g/l for Panel 4.

(f) 50 g/l of $BaSO_4$ of particle size 0.5 to 5.0 μm.

(g) Hardness values are the average of 10 measurements on each specimen using a diamond pyramid indenter with a 25-gram load.

(h) Yield strength was taken at 0.2 percent offset and data are the average of three tests with variations of about ±3 percent within a panel.

(i) Tensile strength was determined with a crosshead rate of 1.27 mm/min and data are the average of three tests with variations of about 3 percent within a panel.

(j) Primary creep strain of 6 to 8 percent for Pb deposits lasted about 10 hours; primary creep strain of 0.5 to 2.0 percent for lead alloys lasted 1 to 2 hours.

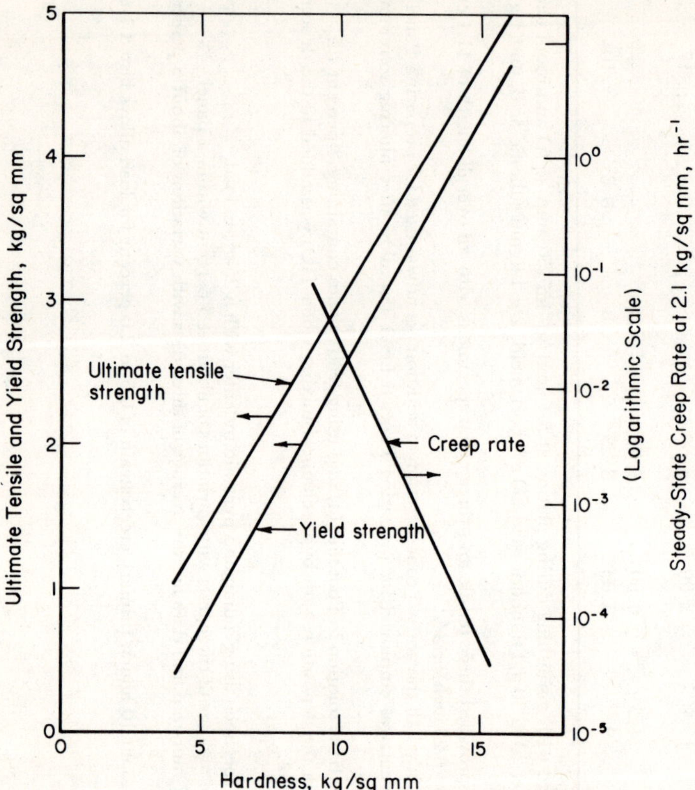

Figure 10.1. Correlation of Strength and Creep Rate with Hardness for Data in Table 10.6 for Electrodeposited Lead and Lead Alloys.

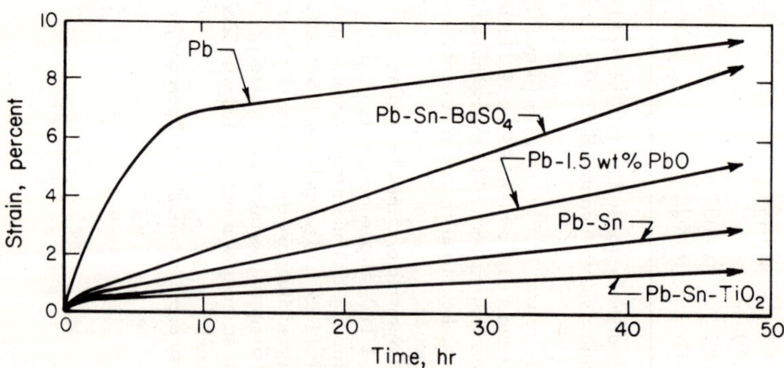

Figure 10.2. Creep Curves for Electroformed Lead, Electroformed Lead Alloys, and Metallurgical Lead Containing 1.5 Percent Lead Oxide at 23 C and 2.1 kg/sq mm (3,000 psi).

The manner in which the carbon was incorporated in the deposits was not discussed by the investigators. No second-phase particles were detected during examination at a magnification of 50,000. Grain size was in the range of 5 to 10 μm. The aroma of coumarin was detected when electrodeposits were melted with a torch. All of the deposits listed in Table 10.6 also contained 0.10 $\pm$ 0.05 percent oxygen. Thus at least part of the carbon evidently existed as occluded coumarin. The hardening effect is more likely due to the inclusion of a metallic compound of carbon, however.

Codeposition of tin ($<$ 0.2 percent) increased the carbon content of the deposit. The increased hardness of the lead-tin deposits in comparison with the lead deposits reflects the hardening effect associated with increasing the carbon content. Alloy deposits with the same tin content (0.11 percent) differed in hardness and carbon content when they were deposited in solutions containing different amounts of the addition agents.

Tensile strength was improved, while ductility was reduced slightly by codepositing 0.11 to 0.16 percent tin and increasing the carbon from about 0.04 to 0.07 percent. An increase in the tin content to 0.44 or 0.53 percent caused a large decrease in ductility, and tensile strength was reduced to a level only slightly greater than that for the tin-free deposit containing 0.044 percent carbon.

Effect of Barium Sulfate and Titanium Dioxide

Barium sulfate particles dispersed in the lead deposit did not contribute to strengthening, but caused a loss in ductility. However, hardness appeared to be increased by occlusion of barium sulfate particles, above that expected from the effect of the carbon content. Barium sulfate, probably because of the larger particle size, did not produce the degree of improvement in creep that titanium dioxide did.

Dispersing titanium dioxide particles (0.4 to 0.5 percent) in the lead-tin alloy increased hardness and strength above that attributable to the carbon content. Elongation was reduced to 7 or 8 percent. The carbon content of the deposit was reduced about 50 percent, which was attributed to the adsorption of the organic addition agents on the titanium dioxide dispersed in the plating bath.

References

(1) Brenner, A., *Electrodeposition of Alloys, Principles and Practice, Vol. II*, Academic Press, New York (1963).

(2) Buss, G., "Properties and Corrosion Protection of Electrodeposited Coatings and Anodized Aluminum," *Metall, 17* (6), 567–571 (1963).

(3) Machu, W., "Brush Plating for Localized Metal Deposition (Dalic Process)," *Metallwaren-Industrie und Galvanotechnik, 49* (11), 467–481 (1958). Cf. Hughes, H. D., *Trans. Inst. Metal Finish., 33*, 424–439 (1956).

(4) Rabinowicz, E., Imai, M., and Santos, V., "Friction and Wear at Elevated Temperatures," *Report No. RTD-TDR-63-4272, Part 5, Contract AF 33(616)-7648*, Surface Laboratory, Massachusetts Institute of Technology (December 1963), 37 pp. AD 431 818.

(5) Graham, A. K., and Pinkerton, H. L., "Properties and Comparative Performance of Electrodeposited Terne Alloys," *Plating, 54*, 367–377 (1967).

(6) Dini, J. W., and Helms, J. R., "Properties of Electroformed Lead," *Metal Finishing, 67* (8), 53–55 (1969).

(7) Raub, E., and Blum, W., "The Electrodeposition of Lead-Tin Alloys," *Metalloberflaeche, 9A* (4), 54–57 (1955). Brutcher Translation No. 3631.

(8) *Metals Handbook, Vol. I,* 8th ed., American Society of Metals (1961).

(9) Heussner, C. E., Balden, A. R., and Morse, L. M., "Some Metallurgical Aspects of Electrodeposits. II," *Plating, 35* (7), 719–723, 768 (1948).

(10) Hothersall, A. W., "Stress in Electrodeposited Metals," *J. Inst. Metals, Symposium on Internal Stresses in Metals and Alloys, Preprint No. 1082* (1947), 12 pp.

(11) Popereka, M. Ya., "Internal Stresses in Lead Deposits," *Russian Journal of Physical Chemistry, 39* (6), 705–709 (1965); *Zh. Fiz. Khimii, 39* (6), 1321–1327 (1965).

(12) Stareck, J. E., Seyb, E. J., and Tulumello, A. C., "The Effect of Different Chromium Deposits on the Fatigue Strength of Hardened Steel," *Plating, 42* (11), 1395–1402 (1955); *Proceedings Am. Electroplaters' Soc., 42,* 129–136 (1955).

(13) Dini, J. W., and Helms, J. R., "Electrodeposition of Lead-Antimony Alloys," *Journal Electrochem. Soc., 117,* 269–272 (1970).

(14) Kalyuzhnaya, P. F., Pimenova, K. N., and Sobko, V. B., "Electrodeposition and Some Physicochemical Properties of Coatings Made of Lead-Manganese Alloy," *Ukr. Khim. Zh., 36,* (5), 464–468 (1970).

(15) Mohler, J. B., and Sedusky, H. J., "Alloys by Electrodeposition," *Metal Finishing, 45* (12), 65–70 (1947).

(16) Fedot'ev, N. P., and Vyacheslavov, P. M., "The Phase Structure of Binary Alloys Produced by Electrodeposition," *Plating, 57* (7), 700–706 (1970).

(17) Riedel, W., "Structure and Properties of Electrolytically Deposited Tin-Lead Alloys," *Metalloberflaeche, 23* (2), 42–44 (1969).

(18) Solov'ev, N. A., Merkulova, N. V., Asrhinov, V. D., and Popov, A. M., "Soft Galvanic Antifriction Coating on Bearing Liners," *Avtomob, Prom., 32* (4) 35–38 (1966).

(19) Vandervoort, R. R., Raymond, E. L., Wiesner, H. J., and Frey, W. P., "Strengthening of Electrodeposited Lead and Lead Alloys. II. Mechanical Properties," *Plating, 57,* 362–368 (1970).

(20) Wiesner, H. J., Frey, W. P., Vandervoort, R. R., and Raymond, E. L., "Strengthening of Electrodeposited Lead and Lead Alloys. I. Process Development," *Plating, 57* (4), 358–361 (1970).

Chapter 11

Manganese

Manganese is produced by electrowinning the metal from sulfate or chloride solutions. (1–4) Electrodeposited coatings of manganese and its alloys have been studied for steel products requiring protection from corrosion. Like zinc and cadmium, manganese protects steel sacrificially. In outdoor weathering, 12.5-μm-thick deposits protected steel for 2 years. (4) The manganese coating was darkened and discolored. Although manganese electrodeposits corrode rapidly in sea water, about 20 times faster than zinc, the application of a chromate film slightly reduced the corrosion rate of manganese. (5) In a moist atmosphere of hydrogen sulfide or carbon dioxide, manganese electrodeposits containing a small amount of selenium were better than zinc for resisting corrosion.

Relatively pure manganese electrodeposits reduced the tensile strength of the steel wire substrate and failed to protect the steel during polarization in a dilute sulfuric acid solution. (6) However, deposits produced in solutions containing 0.1 g/l selenic acid or 1.0 g/l telluric acid had no harmful effect on the tensile strength of the steel, which was attributed to the improved density obtained by codepositing selenium or tellurium and the improved protection against corrosion obtained with the denser coating.

The corrosion rate of manganese-iron alloy electrodeposits in a concentrated (8 N) solution of sodium hydroxide was decreased by adding 0.1 to 0.2 g/l selenic acid to a sulfate-citrate bath with a pH of 4 to 6. (7) The effect was attributed to the grain refinement obtained by codepositing selenium.

Manganese deposits containing about 25 percent zinc were better than unalloyed zinc for protecting steel from corrosion in accelerated tests. (8) Attempts to codeposit more than about 10 percent iron or nickel were unsuccessful.

Additions of selenic acid or sulfur dioxide to the acid solutions employed for electrowinning or electrodepositing manganese improve the cathode efficiency and the smoothness and density of the deposit. To prevent dissolution of the manganese during electrowinning, a diaphragm is used to assist pH control of the solution adjacent to the cathode and isolate impurities and/or anodic oxidation products that otherwise promote manganese dissolution. A diaphragm to separate the anolyte and catholyte also has been employed during investigations of manganese and its alloys as a protective coating.

Property Data

The density of electrolytic, alpha or cubic manganese is 7.42-7.43 g/cu cm. (9) Degassing to remove hydrogen increased the density to 7.48 g/cu cm. The gamma, tetragonal, form of electrolytic manganese has a density of 7.18-7.20 g/cu cm. The electrical resistivities of the cubic and tetragonal forms of electrolytic manganese are 189 and 45 microhm-cm, respectively.

Gamma tetragonal manganese is softer than the alpha or cubic form, which is a hard, brittle material. The nature of the substrate affects the structure of the deposit. Gamma manganese has been deposited on iron, cobalt, nickel, and zinc; but after an addition of sulfur dioxide to a sulfate bath with a pH of 6 to increase efficiency and refine grain size, only the brittle alpha form of manganese was deposited. (10) In any case, gamma manganese transforms to the alpha, cubic form at room temperature in a few days or weeks, although gamma manganese deposits have been preserved for as long as a year by refrigeration. (4)

The microhardness of electrodeposited manganese ranges from 430 to 1120 kg/sq mm, depending on the addition agent employed to refine grain and the sulfur content of the deposit. Table 11.1 data show that hardness increased from 430 to 1120 kg/sq mm as the sulfur content of the deposits produced with additions of sulfur dioxide increased from 0.17 to 0.4 percent.

A tensile stress of 3 kg/sq mm was reported for manganese deposited in a manganese sulfate-ammonium sulfate bath containing 0.3 g/l sulfur dioxide when the pH was 4.2. (14) Stress was increased to 4.6 kg/sq mm when the pH was reduced to 3.0. An increase in stress has been associated with the transformation of gamma to alpha manganese. (15) On the other hand, alpha manganese deposited directly in solutions with a pH < 4 were reported to have little internal stress in comparison with alpha manganese deposited directly in solutions with a pH > 4. (16) The stress remaining in manganese deposited in manganese sulfate-ammonium sulfate-glycerol solutions after stress relief by cracking was reported to range from 5 to 20 kg/sq mm. (17) Additions of glycerol to this type of bath accelerate the transformation of gamma to alpha manganese. (18)

Structure and Purity

Electrolytic gamma manganese consists of an interstitial solution of hydrogen in manganese. (18) As noted previously, gamma manganese transforms, usually in a few hours, to the brittle alpha form. An increase in stress and cracking is associated with this transformation. (19) The hydrogen content of the deposit also is reduced during this transformation, because the solubility of hydrogen is considerably smaller in alpha manganese. The lattice constant of electrolytic alpha manganese is reduced from 8.902 to 8.894 angstroms by degassing. (9) Heating to 125 C evolves a great deal of hydrogen, and all of the hydrogen can be completely expelled at 300 C. (17)

The ductile, gamma (tetragonal) form of manganese deposited at low current

TABLE 11.1

Hardness Data for Electrodeposited Manganese

Solution	Addition Agent	pH	Temperature, C	Current Density, amp/sq dm	Micro-hardness, kg/sq mm	Reference
Manganese sulfate and ammonium sulfate	Sulfur dioxide (0.17% S in deposit)	4.0	20	10	430	11
Manganese sulfate, ferrous ammonium sulfate, and ammonium citrate	Glycine and selenic acid	4–6	20–25	2–5	531[a]	7
Manganese sulfate and ammonium sulfate	Glycerol	7.0	15–20	10–15	535	12
Manganese sulfate and ammonium sulfate	Citric acid and hydroxylamine hydrochloride	5.0	20–32	23	550–575	13
Manganese sulfate and ammonium sulfate	Sulfur dioxide[b] (0.21% S in deposit)	4 0	20	25	760	11
Ditto	Sulfur dioxide[b] (0.26% S in deposit)	4 0	20	25	980	11
"	Sulfur dioxide[b] (0.4% S in deposit)	4.0	20	25	1120	11

[a] A manganese-iron alloy.

[b] The sulfur content was proportional to the sulfur dioxide concentration in solution.

density (4 amp/sq dm) in a neutral, 30 C manganese sulfate-ammonium sulfate solution free of sulfur dioxide contains 0.003 to 0.008 percent sulfur. (20) Alpha (cubic) manganese deposited in similar solutions containing sulfur dioxide and hydrogen sulfide had a sulfur content of at least 0.04 percent.

The grain-refining effects of codepositing small amounts of selenium or sulfur have been noted. The use of ultrasonics also is reported to produce fine-grained deposits in electrolytic manganese. (21) However, ultrasound reduced adhesion, especially for deposits obtained at high current densities.

Codepositing 2 to 3 percent copper stabilized the gamma form, at least for several months. (22) Smooth gray alloy deposits with a thickness of 10 to 20 μm were obtained at 22 to 32 amp/sq dm in a solution containing 0.73 M/l manganous sulfate, about 0.005 M/l copper sulfate, 0.9 M/l ammonium sulfate and 1 g/l hydroxylamine sulfate. Diethylene triamine was added to adjust pH to the range of 7.2 to 7.4. Cathode efficiency was 37 percent for a solution temperature of 30 C.

References

(1) Hampel, *Rare Metals Handbook*, Reinhold Publishing Company (1954).

(2) Jacobs, J. H., and Churchward, P. E., "Electrowinning of Manganese From Chloride Electrolytes," *Trans. Electrochem. Soc., 94*, 108–121 (1948).

(3) Carosella, M. C., and Fowler, R. M., "A New Commercial Process for Electrowinning Manganese," *Journal Electrochem. Soc., 104*, 352–356 (1957).

(4) Dean, R. S., *Electrolytic Manganese and Its Alloys*, Reinhold Press (1952).

(5) Gofman, N. T., Kurashvili, M. I., and Agladze, R. I., "Use of Manganese to Increase the Chemical Stability of Coatings," *Elektrokhim. Margantsa, 4*, 125–141 (1969).

(6) Shul'gina, N. P., and Polukarov, M. N., "Effect of Tellurium and Selenium on the Hydrogen Charging of Steel During Manganese Electroplating," *Uch. Zap. Perm. Gos. Univ., No. 159*, 104–111 (1966).

(7) Kalyuzhnaya, P. F., Sobko, V. B., and Pimenova, K. N., "Electrodeposition Conditions and Some Physicochemical Properties of Coatings Formed by a Manganese-Iron Alloy," *Ukr. Khim. Zh., 36* (4), 356–360 (1970).

(8) Tripler, A. B., Bride, J. E., Gurklis, J. A., and Faust, C. L., "An Investigation of Electrodeposited Alloys for Protection of Steel Aircraft Parts," *Report No. 5692, Supplement 2*, Materials Lab., Wright Air Development Center (1952).

(9) Naylor, B. F., "The Heat Content of Manganese at High Temperatures," *Journal Chemical Physics, 13*, 329–332 (1945).

(10) Gamali, I. V., Danilov, F. I., and Stender, V. V., "Dimensional Correspondence of Crystal Lattice in the Electrodeposition of Manganese," *Zh. Prikl. Khimii, 37*, 337–342 (1964).

(11) Sanzharovskii, A. T., "Effect of Sulphur Dioxide on the Electrolytic Deposition of Manganese," *Russian Journal of Physical Chemistry, 35*, 8–12 (1961); *Zh. Fiz. Khimii, 35*, 20–25 (1961).

(12) Terekhov, I. I., Reĭkhshtadt, A. K., and Ivanova, A. N., "Electrolytic Deposition of Mn From Sulfate Solutions," *Trudy Vtoroĭ Konferentsiĭ Korrozii Metallov, 2*, 237–252 (1943).

(13) Bell, W. A., "A Laboratory Technique for the Electrodeposition of Manganese on Other Metals," *Trans. Inst. Metal Finish., 31*, 466–475 (1954).

(14) Vagramyan, A. T., and Solov'eva, Z. A., *Technology of Electrodeposition*. English Translation, Robert Draper Ltd., Teddington (1961), 398 pp.

(15) Nishihara, K., Murachi, M., Sakamoto, Y., Nishihara, Jr., M., "Internal Stresses in Electrodeposited Chromium and Manganese," *Interfinish '68 Tagungsberichtsband*, Deutsche Gesellschaft für Galvanotechnik, E. V., Düsseldorf, Germany, 118–123 (1968).

(16) Sanzharovskii, A. T., "The Role of SO_2 in Electrolysis of Manganese," *Doklady Moskov, Sel'skokhoz. Akad. im. K. A. Timiryazeva, Nauch. Konf., No. 32*, 512–517 (1958).

(17) Popova, O. S., and Gorbunova, K. M., "The Structure and Properties of Electrolytic Manganese," *Zh. Fiz. Khimii, 32*, 2020–2028 (1958).

(18) Moiseev, V. P., and Popova, O. S., "X-Ray Investigation of Electrolytic Manganese Deposits," *Zh. Fiz. Khimii, 33*, 2183–2189 (1959).

(19) Grube, G., "The Electrodeposition of Manganese From Aqueous Solutions," *Reichsamt Wirtschaftsausbau, Chem. Ber., Prüf.-Nr. 15*, 33–40 (1942). PB 52010.

(20) Schlain, D., and Prater, J. D., "Electrodeposition of Gamma Manganese," *Trans. Electrochem. Soc., 94*, 58–73 (1948).

(21) Kenahan, C. B., and Schlain, D., "Effects of Ultrasonics on Electrolytic Deposition of Manganese and Manganese Dioxide From Sulfate Electrolytes," *Report of Investigation 6073*, U.S. Bureau of Mines (1962), 26 pp.

(22) Graham, A. K., "Manganese Coatings," Graham, Crowley, and Associates, *Report Nos. 3 to 9, Contact NOa(s) 9930*, U.S. Bureau of Aeronautics (December, 1948 to July, 1949).

Chapter 12

Nickel

Nickel plating is an important electrodeposition process for preserving steel, zinc, and other basis metals from corrosion. Nickel consumption for such purposes exceeds the consumption of any other metal by electroplating. Combined with a thin overlay of chromium, nickel is the most effective electroplated coating identified for preserving a decorative appearance for extended periods of exposure to corrosive environments. Nickel electrodeposition also is popular for non-decorative, corrosion-protection applications and for electroforming.

Bright nickel electroplated for decorative uses differs appreciably from non-decorative nickel deposits. The high sulfur content (>0.05 percent) of the bright nickel reduces its ductility and resistance to corrosion. Yet these disadvantages are outweighed by the appreciable cost saving obtained by using bright nickel in place of dull nickel, which must be buffed or polished to achieve a decorative finish.

Brightness is induced with organosulfur compounds that decompose at cathode surfaces, forming very small particles of nickel sulfide which refine grain size at least two orders of magnitude. The selection of brightener additions to the plating bath affects the ductility of the deposit and internal stress. Some organic additives such as saccharin induce a compressive stress, which is desirable for avoiding a reduction in the fatigue strength of nickel-plated steel. Nickel coatings stressed in tension invariably reduce the fatigue strength of steel. Thus, stress reducers frequently are added to solutions used for plating dull nickel, to induce a slight compressive stress and avoid the undesirable consequences of a high tensile stress. For electroforming purposes, a sulfur content of 0.01 to 0.015 percent is desirable to increase strength. However, 0.02 percent sulfur, or more, causes notch sensitivity. The sulfur content of electroplated nickel influences many properties.

Also important are oxides and hydrated nickel compounds sometimes occluded in nickel deposited at high current densities in high pH solutions. Such inclusions reduce ductility and strength. For coatings to be used at high temperatures, these impurities must be minimized or they will decompose to form gas. Metallic impurities in the plating salts or the anodes and organic impurities introduced inadvertently or by the decomposition of additives also influence the properties of the deposits.

Grain size and impurities, which have profound effects on the properties of electrodeposited nickel, are influenced by changes in solution composition,

TABLE 12.1

Physical Properties of Nickel Coatings

Type of Plating Bath	Density, g/cu cm	Specific Heat, cal/g/C	Thermal Expansion Coefficient, 10^{-6}/C	Thermal Conductance, cal/cm/sec-cm^2/C	Reference
Acetate	8.90	—	13.6 to 17.0	—	1
Bright (organic)	8.86	0.1083	Ditto	—	1, 2
Bright cobalt alloy	8.90	—	—	—	1
Chloride, 2 N Ni	8.90	—	13.6 to 17.0	—	1, 2, 5
Chloride, 4 N Ni	8.86	—	Ditto	—	1
Electroless[(a)]	7.85 to 8.2	—	13 to 14.5	0.0105 to 0.0135	7
Fluoborate	8.91	—	—	—	1
Sulfamate	8.93	—	13.6 to 17.0	—	2
Watts	8.90 or 8.91	0.1073[(b)]	14.6 to 17.2	0.195 to 0.26	1–6
Watts with $(NH_4)_2SO_4$	8.90	—	Ditto	—	1, 5
Pure nickel[(c)]	8.902	0.1125	13.3	0.22	205

[(a)] Nickel containing 6 to 10 percent phosphorus.

[(b)] Specific heat after annealing at 1000 C ranged from 0.1112 to 0.1222 cal/g/C.

[(c)] 99.95 percent Ni + Co, electrolytic and carbonyl. (205)

temperature, and agitation, and by the current density. Extensive published information on the properties of nickel, summarized herein, can be used for selecting conditions that result in the properties required or desired for specific applications. Electroplating processes can often be tailored for optimizing the performance of the nickel coatings.

Physical Properties

Density data have been reported for nickel electrodeposited in several solutions, as summarized in Table 12.1. However, data for other physical properties, also summarized in Table 12.1, are relatively meager.

Density

The density of nickel coatings ranges from 8.85 to 8.93 g/cu cm, as a rule, which approaches or exceeds the value for pure electrolytic or carbonyl nickel at 8.902 g/cu cm.* A high density of 8.91 to 8.93 g/cu cm was reported for 0.7-mil or

* 99.95 percent Ni + Co, electrolytic and carbonyl. (205)

thicker nickel deposited in a Watts bath, (1–5) a Watts bath containing sodium sulfate, (1) fluoborate baths, (1) and sulfamate solutions. (2) An intermediate density of 8.85 to 8.89 g/cu cm was obtained for nickel deposited in a Watts bath containing ammonium sulfate, (1,5) a diluted Watts bath (1.0 N Ni), (1) a chloride solution, (1,2,5) an acetate (1) bath, and a bright bath developed for plating bright nickel-cobalt alloy. (1) Nickel deposited in Watts baths containing organic brighteners or a 4 N nickel chloride solution exhibited a low density of < 8.86 g/cu cm. (1)

Solution temperature does not seem to affect density appreciably. Annealing at 1000 C reduced the density of Watts-type nickel from 8.88 to 8.58, or from 8.85 to 7.95 g/cu cm for deposits from a Watts bath at pH 3 and 5, respectively. (1,5) Table 12.2 cites supplemental data on changes in density as a result of heat treatment. Annealing at 1050 C caused a decrease in density of only 0.6 percent for nickel deposited in a sulfamate solution with a pH of 4.3. Such a heat treatment decreased the density of other deposits as much as 7.8 percent. These changes probably were associated with inclusions of hydrated salts or oxides, which decomposed to make steam. The gas accumulated at grain boundaries expanded the metal.

Specific Heat

The specific heat of nickel deposited in bright nickel solutions is about 1 percent higher than nickel deposited in a Watts bath. The average value for the range of 0 to 94 C is 0.1083 ± 0.0002 cal/g/C for bright nickel. (1) Earlier data showed values

TABLE 12.2

Density of Nickel Coatings before and after Annealing (2)

	Density, g/ml		
Bath Type	Before Annealing	After Heating at 1050 C	Change, percent
Watts, pH 4.4[a]	8.90	8.59	−3.5
Watts, pH 2.5[a]	8.90	8.24	−7.8
Watts, pH 3.0[b]	8.88	8.81	−0.8
Sulfamate, pH 4.3[a]	8.93	8.88	−0.6
Sulfamate, pH 2.5[a]	8.93	8.46	−5.6
Chloride[c]	8.91	8.80	−1.2

[a] Wetting agent in bath.

[b] No wetting agent or peroxide in bath.

[c] Sample supplied by the National Bureau of Standards.

TABLE 12.3

Coefficients of Thermal Expansion (1,2)

Temperature Range, C	Expansion Coefficient, 10^{-6}/C Watts Bath	Sulfamate Bath
20 to 200	14.6	13.6
20 to 400	16.2	14.8
20 to 600	16.8	15.6
20 to 800	17.2	16.2
20 to 1000	17.2	17.0

of 0.1112 at O C and 0.1222 at 100 C for annealed cathode nickel. (5) Pure electrolytic or carbonyl nickel has a reported specific heat of 0.1125 cal/g/C at 100 C.*

Thermal Expansion and Thermal Conductivity

Expansion coefficients for nickel deposited in sulfamate solutions were slightly lower than the coefficients for Watts-type nickel (Table 12.3). Thermally unstable impurities such as hydrated salts or oxides which are occluded in nickel electrodeposits cause a permanent expansion during heating. The degree is proportional to the impurity content of the coatings.

A thermal conductance ranging from 0.195 to 0.26 cal/cm/sec-cm²/degree C was reported for electrodeposited nickel. (1,4,6) By comparison, electroless nickel containing 6 to 10 percent phosphorus shows much lower values, from 0.0105 to 0.0135 cal/cm/sec-cm²/degree C. Its coefficient of thermal expansion is 13×10^{-6} per degree C. (7)

Electrical Resistivity

The electrical resistivity of electrodeposited nickel at 20 C generally varies from 7.4 to 11.49 microhm-cm, depending on minor amounts of other metals (impurities) and incorporated foreign materials from organic additives. These values compare with wrought and cast nickel, as follows: (6)

Form	Resistivity, microhm-cm at 20 C
"A" Nickel (205)	9.5
"D" Nickel (205)	14
Cast Nickel	21

* 99.95 percent Ni + Co, electrolytic and carbonyl. (205)

Resistivity increases with decreasing density of nickel. For example, a resistivity of 7.8 microhm-cm was reported for nickel with a density of 8.9 g/cu cm, whereas a resistivity of as high as 35 microhm-cm was measured for nickel with a density of 8.45 g/cu cm, which was deposited in an uncommon, 25 C ammoniacal solution with a pH of 8.5. (1) This high-resistivity nickel exhibited high hardness and tensile strength.

Table 12.4 lists resistivities for nickel deposited at 5 amp/sq dm in several different solutions heated to 55 C and adjusted in pH to 3.0. Nickel from a cobalt-free Watts bath was lowest in resistivity (7.44 microhm-cm), whereas a bright nickel containing about 1.1 percent cobalt exhibited a resistivity of 11.49 microhm-cm. Other deposits containing 1 or 2 percent cobalt showed a lower resistivity of about 8.5 microhm-cm, however. Electrical resistivity increased from 8.6 to 16.3 microhm-cm when 33 g/l of cobalt chloride was added to a sulfamate solution. (8)

Increasing thickness from 12.5 to 50 μm (0.5 to 2 mils) reduced the resistivity of a Watts nickel deposit from 8.08 to 7.8 microhm-cm. Further increase in thickness to 380 μm had little or no effect on resistivity. (9)

As a rule, resistivity decreases with increasing current density for a Watts solution at 65 C, as shown in Figure 12.1. (1) Minimum resistivity was obtained at current densities of 5 to 25 amp/sq dm.

Variations in the solution pH in the range of 1.5 to 5.0 had little effect (e.g., from 7.5 to 7.4 microhm-cm for nickel deposited at 5 amp/sq dm in 55 C baths). On the other hand, increasing the pH of a cobalt-free, nickel chloride solution increased resistivity from 7.65 to 7.85 microhm-cm.

An increase in nickel concentration from 2 to 4 N increased resistivity about 10 percent for Watts-type nickel. (1) Operation at 20 C resulted in a high resistivity of 8.6 microhm-cm, but a change from 30 to 55 C had no significant influence on resistivity. On the other hand, a bath temperature of 70 to 85 C resulted in resistivities of 7.9 to 8.3 microhm-cm.

TABLE 12.4

Electrical Resistivities of Electrodeposits from Various Solutions (1,8)

Type of Bath	Resistivity, microhm-cm at 20 C	
	Before Annealing	Annealed at 1000 C
Acetate	9.18	9.34
Chloride	8.24	8.64
Chloride, cobalt-free	7.71	—
Co-bright	11.49	9.78
Fluoborate	8.34	8.37
Organic bright	10.00	8.42
Sulfamate	8.6	—
Watts	7.76	7.66
Watts, cobalt-free	7.44	—
Watts plus Na_2SO_4	7.81	7.70
Watts plus $(NH_4)_2SO_4$	7.89	8.01

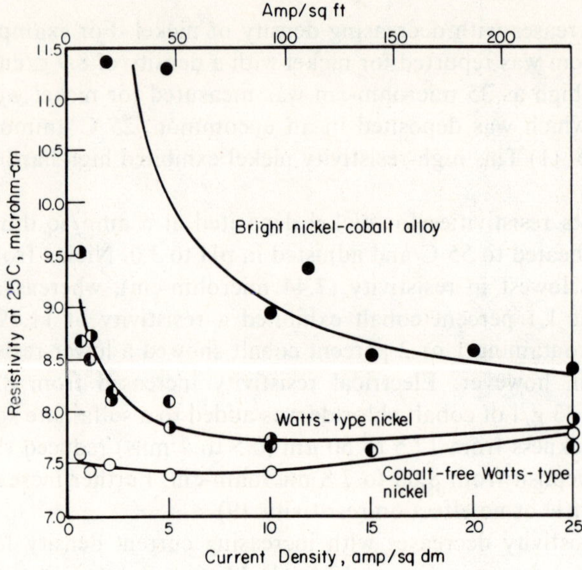

Figure 12.1. Resistivity of Watts-Type Nickel as a Function of Current Density. (1)

The addition of 0.02 g/l trisulfonated naphthalene increased resistivity from 8.6 to 12.4 microhm-cm at 18 C for a Watts-type deposit obtained from a solution also containing 6 g/l sodium fluoride and 110 g/l sodium chloride at pH 5.8. (10)

Increasing concentrations of nickel fluoborate, from 220 to 440 g/l in a bath operated at pH 2.0 to 2.5 and at 55 C, decreased resistivity approximately 25 percent. The reported values were lower than those reported for nickel deposited in Watts-type and chloride baths. (11)

Electrical resistivity values from 30 to 65 microhm-cm have been reported for electroless nickel deposits containing 6 to 10 percent phosphorous. (7)

Magnetic Properties

The magnetic properties of nickel coatings vary with solution composition and type and amount of organic additives. For example, bright deposits in general show about twice the coercive force and half the permeability of Watts-type deposits. Operating variables also alter the magnetic properties of nickel coatings. Comparative values of coercive force (H_c, in oersteds), residual induction (B_r, in gausses), and saturation magnetization (B_m, in gausses) for electrodeposited, chemically reduced, and metallurgical nickel are given in Table 12.5. (7)

The coercive force of thin electrodeposited nickel layers has been measured as a function of coating thickness. A thickness above 1 μm showed a nearly constant coercive force of 70 oersteds. Below 1 μm, the coercive force increased rapidly to a maximum of 220 oersteds at a thickness of 150 Å. (12)

TABLE 12.5

Magnetic Property Data (7)

Magnetic Property	Electrodeposited Nickel	Chemically Reduced Nickel				Metallurgical Nickel[b]
		From Acid Solutions		From Alkaline Solutions		
		Initial Stage	Heat Treated[a]	Initial Stage	Heat Treated[a]	
Coercive force, oersteds	40 to 120	Nonferromagnetic	140	4.1	140	3.4
Residual induction, gausses	200 to 4500	Nonferromagnetic	850	45	3000 to 3300	—
Saturation magnetization, gausses	~7000	Nonferromagnetic	1500 to 1700	140	3400 to 3700	6100

(a) 400 C for 1 hour.

(b) "A" Nickel, 99.4 percent Ni + Co. (3,205) Initial permeability 110 gausses. Maximum permeability 600 gausses.

The coercivity of Watts nickel, 17.4 oersteds at 55 C, was increased with lower operating temperatures and lower pH values. Thus, H_c was 51.0 oersteds for nickel deposited in a bath at 30 C and pH 1.5. Organic bright nickels had a coercivity of 53 oersteds, or more. (1)

Coercivity of nickel increased as the nickel chloride content was increased for deposits at 1.6 amp/sq dm: (13)

Nickel Chloride Concentration, g/l	H_c, oersteds
30	48
45	56
75	60
125	74
160	76

A similar relationship was found for cobalt deposits, which also showed a decrease from 38 to 8 oersteds with additions of up to 1.6 g/l saccharin.

Increase in current density and pH increased the coercive force of nickel electrodeposited in a sulfate bath, as a result of the increase in internal stress of the deposit. When the pH increased beyond the value of the hydrate formation, the structure of the deposit changed, with a resultant decrease in coercive force. (14)

A soft nickel-20 percent iron alloy with a high permeability and saturation magnetization and a low coercivity of 0.5 to 2 oersteds has been deposited at 0.6 amp/sq dm, at room temperature and pH 2.8 in a solution containing 88 g/l $NiSO_4$, 2.5 g/l $FeSO_4$, 15 to 80 g/l boric acid, 12 g/l NaCl, and 0.2 g/l sodium lauryl sulfate. The values compared favorably with other magnetic alloys, as for example, cobalt-nickel (50 g/l Co^{2+} + 50 g/l Ni^{2+}), which had a coercivity of 200 oersteds and a remanence of about 8000 gausses. Superimposed alternating current increased H_c from 200 to 300 oersteds, but reduced remanence to 4500 or 5000 gausses. Other data include coercivities from 200 to 600 oersteds for cobalt-phosphorous alloys. (15)

The permeability of nickel deposited in a Watts bath containing 2 g/l saccharin dropped to nearly zero as a result of additions of either 10 mg/l zinc or copper to the solution. Larger additions slightly increased permeability. With concentrations of less than 50 mg/l iron in the bath, magnetic permeability was little affected, but increased strongly with higher concentrations. (16)

The magnetic induction of nickel and iron films measured by double-exposure electron diffraction of the ferromagnetic specimen over gold measured 7000 and 3000 gausses, which is in agreement with the known saturation inductions. (17) The magnetic anisotropy constants K_1 and K_2 for electrodeposited nickel films 500 Å thick were reported as $(-2.1 \pm 0.2) \times 10^6$ and $(-1.8 \pm 0.4) \times 10^6$ ergs/sq cm, respectively. (18)

Impurities in Nickel Deposits

Hydrogen, oxygen, carbon, sulfur, and chloride are impurities in nickel deposits that adversely affect physical properties (density and resistivity) with increasing concentration in the following ranges:

Hydrogen	0.001 to 0.01
Oxygen	0.003 to 0.046
Sulfur	0.001 to 0.08
Carbon	0.02 to 0.1
Chloride	0.005 to 0.025

Sources for hydrogen in nickel include: hydrogen adsorption on the surface, inclusions of basic nickel salts or nickel hydroxide, or solution of atomic hydrogen in nickel. (19) The concentration is relatively small, 0.001 to 0.01 percent, with few exceptions. Elementary hydrogen has been reported at 0.13 ± 0.02 ml/g of metal, and is independent of current density and pH from 2.5 to 5, but dependent on temperature during nickel deposition in a Watts-type bath containing sodium chloride in a diaphragm cell. (20)

The oxygen content may vary from 0.003 to 0.046 percent depending on solution composition. (1) The lowest oxygen content in this range was obtained in nickel deposited in a concentrated Watts bath, and the highest was found in nickel deposited in a concentrated chloride bath. A Watts bath with a 2.0 N nickel concentration produced deposits with about 0.023 percent oxygen when the current density, temperature, and pH were adjusted to 5 amp/sq dm, 55 C, and 3.0, respectively. Nickel produced in other solutions, including bright plating solutions, showed oxygen contents from 0.005 to 0.016 percent. A very low or high current density or an increase in the pH to 5 increased the oxygen content. A decrease in pH to 1.5, an increase in solution temperature, and an increase in nickel ion concentration tended to decrease the oxygen content in the deposit.

The sulfur content in Watts-type nickel is generally less than 0.005 percent. Bright nickel solutions with sulfur-containing addition agents show concentrations from 0.05 to 0.15 percent. As much as 0.4 percent was found in some coatings, however. The nickel sulfide inclusions reduce corrosion resistance and induce brittleness. The amount of sulfur contamination is related to the concentration and type of organosulfur addition agents. However, the concentration of other solution constituents influences the sulfur content of the nickel. For example, the addition of 2 g/l sodium lauryl sulfate to a Watts bath reduced the sulfur content in the deposit from 0.013 to 0.006 percent. (19)

Carbon content ranges from 0.02 to 0.1 percent and depends solely on concentrations of organic additives in solution. (19)

Chlorides have also been found in nickel deposits. (1,19) The chloride content can range from 0.005 to 0.025 percent and is believed to come from inclusions of basic nickel salts.

Metallic impurities sometimes occur in nickel deposits. Their effects on corrosion resistance and mechanical properties are discussed later.

Corrosion Resistance

Watts nickel deposits exhibit good corrosion resistance. (4) Brightener additions which induce sulfide inclusions reduce corrosion resistance. (21) Nickel coatings on steel deposited in a Watts bath which contained benzene- or naphthalene- sulfonic acids corroded much faster in hydrochloric acid solution with a pH of 3 than deposits from a simple Watts bath. (22)

Ten ppm of copper in bright plating baths caused a 20 percent, and 25 ppm a 50 percent, loss in the salt spray corrosion resistance of steel plated with nickel. With 50 ppm, the loss in corrosion resistance amounted to 60 percent. (23) Increases in zinc concentration improved corrosion resistance, however. (24)

Additions of 10, 50, and 100 mg/l zinc, copper, and iron to a nickel bath containing 2 g/l saccharin as brightener at a pH of 4, 55 C, and a current density of 3 amp/sq dm were correlated with the potential changes of the nickel deposits. Whereas iron additions had little effect on the potential, copper and zinc caused a significant shift in the potential toward reduced corrosion resistance. For example, a potential change from −260 to −550 millivolts in the potential of nickel deposits due to increasing zinc concentration has been interpreted as loss of corrosion resistance. (25) There was no change in the salt spray resistance of coatings produced in baths containing up to 200 ppm of iron. (25,26)

Rain-erosion of Watts and bright nickel is dependent on hardness, internal stress, and thickness of the coating, and the hardness and pretreatment of the basis metal. (16)

Porosity

Porosity in nickel deposits is dependent on thickness of the coating and finish of the basis metal. For example, 5 μm nickel is nonporous on polished smooth copper, whereas steel finished with 240-grit paper requires 25 to 30 μm nickel to eliminate porosity. (19) For deposits from a sulfate-chloride solution, minimum porosity was observed when the pH of solution adjacent to the cathode was 7.6 to 7.7. The solution contained 260 g/l $NiSO_4 \cdot 7H_2O$, 20 g/l $Na_2SO_4 \cdot 10H_2O$, 45 g/l NaCl, and 21 g/l H_3BO_3. (27)

Hardness

Nominal hardness values for dull nickel deposits frequently range from 140 to 250 kg/sq mm (4) or 110 to 246 kg/sq mm. (28) Higher values of 500 kg/sq mm, (29) 530 kg/sq mm, (30) and 800 kg/sq mm (31) have been reported for nickel from undisclosed solutions, however. Bright nickel is usually harder than dull nickel, as indicated by the following comparison: (32,33)

Bath	Indentation Hardness, kg/sq mm	Scratch Hardness, kg/sq mm
Watts nickel	338 to 371	316 to 355
Bright nickel	463 to 622	286 to 433

Hardness measurements seem to be greatly dependent on the applied load and type of indentor, particularly for bright nickel deposits. These relationships have been discussed in a number of reviews dealing with the properties of electrodeposited nickel. (34–38)

Nickel coatings with a thickness of 0.15 to 0.5 mm (0.006 to 0.020 inch) on mild steel and deposited from a Watts bath at 63 to 68 C with a hardness of 140 to 160 kg/sq mm can be roll-formed, die-formed, flame-cut, sheared, and welded. (39) A hardness of 200, 300, and 350 kg/sq mm has been reported for electroformed nickel with an ultimate strength of 56, 112, and 140 kg/sq mm (80,000, 160,000, and 200,-000 psi), respectively. (40) A hardness of 110 kg/sq mm corresponded to an ultimate strength of 42 kg/sq mm (60,000 psi), and a hardness of 210 kg/sq mm to a strength of 78 kg/sq mm (110,000 psi) for Watts-type nickel. (41)

Application of ultrasound at 30 kc/sec and 0.3 w/sq cm to a nickel sulfate electrolyte reduced the amount of included hydrogen and the hardness. (42) For thin coatings from 5 to 10 μm, the deposits obtained in undisclosed solutions had hardness values from 350 to 440 kg/sq mm with or without ultrasound. (43) The hardness of nickel deposited at 2 to 10 amp/sq dm in a Watts nickel bath at 55 C with a pH of 1.5 also was unchanged with ultrasonics. (44).

Whereas the hardness of nickel deposited at 18.6 amp/sq dm in a 60 C chloride bath was 290 kg/sq mm, periodically reversing the current with a frequency of 500 c/s, an anodic to cathodic current ratio of 1, and pulse times of 75 percent cathodic and 25 percent anodic, reduced hardness to 145 kg/sq mm (50-g load). (45)

Superimposed alternating current at a magnitude three times greater than the direct current decreased the hardness from 250 to 150 kg/sq mm of deposits from an electrolyte containing 200 g/l nickel sulfate, 120 g/l sodium sulfate, 25 g/l sodium chloride, and 30 g/l boric acid. (46) Low-frequency oscillations produced by magnetostrictive methods and high-frequency ultrasonic oscillations of the cathode normal to the surface increased the hardness of the deposits, but oscillations parallel to the cathode reduced hardness. (47) Hard deposits were produced by plating very thin successive layers of nickel from a sulfamate bath onto a rotating cathode. Hardness decreased from 350 to 240 and to 200 kg/sq mm with increasing coating thicknesses from 100 Å, to 1000 Å and 10,000 Å. (48) The hardness of nickel deposited by brush plating was 540 kg/sq mm. (50, 51) Methods for measuring hardness of nickel coatings have been recommended. (52)

Effect of Solution Composition

Soft deposits, below 150 kg/sq mm, were reported for nickel deposited in fluoborate, sulfamate, and Watts-type nickel baths, but some conditions resulted in hardness values as high as 650 kg/sq mm. Deposits from acetate and chloride solutions were intermediate in hardness, usually from 200 to 360 kg/sq mm. Hardness ranges for nickel from these and other solutions are listed in Table 12.6.

Table 12.7 details hardness data for nickel deposited in chloride baths. A high nickel concentration (4 N) or a high pH (5.8) shifted hardness from the range of

TABLE 12.6

Hardness of Nickel Deposited in Several Types of Plating Baths

Type of Bath	pH	Current Density, amp/sq dm	Solution Temp, C	Hardness, kg/sq mm	Reference
Acetate	4.5	5	50	335	1
Chloride	1.5 to 5.8	2.5 to 21.6	30 to 70	181 to 360	1, 53–62
Electroless NiP	~5	—	82 to 93	355 to 1200	7, 61
Electroless NiB	~5	—	82 to 93	705 to 1200	61
Fluoborate	1.0 to 4.0	3 to 15	30 to 70	125 to 300	1, 11, 53, 57, 60, 63–67
Pyrophosphate	9.5	0.1 to 6	20 to 80	240 to 600	68, 69
Sulfamate	3.5 to 4.5	2.2 to 30	49 to 57	140 to 650	53, 56, 57, 59, 61, 65, 70–72
Sulfate[(a)]	1.2 to 5.8	2.5 to 5.4	30 to 70	160 to 450	10, 54, 55, 57, 58, 61, 62, 70, 73
Watts, 1.0 N Ni	3.0	5	55	170	1, 5
Watts, 1.1 to 2.5 N Ni	3.0 to 5.2	0.75 to 10.8	40 to 70	130 to 560	1, 5, 53–55, 63, 70, 71
Watts, 2.6 to 3.5 N Ni	1.4 to 4.8	2.7 to 10.8	45 to 70	113 to 560	53, 54, 56–58, 73, 75
Watts, 3.6 to 4.0 N Ni	1.1 to 4.2	5 to 15	30 to 70	220 to 370	1, 4, 64
Watts with cobalt sulfate	3.7 to 5	2 to 7	40 to 65	202 to 509	1, 53, 63
Watts with organic additives	3 to 5	2 to 5	45 to 55	330 to 546	1, 59–61, 63, 71, 73

[(a)] Bath contained $(NH_4)_2SO_4$, NH_4Cl, Na_2SO_4, and/or NaCl.

[(b)] Bath contained 25 g/l $Na_3C_6H_5O_7$, 10 g/l NH_4Cl, 8 g/l $MgSO_4$, and ~0.3 percent wetting agent.

TABLE 12.7

Hardness of Nickel Deposits as a Function of Solution Composition—Chloride Solutions

Solution Composition, N		pH	Current Density, amp/sq dm	Temp, C	Hardness, kg/sq mm	Reference
$NiCl_2 \cdot 6H_2O$	H_3BO_3					
2.0	0.5	3.0 to 5.0	5	55	244 to 300	1
2.10	1.21	1.5 to 2.0	2.7 to 21.6	60	230 to 260	70
2.42	1.82	3.0	4.3	NR	181 to 232	56
2.42[a]	1.45	2.0	2.7 to 10.8	50 to 70	200 to 350	57
2.42	1.45	5.4 to 5.8	NR	NR	230 to 360	53
2.42	1.45	2.0	2.5 to 10	60	230 to 260	58
2.77	1.45	2	10.8	60	240	54
2.77	1.45	2 to 2.5	5 to 10	60	260	55
4.0	0.5	3.0	5.0	55	324	1
0.34[b]	—	7.8 to 8.2	10 to 20	78 to 82	~300	76
1.18[c]	—	7.5 to 8.0	10	75 to 80	550 to 650	49, 77

[a] The solution also contained 24.8 g/l NH_4Cl.
[b] The solution also contained 300 g/l $(NH_4)_2C_2O_4$, 3 to 5 g/l NH_4Cl, and 15 g/l NaF.
[c] The solution also contained 300 g/l $(NH_4)_2C_2O_4$.

TABLE 12.8

Hardness of Nickel Deposits as a Function of Solution Composition—Fluoborate Baths

Solution Composition, N		pH	Current Density, amp/sq dm	Temp, C	Hardness, kg/sq mm	Reference
$Ni(BF_4)_2$	H_3BO_3					
1.90	1.45	2.5 to 4.0	5	32	243 to 280	11
1.90	1.45	2.5	8	54	204	11
2.0	1.45	3.0	5	55	174	63
2.0	0.5	3.0	5	55	174	1
2.54[a]	1.45	2.0 to 3.5	4.3 to 10	40 to 60	183	53
2.54	1.45	2.6 to 2.7	3	54	126 to 131	67
2.54	1.45	2.5 to 3.5	8	54	159 to 183	11
2.45[b]	1.45	1.0	5 and 15	30 to 70	260 to 300	64
3.45	1.45	2.0	8	54	164	11
2.54[c]	1.45	2.5	8	54	466	11

[a] The solution contained 3.7 to 37 g/l HBF_4.
[b] The solution also contained 30 g/l $NiCl_2$.
[c] The solution also contained 1 g/l saccharin.

TABLE 12.9

Hardness of Nickel Deposits as a Function of Solution Composition—Sulfamate Baths

Solution Composition, N							
$Ni(SO_3 \cdot NH_2)_2$	$NiCl_2 \cdot 6H_2O$	H_3BO_3	pH	Current Density, amp/sq dm	Temp, C	Hardness, kg/sq mm	Reference
2.4	0.25	1.45	3.5 to 4.2	2.2 to 10	50	To 200	70
2.6	0.25	1.45	4.5	3	50	220	71
2.7	0.38	1.09	3.5 to 4.2	4.3	49	140 to 190	53
3.2	—	1.45	4.5	4	57	185 to 207	56
1.9 to 5.7[(a)]	0.05 to 0.25	1.45 to 2.2	NR	NR	NR	190	65
2.4[(b)]	0.25	1.45	3.5	2.2 to 10	50	525	70

[(a)] 0.75 to 1.13 g/l benzene disulfonic acid and 0.1 to 0.2 percent by volume of an antipitting agent were added to this solution.
[(b)] 7.5 g/l naphthalene trisulfonic acid was added to this solution.

TABLE 12.10

Hardness of Nickel Deposits as a Function of Solution Composition—Watts-Type Baths

Solution Composition, N								
Total Ni	$NiSO_4$	$NiCl_2$	H_3BO_3	pH	Current Density, amp/sq dm	Solution Temp, C	Hardness kg/sq mm	Reference
1.0	0.75	0.25	0.5	3.0	5	55	170	1, 5
1.5	0.76	0.74	2.0	1.5	10.8	45	205	54
2.0[a]	1.5	0.5	0.5	3.0	5	55	150	1, 5
2.03 to 2.71	1.6 to 2.0	0.43 to 0.71	1.19 to 1.57	4.8 to 5.4	0.75 to 4.3	40 to 45	480 to 560	75
2.12	1.6	0.52	1.45	3.0	5	55	150	63
2.21	1.82	0.39	1.45	4.5 to 5.2	3.8 to 7.6	50	150 to 200	70
2.21	1.82	0.39	1.45	4.5 to 6.0	2.2 to 10.8	46 to 71	130	54
2.67	2.28	0.39	0.55	2.0 and 4.75	—	—	113 and 329	73
2.75	2.5	0.25	1.45	2 to 2.5	4 to 8	60	150	55
2.76	2.5	0.26	1.45	2.0	2.5 to 10	60	140 to 160	58
2.80	2.28	0.52	0.55	3.0	4.3	60	125 to 135	56
2.89	2.5	0.39	0.55	1.5 to 4.5	—	—	140 to 160	53
3.02	2.5	0.52	1.45	1.5 to 4.5	2.7 to 10.8	46 to 71	120	54
3.07	2.28	0.79	0.55	2.0	2.3 to 6.5	—	141 to 187	73
3.17	2.74	0.43	2.0	1.4 to 4.5	2.7 to 10.8	46 to 60	140 to 250	57
3.85	3.2	0.65	1.45	1.1 to 4.2	5 and 15	30 to 70	220 to 370	64
4.0	3.0	1.0	0.5	3.0	5	65	253	1, 5
2.0 to 2.2[b]	1.5 to 1.86	0.35 to 0.5	0.5 to 1.5	3.0 to 5.0	2 to 7	40 to 65	320 to 540	1, 10, 59–61, 63, 71, 73
1.82 to 2.08[c]	1.7 to 1.82	0.25	0.5 to 1.45	3.7 to 4.7	4.3 to 7	40 to 65	202 to 509	1, 53, 63
1.3 to 2.1	1.3 to 2.1	None	1.5	1.2 to 5.9	5	45 to 60	160 to 450	57, 62
0.6 to 2.3[d]	0.6 to 2.3	None	0.7 to 2.2	1.2 to 5.0	0.9 to 20	25 to 70	225 to 500	53–55, 58, 61 69, 70, 78
1.1[e]	1.1	None	1.0	5	—	—	180 to 300	10

[a] 147 kg/sq mm with an addition of 0.5 N Na_2SO_4 and 170 kg/sq mm with an addition of 0.5 N $(NH_4)_2SO_4$.

[b] Bright Watts nickel solutions containing organic additives.

[c] Watts nickel solutions containing 0.01 and 0.02 N $CoSO_4$, $(NH_4)_2SO_4$, and CH_2O.

[d] Nickel sulfate solutions containing 0.24 and 0.47 N NH_4Cl.

[e] Nickel sulfate solution 0.75 N Na_2SO_4 and 2 N NaCl.

about 200–250 kg/sq mm to > 300 kg/sq mm. The addition of ammonium oxalate, which increased the pH to 8.0, increased hardness to about 600 kg/sq mm.

The addition of nickel chloride (64) or saccharin (11) to fluoborate solutions increased hardness from the range of 125–200 kg/sq mm to about 300 and 460 kg/sq mm, respectively. Data for fluoborate baths are detailed in Table 12.8.

Nickel deposited in sulfamate solutions exhibited hardness values of 140 to 220 kg/sq mm, as a rule. The addition of 7.5 g/l naphthalene trisulfonic acid increased hardness to 575 kg/sq mm. (70) Table 12.9 includes hardness data for nickel deposited in nickel sulfamate solutions.

As shown in Table 12.10, the hardness of nickel deposited in Watts-type baths containing 1.0 to 3.0 N nickel usually ranges from 130 to 200 kg/sq mm, except for deposits produced in solutions with a pH of > 4.8, when hardness reached 560 kg/sq mm. When the nickel concentration was 4.0 N in a bath with a pH of 3.0, a hardness value of 250 kg/sq mm was recorded. Additions of cobalt sulfate increased hardness to the range of 202 to 509 kg/sq mm. The harder deposits in this range were obtained at 4.3 to 10 amp/sq dm with a cobalt sulfate concentration of 2.6 g/l in the 40 to 65 C solutions with a pH of 3.7 or 4.7. The cobalt content of the deposits was about 1 percent.

Effect of Impurities

Impurities in nickel deposits may be (a) metallic, (b) decomposition products of organic additives, and (c) gaseous or anionic inclusions. Some metallic impurities such as cobalt, iron, and lead increase hardness, but others show little or no effect, as indicated in Table 12.11.

The change in hardness with the addition of chromium ions depended on their concentration and the pH of the plating bath, as follows: (82)

	Change in Hardness, percent					
	Watts Nickel				Bright Nickel	
	pH of 2.0		pH of 2.5		pH of 3.2	
Cr Concentration, mg/l	Cr^{3+}	Cr^{6+}	Cr^{3+}	Cr^{6+}	Cr^{3+}	Cr^{6+}
25	−16	−12	38	7	−29	0.5
75	−15	−10	97	—	−12	−1.5
125	−15	—	—	—	−3	—

Effect of Current Density

In most instances, the hardness of nickel electrodeposits increases as current density decreases from 2 to 0.25 amp/sq dm. (1,19) For example, Watts nickel deposits obtained at a current density of 2 amp/sq dm exhibited a hardness of 138 or 139 kg/sq mm, whereas the hardness of nickel deposited at lower current densities ranged from 178 to 322 kg/sq mm. With current densities of 10 amp/sq dm or more, hardness was in the range of 186 to 234 kg/sq mm. Table 12.12 includes supplemental data showing harder nickel reported elsewhere. Increases in hardness

TABLE 12.11

Changes in Hardness with Metallic Impurities

Impurity	Concentration, g/l	Type of Bath	Hardness Change, percent	Reference
Al	0.010 to 0.100	Watts	Small	79
Ca	≤ 4.0	Sulfamate	None	80
Cd	≤ 1.0	Sulfamate	Increase	80, 81
Co	>7.0	Sulfamate	Increase	80
Cr^{3+}	0.025 to 0.075	Watts		
		pH 2	−15	82
		pH 5.2	+38 to +97	82
		Bright	−3 to −29	82
	0.010	Sulfamate	None	80
Cr^{6+}	0.025 to 0.075	Watts		
		pH 2	−10 to −12	82
		pH 5.2	+7	82
		Bright	0.5 to −1.5	82
	0.080	Sulfamate	Increase	80
Cu	0.050	Watts		
		pH 2.2	Significant increase	24
	≤ 1.0	Sulfamate	None	80
Fe	2.5 to 200×10^{-3}	Watts	Increase	26
	0 to 0.9	Sulfamate	Increase[a]	80
H and O		Sulfate and Chloride	Linear increase	27, 83
Mg	12.16	Watts	None	84
		Chloride	None	84
	24.32	Watts	+10.4	84
		Chloride	+9.8	84
	≤ 5.0	Sulfamate	None	80
Mn	0.7	Sulfamate	Increase	80, 81
	0.7 to 1.0	Sulfamate	Decrease	80, 81
Na	≤ 1.0	Sulfamate	None	80
Pb	0.60	Sulfamate	25	80, 81
Zn	0.010 to 0.300	Watts, pH 2.2		
		pH 5.2	Increase	25
		Bright	Increase	25
		Alloy-Ni	Decrease	25
NO_3	>0.5	Sulfamate	Increase	80, 81
PO_4		Sulfamate	Small	80, 81

[a] Hardness increased by 50 kg/sq mm for each 0.25 g/l of Fe^{2+}.

TABLE 12.12

Hardness Values of Watts-Type Nickel Deposits with Change in Current Density

Current Density, amp/sq dm	pH	Solution Temp, C	Concentration, g/l $NiSO_4 \cdot 7H_2O$	$NiCl_2 \cdot 6H_2O$	Hardness, kg/sq mm	Reference
0.25	3.0	55	210	60	322	1
0.5	3.0	55	210	60	178	1
1.0	3.0	55	210	60	194	1
2.0	3.0	55	210	60	138	1
2	5.7	54	180	25	375	85
3 to 5	2.0	50	250	0	320	87
5	3.0	55	210	60	142	1
5 to 15	1.1	50	~400	~150	245 to 275	64
5	2.0	71	~300	~45	275 to 305	86
8	5.7	54	180	25	475	85
10.0	3.0	55	210	60	186	1
7 to 20	1 to 2	60	250	0	220 to 300	87, 88
25	3.0	55	210	60	234	63
300 to 400[a]	2.0	71	~300	~45	340 to 360	86

[a] Electrolyte solution was pumped at a rate of 750 ft/min.

have been associated with grain refinement and an increased amount of inclusions. (19)

Hardness is said to be controllable by varying the current density during deposition in a sulfamate bath. (89) Nickel was hardest for deposits at 6 amp/sq dm from a bath containing 330 g/l nickel sulfamate: (90)

Temperature, C	Current Density, amp/sq dm	Hardness, kg/sq mm
40	4	380
	6	390
	8	360
50	4	460
	6	500
	8	490
60	4	550
	6	580
	8	580

The above values are harder than those generally reported for sulfamate nickel (Table 12.13), probably because the harder nickel contained sulfur induced by the decomposition of sulfamate ions.

Current density shifts caused large differences in hardness values (100 to 150 kg/sq mm) for nickel deposited in a Watts-type solution containing 0.13 M nickel fluoborate. (94) However, at a solution temperature of 65 C and a pH of 3.0 or 3.5, hardness values varied only by 22 units. Deposits from an all-fluoborate bath increased slightly in hardness from 260 to 275 kg/sq mm when the current density was increased from 5 to 15 amp/sq dm at 50 C. (64) An increase to 300 kg/sq mm was noted when 30 g/l nickel chloride was added to the fluoborate bath.

The hardness of nickel deposited in 0.5 M nickel pyrophosphate solutions containing 33 g/l ammonium citrate increased from 380 to 450 kg/sq mm as the current density was increased from 1 to 6 amp/sq dm. (68) Current density did not affect hardness (550 to 650 kg/sq mm) for nickel deposited from a sulfate-ammonium oxalate electrolyte, however. (49)

Effect of Solution Temperature

An increase in solution temperature generally decreases the hardness of nickel deposits. With a bath containing 180 g/l nickel sulfate, 25 g/l nickel chloride, and 30 g/l boric acid, hardness decreased from 430 kg/sq mm at 43 C to 350 kg/sq mm at 60 C with a solution pH of 5.4 to 6.0. (85) A temperature decrease to 30 C increased hardness of nickel from other baths. (1) With a bath containing 240 g/l nickel sulfate, 20 g/l potassium chloride, and 30 g/l boric acid, hardness decreased from 370 to 200 kg/sq mm (55-g load) when the bath temperature was raised from 10 to 55 C. (95) The decrease in hardness with increasing temperature was more pronounced for deposits obtained at pH 2 than at pH 5. An increase from 30 to 77 C lowered the hardness from 220 to 110 kg/sq mm when the pH was 2 and from 140 to 120 kg/sq mm when the pH was 5. (96) With solutions of 1 M nickel sulfate or nickel chloride, hardness also is affected by temperature. The highest hardness was obtained at 20 C. (97) Hardness of nickel deposits obtained in a sulfate bath containing magnesium salts was reported to decrease when the temperature increased above 50 C. (98)

TABLE 12.13

Hardness Values of Sulfamate Nickel Deposits with Change in Current Density

Current Density, amp/sq dm	Solution Temp, C	Nickel Sulfamate Concentration, g/l	Hardness, kg/sq mm	Reference
1 to 3	60	280	260 to 319	91
2.2	NR*	600	240 to 270	92
>2.2	NR*	600	210 to 230	92
4 to 8	20	600	180 to 250	90
6	40 to 60	330	390 to 580	90
10	20	600	280	90
25	NR*	600	225	93
40	20	600	210	90

* Not reported.

TABLE 12.14

Effect of pH on Hardness of Nickel Deposited in a Watts Bath (101, 102)[a]

pH	Vickers Hardness	
	As Deposited	Annealed[b]
2.0	138	63
3.0	140	66
4.0	130	58
5.0	150	68
5.3	215	77

[a] Bath composition:

$NiSO_4 \cdot 6H_2O$	325 g/l
$NiCl_2 \cdot 6H_2O$	45 g/l
H_3BO_3	30 g/l
54 C	
5 amp/sq dm.	

[b] At 430 C, 15 minutes.

For deposits from a solution containing 1.15 M $NiSO_4$, 0.13 M $Ni(BF_4)_2$, 0.06 M $NiCl_2$, and 0.6 M H_3BO_3, hardness decreased with increasing temperature from 43 to 60 C, as a rule. (94) With an all-fluoborate bath, hardness changed from 295 to 275 to 270 kg/sq mm at 30, 50, and 70 C, respectively. (64)

For deposits at 3 amp/sq dm from a sulfamate solution containing 280 g/l nickel sulfamate with a pH of 4.5, the hardness decreased from 567 to 310 kg/sq mm as the temperature was raised from 18 to 60 C. (91) With an increase in temperature from 30 to 40 C of another solution with a pH of 4.0, hardness decreased from 204 to 168 kg/sq mm and then held nearly constant up to 60 C. (99) Similar data were obtained for temperature changes from 32 to 50 C at a current density of 4 amp/sq dm. (53, 100) Minimum hardness was reported to occur at 39 C with a more pronounced increase at lower than at higher temperatures. (81)

Effect of Solution pH

A low pH (1.5 or 2.0) or a high pH (> 5.0) results in harder deposits, in comparison with intermediate pH values. For example, nickel deposited at 5 amp/sq dm in Watts-type solutions at 55 C with a pH of 1.5 and 5.5 was harder than nickel deposited at a pH of 3.0 because of an increase in oxygen content from about 0.01 to 0.02 or 0.04 percent. The respective hardness values were 200 to 240 at pH 5.5 and 160 to 170 kg/sq mm at pH 1.5, as compared with 150 kg/sq mm at pH 3.0. (1)

With a low-concentration Watts bath (180 g/l $NiSO_4$, 25 g/l $NiCl_2$, and 30 g/l H_3BO_3) at 43 C, hardness increased from 430 to 510 kg/sq mm as the pH was increased from 5.0 to 6.0. (85) A hardness (10-g load) of 220 kg/sq mm for nickel deposited in a 30 C Watts solution with a pH of 2.0 compared with 140 kg/sq mm at pH 5.0. (96) At 55 C, hardness increased from 120 to 195 kg/sq mm when the pH was increased from 2.0 to 5.3. Raising the pH of Watts-type nickel bath from 2.0 to 5.0 showed little difference in hardness, but increasing pH from 5.0 to 5.3 increased hardness from 150 to 215 kg/sq mm (Table 12.14). (101,102)

With a 2 N $NiCl_2$, 0.5 M H_3BO_3 solution at 55 C, the hardness increased from 255 to 300 kg/sq mm when the pH was raised from 3.0 to 5.0, while the oxygen content increased from 0.01 to 0.1 percent. (1) Bright nickel deposits showed appreciable increases in hardness when solution pH was increased from 4.0 to 5.0. (103) A linear increase was established between hardness and catholyte pH for a solution consisting of 260 g/l $NiSO_4$, 20 g/l Na_2SO_4, 45 g/l NaCl, and 21 g/l H_3BO_3. (27)

An increase in the pH of a 50 C sulfamate bath from 2.5 to 5.5 increased hardness from 140 to 410 kg/sq mm. (104) Other investigators have reported the same trend for the pH range of 2 to 4.5. (91,99) With a sulfamate bath operated at 28 C, the hardest deposits were obtained at pH 5.0 to 5.2. (81) A nickel sulfate-ammonium oxalate bath produced coatings for which hardness was independent of solution pH. (49)

Effect of Organic Additives

Many organic additives increase the hardness of nickel electrodeposits. Table 12.15 shows typical hardness values reported for several common addition agents.

Coumarin, used primarily as a leveling agent for depositing semibright nickel, hardens nickel to about 575 kg/sq mm, in comparison with 220 kg/sq mm for nickel deposited in the Watts-type bath containing no organic addition agent. Additions of naphthalene disulfonic acid and saccharin for reducing stress are similar to coumarin in their effect on hardness, as indicated in Table 12.15. Hardness values above 600 kg/sq mm have been reported for nickel deposited in solutions containing sulfonated naphthalene or toluene. (10,64,116) Benzene disulfonic acid is also a hardener, but appears to have less effect on hardness than naphthalene trisulfonic acid. (74,105,117) These and other compounds containing sulfur consistently increase hardness. For example, a thiourea addition to a sulfate-chloride bath, which contributed 0.69 percent sulfur to the deposit, increased hardness from 210 to 570 kg/sq mm. (115)

A combination of butynediol, saccharin, and formalin additions to a Watts-type bath for depositing bright nickel raised hardness to 600 kg/sq mm. (118) Hardness ranged from 350 to 650 kg/sq mm for nickel deposited in a Watts-type bath containing 40 g/l sodium saccharinate and 0.5 g/l alkane monosulfonic acid. (119) Very hard deposits, 700 kg/sq mm, were cited for nickel deposited in a solution containing nickel sulfate, sodium chloride, coumarin, chloramine-B, and pyridine. (107,120)

TABLE 12.15

Effect on Hardness of Organic Addition Agents

Agent	Concentration, g/l	Type of Bath	Hardness, kg/sq mm No Agent	Hardness, kg/sq mm With Agent	Reference
Benzene disulfonic acid	1.5	Sulfamate	—	260	105
Benzene disulfonic acid	0.6	Watts	280	380	74
Butynediol	0.2	Watts	220	460	106
Butynediol	0.5	Watts	220	620	106
Chloral hydrate and naphthalene disulfonic acid	1.0 (0.6)	Watts	280	570	74
Coumarin	1.0	Watts	220	574	103
Coumarin	3.0	Watts	220	578	103
Coumarin	3.0	Watts	220	580	106
Coumarin, trichloromethyl	1.5	Watts	—	700	107
Cystine	0.5 to 0.8	Watts	250	450	108
Glycine	20	Sulfate-chloride	—	380	109
Naphthalene disulfonic acid	5.0	Fluoride	—	550	110
Ditto	3.75	Sulfamate	190	270	8, 53
"	7.5	Sulfamate	190	475	8, 53
"	1.0	Sulfate-fluoride	374	695	10
"	4.6	Sulfate-chloride	—	520	111
"	1.0	Watts	280	380	74
"	8.0	Watts	250	650	64
Saccharin	0.5 to 1.5	Fluoborate	256	580	112
Saccharin	0.5	Sulfamate	—	525	113
Saccharin	0.03	Watts	220	520	114
Saccharin	3.0	Watts	220	580	106
Thiourea	0.2	Sulfate-chloride	210	570	115
p-toluene sulfonamide	1.0	Watts	280	340	74
p-toluene sulfonamide	2.0	Sulfamate	—	538	120

Wetting agents such as sodium lauryl sulfate generally have little or no effect on hardness. For example, no change in hardness was detected when sodium lauryl sulfate was added to a nickel sulfamate solution containing saccharin. (81) On the other hand, a reduction in hardness from 340 to 190 kg/sq mm was reported after sodium lauryl sulfate was added to a Watts bath. (121)

The equation $V = [500A + 800(F - 0.128)] - pH/(CD)^{1/3}$ has been proposed to correlate hardness (V), the ratio of sodium formate (F) to total nickel in the bath, the agitation factor (A), pH, and current density for deposits obtained in a modified Watts bath. (122)

Effect of Heat Treatment

Annealing reduces hardness of all pure nickel deposits. Hardmess values of nickel from fluoborate solutions annealed at 760 C were about half the values of the unannealed metal (64 to 71 relative to126 to 131 kg/sq mm). (67) Similar values were measured for annealed and unannealed deposits from a Watts bath, as shown in Table 12.14. The annealing temperature for these Watts-type deposits was 430 C (800 F). However, nickel with an initial hardness of 200 to 250 kg/sq mm exhibited hardness values of 170 to 215 kg/sq mm after annealing at 400 C. (1) The hardness of nickel deposited in a chloride bath was reduced from the range of 300 to 360 to the range of 275 to 325 kg/sq mm after heating at 400 C. A bright nickel with an initial hardness of 496 kg/sq mm was reduced to 216 kg/sq mm with such a heat treatment. Other bright nickels ranging in initial hardness from 572 to 823 kg/sq mm exhibited a hardness of 383 to 605 kg/sq mm after 400 C heating. Heat treating at 1000 C reduced the hardness of the initially soft Watts-type nickel deposits to 30 or 40 kg/sq mm. The initially hard bright nickel plates were reduced in hardness to < 95 kg/sq mm after annealing at 1000 C.

A deposit containing carbon, deposited in a nickel sulfate - ammonium oxalate bath, was increased in hardness from 550 to 900 kg/sq mm by heating at 300 C for an hour. (49) In another report citing the same kind of solution, the hardness of the nickel deposit was increased from 500 to 900 kg/sq mm by heat treating at 300 C, but heating at > 350 C softened the deposit. The increase in hardness caused by heating at 300 C probably was a result of the precipitation of carbides. (127)

Hardness at High Temperature

Nickel with an initial hardness of 340 kg/sq mm, deposited at 5 amp/sq dm in a nickel chloride bath at 55 C, exhibited a hardness value of only 50 kg/sq mm, measured at 600 C. (123) At 800 C, hardness was further reduced to only 25 kg/sq mm. Nickel alloy containing 17 or 25 percent cobalt was slightly harder than nickel at elevated temperatures. (124) Cobalt electrodeposits also are slightly harder than nickel at high temperatures. (123) Chromium and cobalt-tungsten alloy electrodeposits retained considerably higher hardness at high temperatures.

Strength and Ductility

Reports on the tensile strength of nickel cite values from 35 to > 140 kg/sq mm (50,000 to > 200,000 psi). The low-strength nickel was the most ductile with an elongation of 30 to 35 percent, whereas the highest strength nickel exhibited an elongation of only 1.0 percent. Data for nickel electroformed in 11 types of solutions are summarized in Table 12.16.

Thin nickel deposited in a Watts-type bath was stronger and less ductile than thicker deposits. (1,9) As thickness increased from 12.5 μm (0.5 mil) to 350 μm (14 mils), strength decreased from 86 to 42 kg/sq mm (122,000 to 60,000 psi). (1) Elon-

TABLE 12.16

Strength and Elongation of Electrodeposited Nickel

Type of Solution	Ultimate Tensile Strength		Yield Strength		Elongation, percent	Reference
	kg/sq mm	psi	kg/sq mm	psi		
Acetate[a]	103	147,000	No data	No data	4.0	1
Bright (cobalt)[b]	145	206,000	Ditto	Ditto	4.0	1
Bright (organic)[c]	149	212,000	—	—	5.0	1
Chloride[d]	49	70,000	No data	No data	21	58
Chloride[e]	76 to 87	108,000 to 124,000	65	92,000	8.0	56
Fluoborate[f]	39	56,000	—	—	32	67
Fluoborate[g]	53	74,500	37	52,700	16.6	11
Fluoborate + saccharin[h]	141	203,000	84	120,000	1.0	11
Sulfamate[i]	56 to 67	79,000 to 96,000	32 to 48	46,000 to 68,000	9 to 14	125
Sulfamate[j]	78 to 84	111,000 to 119,000	52 to 55	74,000 to 79,000	7 to 14	56
Sulfamate + organic agent[k]	165	235,000	No data	No data	1.0	126
Sulfate[l]	39	56,000	Ditto	Ditto	31.0	1
Sulfate-chloride[m]	107	152,000	"	"	5 to 8	53, 55
Watts[n]	35	50,000	"	"	30	55, 58
Watts[o]	39	56,000	"	"	28.0	1, 54, 63
Watts[p]	40 to 44	57,000 to 64,000	22	32,000	27 to 35	56
Watts[q]	91 to 105	130,000 to 150,000	No data	No data	5 to 7	75

	Nickel Conc., N	pH	Other Constituents	Current Density, amp/sq dm	Temp, C
(a)	2.2	4.5	$NiCl_2$ and $Ni(C_2H_3O_2)_2$	5	50
(b)	4.2	3.7	$CoSO_4$ (0.02 N), $Ni(HCO_2)_2$ (0.6 N), and NH_4SO_4 (0.01 N)	5	65
(c)	2.0	3.0	$NiCl_2$ (0.5 N), H_3BO_3 (0.5 N), 7.5 g/l benzene sulfonate, etc.	5	55
(d)	2.5	2.0	H_3BO_3 (0.5 M)	8 to 10	60
(e)	2.5	3.0	H_3BO_3 (0.6 M)	4	60
(f)	2.6	2.7	H_3BO_3 (0.5 M)	3	54
(g)	2.6	2.5	H_3BO_3 (0.5 M) and HBF_4 (0.3 M)	7.5	55
(h)	2.6	2.5	H_3BO_3 (0.5 M), HBF_4 (0.3 M), plus 1.0 g/l saccharin	7.5	55
(i)	2.2	3.0 to 3.7	$NiCl_2$ (0.1 M) and H_3BO_3 (0.6 M)	2.1	38
(j)	3.3	4.5	H_3BO_3 (0.5 M)	4	58
(k)	No data	4.0	Saccharin or other stress-reducing agent	5 to 6	46
(l)	2.0	3.0	H_3BO_3 (0.5 N) and Na_2SO_4 (0.5 N)	5	55
(m)	1.7	5.4 to 5.8	H_3BO_3 (0.5 M)	No data	—
(n)	2.0 to 2.5	2.0 to 2.5	$NiCl_2$ (0.12 M) and H_3BO_3 (0.5 M)	8 to 10	60
(o)	2.0	3.0	H_3BO_3 (0.5 M)	5	55
(p)	2.3	3.0	$NiCl_2$ (0.25 M) and H_3BO_3 (0.6 M)	4	60
(q)	2.2	4.8 to 5.4	$NiCl_2$ (0.25 M) and H_3BO_3 (0.5 M)	4	40 to 45

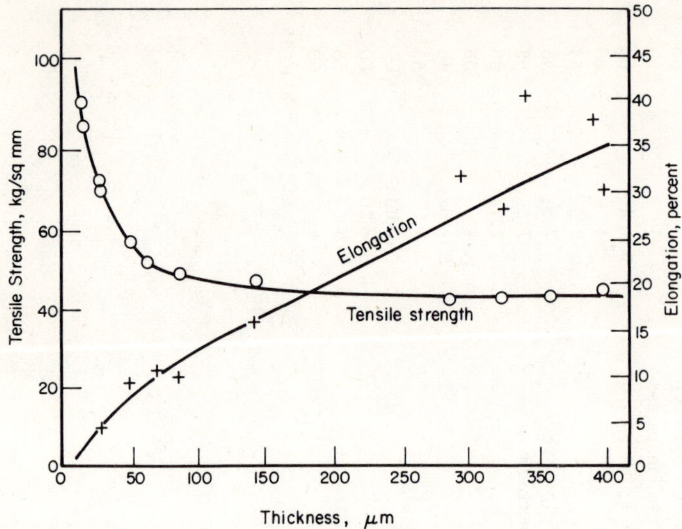

Figure 12.2. Variation of Tensile Strength and Elongation with the Thickness of Nickel Deposited at 5 amp/sq dm in a Watts-Type Bath at 55 C with a pH of 3.0. (1)

gation increased from about 10 to > 30 percent. Figure 12.2 shows the effect of thickness on elongation and strength. The data in Table 12.16 refer to deposits with a thickness of at least 175 μm (7 mils), which were about equal to thicker deposits in tensile strength.

Coarse-grained nickel deposited in fluoborate and Watts-type baths at 54 and 55 to 60 C, respectively, had a low tensile strength of 35 to 44 kg/sq mm (50,000 to 64,000 psi), as a rule, and an elongation of 27 to 35 percent. However, a tensile strength as high as 105 kg/sq mm (150,000 psi) and an elongation of 5 to 7 percent were reported for Watts-type nickel deposited in a bath operated at a lower than normal temperature (40 to 45 C) with a higher than normal pH (4.8 to 5.4).

Nickel with a tensile strength of > 140 kg/sq mm (> 200,000 psi) has been deposited in bright nickel baths and sulfamate plating solutions, which contained organosulfur compounds to brighten the plate and/or reduce stress. The minimum sulfur content for depositing such high-strength nickel is approximately 0.01 percent. (126) In comparison, deposits from sulfamate solutions which contained < 0.005 percent sulfur ranged in tensile strength from 42 to about 85 kg/sq mm (60,000 to 120,000 psi).

Intermediate values of tensile strength were reported for nickel from acetate, chloride, and sulfate-chloride baths. Such deposits exhibited tensile strengths of 49 to 107 kg/sq mm (70,000 to 153,000 psi) and elongations of 4 to 21 percent. Low-sulfur bright nickel-cobalt alloys deposited in a sulfate solution containing nickel formate had a tensile strength of 145 kg/sq mm (206,000 psi) and an elongation of 16.6 percent. (1)

TABLE 12.17

Effect of Nickel Fluoborate Concentration on Tensile Strength and Elongation (11)

Nickel Fluoborate Concentration, N[a]	Tensile Strength		Yield Strength		Elongation, percent
	kg/sq mm	psi	kg/sq mm	psi	
1.9	57	81,400	41	58,300	14.4
2.8	52	74,500	37	52,700	13.0
3.8	38	54,800	28	40,300	20.4

[a] The solution contained 0.5 M H_3BO_3 and had a pH of 2.0 to 2.5. The temperature was 55 C and the current density was 7.5 amp/sq dm.

Effect of Changes in Concentrations of Nickel Salts

Watts-type nickel deposited in 1 N nickel baths was about 10 percent stronger than deposits from 2 N solutions. (1) However, nickel deposited in 4 N solutions was about 45 percent stronger, when the solution pH was adjusted to 3.0, or 90 percent stronger when the pH was 5.0. Such nickel with a tensile strength of 78.5 kg/sq mm (112,000 psi) had an elongation of only 2 percent, in comparison with 25 percent for 42 kg/sq mm (60,000 psi) nickel.

An increase in the nickel concentration of a fluoborate bath reduced tensile and yield strengths of electrodeposited nickel and increased elongation, as detailed in Table 12.17. (11)

TABLE 12.18

Effect of Chloride-Sulfate Ratio on Strength and Ductility of Nickel Deposited in a 0.5 N Nickel (Watts) Solution (1,9)

Ratio of $NiCl_2$ to $NiSO_4$[a]	Tensile Strength				Elongation, percent	
	pH = 3		pH = 5			
	kg/sq mm	psi	kg/sq mm	psi	pH = 3	pH = 5
1:3	42	60,000	46	65,000	30	26
2:2	52	75,000	77	110,000	26	9
3:1	63	90,000	88	125,000	16	7
4:0	70	100,000	105	150,000	15	3

[a] The solution contained 0.5 M H_3BO_3 and was operated at 5 amp/sq dm at 55 C.

Doubling the nickel chloride concentration from 45 to 90 g/l in a Watts bath increased the ultimate strength slightly from 38 to 41 kg/sq mm (54,800 to 58,300 psi). (73) For a dilute nickel solution containing 0.5 N total nickel, an increase in nickel chloride showed greater increases in tensile strength, as shown in Table 12.18. (1,9)

Effect of Cationic Additions

The effect of bath impurities on the strength of nickel coatings has drawn surprisingly little attention. A 1 N addition of magnesium had no effect on the strength of nickel deposited in a Watts or chloride bath, but 2 N additions increased the strength of Watts-type deposits by 15 percent and chloride-bath deposits by 6 percent. (84) Elongation decreased 23 percent for the Watts nickel and 31 percent for the chloride nickel.

Nickel deposited in solutions containing 0.5 N sodium or potassium ions showed no effect for these on the mechanical properties of Watts-type nickel. (1) Small concentrations of cobalt ions, which codeposited a maximum of 1.4 percent cobalt, had only a slight influence on tensile strength and elongation. On the other hand, additions of ammonium ions harden nickel and increase its tensile strength. Nickel from a bath with 0.5 N ammonium sulfate had a tensile strength of 54 kg/sq mm (77,000 psi) and an elongation of only 17 percent, in comparison with Watts-type nickel with a tensile strength of 39 kg/sq mm (56,000 psi) and an elongation of 28 percent with the same temperature and current density conditions.

The effect of iron as an impurity in four types of nickel plating solutions has been investigated. There was a slight decrease in ductility of deposits from Watts and bright organic baths as the iron concentration of the solution increased to 200 mg/l. (26)

Chromium additions (Cr^{3+} and Cr^{6+}) reduced the ductility of Watts nickel deposits; 25 mg/l was sufficient to cause an appreciable reduction in ductility, especially in a high pH bath of 5.2, although no quantitative elongation data were reported. (82)

Ten mg/l copper added to Watts-type baths caused a slight loss of ductility in the deposits, which decreased further with concentrations of 25 and 50 mg/l copper. Deposits were unusable when the copper concentration reached 100 mg/l. (24)

No appreciable change was noted in the ductility of Watts bright nickel when zinc was present up to 150 mg/l. Ductility, however, decreased with addition of increasing concentrations of zinc to a Watts bath without additives. (25)

Additions of 20 to 300 mg/l manganese to a low pH (2.2) Watts-type solution appeared to improve ductility slightly. (128)

Effects of Changes in Current Density

With few exceptions, changes in current density seem to have only slight effects on strength and elongation. (19) A 38 percent decrease in the tensile strength and an

18 percent increase in the elongation of Watts-type nickel was noted when the current density was increased from 1 to 2 amp/sq dm. (1) No change in strength occurred for nickel deposited at higher current densities to 5 amp/sq dm.

Doubling the current density from 1.5 to 3.0 amp/sq dm with a Watts bath had little effect on the tensile strength of nickel deposited at 25 C and 50 C, whereas nickel deposited at 60 C showed a slight tensile strength increase from 49 to 56 kg/sq mm. (129) Another investigator showed no significant change in strength for nickel deposited at current densities in the range of 2.2 to 6.5 amp/sq dm. (105)

A slight decrease (< 10 percent) in the strength of nickel from a 49 C sulfamate bath was reported when the current density was increased from about 1.5 to 6.6 amp/sq dm. (130) For nickel deposited at other temperatures, similar slight differences occurred when the current density was increased or decreased.

Nickel deposited in a 60 C bath containing 350 g/l nickel chloride and 40 g/l boric acid at pH 3 showed a reduction in elongation from 15 to 3 percent with increasing current densities from 5.4 to 45.5 amp/sq dm. (45) Elongation varied irregularly between 3 and 33 percent for the deposits from the same solution when a variable pulsed current of 22 amp/sq dm was applied.

Effect of Solution Temperature

As a general rule, the tensile strength of nickel tends to decrease while ductility increases as the temperature of a Watts-type solution is increased. (1,96) Data illustrating such trends are given in Table 12.19. Nickel from chloride baths with a pH of 1.5 or 5.0 also followed this pattern. On the other hand, nickel deposited in a chloride bath with a pH of 3.0 exhibited a maximum strength when the temperature of the solution was 55 C and lower strengths for nickel deposited in 30 or 80 C solutions. In the case of nickel deposited in a sulfamate bath, the tensile strength was higher for a temperature of 49 C, in comparison with temperatures of 38 or 60 C. However, nickel deposited at a temperature of 38 C exhibited the best ductility. (130)

Effect of pH

Increasing solution pH from 3.0 or 4.0 to 5.0 or > 5.0 consistently increases strength and reduces ductility, as shown in Table 12.20. An increase in pH from 5.0 to 5.3, 5.6, or 6.0 also increased strength and reduced ductility. These trends were noted for deposits from chloride, fluoborate, sulfamate, sulfate, and Watts-type baths. (1,9,10,73,96,101,130) Changes in pH from 3.0 to 4.0 had little or no influence on strength or ductility.

In the case of nickel deposited in a chloride bath, decreasing the pH from 3.0 to 1.5 reduced tensile strength. (1) The reverse trend was reported for Watts-type nickel deposited in 2 N nickel baths, (1,9) but another investigator observed no significant difference in the strength or ductility of Watts nickel when the pH was reduced from 3.0 to 2.0 in a 2.9 N nickel solution. (101) Data are detailed in Table 12.20.

TABLE 12.19

Effect of Increasing Temperature on the Strength and Ductility of Nickel

Type of Bath	Temperature, C	pH	Current Density, amp/sq dm	Ultimate Tensile Strength		Yield Strength		Elongation, percent
				kg/sq mm	psi	kg/sq mm	psi	
Chloride[a]	55	1.5	5	65	92,000	—	—	15
	80	1.5	5	62	88,000	—	—	18
Chloride[a]	30	3.0	5	62	88,000	—	—	14
	55	3.0	5	72	102,000	—	—	14
	80	3.0	5	67	96,000	—	—	5
Chloride[a]	30	5.0	5	96	137,000	—	—	5
	55	5.0	5	95	135,000	—	—	6
	80	5.0	5	71	101,000	—	—	6
Sulfamate[b]	38	3.5	2.4	51	73,000	—	—	14
	49	3.5	1.9	76	108,000	58	83,000	4
	60	3.5	2.4	73	104,000	54	76,000	3
	38	3.5	6.5	49	70,000	32	45,000	16
	49	3.5	6.6	70	100,000	54	76,000	8
	60	3.5	6.0	51	73,000	35	50,000	3

Watts[c]	30	2.0	4 or 5	65	93,000	—	—	15
	55	2.0	4 or 5	37	52,000	—	—	32
	78	2.0	4 or 5	35	50,000	—	—	35
	30	5.0	4 or 5	47	67,000	—	—	23
	55	5.0	4 or 5	44	63,000	—	—	26
	78	5.0	4 or 5	40	57,000	—	—	27
Watts[d]	30	3.0	5	53	75,000	—	—	15
	55	3.0	5	39	56,000	—	—	28
	80	3.0	5	55	78,000	—	—	28
	30	1.5	5	74	105,000	—	—	11
	55	1.5	5	47	67,000	—	—	28
	80	1.5	5	47	67,000	—	—	17
Watts[e]	25	2.5	3	97	138,000	—	—	—
	50	2.5	3	62	88,000	—	—	—
	60	2.5	3	56	80,000	—	—	—

Footnote	Nickel Concentration, N	Other Constituents	Reference
(a)	2.0	0.5 M H_3BO_3	1
(b)	4.3	Saturated in H_3BO_3	130
(c)	2.8	0.5 M H_3BO_3 and 0.4 N $NiCl_2$	96
(d)	1.5	0.5 N $NiCl_2$ and 0.5 M H_3BO_3	1
(e)	Not disclosed	No data	129

TABLE 12.20

Effect of Solution pH on the Strength and Ductility of Electrodeposited Nickel

Type of Bath	pH	Current Density, amp/sq dm	Tensile Strength kg/sq mm	Tensile Strength psi	Elongation, percent
Chloride[a]	1.5	5	65	92,000	15
	3.0	5	72	102,000	14
	5.0	5	95	135,000	6
Fluoborate[b]	3.0	5	44	62,000	30
	4.5	5	49	69,000	22
Sulfamate[c]	3.5	6.6	70	100,000	8
	4.0	6.2	60	85,000	4
Sulfate[d]	1.5	5	58	82,000	20
	3.0	5	46	66,000	20
	5.0	5	73	104,000	6
Sulfate[e]	5.0	3	35	50,000	—
	6.0	3	56	80,000	—
Watts[f]	1.5	5	53	75,000	—
	4.0	5	42	60,000	—
	5.5	5	67	95,000	—
Watts[g]	2.0	No data	38	54,500	22.5
	4.75	No data	62	87,800	4.0
Watts[h]	2.0	4 or 5	37	52,000	32
	5.0	4 or 5	44	63,000	26
	5.3	4 or 5	62	88,000	20
Watts[i]	2.0	5	46	65,000	31
	3.0	5	46	65,000	28
	4.0	5	42	60,000	35
	5.0	5	51	72,000	25
	5.3	5	69	98,000	21
Watts[j]	1.0	5	49	69,000	26
	1.5	5	47	67,000	28
	3.0	5	39	56,000	28
	4.0	5	46	66,000	25
	5.0	5	42	59,000	25
	5.5	5	60	85,000	25
	5.6	5	71	101,000	5

Footnote	Nickel Concentration, N	Other Constituents	Temperature, C	Reference
(a)	2.0	0.5 M H_3BO_3	55	1
(b)	2.0	0.5 M H_3BO_3	55	1
(c)	4.3	Saturated in H_3BO_3	50	130

TABLE 12.20 (Continued)

Footnote	Nickel Concentration, N	Other Constituents	Temperature, C	Reference
(d)	2.0	0.5 M H_3BO_3	55	1
(e)	1.2	0.3 M H_3BO_3 and 0.75 M Na_2SO_4	55	10
(f)	2.0	0.5 N $NiCl_2$ and 0.5 M H_3BO_3	55	9
(g)	2.5	0.2 N $NiCl_2$ and 0.6 M H_3BO_3	No data	73
(h)	2.9	0.4 N $NiCl_2$ and 0.5 M H_3BO_3	55	96
(i)	2.9	0.4 N $NiCl_2$ and 0.5 M H_3BO_3	55	101
(j)	2.0	0.5 N $NiCl_2$ and 0.5 M H_3BO_3	55	1

Effect of Organic Addition Agents

Organic addition agents containing sulfur increase tensile strength (and hardness) and reduce ductility. Table 12.21 shows typical data on such effects. Trisulfonated naphthalene had a profound influence. Only 0.02 g/l increased tensile strength from 61.5 to 175 kg/sq mm (87,500 to 250,000 psi) for nickel deposited in nickel sulfate-sodium chloride-boric acid solution. (10) Nickel from a sulfamate bath was increased in strength from 56 to 98 kg/sq mm (80,000 to 140,000 psi) with an 8 g/l addition of sulfonated naphthalene. (131) Elongation of this nickel was only 1.0 percent.

Nickel deposited in solutions containing organosulfur compounds ordinarily contains at least 0.01 percent sulfur. With such a sulfur content, tensile strength is usually about 150 kg/sq mm (214,000 psi). (126) Higher sulfur contents had no effect on tensile strength. For example, a Watts-type nickel deposit containing 0.14 percent sulfur (and 0.09 percent carbon) exhibited a tensile strength of 150 kg/sq mm (214,000 psi) and elongations of 3 to 4 percent. (133)

Bright nickel deposited in proprietary solutions ranged in tensile strength from 91 to 163 kg/sq mm (130,000 to 230,000 psi). (1) Elongations ranged from 1 to 6 percent. Other bright nickel deposits exhibited elongations of 0.5 to a maximum of 4.5 percent, depending on the carbon and oxygen contents of the deposit. (133) The least ductile deposit contained 0.1 percent carbon and 0.09 percent oxygen, in comparison with 0.05 and 0.02 percent carbon and oxygen, respectively, for the deposits with a 2.0 to 4.5 percent elongation. Even lower elongations of 0.40 to 0.47 percent were reported for some bright nickel deposits. (134)

Additions of coumarin (which does not contain sulfur) as a leveling agent for semibright nickel reduced the elongation of Watts-type nickel from 29.5 to 23 percent. (135) The coumarin concentration was 0.07 g/l. During bath aging, which formed coumarin reduction products, elongation was further reduced to 4 percent. From another report, higher concentrations of coumarin, such as 1 and 3 g/l, reduced elongation to only 5 and 1 percent, respectively. (103) Butynediol (0.2 g/l) reduced elongation to 4 percent. With a combination of butynediol and 3 g/l of acetic acid, an elongation of 9 percent was reported.

The tensile strength of notched nickel specimens is sometimes appreciably lower than the strength of unnotched specimens. Whereas the ratio of the tensile strengths

TABLE 12.21

Effect of Organic Addition Agents on Strength and Ductility

Agent	Type of Bath	Temp, C	Cathode Current Density, amp/sq dm	Ultimate Tensile Strength		Yield Strength		Elongation, percent
				kg/sq mm	psi	kg/sq mm	psi	
None	Fluoborate[a]	55	7.5	52	74,500	37	52,700	16.6
Saccharin, 1 g/l	Ditto	55	7.5	142	203,000	85	120,000	1.0
None	Sulfamate[b]	49	3.0	56	80,000	35	50,000	8
Naphthalene trisulfonic acid, 8 g/l	Ditto	49	3.0	98	140,000	77	110,000	1.0
Naphthalene trisulfonic acid, 8.7 g/l, and 1 g/l coumarin	"	49	3.0	119	170,000	98	140,000	1.0
None	Sulfamate[c]	49	3.0	77	110,000	49	70,000	7
Naphthalene trisulfonic acid, 8 g/l	Ditto	49	3.0	105	150,000	77	110,000	1.0

Dibenzene sulfonic acid, 1.5 g/l	Sulfamate[d]	50	2.2 to 6.5	53	75,500	—	—	12
None	Sulfate[e]	45	1.0	61.5	87,500	—	—	—
Trisulfonated naphthalene, 0.02 g/l	Ditto	45	1.0	175	250,000	—	—	—
Trisulfonated naphthalene, 1.0 g/l	"	45	1.0	98	140,000	—	—	—
None	Watts[f]	55	5.0	39	56,000	—	—	28
Nickel benzene sulfonate, 7.5 g/l, and triamine tolylphenyl methane chloride, 5–10 mg/l	Ditto	55	5.0	149	212,000	—	—	5.0

[a] 2.6 N Ni bath containing 0.5 M H_3BO_3 with a pH of 2.5. (11)
[b] 2.9 N Ni bath containing 0.5 M H_3BO_3 and 0.02 M $MgCl_2$ with a pH cf about 4.0. (131)
[c] 2.6 N Ni bath containing 0.5 M H_3BO_3 and 0.06 N $NiBr_2$ with a pH of about 4.0. (131)
[d] 4.6 N Ni bath containing 0.18 N $NiCl_2$ and 0.5 to 0.75 M H_3BO_3 with a pH of 4.0 to 4.2. (105)
[e] 0.8 N Ni bath containing 1.9 M NaCl, 0.15 M NaF, and 0.4 M H_3BO_3 with a pH of 5.8. (10)
[f] 1.5 N Ni bath containing 0.5 M H_3BO_3 and 0.5 N $NiCl_2$ with a pH of 3.0. (1)

of notched and unnotched specimens is > 1.0 when the nickel contains < 0.01 percent sulfur, (126,136) this ratio was reduced to < 0.75 when the sulfur content exceeded 0.017 percent. Thus the sulfur content should be less than 0.017 percent for electroformed nickel used for structural purposes.

Effect of Chromium Overlay Deposits

With a 0.25 to 0.5-μm chromium overlay deposit, the tensile strength of dull nickel electroformed in Watts-type, fluoborate, and chloride baths was not affected significantly. (137) Watts-type nickel increased slightly in strength from 35 to 41 kg/sq mm (50,000 to 58,000 psi) with such a chromium overlay. (137) Ductility was not affected. On the other hand, some bright nickel deposits were reduced appreciably in tensile strength when they were chromium plated. An initial strength of 119 kg/sq mm (170,000 psi) was reduced to only 39 kg/sq mm (56,000 psi) with the chromium overplate, for example. The ductility of these bright nickel foils also was reduced when they were chromium plated. The decrease in ductility ranged from 16 to 57 percent.

Bright nickel foils electroformed in a bath containing sulfonated naphthalene and formaldehyde were appreciably reduced in strength and ductility by chromium plating, whereas bright nickel deposited in a bath containing chloral hydrate and p-toluene sulfonamide was only slightly reduced in these properties. (137) The effects of the chromium persisted after several days of aging. Thus, hydrogen embrittlement does not appear to have significance in the data developed with chromium overlays. No obvious explanation of the variable effects of chromium plating on strength and ductility can be offered.

Charging electrodeposited nickel with hydrogen by a cathodic process in a cyanide solution decreased the tensile strength and ductility of dull and bright deposits. (138) Decreases of 35 to 55 percent in strength and 65 to 75 percent in ductility were reported. However, the effects of hydrogen charging were time dependent. Recovery of initial strength and ductility of Watts-type nickel occurred within 3 or 6 hours after charging, whereas a bright nickel deposited in a bath containing chloral hydrate and p-toluene sulfonamide required 23 hours to recover its original strength and ductility; 24 hours after charging a semibright nickel produced with an addition of p-toluene sulfonamide, strength and ductility were only partially restored. Periodically reversing the current for plating the bright nickel avoided any harmful effect of hydrogen charging.

Effect of Annealing

Annealing in a vacuum at 400 C for 1 hour reduced the tensile strength of ductile nickel electroformed in a Watts-type bath from about 39 to 37 kg/sq mm (from 56,000 to 53,000 psi). (1) Nickel deposited in acetate, chloride, fluoborate, sulfate, and Watts solutions, which exhibited initial strengths of 53 to 143 kg/sq mm

TABLE 12.22

Tensile Strength and Ductility of Annealed, Electroformed Nickel[a]

Plating Bath	Annealing Temperature, C	Ultimate Tensile Strength				Elongation, percent		Reference
		Before Annealing		After Annealing				
		kg/sq mm	psi	kg/sq mm	psi	Before Annealing	After Annealing	
Acetate	400	103	147,000	48	68,000	4	43	1
	1000	103	147,000	26	37,000	4	23	1
Chloride	400	71.5	102,000	59	84,000	14	25	1
	1000	71.5	102,000	27	39,000	14	12	1
Sulfamate	400	88	125,000	65	93,000	18	18	124
	600	88	125,000	46	65,500	18	43	124
Watts	400	39	56,000	37	53,000	28	39	1
	1000	39	56,000	25	35,000	28	17	1
Bright, Watts[b]	400	149	212,000	25	35,000	5	0	1
	1000	149	212,000	14	20,000	5	3	1

[a] One-hour heat treatments.

[b] The organic brighteners were nickel benzene disulfonate and triamine tolyldiphenyl methane.

(75,000 to 205,000 psi) was reduced in strength to the range of 42 to 56 kg/sq mm (60,000 to 80,000 psi) with the same heat treatment. Elongation was increased from as little as 4 to the range of 24 to 32 percent with this heat treatment.

Annealing at 400 to 600 C reduced the tensile strength of nickel electroformed in a sulfamate bath, as shown in Table 12.22. The sulfamate nickel was slightly harder than the nickel deposits from acetate, chloride, or Watts-type solutions, after annealing at 400 C.

The tensile strength of nickel deposited in many solutions was reduced to a maximum of 28 kg/sq mm (40,000 psi) by annealing at 1000 C. (1) Elongation ranged from about 17 to 23 percent after this heat treatment, except for some deposits which exhibited a lesser elongation.

Annealing at intermediate temperatures of 650 or 815 C maximized the ductility of Watts nickel. (101) An elongation of about 50 percent was reported for this annealed, electrodeposited nickel. However, an elongation of only 30 to 35 percent was reported for nickel heated at 870 C.

Whereas nickel containing 0.001 percent sulfur was improved in ductility (from 8 to 12 or 14 percent in elongation) by heat treating at 620 C, nickel containing 0.002 percent sulfur was either unchanged or reduced in ductility with such a heat treatment. (131) Elongations for the nickel containing 0.002 percent sulfur ranged from < 1 to 4 percent before annealing and < 1 to 2 percent after annealing at 620 C. The deposits were obtained in nickel sulfamate baths containing naphthalene trisulfonic acid.

Bright nickel deposited in a solution containing an organosulfur compound was reduced in tensile strength and ductility by heat treating at 400 or 1000 C. (1) A typical deposit was reduced in strength from 147 kg/sq mm (210,000 psi) to only 14 kg/sq mm (20,000 psi) by heat treating at 400 or 1000 C. Elongation was reduced from 2 to 1 percent by the 400 C heat treatment and increased to 3 percent by the 1000 C heat treatment.

Strength and Ductility at Subzero and High Temperatures

Figure 12.3 shows the strength and ductility of nickel electroformed in chloride, sulfamate, and Watts-type solutions, measured at temperatures from −195 to 870 C. (56) The nickel specimens were deposited in solutions containing no organosulfur stress-reducing compounds. At −195 C, nickel deposited in chloride and sulfamate baths exhibited an ultimate tensile strength of > 100 kg/sq mm (>140,000 psi), in comparison with a value of about 56 kg/sq mm (80,000 psi) for Watts-type nickel and annealed, wrought nickel. At high temperatures above 650 C, each type of nickel exhibited an ultimate tensile strength of < 8.5 kg/sq mm (< 12,000 psi).

An increase in ductility as the measuring temperature is decreased from 20 or 25 to −210 C has been observed by several investigators. (56,126) An increase in tensile strength also occurs at subzero temperatures. However, the increase in the strength of unnotched tensile specimens was greater than the increase in the strength of notched specimens at subzero temperatures, as follows: (136)

Test Temperature, C	Ultimate Tensile Strength, kg/sq mm		Tensile Strength Ratio of Notched and Unnotched Specimens
	Unnotched	Notched	
25	77	105	1.36
−195	112	140	1.25
−253	123	126	1.03

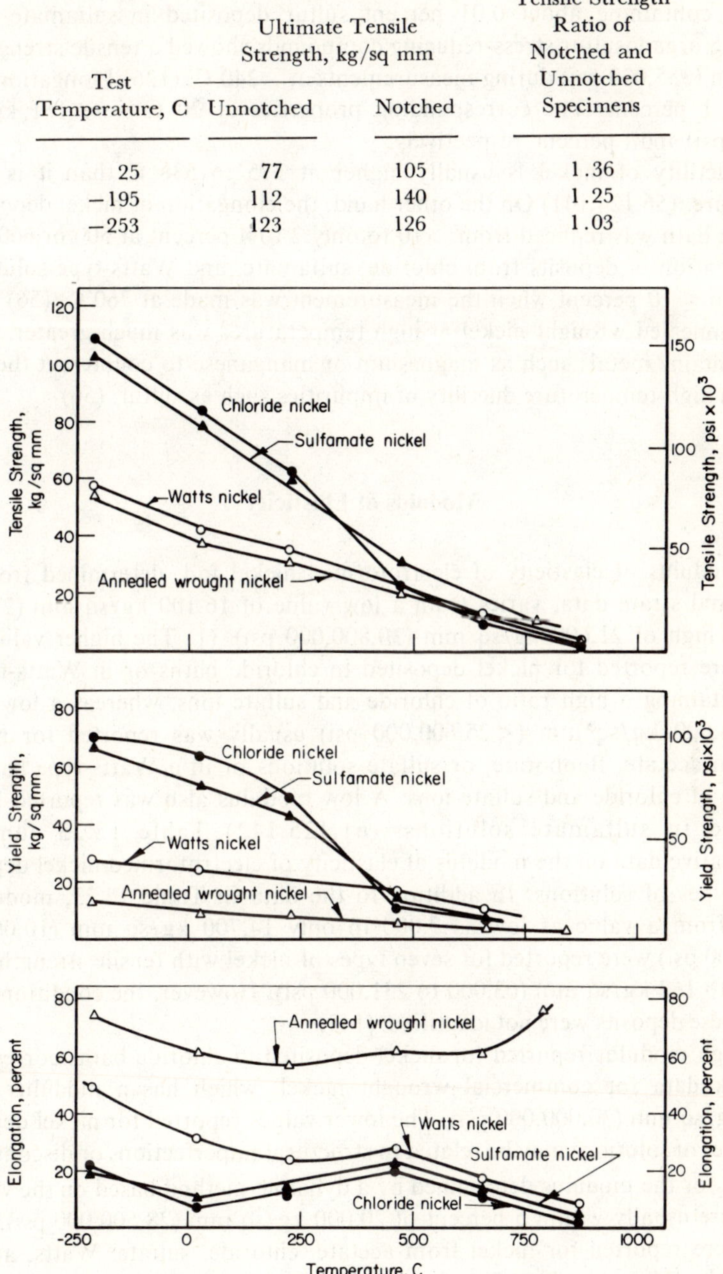

Figure 12.3. Strength and Ductility of Electroformed Nickel at Subzero and Elevated Temperatures. (56)

These nickel deposits were obtained in a nickel sulfamate bath containing an organosulfur stress reducer.

Nickel containing about 0.01 percent sulfur deposited in sulfamate solutions containing organosulfur stress-reducing compounds showed a tensile strength of 250 kg/sq mm (355,000 psi) during measurements at −240 C. (126) Elongation at −240 C was < 1 percent. The corresponding properties at 20 C were 161 kg/sq mm (230,000 psi) and 1 percent, respectively.

The ductility of nickel is usually higher at 375 to 538 C than it is at room temperature. (56,124,141) On the other hand, the elongation of nickel deposited in a sulfamate bath was reduced from >10 to only 2 to 4 percent at 500 or 600 C. (124) The elongation of deposits from chloride, sulfamate, and Watts-type solutions was reduced to < 10 percent when the measurement was made at 760 C. (56) The ductility of annealed wrought nickel at high temperatures was much greater. Wrought nickel contains metals such as magnesium or manganese to counteract the harmful effects on high-temperature ductility of impurities such as sulfur. (56)

Modulus of Elasticity

The modulus of elasticity of electroformed nickel foil, determined from tensile loading and strain data, varies from a low value of 16,100 kg/sq mm (23,000,000 psi) to a high of 21,600 kg/sq mm (30,800,000 psi). (1) The higher values in this range were reported for nickel deposited in chloride baths or in Watts-type solutions containing a high ratio of chloride and sulfate ions, whereas a low modulus below 18,000 kg/sq mm (< 25,600,000 psi) usually was reported for nickel deposited in acetate, fluoborate, or sulfate solutions or in a Watts-type bath with a low ratio of chloride and sulfate ions. A low modulus also was reported for nickel deposited in sulfamate solutions. (61,125,142) Table 12.23 summarizes representative data on the modulus of elasticity of electroformed nickel deposited in several types of solutions. In addition to the data in Table 12.23, moduli values ranging from a value as low as 7,000 to only 14,700 kg/sq mm (10,000,000 to 21,000,000 psi) were reported for seven types of nickel with tensile strengths ranging from 44 to 162 kg/sq mm (63,000 to 231,000 psi). However, the conditions for producing these deposits were not identified. (143)

The high modulus reported for nickel deposited in chloride baths corresponds to handbook data for commercial wrought nickel, which has a modulus of about 21,000 kg/sq mm (30,000,000 psi). The lower values reported for nickel deposited in other types of solutions may be related to structural imperfections or discontinuities.

Values for the modulus determined by a dynamic method based on the velocity of sound were usually within 5 percent of 20,000 kg/sq mm (28,500,000 psi). (1) Such values were reported for nickel from acetate, chloride, sulfate, Watts, and bright nickel baths. Thus data based on the dynamic measuring method show little significant difference in the modulus of nickel deposited in these different types of solutions.

TABLE 12.23

Young's Modulus of Elasticity for Electroformed Nickel

Type of Bath	Solution Temperature, C	Current Density, amp/sq dm	Modulus of Elasticity	
			kg/sq mm	psi × 10^6
Acetate[a]	50	5	15,900	22.8
Bright[b]	65	5	18,100	25.8
Bright[c]	50	4	18,300	26.1
Chloride[d]	30	4	21,400	30.5
Chloride[e]	55	5	21,600	30.8
Fluoborate[f]	55	5	14,300	20.4
Sulfamate[g]	60	30	15,900	22.7
Sulfamate[g]	60	10	14,900	21.3
Sulfamate[h]	45	4	17,500	25.0
Sulfamate[i]	38	2	15,200	21.5
Sulfate[j]	55	5	16,500	23.5
Watts[k]	30	5	16,600	23.8
Watts[k]	55	5	16,500	23.6
Watts[k]	80	5	19,100	27.3
Watts[l]	55	5	17,500	25.0
Watts[m]	55	5	18,800	26.8
Watts[n]	50	4	17,100	24.4

Footnote	Nickel Concentration N	Ratio of $NiCl_2$ to $NiSO_4$	Supplementary Salts	pH	Reference
(a)	2.2	—	1.1 N $NiCl_2$	3.0	1
(b)	2.55	0.14	0.25 N $NiCl_2$, 0.5 M H_3BO_3, 0.6 N $Ni(HCO_2)_2$, 0.02 N $CoSO_4$, etc.	3.7	1
(c)	1.4	0.17	0.2 N $NiCl_2$, 0.5 M H_3BO_3, 4 g/l saccharin	4.5	61
(d)	0.95		0.4 M H_3BO_3	2.0	61
(e)	2.0		0.5 M H_3BO_3	3.0	1, 9
(f)	2.0	—	0.5 M H_3BO_3	3.0	1
(g)	4.8	—	0.04 N $NiCl_2$, 0.65 M H_3BO_3	4.0	61
(h)	No data	—	—	—	142
(i)	2.35	—	0.24 N $NiCl_2$, 0.6 M H_3BO_3	3.3 ± 0.3	125
(j)	2.0	0.0	0.5 M H_3BO_3	3.0	1
(k)	2.0	0.33	0.5 N $NiCl_2$, 0.5 M H_3BO_3	3.0	1, 9
(l)	2.0	1.0	1.0 N $NiCl_2$, 0.5 M H_3BO_3	3.0	1, 9
(m)	2.0	3.00	1.5 N $NiCl_2$, 0.5 M H_3BO_3	3.0	1, 9
(n)	2.2	0.21	0.38 N $NiCl_2$, 0.5 M H_3BO_3	4.5	61

TABLE 12.24

Fatigue Strength of Electroformed Nickel (1)

Type of Bath	Solution Temperature, C	Current Density, amp/sq dm	pH	Fatigue Strength (10^7 Cycles)		Tensile Strength, kg/sq mm	Endurance Ratio (Fatigue/Tensile)
				kg/sq mm	psi × 10^3		
Acetate[(a)]	50	5	4.5	56	80	103	0.544
Bright[(b)]	55	2	3.5	69	98	119	0.576
Bright[(c)]	65	5	3.7	81	116	144	0.564
Chloride[(d)]	55	5	5.0	28	40	95	0.296
Watts[(e)]	30	5	3.0	25	36	53	0.480
Watts[(e)]	55	5	3.0	21	30	39	0.535
Watts[(e)]	55	5	5.0	24	35	41	0.593
Watts with high chloride[(f)]	55	5	3.0	28	40	64	0.434
Commercial sheet (10^8 cycles)	—	—	—	18	25	50	0.347
Ditto	—	—	—	28	40	85	0.333

Footnote	Nickel Concentration, N	Supplementary Salts
(a)	2.2	1.1 N $NiCl_2$
(b)	2.0	0.5 N $NiCl_2$, 0.5 M H_3BO_3, 7.5 g/l nickel benzene disulfonate, etc.
(c)	2.55	0.25 N $NiCl_2$, 0.5 M H_3BO_3, 0.6 N $Ni(HCO_2)_2$, 0.02 N $CoSO_4$, and formaldehyde
(d)	2.0	0.5 M H_3BO_3, 0.5 N $NiCl_2$
(e)	2.0	0.5 M H_3BO_3, 0.5 N $NiCl_2$
(f)	2.0	1.5 N $NiCl_2$, 0.5 M H_3BO_3

Fatigue Strength

Although the fatigue strength (endurance limit) of electroformed nickel varied from a low value of 21 kg/sq mm (10^7 cycles) to a high of 81 kg/sq mm, the endurance ratio—the ratio of the fatigue strength and tensile strength—usually was > 0.48, as shown in Table 12.24. (1) Nickel deposited in a chloride bath or in a high chloride - low sulfate solution exhibited lower endurance ratios of only 0.3 and 0.43, however, in comparison with endurance ratios of 0.33 or 0.35 for commercial wrought nickel.

The fatigue strength of bright nickel deposited in two solutions exceeded 65 kg/sq mm (93,000 psi), whereas nickel deposited in chloride or Watts-type baths exhibited a fatigue strength of < 35 kg/sq mm (< 50,000 psi). An intermediate value of 56 kg/sq mm (80,000 psi) was reported for nickel deposited in an acetate bath. No data have been reported for nickel deposited in sulfamate solutions.

Internal Stress

Stress in nickel electrodeposits ranges from 10 kg/sq mm (14,000 psi) in compression to > 70 kg/sq mm (> 100,000 psi) in tension, depending on the solution composition and operating conditions. The nature of the substrate, the solution composition and temperature, the cathode current density, the degree of solution agitation, thickness of deposit, and other factors influence stress in electrodeposited nickel. A high tensile stress may cause cracking in the deposit when electroplated articles are subjected to a mechanical strain. In electroformed nickel, a high tensile stress causes changes in dimensions and warpage when the electroformed article is separated from the mandrel. High tensile stress reduces the fatigue strength of nickel-plated steel. Bright nickel highly stressed in tension (7 to 25 kg/sq mm or 10,000 to 35,000 psi) accelerated corrosion outdoors of copper-, nickel-, and chromium-plated die castings, in comparison with compressively stressed bright nickel (−3.5 kg/sq mm or −5,000 psi). (144)

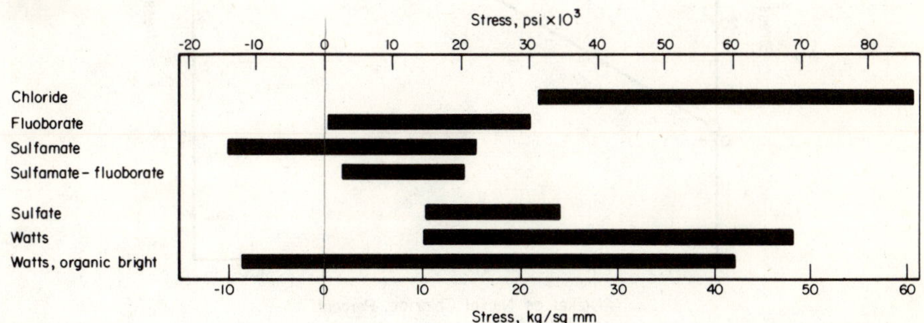

Figure 12.4. Ranges in Stress Reported for Nickel from Seven Popular Types of Plating Baths.

Figure 12.4 shows the ranges of stress reported for nickel from seven types of plating baths. In each case, stress values span a broad range. However, the ranges reported for nickel deposited in Watts-type solutions with brighteners are broader than the ranges for nickel from other solutions. Sulfamate baths that sometimes contained fluoborate or chloride ions yielded nickel with a lower tensile stress (below 16 kg/sq mm or 23,000 psi) than chloride or Watts-type nickel baths containing chloride ions.

Stress in nickel deposited at 3.3 amp/sq dm in four different 1.0 N nickel baths at 25 C (which were all buffered with 0.5 M boric acid and adjusted in pH to 4.0) was as follows: (145,146)

Solution	Stress	
	kg/sq mm	psi
Sulfamate	6	8,500
Fluoborate	12	17,000
Sulfate	16	22,800
Chloride	23	33,000

Thin nickel deposits below 12.5 μm (0.5 mil) tend to be more highly stressed in tension than thicker deposits. For example, stress in a Watts-type nickel decreased from 24 to 13 kg/sq mm (34,000 to 18,500 psi) as thickness was increased from 10 to 30 μm. (147) Another report showed a decrease from 20 to 15 kg/sq mm (28,500 to 21,500 psi) as thickness was increased from 1 to 30 μm. (148) Nickel deposited in

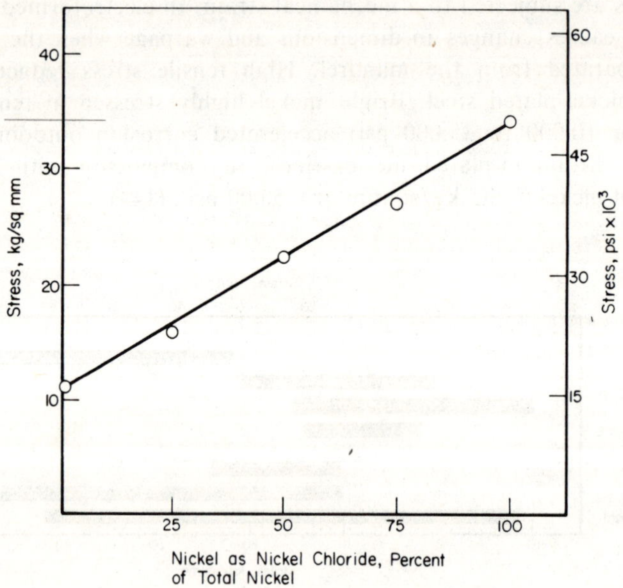

Figure 12.5. Stress in Nickel as a Function of the Nickel Chloride/Nickel Sulfate Concentration Ratio in a Boric-Acid-Buffered Bath Operated at 55 C with a Current Density of 5 amp/sq dm. (1)

a sulfamate bath decreased in stress from about 2 to only 0.7 kg/sq mm (3,000 to 1,000 psi) as thickness increased from < 2.5 to 12.5 μm (from < 0.1 to 0.5 mil). (149) Very thin nickel films (0.1 to 0.8 μm) from a sulfamate bath exhibited a high tensile stress ranging from 18 to 28 kg/sq mm (26,000 to 40,000 psi). (150) Stress decreased as thickness increased.

Little or no change in stress occurs after the thickness of the deposits exceeds 30 μm (1.2 mils). The decrease in stress reported for increasing thickness from 1 to about 30 μm has been associated with substrate effects. As an example of the influence of the substrate, stress in nickel was increased 50 percent when coarsely crystalline copper replaced a polished, fine-grained copper surface. (149)

Most of the stress values reported in the literature were determined by plating on the outside diameter of spiral helices which expanded or contracted during plating due to stress in the deposits. (151) Values developed by the rigid-strip method or the Stressometer tend to be lower. (152) In the case of highly stressed nickel deposited in a high-chloride Watts nickel bath, stress developed in deposits on a rigid-strip was 42 percent lower than the stress in deposits on spiral helices. An intermediate value was reported for stress in nickel from the same bath as determined with the Stressometer. A total of ten methods for determining stress in electrodeposited metals was compared in another report. (153)

Effects of Changes in Solution Composition

A slight increase in stress was reported for Watts-type nickel as nickel concentrations of solutions were increased from 1.0 to 4.0 N. (1) However, another report showed no effect as a result of a nickel concentration increase from 1.8 to 3.1 N in a Watts bath. (147)

A shift from zero stress to a compressive stress of 8.5 kg/sq mm (12,000 psi) was cited on one report as the nickel concentration in a sulfamate bath was increased from 1.5 to 3.0 N. (93) The bath was operated at 60 C with a current density of 5.4 amp/sq dm. According to another report, increasing the nickel concentration from 3.0 to 3.4 N in a nickel sulfamate bath decreased the tensile stress or induced a compressive stress, as follows: (19,154)

Nickel Concentration, N	Solution Temperature, C	Stress	
		kg/sq mm	psi
3.0	30	15.5	22,000
3.4	30	14.0	20,000
3.0	40	9.0	13,000
3.4	40	2.5	3,500
3.0	50	5.5	8,000
3.4	50	−1.0	−1,400

Figure 12.5 shows the effect of replacing some or all of the nickel sulfate in a boric-acid-buffered bath with nickel chloride. The tendency for increased stress with

an increasing concentration of nickel chloride in Watts-nickel baths has also been noted in several other reports. (147,155–158) An increase in the concentration of sodium chloride in a nickel sulfate - boric acid bath also increased stress, as follows: (159)

Sodium Chloride Concentration, M	Stress kg/sq mm	psi
1.0	11	16,000
2.0	20	28,500
3.0	20	28,500

Similar results were reported for additions of sodium chloride to a nickel sulfate - sodium sulfate solution. (10) However, another report showed higher stress in nickel deposited at 2 amp/sq dm in a 54 C, nickel sulfate - sodium sulfate bath, as follows: (204)

Sodium Chloride Concentration, M	Stress kg/sq mm	psi
0.17	14	20,000
0.70	20	29,000
0.85	24	34,000
1.4	29	41,000
1.7	37	53,000

Potassium chloride additions increased stress to a greater degree than sodium or nickel chloride. (156) Magnesium or ammonium chloride increased stress more than potassium chloride. A combination of magnesium sulfate, sodium chloride, and sodium fluoride increased stress to about 28 kg/sq mm (40,000 psi). (98) Fluoborate ion additions to a Watts-type bath had no effect on stress. (94)

Chloride additions to nickel sulfamate baths also increase stress. (53,160,161) For example, one report showed a 300 percent increase in stress for nickel deposited at 5 amp/sq dm after a 0.25 M addition of nickel chloride and a 60 percent increase for nickel deposited at 10 amp/sq dm, as follows: (160)

Halide Concentration, M	Current Density, amp/sq dm	Stress kg/sq mm	psi
None	5	4.2	6,000
	10	9.7	13,800
Chloride, 0.25	5	12.8	18,200
	10	15.6	22,200
Bromide, 0.11	5	8.1	11,500
	10	9.5	13,500

A 0.11 M addition of nickel bromide had less influence on stress than a 0.25 M nickel chloride addition.

The addition of 0.13 M nickel chloride increased stress in nickel deposited at 2.0 to 5.0 amp/sq dm in a nickel fluoborate bath from 11 or 12 kg/sq mm (16,000 or 17,000 psi) to 18 or 20 kg/sq mm (25,500 or 28,500 psi). (155) According to another report, an increase in the concentration of nickel chloride from 0.065 to 0.13 M in a nickel fluoborate bath increased stress from 11 to 18 kg/sq mm (from 16,000 to 25,500 psi). (64)

Unbuffered sulfate-chloride baths operated at current densities from 1 to 5 amp/sq dm produced nickel that was much higher in stress than similar solutions buffered with boric acid or another buffering agent. (162) Citric acid was a better buffer than boric acid with respect to minimizing stress, however, as indicated by the following comparative stress values:

Buffer	Stress	
	kg/sq mm	psi
None	30 to 45	43,000 to 64,000
Boric acid	12 to 19	17,000 to 27,000
Citric acid	6 to 9	8,500 to 13,000
Acetic acid	10 to 16	14,000 to 23,000

In another report, supplementing boric acid with acetate ions reduced stress from the range of 9 to 19 kg/sq mm (13,000 to 27,000 psi) to only 6 kg/sq mm (8,500 psi) in nickel deposited at 2 amp/sq dm in a 45 C solution. (156)

Effect of Impurities

Several metallic impurities increase stress in electrodeposited nickel. Their effects in sulfamate solutions are detailed in Table 12.25. Hexavalent chromium ions had the greatest effect. (163) Next in order of influence was trivalent chromium. Other metallic impurities had slight effects, but all, except manganese and copper, increased tensile stress.

Among the anionic impurities investigated in sulfamate baths, all of the halides increased tensile stress, but the effect of iodide ions was greatest. Sulfide, sulfite, and dithionite ions shifted stress to compressive values, as shown in Table 12.25. The effects of adding these sulfur-containing compounds (which may be similar to the products of oxidation of sulfamate ions at passive anodes) are similar to the effects of operating sulfamate baths with insoluble anodes. For example, nickel deposited in a bath operated with platinized titanium anodes exhibited a compressive stress of −5.5 kg/sq mm (−7,800 psi), whereas nickel from a similar solution operated with sulfur-containing nickel anodes was stressed in tension (3.5 kg/sq mm or 5,000 psi). (164) The sulfur contents of these nickel deposits were 0.026 and 0.003 percent, respectively.

At 3.0 and 0.9 g/l, respectively, zinc and cadmium additions to Watts-type nickel baths increased stress considerably from 17 to 31 kg/sq mm (from 24,000 to 44,000 psi). (165) Smaller additions of 0.05 g/l zinc or copper had little effect on stress.

TABLE 12.25

Effect of Impurities on Stress of Sulfamate Nickel Deposits (163)

Ions	Concentration, g/l	Stress, kg/sq mm	Change in Stress,[a] kg/sq mm per ppm
Ammonium	0.4	4.25	0.006
Chromium (3+)	0.1	7.5	0.058
Chromium (6+)	0.002	9.0	3.75
Cobalt	0.4	4.0	0.006
Copper (2+)	0.1	1.75	0
Iron (2+)	0.4	2.5	0.002
Lead	0.4	3.5	0.004
Magnesium	0.5	4.2	0.005
Manganese (1+)	0.1	1.75	0
Manganese (2+)	0.2	1.75	0
Sodium	1.0	2.0	0.0003
Tin (2+)	0.4	2.25	0.001
Tin (4+)	0.2	4.75	0.015
Zinc	0.1	2.3	0.006
Bromide	8.0	5.0	0.0004
Chloride	8.0	6.0	0.0008
Fluoride	4.0	6.0	0.001
Iodine	0.05	8.5	0.135
Nitrate	1.0	6.0	0.058
Phosphate	1.0	10.0	0.008
Sulfate (SO_4^{2-})	0.5	3.75	0.004
Sulfide (S^{2-})	0.2	−12	−0.068
Sulfite (SO_3^{2-})	0.2	−13	−0.074
Dithionite ($S_2O_4^{2-}$)	0.2	−8.0	−0.049

[a] The solution contained 330 g/l nickel sulfamate and 30 g/l boric acid and was operated at pH = 4.0, 50 C, and 2.75 amp/sq dm. Stress was 1.75 kg/sq mm when the solution contained no impurity.

(147) However, 0.1 g/l copper, iron (II), or zinc in a bath with a pH of 4.75 raised the stress from 6.85 to 11.8, 14, and 17 kg/sq mm, respectively. (158) Similar stress values were measured when the pH was 2.0. The addition of 0.2 g/l hydrogen peroxide increased the stress to 33 kg/sq mm (47,500 psi), and the addition of 3 g/l cobalt sulfate increased the stress to 9.5 kg/sq mm (13,600 psi), in comparison with 8.5 kg/sq mm (12,200 psi) stress for nickel deposited in a purified bath.

Removal of iron by purification has been reported to lower stress. (81,163,166) Solution treatment with hydrogen peroxide (to oxidize iron) and with activated carbon (to absorb organic impurities) reduced stress in nickel from a Watts-type bath from 22.4 to 10.5 kg/sq mm (from 31,000 to 15,000 psi). (19,151) Nickel deposited in purified Watts-type solutions exhibited less stress than nickel from impure solutions, as shown in Figure 12.6. (167)

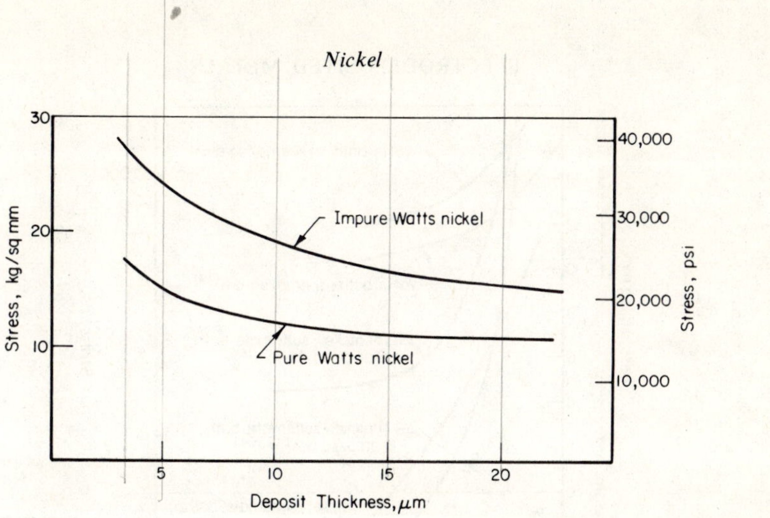

Figure 12.6. Stress of Nickel Electrodeposits from Pure and Impure Watts-Nickel Solutions. (167)

Continuous purification of sulfamate solutions to remove metallic impurities by low-current density electrolysis and organic impurities with activated carbon, combined with operation at 70 C with an anode current density of 0.5 to 1.08 amp/sq dm, resulted in stressfree deposits with a cathode current density of 32.4 amp/sq dm. (168)

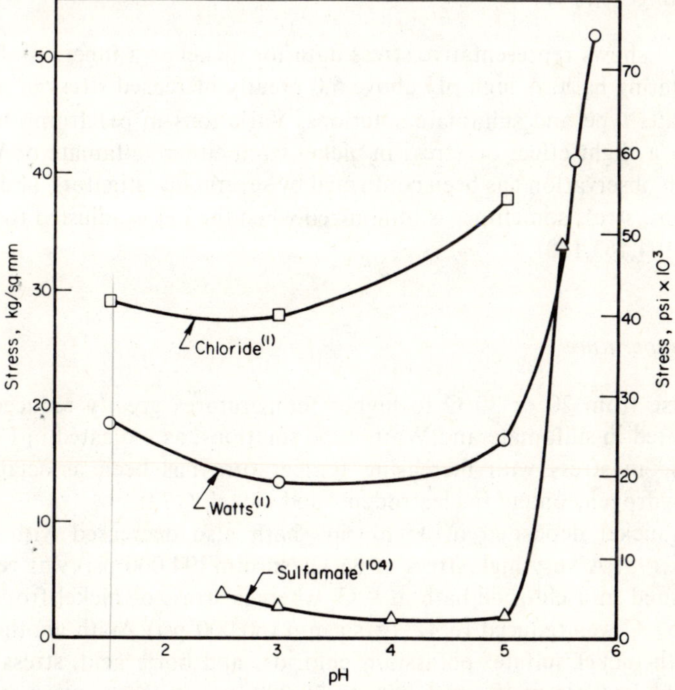

Figure 12.7. Effect of pH of Plating Bath on Stress of Nickel Deposits.

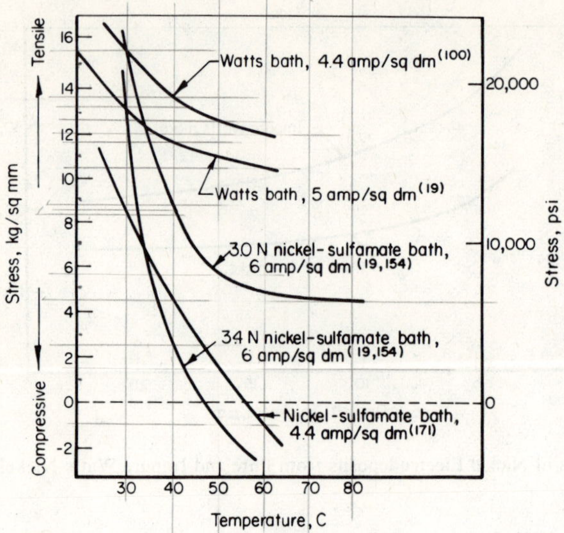

Figure 12.8. Effect of Solution Temperature on Stress of Deposits Obtained in Watts-Type and Sulfamate Baths.

Effect of Changes in pH

Figure 12.7 shows representative stress data for nickel as a function of the pH of the electroplating bath. A high pH above 5.0 greatly increased stress of nickel from both the Watts-type and sulfamate solutions. Variations in pH from about 2.5 to 4.5 had only a slight effect on stress in nickel from either sulfamate or Watts-type baths—which observation has been confirmed by several investigators. (1,104,169) In these solutions, stress sometimes is minimized when the pH is adjusted to 3.5 or 4.0, however, (1,104,163,170)

Effect of Temperature

An increase from 20 or 30 C to higher temperatures greatly reduced stress of nickel deposited in sulfamate and Watts-type solutions, as indicated in Figure 12.8. The decrease in stress with increasing temperature has been associated with a decreasing hydrogen content in electrodeposited nickel. (172)

Stress of nickel deposited in a chloride bath also decreased with increasing temperature. (95) A very high stress of 66 kg/sq mm (94,000 psi) was reported for nickel deposited in a chloride bath at 9 C, whereas stress of nickel from the same solution at 55 C was reduced to 42 kg/sq mm (60,000 psi). With another solution prepared with nickel sulfate, potassium chloride, and boric acid, stress decreased from 53 to 14 kg/sq mm (from 76,000 to 20,000 psi) as the solution temperature increased from 10 to 55 C.

Effect of Current Density

As a general rule, increasing current density increases stress in nickel deposited in several solutions, as shown in Figure 12.9. Many research reports confirm this trend. (8,19,45,66,93,94,154,159,163,165,172) However, some investigators reported exceptions. For example, data for a high chloride-Watts bath did not follow this pattern. (1) Some data showed little change in stress as a function of current density from about 3 to 15 amp/sq dm for Watts nickel, (1,147,158) nickel deposited in a sulfamate bath, (90,161) or nickel deposited in a sulfamate-chloride solution containing naphthalene trisulfonic acid. (174)

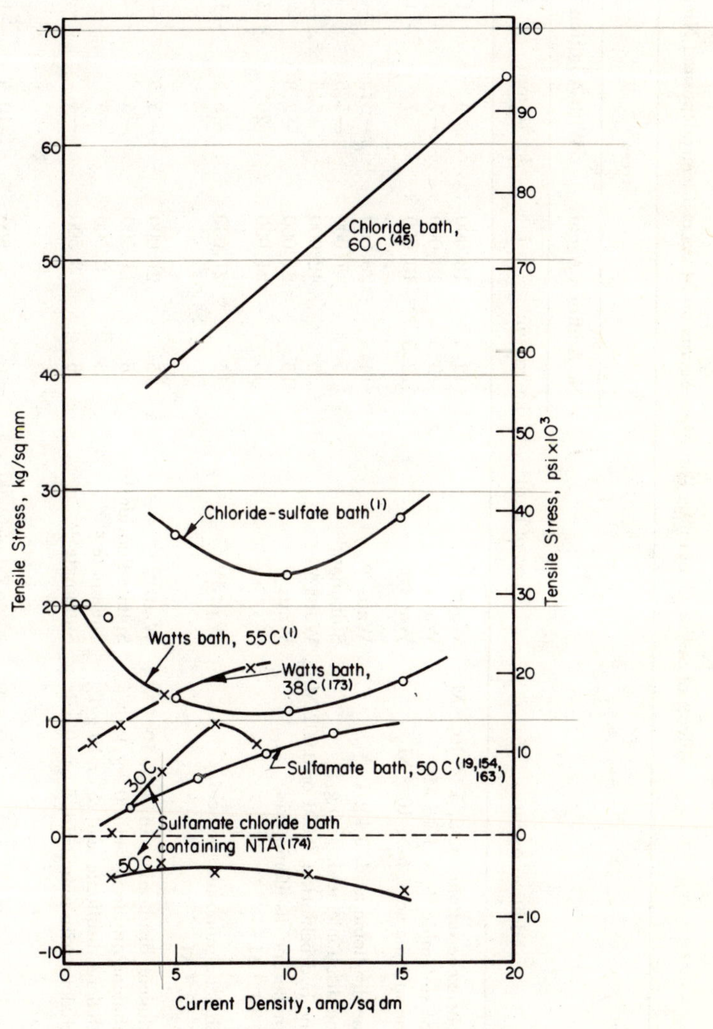

Figure 12.9. Effect of Current Density on Stress of Nickel Electrodeposits.

TABLE 12.26

Effects of Sulfur-Containing Addition Agents on Stress of Electrodeposited Nickel

Addition Agent	Type of Bath	Stress					
		No Addition Agent		With Addition Agent		Change in Stress	
		kg/sq mm	psi	kg/sq mm	psi	kg/sq mm	psi
Aminobenzene sulfonic acid, 5×10^{-3} M	Watts[a]	12.5	18,000	9	13,000	−3.5	−5,000
Benzene sulfamide, 5×10^{-3} M	Watts[a]	12.5	18,000	−7	−10,000	−19.5	−28,000
Benzene sulfonic acid, 5×10^{-3} M	Watts[a]	12.5	18,000	7	10,000	−5.5	−8,000
Benzene disulfonic acid, 0.4 g/l	Watts[b]	14	20,000	−3.5	−5,000	−17.5	−25,000
Benzene disulfonic acid, 33.5 g/l	Sulfamate[c]	5	7,000	−5.5	−8,000	−10.5	−15,000
Cysteine hydrochloride, 0.001 or 0.002 M	Watts[d]	30	42,000	−5.5	−8,000	−35.5	−50,000
Naphthylamine disulfonic acid,	Watts[e]	15	21,000	−2.8	−4,000	−17.8	−25,000
4×10^{-3} M		15	21,000	26*	37,000*	11	16,000
Naphthalene monosulfonic acid, 4×10^{-3} M	Watts[e]	15	21,000	−8.5	−12,000	−23.5	−33,000
Naphthalene disulfonic acid, 4×10^{-3} M	Watts[e]	15	21,000	−5.5	−8,000	−20.5	−29,000
Naphthalene disulfonic acid, 0.2 g/l	Sulfate-fluoride[f]	8	11,500	−5	−7,000	−13	−18,500
Naphthalene disulfonic acid, 5 or 10 g/l	Sulfamate[g]	6	8,500	−5	−7,000	−11	−15,500
Naphthalene trisulfonic acid, 5 or 10 g/l	Sulfamate[g]	6	8,500	−4	−6,000	−10	−14,500
Naphthalene trisulfonic acid, 0.5 to 1 g/l	Sulfate-fluoride[h]	6	8,500	−10.5	−15,000	−16.5	−23,500
Naphthalene trisulfonic acid, 2.3 g/l	Sulfamate[k]	1.4	2,000	−2.1	−3,000	−3.5	−5,000
Naphthalene trisulfonic acid, 4×10^{-3} M	Watts[e]	15	21,000	−2.8	−4,000	−17.8	−25,000
Naphthol monosulfonic acid, 4×10^{-3} M	Watts[e]	15	21,000	32	45,000	17	24,000
Saccharin, 1 g/l	Watts[j]	8	11,250	−8	−11,250	−16	−22,500
Saccharin, 3 g/l	Sulfamate[g]	6	8,500	−10	−14,000	−16	−22,500
Saccharin, 2 g/l	Sulfamate[c]	5	7,000	−12	−17,000	−17	−24,000

Saccharin, 2 to 4 g/l	Sulfamate[k]	9	13,000	−12	−17,500	−21	−30,500
Saccharin, 4×10^{-3} M	Watts[d]	30	42,000	−7	−10,000	−37	−52,000
Saccharin, 4 g/l	Watts[l]	44	62,000	2	2,700	−42	−59,300
Saccharin, 5×10^{-4} M	Watts[m]	21	30,000	−5	−7,000	−26	−37,000
p-toluene sulfonamide, 2 g/l	Sulfamate[n]	6	8,500	0	0	−6	−8,500
p-toluene sulfonic acid, 3 g/l	Sulfamate[g]	6	8,500	−9	−13,000	−15	−21,500
p-toluene sulfonic acid, 30 g/l	Sulfamate plus 30 g/l $NiCl_2$[g]	17	24,000	−4	−5,500	−21	−29,500
p-toluene sulfonic acid, 4×10^{-3} M	Watts[d]	30	42,000	3.5	5,000	−26.5	−37,000
Thioacetamide, 1×10^{-3} M	Watts[d]	30	42,000	−2.0	−3,000	−32	−45,000
Thiosemicarbazide, 0.3×10^{-3} M	Watts[d]	30	42,000	−5.5	−8,000	−35.5	−50,000
Thiourea, 8×10^{-3} M	Watts[o]	20	28,500	−25	−35,500	−45	−64,000
Thiourea, 5×10^{-3} M	Watts[o]	20	28,500	−50	−71,000	−70	−99,500

Footnote	Temperature, C	pH	Current Density, amp/sq dm	Reference
(a)	20	4.0	2	181
(b)	50	5.0	5	117
(c)	49	4.0	19	163
(d)	55	4.0	4	182
(e)	40	4.0	4	183
(f)	35	5.8	1	184
(g)	50	3.5	6	90
(h)	45	5.8	1	10
(i)	49	4.0	3	185
(j)	50	4.0	4	114
(k)	50	4.0	20	171
(l)	55	4.2	2.5	151
(m)	55	4.0	4	186
(n)	18	4.5	3	91
(o)	50	2.6	3	187

* Value after several deposits were produced in the bath.

TABLE 12.27

Sulfur Content of Nickel Deposited in Watts-Type Solutions Containing Organosulfur Addition Agents[(a)]

Addition Agent[(b)]	Sulfur Content, percent[(c)]	Carbon Content, percent[(c)]
Benzene sulfonic acid	0.019	0.01
Naphthalene monosulfonic acid	0.026, 0.056	0.01
Naphthalene disulfonic acid	0.032, 0.035	0.01
Naphthalene trisulfonic acid	0.034	—
Naphthylamine monosulfonic acid	0.10, 0.18	0.39, 0.68
Naphthylamine disulfonic acid	0.11	0.04, 0.38
Naphthol monosulfonic acid	0.026, 0.057	0.01
Naphthol disulfonic acid	0.018, 0.037	0.01
Saccharin[(d)]	0.03	—

(a) Watts nickel deposited at 4 amp/sq dm in a Watts bath at 55 C with a pH of 4.0. (182)

(b) The concentration of all addition agents (except saccharin) was 4×10^{-3} M. (182)

(c) Sulfur and carbon contents for nickel deposited in a similar bath containing no addition agent were 0.002 and 0.01 percent, respectively. (182)

(d) The concentration of saccharin was 5×10^{-4} M. (186)

Nickel deposited in a sulfate - boric acid - potassium chloride bath with a pH of 2.5 increased in stress with increasing current density. (159) On the other hand, nickel from a sulfate - boric acid - sodium chloride bath with a pH of 5.2 was only slightly changed in stress with changes in current density. (175)

Pulsating current with frequencies from 10 to 500 cycles/second generally reduced the stress of nickel deposited in a chloride bath. (45) Stress was minimized when the anode and cathode area ratio was 1.0 or > 1.0. Each anodic pulse ranged from 15 to 30 percent of the total cycle time, as a rule. Cathode current density was maintained at 20 amp/sq dm.

Superimposing 50-cycle alternating current on the direct current slightly reduced stress, according to one report. (176) Another stated that stress was reduced to only 1.4 kg/sq mm (2,000 psi) when 60-cycle alternating current was superimposed with a peak voltage three times greater than the direct-current voltage. (177–179) Ultrasonic agitation reduced stress from the range of 15 to 20 kg/sq mm (21,500 to 28,000 psi) to between 7.8 and 12 kg/sq mm (11,000 to 17,000 psi) in the case of nickel deposited in sulfate solutions. (180)

Effect of Addition Agents

Saccharin, sulfonated naphthalene, p-toluene sulfonamide, and other sulfur-containing organic compounds are well-known stress-reducing agents for electrode-

TABLE 12.28

Effects of Non-Sulfur-Containing Organic Addition Agents on Stress of Electrodeposited Nickel

Addition Agent	Type of Bath	Stress					
		No Addition Agent		With Addition Agent		Change in Stress	
		kg/sq mm	psi	kg/sq mm	psi	kg/sq mm	psi
Acetamide, 4×10^{-3} M	Watts[a]	30	43,000	10.5	15,000	−19.5	−28,000
Acetone, 4×10^{-3} M	Watts[a]	30	43,000	12	17,000	−18	−26,000
Adipic acid, 4×10^{-3} M	Watts[a]	30	43,000	18	26,000	−12	−17,000
Butynediol, 0.2 g/l	Watts[b]	12.5	18,000	36	51,000	23.5	33,000
Butynediol, 0.2 g/l	Watts[c]	10	14,000	35	50,000	25	36,000
Butynediol, 0.5 g/l	Watts[c]	14	20,000	45	64,000	31	44,000
Chloroamine, 1 g/l and quinoline, 0.03 g/l	Watts[d]	17	24,000	−7	−10,000	−24	−34,000
Chlorohydrate, 3×10^{-4} M	Watts[a]	30	43,000	15.5	22,000	−14.5	−21,000
Chlorohydrate, 4×10^{-4} M	Watts[a]	30	43,000	22.5	32,000	−7.5	−11,000
Coumarin, 4×10^{-4} M	Watts[a]	30	43,000	1.4	2,000	−28.6	−41,000
Coumarin, 1 g/l	Watts[b]	12.5	18,000	8.4	12,000	−4.1	−6,000
Coumarin, 0.5 or 1.0 g/l	Watts[c]	14	20,000	8.4	12,000	−5.6	−8,000
Glycine, 4×10^{-3} M	Watts[a]	30	43,000	11	16,000	−19	−27,000
Pyruvic acid, 2.25 g/l	Watts[e]	15	21,300	0	0	−15	−21,300
Semicarbazide, 4×10^{-3} M	Watts[a]	30	43,000	12	17,000	−18	−26,000
Succinamide, 4×10^{-3} M	Watts[a]	30	43,000	7	10,000	−23	−33,000
Sucrose, 4×10^{-3} M	Watts[a]	30	43,000	10.5	15,000	−19.5	−28,000
Urea, 4×10^{-3} M	Watts[a]	30	43,000	14	20,000	−16	−23,000

Footnote	Temp, C	pH	Current Density, amp/sq dm	Reference
(a)	55	4.0	4	182
(b)	50	4.0	3	103
(c)	50	4.0	3	106
(d)	25	4.0	1	165
(e)	60	4.0	4	188

posited nickel. Table 12.26 summarizes representative data from the literature on the effects of such addition agents. Large shifts in stress, usually from tension to compression, were observed when benzene sulfonic acid, cysteine hydrochloride, naphthalene sulfonic acids, saccharin, p-toluene sulfonic acid, thioacetamide, or thiosemicarbazide was added to Watts-type or sulfamate solutions. All of these compounds contribute sulfur to the deposit. Table 12.27 shows the sulfur contents of typical deposits from solutions containing several of these addition agents.

Changes in the concentration of benzene sulfonic acid, naphthalene sulfonic acids, saccharin, and p-toluene sulfonic acid had relatively little effect on stress. However, considerable sensitivity to concentration changes has been reported for allylthiourea, cysteine hydrochloride, naphthylamine disulfonic acid, thioacetamide, and thiosemicarbazide. (182) In typical cases, a minimum (compressive) stress was observed when the concentration of these addition agents was below 0.5×10^{-3} M. Increasing tensile stress was reported with concentrations of 2 to 4×10^{-3} M. The same pattern was exhibited with thiourea additions.

Table 12.28 cites several stress-reducing agents that do not contribute sulfur to the nickel deposit. Acetamide, coumarin, succinamide, and sucrose are examples. Literature data indicate that these agents tend to reduce tensile stress increasingly as their concentration is increased in Watts-type solutions operated at 55 C. (182) No relationship between the chemical properties of addition agents and their effect on stress could be established.

Supplementing the agents cited in Table 12.28, phthalamide (like coumarin) was reported to reduce stress, (189) but no quantitative data were disclosed. Individual additions of butynediol and quinaldine increased stress in nickel deposited in a Watts-type solution. The addition of 0.3 to 0.75 ml/l formalin slightly increased stress in a Watts bath containing coumarin. (106)

The net effect of mixtures of two addition agents on stress was intermediate between their individual effects. (182) Although an addition of butynediol increased stress of Watts-type nickel, a combination of 0.1 or 0.2 g/l butynediol and 2 or 5 g/l acetic acid reduced stress from 14 to a maximum of 10.5 kg/sq mm (from 20,000 to 15,000 psi). (106) A combination of butynediol (0.2 g/l), formic acid (3 g/l), and naphthalene trisulfonic acid reduced the stress of Watts nickel from 12.5 kg/sq mm (18,000 psi) to 7.7 kg/sq mm (11,000 psi). (103) A compressive stress of −8.0 kg/sq mm (−11,500 psi) was reported for Watts nickel deposited in a solution containing 2 g/l choloramine-B, 1 g/l coumarin, and 0.03 ml/l pyridine. (107,120) On the other hand, the addition of 1 g/l acetanilid, 2 g/l benzene sulfonamide, and 2 g/l of naphthalene disulfonic acid slightly increased the stress of Watts nickel. (3)

Anionic wetting agents were reported to have little effect on stress. (153) However, some nonionic agents in small concentrations increased stress significantly. (108)

Effect of Stressed Nickel on the Fatigue Strength of Steel

The fatigue strength of steel is reduced when it is electroplated with nickel stressed in tension. The amount of reduction depends on the strength of the steel and the degree of tensile stress. Published data are summarized in Table 12.29. No loss in fatigue strength was observed when a low-strength steel (60 kg/sq mm or 86,000 psi UTS) was plated with compressively stressed nickel. A compressively stressed nickel reduced the fatigue strength of a high-strength steel (147 kg/sq mm or 211,000 psi UTS) 22 percent, in comparison with a 59 percent reduction in the fatigue strength of the same high-strength steel plated with nickel stressed in tension. (191)

By shot peening steel before plating, loss in the fatigue strength of steel plated with nickel stressed in tension was reduced appreciably, in comparison with the loss observed on machined specimens. For example, the fatigue strength of steel with a limit of 33 kg/sq mm (47,000 psi) before plating was reduced to $<$ 21 kg/sq mm ($<$ 30,000 psi) by plating with nickel stressed in tension, whereas shot-peened and nickel-plated steel exhibited a fatigue strength of 29 kg/sq mm (41,000 psi). (194) Peening after plating increased the fatigue limit to 39 kg/sq mm (56,000 psi).

Nickel stressed in tension also reduced the fatigue strength of an aluminum alloy substrate treated with a displacement zinc coating and a copper strike. (198) The reduction was 55 percent in comparison with a gain in fatigue strength of 10 percent for specimens plated with compressively stressed nickel.

Structure

The structure of electrodeposited nickel is dependent on bath composition, the nature and concentration of addition agents, plating conditions, and, to some degree, the structure of the substrate. Of course, the physical and mechanical properties of nickel also depend on these factors.

Low-strength nickel deposited in Watts-type solutions containing no grain-refining agents is typically columnar in structure (Figure 12.10). A decrease in temperature from 55 or 60 C refines the structure to a fibrous condition (Figure 12.11). (1) An increase in pH from 3.0 to 5.0 or an increase in current density from 5 to $>$ 20 amp/sq dm induces a similar fibrous structure. Increasing the nickel chloride concentration or adding ammonium ions also refines grain structure. The changes in conditions that refine the structure also increase strength and hardness and reduce ductility.

Deposits from fluoborate solutions are similar in structure to that of Watts-type electroplates. (1) Acetate and chloride baths tend to yield finer-grained deposits, (1) as do sulfamate solutions. (56)

Many stress reducers and all brighteners refine grain structure appreciably. Naphthalene monosulfonic acid (a stress reducer) induced the structure in Figure 12.12. Figure 12.13 shows the structure of a duplex nickel system consisting of a sulfur-free, semibright nickel and a top layer of a bright nickel deposited in solutions containing an organosulfur addition agent. The banded (lamellar) bright nickel deposits are hard and strong, but exhibit little ductility.

As Table 12.30 shows, the tensile strength and electrical resistivity of electroformed nickel increase with decreasing grain size for a number of deposits obtained in several solutions. Except for the relatively coarse-grained columnar de-

TABLE 12.29

Effect of Stress in Nickel on the Fatigue Strength of Steel

Stress[a]		Fatigue Strength of Unplated Steel, kg/sq mm (5×10^7 to 10^8 cycles)	Change in Fatigue Strength of Steel by Nickel Plating, percent	Nickel Bath			Current Density, amp/sq dm	Nickel Thickness, μm	Reference
kg/sq mm	psi			Type	Temp, C	pH			
13	18,500	60	−18	Watts, purified[b]	50	1.9	3	No data	190
19	27,500	60	−29	Watts, 0.2 g/l Fe[b]	50	1.9	3	No data	190
27.5	39,500	60	−36	Watts, 0.4 g/l Fe[b]	50	1.9	3	No data	190
Tensile	Tensile	77	−59	Watts	—	—	—	25	191
−2.1	−3,000	77	−22	Watts plus brighteners	—	—	—	25	191
0	0	—	11	Watts plus brighteners	—	—	—	25	192
8.5	12,000	—	0	Watts	—	—	—	25	192
10	14,000	—	−11	Watts	—	—	—	25	192
12.5	18,000	—	−19	Watts	—	—	—	25	192
3.5[c]	5,000[c]	44	−10	Sulfate-chloride-fluoride[c]	50	5.7	1	88	193
16	23,000	44	−32	Sulfate-chloride-fluoride	20	5.7	1	88	193
−4.2	−6,000	32	0	Watts plus brighteners	—	—	—	—	194
17.5	25,000	32	−35	Watts	—	—	—	—	194

9	13,000	20.5	17	Sulfamate	55	4.0	2	55	195
17	24,000	20.5	−42	Watts	55	4.5	4	55	195
23	33,000	20.5	−46	Sulfate-ammonium chloride	55	5.6	2	55	195
9	13,000	28.5	12	Sulfamate	55	5.0	2	100	23, 196
15	21,500	28.5	−9	Watts	40	4.5	3	100	23, 196
35	50,000	54	−32	Fluoborate	40	4.5	3	30	23, 196
2.5	3,600	52	−58	Sulfamate plus chloride	50	4.0	4	150	104, 197
8	11,200	52	−63	Watts	25	3.0	2.5	150	104, 197
No data	—	50	−52	Sulfate-chloride-fluoride	22	—	1	10	7
No data	—	26[d]	−38[d]	Sulfate-chloride-fluoride	22	—	1	10	7
Compressive	—	14[e]	10	Watts plus brightener	60	4.5	14	100	198
Tensile	—	14[e]	−55	Watts	45	2.0	43	100	198

[a] A negative sign indicates compressive stress.
[b] Modified Watts bath with a high chloride content (125 g/l $NiCl_2 \cdot 6H_2O$). All specimens baked at 190 C for 3 hours.
[c] Tensile stress was reduced by superimposing alternating current (2.2 or 3.6 amp/sq dm) and raising the bath temperature.
[d] Notched specimen.
[e] Aluminum Alloy 75-T6 substrate with a displacement zinc coating and a copper strike.

Figure 12.10. Columnar Structure of Watts-Type Nickel Deposited in a Bath at 55 C. (250×) (1)

posits, however, true crystal size did not vary with appearance, strength, or resistivity. (1)

Second-phase impurities are responsible for crystalline discontinuities in structure and produce cleavage planes. (199) Sulfide impurities induced with sulfur-containing, stress-reducing agents cause hot shortness (sulfur embrittlement).

Although hexagonal-close-packed crystals have been observed in combination with face-centered-cubic (fcc) nickel in a few electrodeposits, electroplated nickel is

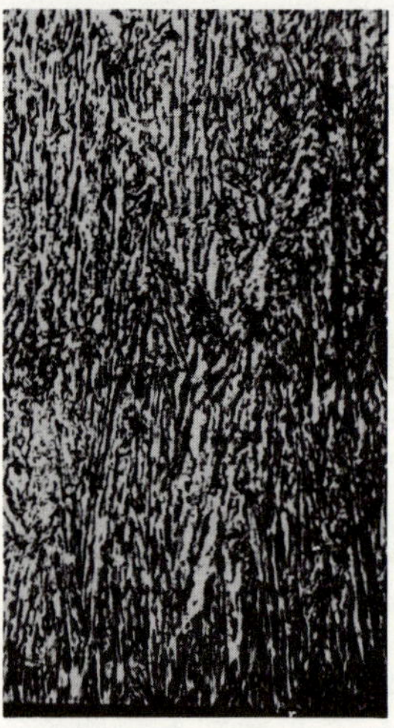

Figure 12.11. Fibrous Structure of Watts-Type Nickel Deposited in a Bath at 80 C. (250×) (1)

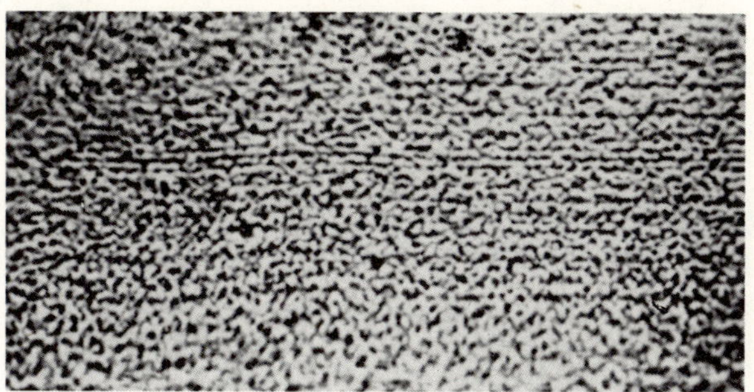

Figure 12.12. Fine-Grained Nickel Deposited in a Watts Bath Containing Naphthalene Sulfonic Acid. (1500×) (183)

typically fcc. (1) Many nickel deposits exhibit a preferred orientation with (100) crystal planes parallel to the surface. However, some bright nickel deposits are randomly oriented or show (100) and (111) planes oriented with the surface. A (210) orientation correlated with good corrosion performance and a (110) orientation with poor performance in one report. (200)

The temperature and pH of a Watts bath influenced orientation. (201) A high temperature of 75 C favored (110) orientation. A random orientation for nickel deposited in a solution with a pH of 2.1 became a (100) orientation when the pH was raised to 3.9 and later to ($10\bar{1}0$) plus (211) orientation at a pH of 5.1. This information referred to 5-μm-thick deposits obtained at 1 amp/sq dm.

Chromium and aluminum ions as impurities in the plating bath influenced orientation. Preferred orientation was shifted from (100) to (110) when 0.4 g/l chromous sulfate was added to the bath. A (110) orientation also was noted when amino acids were added to the plating bath.

The recrystallization temperature for electrodeposited nickel was higher than the heat-treating temperature that reduces hardness. (202) In the case of Watts-type

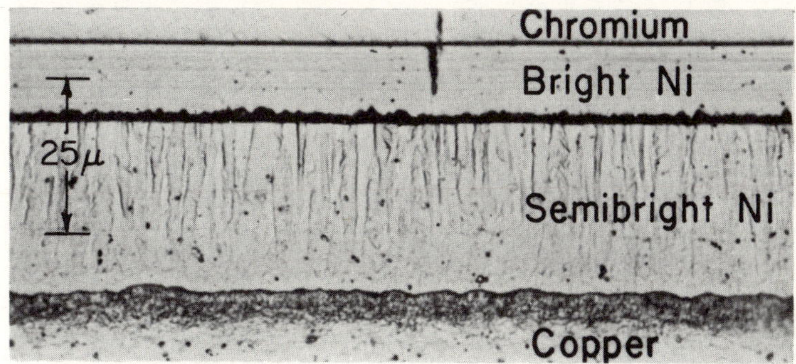

Figure 12.13. Structure of a Duplex System of Semibright and BrightNickel. Note crack propagating from the chromium deposit into the bright Nickel.

TABLE 12.30

Correlation Between Structure and Properties of Nickel Deposits (1)

Type of Bath[a]	Temp, C	Appearance of Deposit	Grain Structure	Crystal Size,[b] μm	Degree of Fibering[c]	Tensile Strength		Resistivity, microhm-cm
						kg/sq mm	psi	
Watts (3.0)	55	Dull	Columnar	0.1	9.2	39	56,000	7.76
Watts (3.0)	30	Dull	Columnar	<0.1	38.6	52	75,000	7.83
Watts with proprietary agent	—	Semibright	Fine grained	0.01	6.6	91	130,000	8.27
Chloride (5.0)	55	Dull	Fibrous	0.01	0.4	95	135,000	8.93
Acetate (4.5)	50	Dull	Fibrous	0.01	5.2	103	147,000	9.18
Watts (3.7)[d]	65	Semibright	Fine grained	0.01	5.5	141	202,000	11.5
Watts, bright (3.5)[e]	55	Bright	Fine grained and lamellar	0.01	2.0	148	212,000	10.0
Watts, plus proprietary brighteners	—	Bright	Fine grained and lamellar	0.01	0.5	155	222,000	11.0

[a] The figures in parentheses are pH values. Current density was 4 to 6 amp/sq dm.

[b] Determined by X-ray diffraction.

[c] Relative intensities of X-ray reflections from (200) and (111) crystal faces. All deposits were oriented with (100) crystal faces parallel to the surface.

[d] The solution contained 0.02 N cobalt, 0.6 N nickel formate, 0.01 ammonium sulfate, and 2.5 g/l formaldehyde.

[e] The solution contained 7.5 g/l nickel benzene disulfamate and 5 mg/l reduced fuchsin.

and other coarsely crystalline deposits, heat treatment at 500 C reduced hardness from 150 to 100 kg/sq mm, but no recrystallization was observed until the temperature reached 700 C. Bright and fine-grained deposits softened from about 625 to 325 kg/sq mm as a result of heating at 400 C. Recrystallization occurred at about 600 C. These data were reported for 150-μm-thick nickel deposits on copper. Another report showed that recrystallization is initiated at lower temperatures. (203) The first observed structural change occurred at temperatures that varied with thickness from 265 C for 10-μm-thick deposits to 310 C for 0.13-μm-thick deposits.

References

(1) Brenner, A., Zentner, V., and Jennings, C. W., "Physical Properties of Electrodeposited Metals. I. Nickel. (3) The Effect of Plating Variables on the Structure and Properties of Electrodeposited Nickel," *Plating, 39* (8), 865–894, 899–927, 933 (1952). Cf. Brenner and Jennings, *Proceedings Am. Electroplaters' Soc., 35*, 31 (1948).

(2) Safranek, W. H., and Schaer, G. R., "Properties of Electrodeposits at Elevated Temperatures," *Proceedings Am. Electroplaters' Soc., 43*, 105–117 (1956).

(3) Matulis, J., and Valentelis, L., "Improvements of Bright Nickel Electrodeposits," *Protsessy Uprochneniya Detalei Mashin, Akad. Nauk SSSR*, 196–199 (1964).

(4) Squitero, A. D., "Designing Electroformed Parts," *Machine Design, 35* (11), 223–226 (1963).

(5) Brenner, A., and Jennings, C. W., "Physical Properties of Electrodeposited Metals. I. Nickel. (1) Literature Survey," *Plating, 35* (1), 52–58 (1948).

(6) Black, G., "Electroplated Coatings," *Materials & Methods, 27* (6), 93–104 (1948).

(7) Gorbunova, K. M., and Nikiforova, A. A., *Physicochemical Principles of Nickel Plating*. Translated by the Israel Program for Scientific Translations, Jerusalem (1963). OTS 63-11003.

(8) Diggin, M. B., "Nickel Plating From the Sulfamate Solution," *Trans. Inst. Metal Finish., 31*, 243–256 (1954).

(9) Brenner, A., and Jennings, C. W., "Physical Properties of Electrodeposited Metals—Nickel," *Proceedings Am Electroplaters' Soc., 35*, 31–50 (1948). Cf. Brenner and Jennings, *Plating, 35* (12), 1228–1231, 1234–1239 (1948).

(10) Fedot'ev, N. P., "Physical and Mechanical Properties of Electrodeposited Metals," *Plating, 53* (3), 309–317 (1966).

(11) Struyk, C., and Carlson, A. E., "Nickel Plating From Fluoborate Solutions," *Plating, 37* (12), 1242–1246, 1263–1264 (1950).

(12) Reimer, L., "Magnetic Properties of Electrolytically Deposited Thin Nickel Layers," *Zeitschrift für Naturforschung, 10a*, 1030–1031 (1955).

(13) Fisher, R. D., "The Influence of Residual Stress on the Magnetic Characteristics of Electrodeposited Nickel and Cobalt," *Journal Electrochem. Soc., 109* (6), 479–485 (1962).

(14) Polukarov, Yu. M., "Study of the Structural and Magnetic Characteristics of Ferromagnetic Metal and Alloy Electrodeposits With Respect to Plating Conditions. I. Nickel," *Zh. Fiz. Khimii, 32* (5), 1008–1015 (1958).

(15) Zentner, V., "Electrodeposited and Electroless Magnetic Alloys for Computers," *Plating, 52* (9), 868–872 (1965); *J. Intern. Applic. Cobalt, Brussels, Belgium*, 152 (June, 1964).

(16) Stahl, G., and Wild, P. W., "Effect of Metal Impurities on Bright Nickel Deposits," *Galvanotechnik, 60* (6), 429–432 (1969).

(17) Yamaguchi, S., "Magnetism of the Electrodeposited Films as Revealed by Electron Diffraction," *Journal Electrochem. Soc., 107* (1), 55–56 (1960).

(18) Wright, J. G., and Goddard, J., "The Lattice Constants and Magnetic Anisotropy Constants of Electrodeposited Single Crystal Films of Hexagonal Close-Packed Nickel," *Philosophical Magazine, 11* (111), 485–493 (1965).

(19) Brugger, R., *Nickel Plating: Bright and Semi-Bright Nickel Plating, Corrosion-Resistance, Heavy Nickel Plating and Electroforming*, Leuze Verlag, Saulgau, West Germany (1967).

(20) Rotinyan, A. L., and Joffe, E. Sh., "Effect of the Electrolysis Conditions on the Gas Content of Electrolytic Nickel," *Tsvetn. Metal., 37* (11), 42–46 (1964).

(21) Belen'kii, M. A., and Lainer, V. I., "Improvement in the Properties of Nickel Coatings," *Vestnik Mashinostroeniya, 48* (6), 58–61 (1968).

(22) Steinebach, E., "Rain-Erosion Behavior of Electrodeposited Coatings," paper presented at the Second Research Conference on Rain Erosion (August, 1967).

(23) Spähn, H., "Deposits of Nickel on Steel Used as a Model for Study of the Influence of Internal Stress on Corrosion Fatigue," *Corrosion, 15* (4), 178–186 (1967).

(24) Ewing, D. T., Rominski, R. J., and King, W. M., "Effect of Impurities and Purification of Electroplating Solutions. I. Nickel Solutions. (4) The Effects and Removal of Copper," *Plating, 37* (11), 1157–1160 (1950).

(25) Ewing, D. T., Brouwer, A. A., Clark, D. D., Owen, C. J., Rominski, R. J., and Werner, J. K., "Effect of Impurities and Purification of Electroplating Solutions. I. Nickel Solutions. (5) The Effects and Removal of Zinc," *Plating, 39* (9), 1033–1037 (1952).

(26) Ewing, D. T., Werner, J. K., Brouwer, A. A., Owen, C. J., and Dow, W. O., "Effect of Impurities and Purification of Electroplating Solutions. I. Nickel Solutions. (7) The Effects and Removal of Chromium," *Plating, 40* (12), 1391–1400 (1953).

(27) Andreev, I. N., Vigalok, A. A., and Gudin, N. V., "Acidity of Catholyte Layer and Properties of Nickel Deposits," *Tr. Kazan, Khim.-Tekhnol. Inst., No. 36*, 145–150 (1967).

(28) Natwick, J. W., "Mechanical Properties of Nickel Foils Made by Electroforming," *Sixth Symposium on Materials of Space Vehicle Use, Seattle, 2*, 1–99 (1963).

(29) Ogburn, F., "Electrodeposited Coatings," *Surface Protection Against Wear and Corrosion*, American Society for Metals, Cleveland, 443–452 (1954).

(30) Ollard, E. A., "Electroforming. Piece Part Production by Electrodeposition," *Metal Industry (London), 70*, 6–8, 51–53, 86–88, 126–128 (1947).

(31) Buss, G., "Properties and Corrosion Protection of Electrodeposited Coatings and Anodized Aluminum," *Metall, 17* (6), 567–571 (1963).

(32) Weiner, R., and Klein, G., "Hardness and Wear-Resistance of Electrodeposits," *Metalloberflaeche, 7B* (1), 1–7 (1955).

(33) Weiner, R., and Klein, G., "Examination of the Cathodic Polarization of Bright Plating Solutions," *Metalloberflaeche, 10* (1), 17–21 (1956).

(34) Castell, H. C., "Nickel and Nickel Alloy Deposits," *Metal Industry (London), 89* (26), 536–540 (1956).

(35) Castell, H. C., "Mechanical Properties of Nickel and Nickel Alloy Electrodeposits," *Métaux (Corrosion-Ind.), 32* (379), 122–131 (1957).

(36) Castell, H. C., "Nickel Plating and Its Industrial Applications," *Métaux (Corrosion-Ind.), 33* (397), 366–376 (1958).

(37) Such, T. E., "The Physical Properties of Electrodeposited Metals," *Metallurgia, 56* (334), 61–66 (1957).

(38) Delforge, P., "Special Applications of Electrodeposits," *Métaux (Corrosion-Ind.), 33* (399), 457–463 (1958).

(39) Bart, S. G., "Electrolytic Nickel-Clad Plate," *Iron Age, 174* (18), 87–89 (1954).

(40) Silverstone, P. C., "Electroforming. Its Use in the Aircraft and Missile Industries," paper presented at the 1964 Western Metal and Tool Conference, Los Angeles, California (March 16–20, 1964), 11 pp.

(41) Natwick, J. W., "Metallurgical Factors in Electroforming," *Proceedings Am. Electroplaters' Soc., 51*, 178–191 (1964).

(42) Trofimov, A. N., "Electrodeposition of Metals in Ultrasound," *Primenenie Ul'traakust. k Issled. Veshchestva, No. 12*, 113–119 (1960).

(43) Roll, A., "The Effect of Ultrasonics on Electrolytic Processes. V. The Effect of Ultrasonics on the Properties of Electrodeposited Metals," *Z. Metall., 42* (9), 271–273 (1951).

(44) Kozan, T. G., "Effect of Ultrasonic Cleaning, Pickling and Plating on the Structure and Adhesion of Electrodeposited Nickel and Chromium," *Plating, 49* (5), 495–505 (1962).

(45) Kendrick, R. J., "The Effects of Alternating Current on the Properties of Nickel Deposits From an All-Chloride Solution," *Trans. Inst. Metal Finish., 44* (2), 78–83 (1966).

(46) Krivtsov, A. K., and Khamaev, V. A., "Effect of Alternating Current on Some Physicomechanical Properties of Nickel Deposits," *Izvestiya Vysshikh Uchebnykh., Zavedenij SSSR, Khimiya i Khimicheskaya Tekhnologiya, 10* (2), 194–197 (1967).

(47) Müller, Fr., and Kuss, H., "The Effects of Using Cathodes Oscillating at Various Frequencies, Particularly in the Ultrasonic Region, on the Electrodeposition of Metals," *Helvetica Chimica Acta, 33* (1), 217–228 (1950).

(48) Greco, V. P., "A Dual Cell Plating Apparatus for Deposition of Multilayer Metal Systems," *Proceedings Am. Electroplaters' Soc., 50*, 125–134 (1963).

(49) Lyakhovich, E. F., "Obtaining Hard Nickel Coatings From Oxalate Electrolytes," *Priborostroenie, No. 2*, 19–21 (1958).

(50) "Improvements in Plating Processes: The Dalic Process," *Product Finishing (London), 6* (9), 77–79 (1953).

(51) Machu, W., "Brush Plating for Localized Metal Deposition (Dalic Process)," *Metallwaren-Industrie und Galvanotechnik, 49* (11), 467–481 (1958). Cf. Hughes, H. D., *Trans. Inst. Metal Finish., 33*, 424–439 (1956).

(52) Elze, J., "Methods of Testing Nickel-Plating Baths and Nickel Coatings," *Metall, 6* (1/2), 5–11 (1952).

(53) Diggin, M. B., "Modern Electroforming Solutions and Their Applications,"Symposium on Electroforming—Applications, Uses and Properties of Electroformed Metals, *ASTM Special Technical Publication No. 318*, 10–31 (1962).

(54) Prine, W. H., "Industrial Nickel Plating in the United States," *Metal Industry (London), 78* (19), 383–388 (1951).

(55) Orbaugh, M. H., "Hard Nickel Plating," *Metal Finishing, 47* (11), 53–55, 59 (1949).

(56) Sample, C. H., and Knapp, B. B., "Physical and Mechanical Properties of Electroformed Nickel at Elevated and Subzero Temperatures," Symposium on Electroforming—Applications, Uses and Properties of Electroformed Metals, *ASTM Special Technical Publication No. 318*, 32 43 (1962).

(57) Spencer, L. F., "Engineering Uses of Plated Coatings," *Metal Finishing, 57* (5), 48–54 (1959).

(58) Benninghof, H., "Heavy- and Hard-Nickel Plating," *Galvanotechnik, 50* (3), 127–131 (1959).

(59) Faust, C. L., "The Challenge Before Us," *Proceedings Am. Electroplaters' Soc., 48*, 17–23 (1961).

(60) Safranek, W. H., "Don't Overlook Electroforming," *Product Engineering, 32* (23), 44–49 (June 5, 1961).

(61) Wiegand, H., and Schwitzgebel, K., "Stresses, Crystal Size and Textures in Electrodeposited and Chemical Nickel Coatings and Their Correlation With Mechanical and Technological Properties of Materials," *Metall, 21* (10), 1024–1038 (1967).

(62) Wesley, W. A., Carr, D. S., and Roehl, E. J., "Nickel Plating With Insoluble Anodes," *Plating, 38* (12), 1243–1250 (1951).

(63) Fishlock, D. J., "Bright Nickel Plating. 4. Deposit Characteristics," *Product Finishing (London), 11* (4), 79–90 (1958).

(64) Fedot'ev, N. P., Yakubovskii, E. S., Kruglova, E. G., and Kosareva, M. K., "Influence of the Physical and Mechanical Properties of Nickel Deposits on the Fatigue Strength of Steel," *Journal Applied Chemistry (USSR), 40* (9), 1897–1902 (1967); *Zh. Prikl. Khimii, 40* (9), 1963–1969 (1967).

(65) Squitero, A. D., "New Developments in All-Nickel Shell Electrotype Plates," *Printing Plates Magazine* (January, 1964).

(66) Winkler, L., "Nickel Electroforming With Special Reference to the Ni-Speed Process," *Metalloberflaeche, 20* (2), 57–61 (1966).

(67) Roehl, E. J., and Wesley, W. A., "Notes on Nickel Plating From a Fluoborate Bath," *Plating, 37* (2), 142–146, 171 (1950).

(68) Panikkar, S. K., and Rama Char, T. L., "Electroplating of Nickel From the Pyrophosphate Bath," *Journal Electrochem. Soc., 106* (6), 494–499 (1959).

(69) Belinskaya, L. S., and Izbekova, O. V., "Some Properties of Nickel Pyrophosphate Baths," *Tr. Kievsk. Politekhn. Inst., 38*, 33–37 (1962).

(70) Oswald, J. W., *Heavy Electrodeposition of Nickel*, International Nickel Company (Mond) Limited, London (1962), 40 pp.

(71) Baeyens, P., "Electroformed Molds and Dies," *International Nickel Company, Electroforming Seminar, New York*, 165–195 (December 2, 1964); *Plating, 53* (5), 591–597 (1966).

(72) Lamb, V. A., and Metzger, Jr., W. H., "Electroforming—A Method for Producing Intricate Shapes," *The Tool Engineer, 33* (2), 55–62 (1954).

(73) Heussner, C. E., Balden, A. R., and Morse, L. M., "Some Metallurgical Aspects of Electrodeposits," *Plating, 35* (6), 554–561; 577–578 (1948).

(74) Read, H. J., and Weil, R., "Grain-Size and Hardness of Nickel Plate as Related to Brightness," *Plating, 37* (12), 1257–1261 (1950).

(75) Hughes, H. D., "Grain Size Control in Electroforming," *Proceedings Am. Electroplaters' Soc., 46*, 326–332 (1959).

(76) Brusentsova, V. N., and Pletnev, D. V., "High-Hardness, Bright Nickel Plating in Oxalic Acid Baths," *Priborostroenie, No. 2*, 19–21 (1963).

(77) Steiger, A. J., "Hard Nickel Plating in Russia. Sulphate-Oxalate Bath," *Metal Finishing, 57* (1), 52–53 (1959).

(78) Vozdvizhenskii, G. S., and Kul'pina, N. V., "Influence of Some Factors of Electrolysis on the Uniformity of Distribution of Nickel Deposits on Complex Shaped Cathodes," *Protective Metallic and Oxide Coatings, Metal Corrosion and Electrochemistry*, edited by N. P. Fedot'ev, Akademiya Nauk SSSR. Translated by the Israel Program for Scientific Translations, Jerusalem, 60–65 (1968).

(79) Ewing, D. T., Smith, A. J., and Dow, W. O., "Effects of Impurities and Purification of Electroplating Solutions. I. Nickel Solutions. (9) The Effect and Removal of Aluminum," *Proceedings Am. Electroplaters' Soc., 43*, 44–46 (1956).

(80) Marti, J. L., and Lanza, G. P., "The Effects of Some Variables Upon the Hardness of Sulfamate Nickel Deposits," *Electrochimica Metallorum, 1* (3), 335–349 (1966).

(81) Marti, J. L., and Lanza, G. P., "Hardness of Sulfamate Nickel Deposits," *Plating, 56* (4), 377–385 (1969).

(82) Ewing, D. T., Werner, J. K., Brouwer, A. A., Owen, C. J., and Dow, W. O., "Effect of Impurities and Purification of Electroplating Solutions. I. Nickel Solutions, (7) The Effects and Removal of Chromium," *Plating, 40* (12), 1391–1400 (1953).

(83) Polukarov, Yu. M., and Semenova, Z. V., "Dependence of the Physical and Mechanical Properties of Electrodeposited Nickel on the Amount of Included Hydrogen," *Navodorozhivanie Metal. Bor'ba Vodordn. Khrupkost'yu*, 150–157 (1968).

(84) Geneidy, A., Koehler, W. A., and Machu, W., "The Effect of Magnesium Salts on Nickel Plating Baths," *Journal Electrochem. Soc., 106* (5), 394–403 (1959).

(85) Pianelli, A., "Heavy Nickel Plating," *Revue du Nickel, 27* (6), 150–156 (1961).

(86) Wesley, W. A., Sellers, W. W., and Roehl, E. J., "Electrodeposition of Nickel at High Current Density," *Proceedings Am. Electroplaters' Soc., 36*, 79–92 (1949).

(87) Doktorina, S. V., "The Properties of Nickel Deposits," *Journal Applied Chemistry (USSR), 36* (9), 1897–1899 (1963); *Zh. Prikl. Khimii, 36* (9), 1955–1958 (1963).

(88) Doktorina, S. V., and Kudryavtsev, N. T., "Nickel Plating at High Current Densities," *Izvestiya Vysshikh Uchebnykh Zavedeniy SSSR, Khimiya i Khimicheskaya Tekhnologiya, 3*, 497–503 (1960).

(89) Rohde, G., "Electroforming of Thin-Walled Nickel Models," *Galvanotechnik, 59* (5), 407–420 (1968).

(90) Kase, H., "Measurement of Internal Stress of Electrodeposited Nickel by Electric Resistance Wire Type Strain Meter," *Journal of the Metal Finishing Society of Japan, 11* (7), 259–263 (1960).

(91) Lainer, V. I., "Deposition of Low-Stress Nickel Layers From Sulfamate Electrolytes," *Russian Engineering Journal, 44* (1), 27–30 (1964); *Vestnik Mashinostroeniya, 44* (1), 32–36 (1964).

(92) Belt, K. C., Crossley, J. A., and Kendrick, R. J., "Properties of Electrodeposits From a Concentrated Nickel Sulfamate Solution," *Interfinish '68 Tagungsberichtsband*, Deutsche Gesellschaft für Galvanotechnik, e. V., Düsseldorf, Germany, 222–228 (1968).

(93) Kendrick, R. J., and Watson, S. A., "Rapid Deposition With Concentrated Sulphamate Nickel Plating Solutions," *International Nickel Company, Electroforming Seminar, New York, 96–106* (December 2, 1964).

(94) McMullen, W. H., and Stoddard, Jr., W. B., "A High Speed Nickel Electroforming Solution," *International Nickel Company, Electroforming Seminar, New York,* 42–61 (December 2, 1964).

(95) Evans, D. J., "The Structure of Nickel Electrodeposits in Relation to Some Physical Properties," *Transactions of the Faraday Society, 54*, 1086–1091 (1958).

(96) Roehl, E. J., "Mechanical Properties of Nickel Deposits," *Monthly Rev., Am. Electroplaters' Soc., 34*, 1129–1140 (1947).

(97) Raub, E., "Electrodeposition of Nickel From Chloride and Sulfate Solutions," *Metalloberflaeche, 9A* (6), 88–93 (1955).

(98) Vagramyan, A. T., and Solov'eva, Z. A., *Technology of Electrodeposition*. English Translation, Robert Draper Ltd., Teddington (1961), 398 pp.

(99) Calderon, E., "Life of Aircraft Parts is Increased, Expensive Components are Salvaged, and Quality is Improved With 'Stress-Free' Plating," *American Machinist, 101* (22), 154–155 (1957); *Automotive Industries, 117* (2), 65, 72 (1957).

(100) Croly, P. B., "New Developments in All-Nickel Shell Electrotype Plates: Effect of Operating Conditions of Plating Solutions on Properties of Nickel Deposits," *Printing Plates Magazine, 50* (10), 3–8 (1964).

(101) Roehl, E. J., "Mechanical Properties of Nickel Deposits. II," *Plating, 35* (5), 452–455, 478 (1948).

(102) Roehl, E. J., "Practical Methods in Heavy Industrial Nickel Plating," *Metal Finishing, 45* (5), 56–59, 71 (1947).

(103) Konishi, S., "Duplex Nickel Plating. A Study of Stress in the Deposits," *Metal Finishing, 63* (4), 67–72 (1965).

(104) Fanner, D. A., and Hammond, R. A. F., "The Properties of Nickel Electrodeposited From a Sulfamate Bath," *Trans. Inst. Metal Finish., 36* (2), 32–42 (1958).

(105) Katz, W., "Electroforming With Nickel," *Metall (Berlin), 21* (6), 580–587 (1967).

(106) Konishi, S., "Stress in Duplex Nickel Deposits," *Journal of the Metal Finishing Society of Japan, 15* (12), 479–484 (1964).

(107) Marchenko, N. A., Borisova, V. A., Kharchenko, E. P., Lipko, S. Kh., and Baksheeva, V. P., "Electrolyte for Deposition of Bright Nickel Platings," *Tekhnol. Organ. Proiz. Nauch.-Proiz. Sb., 2* (38), 84–85 (1966).

(108) Jogarao, A., Guruviah, S., and Parameswara Iyer, K., "Study of Cystine as a Nickel Brightener," *Metal Finishing, 64* (1), 82–84 (1966).

(109) Willson, K. S., and DuRose, A. H., "A Semi-Bright Nickel-Plating Process," *Metal Finishing, 47* (2), 55–57 (1949).

(110) Spiro, P., and Wohlgemuth, F., "Improved Nickel-Plating Electrolyte," British Patent 584,977 (January 28, 1947); also German Patent 804,278 (April 19, 1951). Assigned to London & Scandinavian Metallurgical Co. Ltd.

(111) Ostroumov, V. V., "Electrolytic Nickel Deposits Formed in the Presence of Sulfonated Naphthalene," *Journal Applied Chemistry (USSR), 32* (3), 600–606 (1959); *Zh. Prikl. Khimii, 32* (3), 572–578 (1959).

(112) Young, C. B. F., and Strobach, W., "Electrodeposition of Nickel From Fluoborate Solutions," *Metal Finishing, 53* (8), 53–58; (9), 79–85 (1955).

(113) Newell, I. L., "A Study of the Effect of Several Organic Addition Agents on the Hardness and Residual Stress in Nickel Deposits," *Proceedings Am. Electroplaters' Soc., 43*, 101–104 (1956).

(114) Ebuch, K., Yamazaki, S., and Kuroda, A., "Effect of Saccharin on the Stress of Nickel Electroforming," *Journal of the Metal Finishing Society of Japan, 15* (1), 15–18 (1964).

(115) Sutyagina, A. A., and Gorbunova, K. M., "Crystallization of Metals During Electrodeposition in the Presence of Sulfur-Containing Surface-Active Compounds. II. Effect of Conditions of Electrolysis on the Impurities in Nickel Deposits and Their Properties," *Russian Journal Physical Chemistry, 35* (11), 1243–1248 (1961); *Zh. Fiz. Khimii, 35* (11), 2514–2523 (1961).

(116) Carr, D. S., "Electrodepositing Dull Nickel," U.S. Patent 2,842,487 (July 8, 1958). Assigned to Bart Laboratories Company, Inc.

(117) Metzger, Jr., W. H., Krasley, P. A., and Ogburn, F., "Use of Disodium-m-Benzenedisulfonate as a Hardening Agent in a Watts Nickel Bath," *Plating, 47* (3), 285–287 (1960).

(118) Kunik, E. V., "Development and Adoption of Bright Nickel Plating at Automobile Industry Plants," *Blestyashchie Komb. Metal. Pokrytiya, No. 1*, 93–100 (1967).

(119) Schulze, O., "Electrolytic Deposition of Nickel Coatings," East German Patent 35,084 (February 25, 1965).

(120) Marchenko, N. A., Lipko, S. Kh., and Kharchenko, E. P., "Properties of Coatings From Lustrous Nickel-Plating Electrolytes," *Issled. Obl. Galvanotekh, Novocherkassk.*, 96–99 (1965).

(121) Spähn, H., "The Effect of Wetting Agents on the Hardness of Electrodeposited Nickel," *Metalloberflaeche, 13* (8), 259–264 (1959).

(122) Stoddard, Jr., W. B., "Electrodeposition of Nickel," U.S. Patent 2,533,532 (December 12, 1950). Assigned to Champion Paper & Fibre Company.

(123) Brenner, A., "A Microhardness Tester for Metals at Elevated Temperatures," *Plating, 38* (4), 363–366 (1951).

(124) Belt, K. C., Crossley, J. A., and Watson, S. A., "Nickel-Cobalt Alloy Deposits From a Concentrated Sulfamate Electrolyte," *Trans. Inst. Metal Finish., 48*, 133–137 (1970).

(125) Hanson, R. N., DuPree, D. G., and Lui, K., "Structural Electroforming—Applications and Developments," *Plating, 55* (4), 347–355 (1968).

(126) Safranek, W. H., "A Survey of Electroforming for Fabricating Structures," *Plating, 53* (10), 1211–1216 (1966), extract of *RSIC 210*, under Army Contract DA-01-021-AMC-203(Z), Battelle Memorial Institute (October 1, 1964), Revised Edition, 40 pp.

(127) Lackovic, A., "Tough and Bright Nickel Plating in Oxalate Bath," *Zastita Mater., 11* (9), 336–337 (1963).

(128) Smith, A. J., and Rowe, R. J., "Effects of Impurities and Purification of Electroplating Solutions. I. Nickel Solutions. (10) The Effect and Removal of Manganese," *Proceedings Am. Electroplaters' Soc., 43*, 46–49 (1956).

(129) Phillips, W. M., and Clifton, F. L., "Judging the Quality of Plated Parts," *Proceedings Am. Electroplaters' Soc., 35*, 87–102 (1948).

(130) Asher, R. K., and Harding, W. B., "Mechanical Properties of Electroformed Nickel Produced in Sulfamate Solutions," *Plating, 49* (7), 783–788 (1962).

(131) Endicott, D. W., and Knapp, Jr., J. R., "Electrodeposition of Nickel-Cobalt Alloy: Operating Variables and Physical Properties of the Deposits," *Plating, 53* (1), 43–60 (1966).

(132) Baker, R. A., and Christie, N., "A New Electrolytic System for Producing Satin Nickel Deposits," *Trans. Inst. Metal Finish., 47* (3), 80–83 (1969).

(133) Edwards, J., "Properties of Bright Nickel Electrodeposits in Relation to the Period of Service of the Plating Bath. 3. Composition and Microstructure," *Trans. Inst. Metal Finish., 36* (3), 86–92 (1959).

(134) Read, H. J., "The Effects of Addition Agents on Physical and Mechanical Properties of Electrodeposits," *Plating, 49* (6), 602–606 (1962).

(135) Edwards, J., and Levett, M. J., "Preparation of Coumarin Derivatives and Their Evaluation as Additives for Nickel Plating Solutions," *Trans. Inst. Metal Finish., 47* (1), 7–12 (1969).

(136) Bradley, R. C., "Large Electroformed Bulkheads for Space Vehicles," paper presented at Second Aerospace Finishing Symposium, Kansas City, Missouri (January 18, 1963).

(137) Read, H. J., Karchner, G. H., and Patrician, T. J., "The Effects of Chromium Plating on the Mechanical Properties of Electrodeposited Nickel," *Plating, 50* (1), 35–38 (1963).

(138) Read, H. J., and Oles, Jr., E. J., "Influence of Hydrogen on the Strength and Ductility of Electrodeposited Nickel," *Plating, 52* (9), 860–867 (1965).

(139) See Reference (126).

(140) Stephenson, W. B., "Forming Through Electrodeposition," paper presented at the Design Engineering Conference, American Society of Mechanical Engineers, Chicago, Illinois (May 12, 1964).

(141) Ladd, J. C., and Allie, D. L., "Electroforming for the Space Age," *Products Finishing, 26* (4), 50–56 (1962).

(142) Carlson, R. L., Schneider, B. R., and Berke, L., "Electroforming Thin Shells for Experimental Studies," *Materials Research & Standards, 7* (5), 183–188 (1967).

(143) Green, R. E., "Some Applications of Electroforming," *Machinery and Production Engineering, 109* (2817), 1004–1011 (1966).

(144) Durbin, C., "Evaluation of Quality of Electrodeposited Coatings," *Proceedings Am. Electroplaters' Soc., 38*, 119–132 (1951).

(145) Kushner, J. B., "Factors Affecting Residual Stress in Electrodeposited Metals. A Critical Evaluation," *Metal Finishing, 56* (5), 82–87; (7), 52–55, 59 (1958).

(146) Kushner, J. B., "Stress in Electroplated Metals," *Metal Progress, 81* (2), 88–93 (1962).

(147) Konishi, S., "Trial Production of Stress Meter for Electrodeposits," *Journal of the Metal Finishing Society of Japan, 11* (7), 263–268 (1960); "Stress in Nickel Plating," *Ibid.*, 273–276 (1960).

(148) Heinke, G., and Spähn, H., "The Practical Measurement of Internal Stresses With the Spiral Contractometer," *Metalloberflaeche, 20* (5), 215–222 (1966).

(149) Kushner, J. B., "The Role of the Basis Metal in the Production of Stressed Electrodeposits," *Proceedings Am. Electroplaters' Soc., 45*, 28–32 (1958).

(150) Watkins, H., and Kolk, A., "Measurement of Stress in Very Thin Electrodeposits," *Journal Electrochem. Soc., 108* (11), 1018–1023 (1961).

(151) Brenner, A., and Senderoff, S., "A Spiral Contractometer for Measuring Stress in Electrodeposits," *Proceedings Am. Electroplaters' Soc., 35*, 53–76 (1948).

(152) Borchert, L. C., "Investigation of Methods for the Measurement of Stress in Electrodeposits," *Proceedings Am. Electroplaters' Soc., 50*, 44–50 (1963).

(153) Spähn, H., "Stress Conditions in Electrolytic Metal Coatings," *Galvanotechnik Oberflaechenschutz, 8* (9), 194–200 (1967).

(154) Brugger, H., "On Nickel Plating From Sulfamate Solutions," *Galvanotechnik Oberflaechenschutz, 6* (7), 165–171 (1965).

(155) Soderberg, K. G., and Graham, A. K., "Stress in Electrodeposits—Its Significance," *Proceedings Am. Electroplaters' Soc., 34*, 74–96 (1947).

(156) Bilfinger, R., and Strauch, A., "Investigation of Internal Stress in Nickel Deposits From Watts-Type Nickel Solutions," *Wissenschaftliche Zeitschrift der Hochschule für Elektrotechnik, Ilmenau, 8* (4), 327–335 (1963).

(157) Phillips, W. M., and Clifton, F. L., "Stress in Electrodeposited Nickel," *Proceedings Am. Electroplaters' Soc., 34*, 97–110 (1947).

(158) Heussner, C. E., Balden, A. R., and Morse, L. M., "Some Metallurgical Aspects of Electrodeposits. II," *Plating, 35* (7), 719–723, 768 (1948).

(159) Vagramyan, A. T., and Petrova, Yu. S., *The Mechanical Properties of Electrolytic Deposits*, USSR Academy of Sciences Press for Institute of Physical Chemistry, Moscow (1960). English translation by Consultants Bureau (1962), 108 pp.

(160) Searles, H., "The Effect of Bromide Ion on Stress in a Nickel Sulfamate Bath," *Plating, 53* (2), 204–208 (1966).

(161) Barrett, R. C., "Nickel Plating From the Sulfamate Bath," *Proceedings Am. Electroplaters' Soc., 41*, 169–174 (1954); *Plating, 41* (9), 1027–1032 (1954).

(162) Knödler, A., "Internal Stresses in Electrolytic Nickel Deposits," *Metalloberflaeche, 20* (2), 52–56 (1966).

(163) Marti, J. L., "The Effect of Some Variables Upon Internal Stress of Nickel as Deposited From Sulfamate Electrolytes," *Plating, 53* (1), 61–71 (1966).

(164) Klingenmaier, O. J., "The Effect of Anode Efficiency on the Stability of Nickel Sulfamate Solutions," *Plating, 52* (11), 1138–1141 (1965).

(165) Pamfilov, A. V., and Mel'nik, P. M., "Effects of Additives on Internal Stresses in Electrolytic Nickel Deposits," *Journal Applied Chemistry (USSR), 35* (10), 2179–2182 (1962); *Zh. Prikl. Khimii, 35* (10), 2272–2275 (1962).

(166) Weinhardt, R. A., "Nickel Electroforming of Optical Mirrors," *NOTS Technical Memo 786*, Naval Ordnance Test Station, China Lake, California (February 2, 1953).

(167) Kushner, J. B., "A New Instrument for Measuring Stress in Electrodeposits," *Proceedings Am. Electroplaters' Soc., 41*, 188–195 (1954).

(168) Kendrick, R. J., and Watson, S. A., "Plating With Nickel Sulfamate," *Electrochimica Metallorum, 1* (3), 320–334 (1966).

(169) Muñoz del Corral, F., and Rubio Felipe, A., "Internal Stresses in Electrodeposits," *Inst. hierro y acero, 10*, 355–363 (1957).

(170) Fischer, G., "Measurement of Mechanical Stress in Electrodeposits," *Galvanotechnik, 53* (7), 335–340 (1962).

(171) Pawlak, Jr., "A Study of High-Speed Heavy Nickel Deposition," *Products Finishing, 30* (3), 56–66 (1966).

(172) Nishihara, K., and Tsuda, S., "Studies on the Electrodeposition of Nickel (II)," *Bulletin of the Institute for Chemical Research, Kyoto University, 21*, 64 (1950).

(173) Max, A. M., "Application of Electroforming to the Manufacture of Disk Records," Symposium on Electroforming—Applications, Uses and Properties of Electroformed Metals, *ASTM Special Technical Publication No. 318*, 71–85 (1962).

(174) Diggin, M. B., "Nickel Plating From Sulfamate Baths," *Metal Progress, 66* (4), 132–137 (1954).

(175) Balashova, N. N., Bokov, N. S., Gulova, G. P., Glushko, N. A., and Tarasova, E. G., "Influence of Iron on Electrodeposition of Nickel From Sulfate Electrolytes," *Journal Applied Chemistry (USSR), 38* (12), 2770–2772 (1965); *Zh. Prikl. Khimii, 38* (12), 2850–2852 (1965).

(176) Crossley, J. A., Kendrick, R. J., and Mitchell, W. I., "The Structure of Nickel Deposited From an All-Chloride Solution by Square-Wave Alternating Current," *Trans. Inst. Metal Finish., 45* (2), 58–63 (1967).

(177) Marchese, V. J., "Electroforming Low-Stress Nickel," U.S. Patent 2,706,170 (April 12, 1955). Assigned to Sperry Corporation.

(178) Marchese, V. J., "Stress Reduction of Electrodeposited Nickel," *Journal Electrochem. Soc., 99* (2), 39–43 (1952).

(179) Serota, L., "Science for Electroplaters. 65. Variables in the Watts Bath," *Metal Finishing, 59* (5), 76–78 (1961).

(180) Kochergin, S. M., and Vyaseleva, G. Ya., "Investigation of the Structure and Properties of Nickel Electrodeposits Formed in an Ultrasonic Field," *Zh. Fiz. Khimii, 38* (4), 839–845 (1964); *Russian Journal of Physical Chemistry, 38* (4), 456–459 (1964).

(181) Zheiwite, O. S., Bodnevas, A. I., and Matulus, Yu. Yu., "The Influence of Some Aromatic Sulfur-Containing Compounds on the Internal Stresses in Nickel Electrodeposits," *Electrodeposition of Metals - Proceedings of the 10th Lithuanian Conference of Electrochemists*. Translated by the Israel Program for Scientific Translations, National Bureau of Standards and National Science Foundation, Washington, D.C., 19–21 (1970).

(182) Watson, S. A., "The Effect of Some Addition Agents on Stress in Nickel Deposits," *Trans. Inst. Metal Finish., 40*, 41–47 (1963).

(183) Kendrick, R. J., "The Effects of Some Aromatic Sulfonic Acids on the Stress, Structure, and Composition of Electrodeposited Nickel," *Trans. Inst. Metal Finish., 40* (1), 19–27 (1963).

(184) Fedot'ev, N. P., and Pozin, Yu. M., "Internal Stresses in Copper and Nickel Electrodeposits," *Protective Metallic and Oxide Coatings, Metal Corrosion and Electrochemistry*, edited by N. P. Fedot'ev, Akademiya Nauk SSSR. Translated by the Israel Program for Scientific Translations, Jerusalem, 55–59 (1968).

(185) Mattia, M., "Electroforming of Dies and Molds," *Plating, 55* (1), 40–46 (1968).

(186) Edwards, J., "Aspects of Addition Agent Behavior," *Trans. Inst. Metal Finish., 42*, 22–34 (1964).

(187) Stalzer, M., "Instruments for the Measurement of Stress in Electrodeposited Coatings and Description of a Newly Developed Self-Compensating and Recording Instrument," *Metalloberflaeche, 18* (9), 263–267 (1964).

(188) Fischer, A., "Electrodeposition of Stress-Free Metal Deposits," U.S. Patent 3,338,804 (August 29, 1967). Assigned to Kewanee Oil Company.

(189) Zukaite, M., Vegys, J., Bodnevas, A., and Matulis, J., "Variation of Leveling and Internal Stresses of Electrodeposited Nickel in the Presence of Some Brighteners," *Lietuvos TSR Mokslu Akad. Darbai, Ser. B., No. 1*, 23–29 (1967).

(190) Curkin, L. H., and Moeller, R. W., "Stress Effect of Iron Contamination in a Watts-Type Nickel-Plating Solution and Its Correlation With Endurance Limit," *Plating, 41* (10), 1154–1157 (1954); *Proceedings Am. Electroplaters' Soc., 41*, 196–199 (1954).

(191) Stareck, J. E., Seyb, E. J., and Tulumello, A. C., "The Effect of Different Chromium Deposits on the Fatigue Strength of Hardened Steel," *Plating, 42* (11), 1395–1402 (1955); *Proceedings Am. Electroplaters' Soc., 42*, 129–136 (1955).

(192) Hammond, R. A. F., "Some Engineering Aspects of Electrodeposition," *Proceedings Am. Electroplaters' Soc., 51*, 9–20 (1964).

(193) Barklie, R. D. H., and Davies, H. J., "The Effect of Surface Conditions and Electrodeposited Metals on the Resistance of Materials to Repeated Stresses," *Proceedings Inst. Mechanical Engineers, 1*, 731 (1930).

(194) Almen, J. O., "Fatigue Loss and Gain by Electroplating," *Product Engineering, 22* (6), 109–116 (1951).

(195) Spähn, H., "Corrosion Fatigue of Metallic Materials (VII). Internal Stresses in Electrolytically and Chemically Deposited Nickel Coatings and Their Effect on the Properties of Steel Under Corrosion-Fatigue," *Metalloberflaeche, 17* (1), 1–9 (1963). Cf. Spähn, *Technische Rundschau (Switzerland), 58* (27), 33 (1966).

(196) Spähn, H., "The Effect of Internal Stress on the Fatigue and Corrosion-Fatigue Properties of Electroplated and Chemically Plated Nickel Deposits," *Trans. Inst. Metal Finish., 42*, 364 (1964).

(197) Hammond, R. A. F., "Stresses in Hard Chromium and Heavy Nickel Deposits and Their Influence on the Fatigue Strength of the Basis Metal," *Metal Finishing Journal, 7*, 441–449 (1961).

(198) Paige, H., James, J. H., and Williams, F. S., "How Tough Are Nickel and Chromium Electroplates for Aluminum?" *Product Engineering, 24* (12), 162–167 (1953).

(199) Wiggins, F. W., "The Development of Electroforming Techniques for Wind Tunnel Models," *McDonnell Aircraft Corporation, A-479, Final Report, Contract AF 33(657)-11215* (October 15, 1962).

(200) Reddy, A. K. N., "The Relative Corrosion Rates of Nickel Electrodeposits Having Different Textures," *Journal Electrochem. Soc., 110* (10), 1087–1088 (1963).

(201) Banerjee, B. C., and Walker, Jr., P. L., "Effect of Heat Treatment on the Structure of Oriented Nickel Electrodeposits," *Nature, 206* (4986), 816–817 (1965).

(202) Weil, R., Jacobus, Jr., W. N., and DeMay, S. J., "The Effect of Annealing on the Microstructure and Hardness of Some Nickel Electrodeposits," *Journal Electrochem. Soc., 111* (9), 1046–1052 (1964).

(203) Weil, R., Sumka, H. J., and Greene, G. W., "Annealing Behavior of Fine-Grained Nickel Electrodeposits," *Journal Electrochem. Soc., 114* (5), 449–451 (1967).

(204) Rotinyan, A. L., and Kosich, E. S., "Internal Stresses in Cathodic Nickel Deposits," *Journal of Applied Chemistry (USSR), 31* (3), 411–415 (1958); *Zh. Prikl. Khimii, 31* (3), 424–428 (1958).

(205) *Metals Handbook*, 8th Ed., *Vol. 1, Properties and Selection of Metals*, American Society for Metals, Metals Park, Ohio, 1118–1119, 1217 (1961).

Chapter 13

Nickel Alloys

Nickel-Cobalt Alloys

Nickel-cobalt alloys have been studied extensively for magnetic applications. Where hardness is a desirable property, electroforms are finding use as mold inserts for pressure and injection forming of plastics. (1,2) For engineering applications, alloying with cobalt hardens the nickel in a sulfur-free form that is not embrittled on heat-treatment. Using cobalt as a hardener instead of sulfur simplifies control. Moreover, the hard nickel-cobalt deposits are more ductile than those hardened by sulfur-containing additives.

Magnetic Properties

Table 13.1 shows the range of magnetic properties observed for electrodeposited alloys of nickel and cobalt. Thin films with a coercive force (H_c) above 200 oersteds, which is desirable for high-density permanent storage of information, were reported in several publications. (3,4,5,6) In contrast with this trend, cobalt-rich alloy containing less than 25 percent nickel exhibited a coercivity of < 200 oersteds.*

According to one investigator, a nickel content of 25 to 30 percent was reported to be best for obtaining a high coercivity. (5) Alloy containing >30 percent nickel had low coercive forces and residual induction, which also were decreased by introducing small amounts of copper or lead in the solution. Coatings 4 to 7 μm thick were optimum for recording with a low current level and had a low noise level.

A recent patent (6) recommended deposition of nickel-cobalt alloys from sulfate-perchlorate solutions for achieving coercivities of 200 to 500 oersteds and residual inductions of 4000 to 8000 gausses. Current reversal or the addition of complexing anions were harmful for good magnetic characteristics. The deposition solution contained 12 to 50 g/l nickel sulfate ($NiSO_4 \cdot 7H_2O$), 12 to 50 g/l cobalt sulfate ($CoSO_4 \cdot 7H_2O$), and 10 to 40 g/l boric acid (H_3BO_3) and was operated at 20 to 80 C and a current density of 0.5 to 10 amp/sq dm with a pH of 4.8 to 5.9. Perchlorates

* See Table 5.5 in Chapter 5, Cobalt and Cobalt Alloys.

TABLE 13.1

Magnetic Properties of Nickel-Cobalt Alloys

Alloy	Composition	Coercive Force, H_c, oersteds	Residual Induction, B_r, gausses	Bath	Reference
Ni-Co	—	200–300	5000	Chloride-sulfate	3
Ni-Co	15 to 38% Ni	200–260	4000–6000	Sulfate-chloride	4
Ni-Co[a]	25% Ni	275	—	Sulfate	5
Ni-Co	—	200–500	4000–8000	Sulfate-perchlorate	6
Ni-Co-P	—	500–800	4000–5000	Chloride-hypophosphite	7
Ni-Co-P	—	100–800	—	Sulfate-chloride-tartrate-hypophosphite	3
Co-Ni-P	—	600–800	500	Chloride-sulfate-hypophosphite	3
Ni-Co-P	≤3% P, 15 to 38% Ni	600–700	4000–6000	Sulfate-chloride-hypophosphite	4
Ni-Fe-Co	64% Ni, 16% Fe, 20% Co	2–3	—	—	9
Ni-Fe-Co[b]	43–48% Ni, 40–60% Co, 7–11% Fe	2–15	—	Chloride-sulfate	10

(a) 4–7 μm thick.
(b) 1000 Å thick.

of both metals or any combination of sulfates or perchlorates can be used to prepare the solution. Halogen and complex-forming ions should be avoided, according to the patent.

The codeposition of phosphorous with nickel and cobalt increased coercivity to the range of 600 to 800 oersteds, by comparison with 200 to 300 oersteds for nickel-cobalt alloy electrodeposits. (3) The phosphorus content depended on the concentration of sodium hypophosphite in a chloride-sulfate bath with a 1:2 ratio of nickel and cobalt ions. A pH of 4.5 was best for increasing coercivity and avoiding the precipitation of hypophosphites. The current density was 10 amp/sq dm and the temperature was in the range of 40 to 70 C. With these conditions, residual induction was reduced to 500 gausses, in comparison with about 5000 gausses for phosphorus-free nickel-colbalt alloy.

Nickel-cobalt-phosphorus alloy with a high coercivity of 600 to 800 oersteds and a residual induction of 4000 to 6000 gausses was reported for other films containing about 3 percent phosphorus. (4) The electrodeposition solution contained 140 g/l nickel sulfate ($NiSO_4 \cdot 7H_2O$), 80 g/l sodium sulfate ($Na_2SO_4 \cdot 10H_2O$), 20 g/l sodium chloride (NaCl), and 20 g/l boric acid (H_3BO_3); pH was 4.0 to 5.5. It was used at 70 C at a current density of 1 to 2 amp/sq dm. To deposit an alloy containing 10 or 11 percent phosphorus, 8 to 10 g/l sodium hypophosphite ($NaH_2PO_2 \cdot H_2O$) was added to the solution.

Nickel-cobalt-phosphorus alloys having a coercive force ranging from 100 to 800 oersteds are reported to be suitable for deposition as record carriers on aluminum disks. (8) The deposition solution contained 23 to 25 g/l cobalt, 25 to 35 g/l nickel, 0.5 to 2.0 g/l sodium hypophosphite ($NaH_2PO_2 \cdot H_2O$), 25 to 55 g/l ammonium chloride (NH_4Cl), and 5.7 to 11.4 g/l pyrotartaric anhydride.

Codeposition of iron with nickel and cobalt resulted in coercivities of only 15 oersteds, or less. With about 7 to 11 percent iron, coercivity ranged from 2 to 15 oersteds, depending on the cobalt content. (10) An alloy containing 16 percent iron exhibited a coercivity of 2 or 3 oersteds. (9)

Resistivity

Electrical resistivity of nickel-cobalt alloys was reported for the entire composition range for deposits from sulfate solutions. (11) Values at 20 C, which ranged from 9.4 to 13.0 microhm-cm, are reported in Table 13.2. Similar resistivities were also reported at 50 C. The values in Table 13.2 are slightly higher than those for cast and slowly cooled alloys. The latter had resistivities of 10.0 and 9.0 microhm-cm for 40 percent and 66 percent cobalt, respectively.

Tensile Strength and Ductility

The tensile properties of electrodeposited nickel-cobalt alloys are summarized in Table 13.3. Two investigators (13,15) reported tensile strengths from 140 to 192

TABLE 13.2

Electrical Resistivity of Nickel-Cobalt Alloys[(a)]

Cobalt Content, percent	Resistivity at 20 C, microhm-cm
0	8.4
11	11.6
24	11.6
34	13.0
39	11.6
61	10.8
66	10.3
83	9.4
100	8.7

[(a)] Alloys deposited at 1 amp/sq dm in sulfate solutions at 20 or 50 C containing 60 g/l Ni + Co, 20 g/l H_3BO_3, and 20 g/l NaCl. (11)

kg/sq mm (200,000 to 275,000 psi) for alloy containing 40 to 50 percent cobalt. One of these reported a maximum strength for alloy containing 45 percent cobalt. (13) Elongation ranged from 2 to 4 percent. Heat treatment reduced tensile strength. (13) However, the maximum still occurred at 45 percent cobalt, with values of 133 kg/sq mm (190,000 psi), 112 kg/sq mm (160,000 psi), and 98 kg/sq mm (140,000 psi) after heat treatment at 230, 330 and 455 C, respectively. After heating at 630 C, ultimate tensile strength was in the range of 42 to 63 kg/sq mm (60,000 to 90,000 psi), which was less than the 84 kg/sq mm (120,000 psi) measured for unalloyed cobalt similarly heat treated. Alloys containing 45 to 80 percent cobalt and pure cobalt had the same yield strength of about 56 kg/sq mm (80,000 psi) after heating at 630 C.

Ductility increased with decreasing cobalt content, to values ranging from 2 to 4 percent elongation at 45 percent cobalt before heat treatment and 4 to 10 percent after heating at 630 C. The most ductile alloys, having elongations from 3 to 9 percent before heating and 8 to 22 percent after heating at 630 C, contained 20 to 30 percent cobalt. (13)

Hardness

Hardness data for electrodeposited nickel-cobalt alloys are summarized in Table 13.4. Alloys with a hardness ranging from 643 to 703 kg/sq mm were reported for bright deposits obtained at 6 amp/sq dm from a sulfate-boric acid-sodium chloride-formate solution. (17) Bright deposits containing 1.1 percent cobalt deposited at 5 amp/sq dm in a sulfate-chloride bath containing 45 g/l nickel formate and 2.5 g/l

TABLE 13.3

Strength and Elongation of Electrodeposited Nickel-Cobalt Alloys

Cobalt Content, percent	Tensile Strength		Yield Strength		Elongation, percent	Plating Bath	Reference
	kg/sq mm	psi	kg/sq mm	psi			
1	144	206,000	—	—	4	Sulfate-chloride-formate	12
20	126	180,000	56–70	80,000–100,000	3–6	Sulfamate-bromide	13
35	105–133	150,000–190,000	—	—	1–1.5	Watts	14
40	140–154	200,000–220,000	98–112	140,000–160,000	2–4	Sulfamate-bromide	13
45	154	220,000[a]	—	—	2–4	Sulfamate-bromide	13
50	180–192	255,000–275,000	126	180,000	—	Sulfamate	15

[a] Tensile strength was reduced by heat treating for an hour, as follows:

Temperature, C	kg/sq mm	psi
230	133	190,000
330	112	160,000
455	98	140,000
630	42–63	60,000–90,000

TABLE 13.4

Hardness of Nickel-Cobalt Alloys

Cobalt Content, percent	Hardness, kg/sq mm[a]	Solution	Current Density, amp/sq dm	Reference
1.1	509	Sulfate-chloride-formate	5	12
11	320	Sulfate-chloride[b]	1	11
17	370	Sulfamate-chloride	1.1	16
18	260–290	Sulfamate-bromide	5	13
24	360	Sulfate-chloride[b]	1	11
30	300	Sulfamate-bromide	5	13
35	510	Sulfamate-chloride	1.1	16
37	345	Sulfate-chloride[b]	1	2
39	439	Sulfate-chloride[b]	1	11
40	400–450	Sulfamate-bromide	5	13
41	412	Sulfate-chloride[b]	1	2
50	500	Sulfamate-bromide	5	13
57	402	Sulfate-chloride[b]	1	2
61	392	Sulfate-chloride	1	11
66	371	Sulfate-chloride[b]	1	11
No data	511–593[c]	Sulfate-chloride-formate	3–4	17
No data	237–514	Sulfate-chloride-formate	5.4	18
No data	425–525	Sulfamate	2.2	19

[a] For comparison, the hardness of either nickel or cobalt in similar solutions ranged from abou 250 to 280 kg/sq mm.

[b] Chloride derived from NaCl additions. These solutions were operated at 20 C.

[c] Hardness increased to 643–703 kg/sq mm at a current density of 5–6 amp/sq dm.

formaldehyde exhibited a hardness of 509 kg/sq mm. (12) The bath was operated at 65 C with a pH of 3.7.

The hardness of deposits produced with no brightener increases with increasing cobalt content to a maximum of 400 to 510 kg/sq mm with a cobalt content of 35 to 50 percent, depending on the solution and conditions adopted for depositing the alloy. One report showed a maximum of 500 kg/sq mm for alloy containing 50 percent cobalt deposited in a sulfamate-bromide bath. (13) Another showed a maximum of 439 kg/sq mm for alloy containing 39 percent cobalt deposited in a nickel sulfate-sodium chloride solution. (11) A hardness of 510 kg/sq mm was reported for alloy containing 35 percent cobalt, which was deposited in a sulfamate-chloride bath. (16) With a fluoborate bath, hardness reached a maximum for alloy containing 30 percent cobalt. (20) Unalloyed nickel or cobalt deposited in similar solutions exhibited a hardness of 250 to 280 kg/sq mm.

Alloys containing 11 to 66 percent cobalt deposited in sulfate solutions at 20 C were much harder than their cast and slowly cooled counterparts, which exhibited a hardness of only 70 to 100 kg/sq mm. (11) The higher hardness of the electrode-

TABLE 13.5

Comparison of Hardness before and after Heat Treatment of Nickel-Cobalt Alloy Electrodeposits

Cobalt Content, percent	Plating Bath	Hardness, kg/sq mm			Hardness at High Temperatures, kg/sq mm			Reference
		Before Heat Treatment	After Heat Treatment at 300 or 320 C	After Heat Treatment at 600 or 620 C	200 C	400 C	600 C	
10	Sulfamate-bromide	320–360	320	220–230	—	—	—	13
20	Sulfamate-chloride	400	360	200	270[a]	160[a]	70[a]	16
20	Sulfamate-bromide	270–300	280–300	130–160	—	—	—	13
27	Sulfamate-chloride	440	420	260	—	—	—	16
35	Sulfamate-chloride	500	460	280	380	180	80	16
35	Sulfamate-bromide	400	360	220–280	—	—	—	13
50	Ditto	500	500	300	—	—	—	13
75	"	260	280	300	—	—	—	13

[a] Alloy containing 17 percent cobalt.

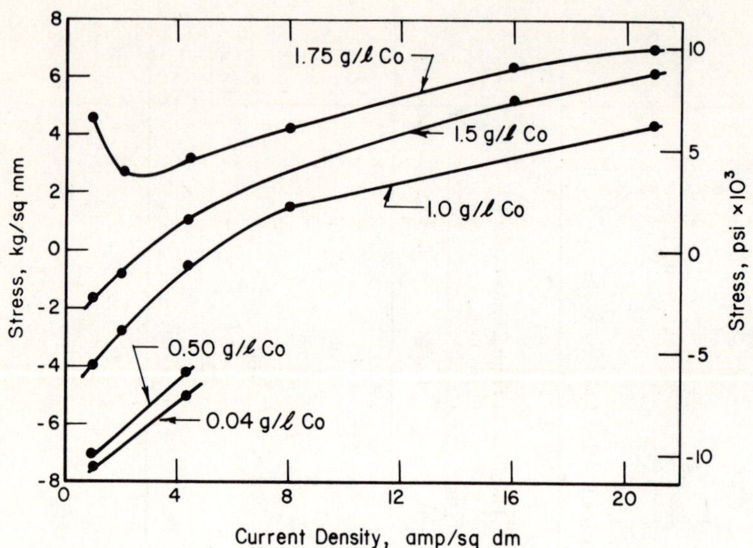

Figure 13.1. Effects of Cobalt Concentration and Current Density on the Stress of Nickel-Cobalt Alloy Deposited in Sulfamate-Chloride Solutions. (16)

posits probably was a result of finer grain size. An increase in the temperature of the sulfate solutions to 50 C, which undoubtedly coarsened grain size, reduced the hardness of the electrodeposited alloys to about 220 kg/sq mm. On the other hand, the nickel-cobalt alloys from sulfamate-chloride or sulfamate-bromide baths at 49 or 60 C exhibited a high hardness, > 400 kg/sq mm. (11,13)

Heat treating at 232 or 332 C had no effect on the hardness of nickel-cobalt alloys deposited in sulfamate-bromide baths, (13) but a slight reduction in hardness was reported for a 300 C heat treatment of alloys deposited in sulfamate-chloride solutions. (16) After heat treatment at 600 or 620 C, hardness of alloys from both baths decreased appreciably, as listed in Table 13.5, to values of 300 kg/sq mm, or less. Table 13.5 also gives hot hardness data for nickel alloy deposits containing 17 and 35 percent cobalt.

Stress

Codeposition of cobalt with nickel increases stress. As a result, stress in alloys deposited in a sulfate-chloride bath was increased from the range of 12 to 25 kg/sq mm (17,000 to 35,000 psi) to as much as 39 kg/sq mm (56,000 psi). (14) Bright alloy containing only 1 or 2 percent cobalt deposited in a solution containing nickel sulfate, nickel chloride, nickel formate, and formaldehyde exhibited a stress of 9.1 to 12.5 kg/sq mm (13,000 to 18,000 psi). (12,18) A bath temperature of 65 C favored a low stress of 9.1 kg/sq mm in this range, in comparison with 11 to 12.5 kg/sq mm for alloy deposited at temperatures of 40 or 50 C. (18)

Stress in alloys deposited in sulfamate-chloride baths was lower than stress in alloys from sulfate-chloride baths, but proportional to the cobalt concentration in the bath (and the cobalt content of the alloy). (16) At a current density of 2 amp/sq dm, stress was compressive (−1 to −7 kg/sq mm or −1400 to −10,000 psi) in deposits produced in sulfamate-chloride solutions containing 0.04 to 1.5 g/l cobalt ions. An increase in the cobalt ion concentration to 1.75 g/l shifted the stress to the tensile form. Increasing current density to 8 or 16 amp/sq dm increased the stress, as shown in Figure 13.1.

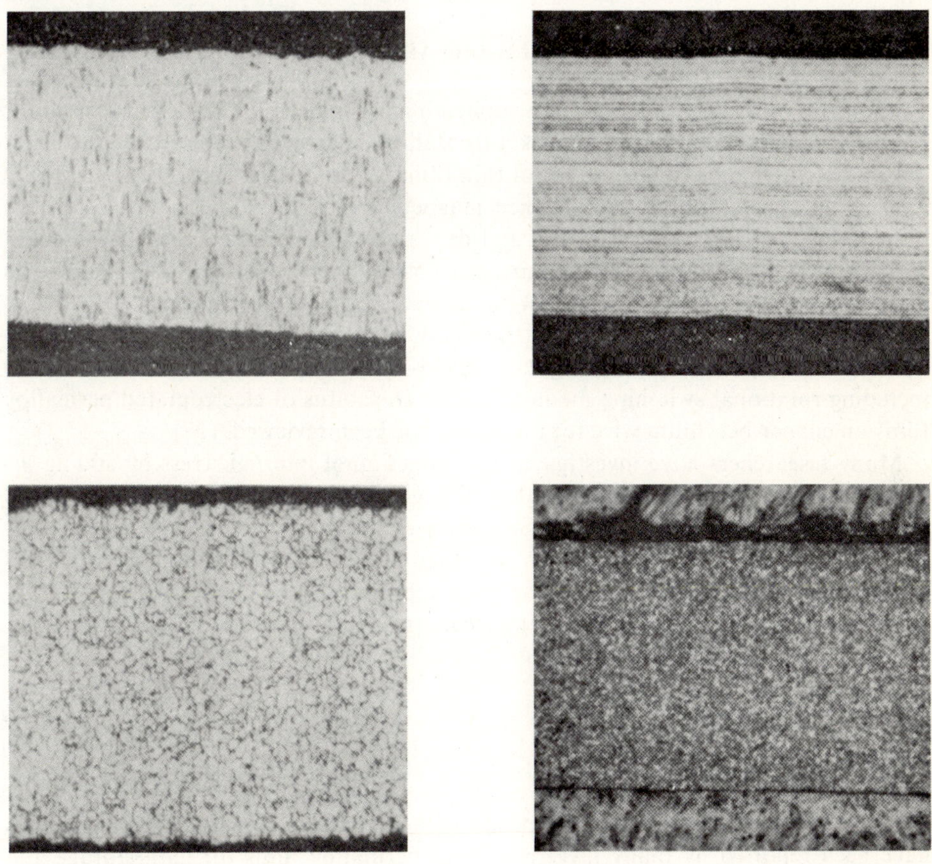

Figure 13.2. Structure of Nickel-Cobalt Alloy Deposits before and after Heat Treatment at 620 C. (13)

(upper left): Alloy Containing 20 percent cobalt, before heat treatment, 300×
(upper right): Alloy containing 47 percent cobalt, before heat treatment, 300×
(lower left): Alloy containing 20 percent cobalt, after heat treatment, 300×
(lower right): Alloy containing 47 percent cobalt, after heat treatment, 300×

Structure

Continuous solid solutions are reported throughout the whole composition range for nickel-cobalt alloys deposited from sulfate solutions. (11) Alloys containing 14 to 72 percent cobalt deposited from a pyrophosphate bath had an fcc structure, but were hcp for nickel contents below 14 percent. (21)

Figure 13.2 shows fine-grained and banded (lamellar) structures for alloy deposits containing 20 and 47 percent cobalt, respectively, which were deposited in a sulfamate bath. Heat treatment at 620 C enlarged the grains as shown at the bottom of Figure 13.2.

Nickel-Iron Alloys

Magnetic recording and switching applications have been the principal stimuli for research on electrodeposition of nickel-iron alloys. Particular interest leans toward optimizing the magnetic properties of thin films. In some cases, magnetic fields are applied during deposition to influence magnetic properties. Special attention has been given to substrate preparation and deposition at controlled cathode potential to prevent composition gradients during deposition. The most suitable nickel-iron alloy for magnetic applications contains about 20 weight percent iron, which corresponds to the composition of the thermally prepared permalloy. Usual applications of the deposits are as magnetic shields (22) or memory elements, (23–25) including rotational switching memories. (26) The status of electroplated permalloy films on copper-beryllium wire for memories has been reviewed. (27)

Many researchers have investigated the reduction of internal stress by adding organic compounds to the plating bath, although one report indicates the beneficial effect of ultrasonic agitation. (28) Saccharin is the most widely used additive for reducing stress. (29–33) A variety of other sulfur-containing organic additives (34–37) and phenol (38,39) also have been reported. Some reports discuss the influence of substrate preparation on internal stress and magnetic properties, (33,36, 38,40,41)

Magnetic Properties

Binary Alloys. Nickel alloy deposits with a coercive force of 1 to 4 oersteds have been reported by many investigators of permalloy films for data storage and retrieval. These deposits contained at least 15 and up to about 25 percent iron. Alloy containing < 15 or > 35 percent iron exhibited higher coercivities. For example, the coercivities of alloys containing 2 and 40 percent iron were 110 and 75 oersteds, respectively. (42) The optimum iron content for minimum coercivity is said to be either 19, (33) 22 (43) or 23 (44) percent.

Alloy with a coercivity of less than 5 oersteds has been deposited in both sulfate and sulfamate solutions. Some of the sulfate baths contained citrate ions. (45, 46)

Another contained phenol, (39) but no organic addition agent (except a wetting agent) was added to other solutions. (26,44,47,48) Some of the sulfamate baths contained saccharin or a similar stress-reducing agent, (29,30) but at least one was operated with no organic addition agent. (49) All solutions were adjusted in pH to the range of 2.2 to 2.8 and all but two were operated at room temperature (20 to 28 C). The exceptions were heated to 50–60 C (39) or 58 to 68 C. (47) Another solution used to deposit low-coercivity nickel-iron alloy was an alkaline pyrophosphate-citrate bath. (50)

Some low-coercivity films had a thickness of only 1,000 to 2,000 angstroms. (10, 39,51) One paper cited decreasing coercivity as thickness increased to 3,000 angstroms. (44) Another reported a minimum coercivity at 5,000 angstroms. (47) Yet a deposit as thick as 500,000 angstroms (50 μm) was reported to have a coercive force of < 1 oersted. (52) Deposition at controlled constant current or controlled constant cathode potential (potentiostatic) has been described for maintaining constant composition, independent of deposit thickness. (41,46,53)

Current densities were usually in the range of 0.5 to 3 amp/sq dm, but two reports of low-coercivity films cited current densities of about 6 amp/sq dm. (26,46) A magnetic field was applied during deposition by several investigators to influence orientation. (25,29,45,46,54)

Representative reports cited a nearly square hysteresis loop with a saturation:residual induction ratio > 0.85. (23–25, 29, 39, 46, 47, 49, 51) The permeability of an alloy deposit containing 20 percent iron was 8,000, in comparison with 300 for unalloyed nickel or iron deposits. (22)

Zero magnetostriction was observed for electrodeposited permalloy films. (31, 52,55,56) For 1000-angstrom-thick films on beryllium-copper wire, magnetostriction varied about 7 percent from zero by raising or lowering the bath pH from about 2 to 6. (56) A 40 percent gain in output voltage was reported for recording on zero magnetostrictive films deposited from a sulfate bath containing $\leq$ 0.7 g/l palladium chloride. (55) Films containing 19 percent iron with zero magnetostriction at 1500 to 2000-angstrom thickness were deposited from a sulfate bath onto a 5000-angstrom layer of chromium that was evaporated in vacuum onto glass. (52) An all-chloride bath containing saccharin for lowering deposit stress produced films that contained 19.5 percent iron and had zero magnetostriction. (31)

Adding 15 or 20 g/l Rochelle salts had a negligible effect on magnetic properties, but increased the bath life during an investigation of alloys having a coercive force of 2.5 to 3.0 oersteds and a squareness ratio of up to 0.92, deposited from a sulfamate solution containing 29 to 31 g/l nickel, 2.5 to 2.7 g/l iron, and 0.5 to 1.0 g/l saccharin (pH 2.5 to 2.7 at 25 to 28 C) at a current density of 3 amp/sq dm while a magnetic field with an intensity of 400 amp/m was applied. (29) Increasing the saccharin concentration decreased the iron content of the alloy deposit. The minimum coercive force was obtained using 0.5 to 1.0 g/l of the additive. Increasing the intensity of the magnetic field did not change the coercive force. However, the residual induction squareness ratio and film anisotropy increased.

In a study of many additives, thiosemicarbazide hydrochloride decreased coercive force (and stress) more than commonly used saccharin or other organosulfur compounds for nickel-iron codeposited from a solution containing 280 g/l nickel sulfate ($NiSO_4 \cdot 7H_2O$), 10 g/l ferrous sulfate ($FeSO_4 \cdot 7H_2O$), and 30 g/l boric acid with a pH of 3.7 at room temperature. (51) The current density was 1 amp/sq dm in a constant magnetic field of 500 oersteds.

In another study, saccharin additions apparently caused a decrease of coercive force for nickel alloy containing 20 percent iron deposited at 25 to 60 amp/sq dm on a 0.2 mm copper wire in a sulfamate solution at 45 C, which contained 86.5 g/l nickel, 20 to 25 g/l ferrous ions, 30 g/l boric acid, 15 g/l sulfamic acid, 2 g/l hydroxylamine hydrochloride, 0.6 g/l saccharin, and 0.05 g/l sodium lauryl sulfonate. (32)

In depositing alloys with a coercive force of 0.62 to 0.72 oersted from another sulfamate solution containing 85 to 87 g/l nickel, 2.7 to 2.8 g/l iron, 3 g/l sodium chloride, and 1.0 g/l saccharin (pH 2.5–3.0 at 20–28 C), increasing the cathodic current density above 3 amp/sq dm increased the coercive force but decreased remanent induction and squareness ratio. (30) Saccharin reduced the coercivity, squareness ratio, and internal stress of the deposits.

TERNARY ALLOYS. The magnetic characteristics of electrodeposited nickel-iron alloys are reported to be improved by codepositing a third element. Examples are antimony, (46,57) arsenic, (58,59) cobalt, (60), copper, (61) molybdenum, (62,63) palladium, (55), phosphorus, (58,64) selenium, (57,65) and tellurium. (66) Sulfate solutions were used with an addition of the third element as an anion, except for copper and cobalt, which were added as cations. The alloys exhibited a low coercivity, usually from about 2 to 8 oersteds, for films ranging in thickness from 200 to 2000 angstroms.

Antimony (0.5 to 5 weight percent) was codeposited with nickel and iron to attain coercive forces in the range of 3.8 to 6.5 oersteds and a squareness ratio of about 1. (46) Current densities were 0.06 to 0.07 amp/sq dm under galvanostatic or potentiostatic conditions. The deposition solution contained 218 g/l nickel sulfate ($NiSO_4 \cdot 6H_2O$), 6.8 to 10.5 g/l ferrous ammonium sulfate [$(NH_4)_2Fe(SO_4)_2 \cdot 6H_2O$], 0.1 to 1.0 g/l potassium antimony tartrate ($KSbC_6H_4O_6 \cdot O_5 \cdot H_2O$), 10 to 40 g/l boric acid, 10 to 40 g/l ammonium chloride, 0.8 g/l saccharin, and 0.4 g/l sodium lauryl sulfonate; the pH was 2.4 to 2.8 at 20 C and the bath was used with platinum anodes. The cathode was subjected to a homogeneous magnetic field (100 oersteds) having force lines parallel to the cathode surface.

With a 0.05 to 0.2 g/l addition of sodium arsenate to a sulfate bath at about 20 C, a current density of 0.6 amp/sq dm deposited ternary alloy with a coercivity of 6 to 8 oersteds. (58) Copper slightly reduced the coercivity of nickel-iron alloys. (61) When investigated for use in computer memories, the ternary alloy containing 23 percent iron and 2.0 percent cobalt exhibited an improved flux level, greater transverse loop linearity, lower anisotropy, and lower coercivity in comparison with the binary, nickel-iron alloy. (60)

Zero magnetostriction was observed for nickel alloys containing 23 percent iron and 10 or 22 percent molybdenum. (62) On the other hand, films containing 15 to

20 percent iron and 15 percent molybdenum were highly magnetostrictive. In another report, the codeposition of molybdenum with nickel and cobalt was facilitated by ultrasonic agitation, which significantly reduced stress. (63) However, the molybdenum content of the deposit was only 4 or 5 percent.

Palladium as ≤ 0.7 g/l sodium chloropalladate was added to a sulfate solution with a Ni^{2+}:Fe^{2+} ratio of 50 to 80:1 for a permalloy deposit with a coercive force of 1.8 oersteds. (55) Codeposition of 2.1 weight percent phosphorus with iron and nickel reduced coercivity from 8.5 to 3.1 oersteds, (64) but an increase in coercivity was noted where as much as 3.1 percent phosphorus was incorporated in a nickel-iron deposit.

The coercive force of alloy containing 13 to 17 percent selenium was lower than that of selenium-free films by 2 to 3 oersteds. (57,65) Selenium was more effective than antimony for lowering coercive force. The films contained 15 to 17 percent iron and up to 10 percent selenium. Hysteresis loops were square, and switching times were 1 to 4 nsec. The reversal of magnetization by rotation required a switching field strength of 8 to 10 oersteds.

SUBSTRATE EFFECTS. Independently of plating variables, substrate type and treatment prior to plating influence coercive force, anisotropy, and squareness ratio for nickel-iron alloy magnetic films. For example, the effect of preparation of beryllium-copper wire on magnetic properties was determined for alloy films containing 15 percent iron. (40) Such films were strongly oriented along the wire axis and had a coercive force of 30 oersteds when deposited directly on the copper alloy. Unoriented films having an easy magnetizing axis in the circumferential direction with coerive forces of 6 of 10 oersteds were obtained by deposition on gold films more than 1000 angstroms thick. Orientation decreased as the gold-film thickness was increased from 0 to 1000 angstroms. Stress-induced magnetic isotropies during epitaxial growth of oriented electrodeposited films was thought to be due to anisotropic stress.

Permalloy films that were electrodeposited onto an organic polymer coating over a metal substrate were found to have reduced coercive force compared to films electrodeposited onto uncoated metal. (67) The decrease in coercive force was attributed to lower surface roughness of polymer-coated substrate. Data were reported for films on plastic and glass, as follows:

Thickness, angstroms	Coercive Force, oersteds	
	On Polymer	On Glass
400	5.8	10.0
800	3.0	4.3
1200	1.9	3.2
2000	1.1	1.6
3000	0.75	0.9

The coercivity of thin nickel-iron electrodeposits as a function of iron content was minimum at 20 percent iron and varied with film thickness and roughness of substrate, as follows: (68)

Thickness, angstroms	Coercivity, oersteds	
	Smooth Polished Surface	Grit Polished Surface
3,000	7	9.5
5,000	4	6.2
7,500	3	4.2
10,000	2	3
20,000	1.2	—

Squareness of the hysteresis loop for alloys containing 12 to 29 weight percent iron was affected by the procedure for smoothing a copper substrate. (33) Square loops were observed when the test magnetic field was parallel with the polishing direction.

Effect of Annealing. Electrodeposited alloys are heat treated to render them magnetically isotropic and reduce stress. Heat treating for 3 hours at 300 C had little influence on the coercivity of 0.7 to 1.2-oersted films containing 28 to 49 percent iron. (68) In another investigation, on the other hand, coercive force decreased as the annealing temperature was increased to 350 C. (69) An increase in coercivity was noted at temperatures above 350 C for films on a copper substrate, but films separated from the substrate continued to exhibit decreasing coercivity at annealing temperatures to 600 C. Heat treating in a vacuum (10^{-4} mm) reduced somewhat the coercivity of a 3.5-oersted film and the squarenss of the hysteresis loop. (39)

Results of thermomagnetic treatment were reported for electrodeposited films containing 1 percent iron. (52) Films deposited on a 500-angstrom-thick chromium layer applied by vacuum evaporation on glass were annealed with a magnetic field of 500 oersteds applied normal to the easy axis of magnetization. Temperature increases up to 200 C produced magnetically isotropic films with a very high angle of dispersion. Dispersion angle increased on annealing at 130 C and the squareness ratio increased, but the films remained anisotropic. Continued annealing at temperatures over 350 C increased parallel coercive force and anistropy field.

Mechanical Properties

Nickel alloy deposits containing 25 to 40 percent iron are about equal in strength to nickel alloy containing 40 to 50 percent cobalt, but stronger than nickel alloys of manganese or tungsten. Typical strength levels are compared in Table 13.6. The nickel-iron and nickel-cobalt alloys exhibit similar ductility.

Alloys deposited from a sulfate solution containing 0.2 to 0.9 g/l ferrous sulfate ($FeSO_4 \cdot 7H_2O$), which resulted in an iron content of about 2 percent, exhibited a tensile strength of 66 to 114 kg/sq mm (94,000 to 162,000 psi), in comparison with 95 to 112 kg/sq mm (136,000 to 160,000 psi) for unalloyed nickel deposited in a similar bath containing no iron. (72) An ultimate tensile strength of >140 kg/sq mm (>200,000 psi) and elongations of 2 or 3 percent were reported for nickel alloys containing 15 to 60 percent iron deposited in a chloride-sulfate bath containing 7.5 g/l sodium 1, 3, 5-naphthalene trisulfonate as a stress reducer. (36) The highest tensile strength was 183 kg/sq mm (257,000 psi). Both strength and elongation were maximum for a solution pH of 2.8 or 2.9. A tensile strength < 80 kg/sq mm (< 114,000 psi) and an elongation < 2 percent were obtained at a pH of only 2.4. Alloy deposits with a tensile strength > 160 kg/sq mm (230,000 psi) contained 25 to 40 percent iron. Solution temperature and current density relationships for deposits with such high tensile strength were as follows:

Bath Temperature, C	Current Density, amp/sq dm
41	5
61	5 or 10
71	10

Deposits obtained at 15 or 20 amp/sq dm in a 76 C solution contained >50 percent iron and had lower tensile strengths.

Annealing at 400 C for 15 minutes reduced tensile strength to < 53 kg/sq mm (75,000 psi) and elongation to < 0.1 percent. Some deposits disintegrated into powder during annealing. On the other hand, alloy deposited in a similar bath that contained no organosulfur stress reducer exhibited high tensile strengths ranging from about 100 to 150 kg/sq mm (140,000 to 210,000 psi) after annealing at 400 C for 15 minutes. However, many deposits obtained in the bath containing no stress reducer were highly stressed and cracked and could not be pulled in tension.

TABLE 13.6

Comparative Strength, Ductility, and Hardness Data for Nickel Alloys

Codeposited Metal	Tensile Strength kg/sq mm	Tensile Strength psi	Elongation, percent	Hardness, kg/sq mm	Reference
None[a]	35–44	50,000–64,000	27 to 35	130 to 320	12
Cobalt (40–50%)	140–192	200,000–275,000	2 to 4	400 to 500	13, 15
Iron (25–40%)	140–180	200,000–255,000	2 to 3	—	36
Manganese (0.5%)	117–131	167,000–187,000	5 to 6	—	70
Tungsten (20%)	78–110	110,000–156,000	0	500 to 600	71

[a] Nickel deposited in Watts-type solutions containing no brightener or stress reducer. See Chapter 12, Nickel.

TABLE 13.7

Sulfur-Containing Organic Additives Which Reduce Internal Stress in Permalloy Electrodeposits (37)

Compound	Functional Group
1. Benzene sulfonic acid	$—SO_3H$
2. Benzene sulfonamide	$—SO_2NH_2$
3. Naphthalene sulfonic acids (1 and 2)	$—SO_3H$
4. Naphthalene disulfonic acid (1,5 and 2,7)	$—SO_3H(2)$
5. Naphthalene trisulfonic acid (1, 3, 5)	$—SO_3H(3)$
6. Naphthol sulfonic acids (1,5 and 2,6 and 2,7)	$—SO_3H$, —OH
7. 1-Nitroso-2-naphthol-3,6-disulfonic acid	$—SO_3H(2)$, —OH, –NO
8. Saccharin	$—CO—NH—SO_2$
9. Hydroquinone sulfonic acid	$—SO_3H$, —OH(2)
10. 4,4-Diamino-2,2-biphenyldisulfonic acid	$—SO_3H(2)$, $—NH_2(2)$

Hardness values for nickel alloy deposits containing >1.0 percent iron were in the range of 300 to 400 kg/sq mm, (73) which is about equal to the hardness of nickel alloy deposits containing 11 to 24 percent cobalt and somewhat harder than unalloyed nickel obtained in similar solutions. Heat treating the nickel-iron alloys at 500 or 800 C reduced their hardness to 180 and 100 kg/sq mm, respectively.

Stress

Iron-salt additions to nickel plating baths increase stress. For example, stress of nickel deposited in a sulfate bath was increased from 13 kg/sq mm (18,500 psi) to 16 kg/sq mm (23,000 psi) with a 0.05 N iron addition and to 35 kg/sq mm (50,000 psi) with a 0.1 N addition. (74) Another investigator reports stress increasing about 2 kg/sq mm for each g/l increment of ferrous ions added to a sulfamate solution. (75)

Saccharin is widely used to reduce internal stress in nickel-iron alloys including permalloy. (29–31,33) An addition of 1.03 g/l saccharin to the solution, a current density of 80 amp/sq dm at a 50 C, and ultrasonic agitation at 22 kc with sufficient intensity to cause cavitation are reported necessary for deposition of stress-free nickel-iron alloy. (28) Initially deposited layers were compressively stressed, but this changed to tensile stress as deposit thickness was increased. Stress in nickel-iron alloy deposits was also reduced by the addition of trisulfonated naphthalene, which also increased tensile strength and ductility. (36)

Table 13.7 lists sulfur-containing additives for sulfate solutions for electrodeposition of permalloy (20 weight percent iron). The additives in an amount less than 0.5 g/l lower internal stress and slightly lower coercive force and anisotropy of field of 200 to 20,000-angstrom-thick films. (37) The electrodeposits were made at 0.3 amp/sq dm in a sulfate solution containing 61.5 g/l nickel, 1.25 g/l ferrous ions,

Figure 13.3. Structure of High-Strength Nickel-Iron Alloy Deposited in a Sulfate-Chloride Bath Containing Trisulfonated Naphthalene with the pH Adjusted to 2.8. 1000× (36)

9.75 g/l sodium chloride, 25 g/l boric acid, and 0.42 g/l sodium lauryl sulfate at a pH of 2.7 to 3.0.

Substrate pretreatment and current density are reported to influence the internal stress in nickel-iron alloy deposits from a sulfate solution containing phenol. (38) Stress was minimized at 5 to 7 kg/sq mm (5 to 7 $\times$ 10^8 newtons/sq meter) when deposition was at a current density of 3 to 4 amp/sq dm on copper, brass, and nickel substrates. Highest and lowest stress were observed for brass and copper cathodes, respectively. Stress also increased with increase of surface area of the substrates due to etching. It reached a maximum of 25 kg/sq mm (25 $\times$ 10^8 newtons/sq meter) for a film deposited at 7 to 8 amp/sq dm and then decreased for higher current densities.

Stressed 5000-angstrom-thick films of nickel-iron alloy containing 25 percent iron and 2 to 3 percent phosphorus were obtained by adding 5 g/l sodium hypophosphite ($NaH_2PO_2 \cdot H_2O$) to the bath. (34) Tensile stress decreased abruptly with increasing amounts of saccharin additions and was constant for concentrations over 0.5 g/l.

Crystal Texture and Structure

A lamellar microtexture observed in nickel-iron deposits consisted of alternate layers of γ-phase solid solution and very thin layers of unidentified material, which might have been nickel sulfide induced by the reduction of a sulfonated addition

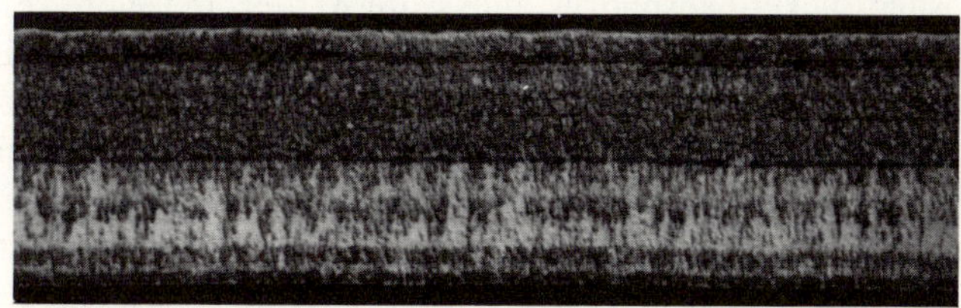

Figure 13.4. Structure of Relatively Low-Strength Nickel-Iron Alloy Deposited in a Sulfate-Chloride Bath Containing Trisulfonated Naphthalene with the pH Adjusted to 2.1. 1000× (36)

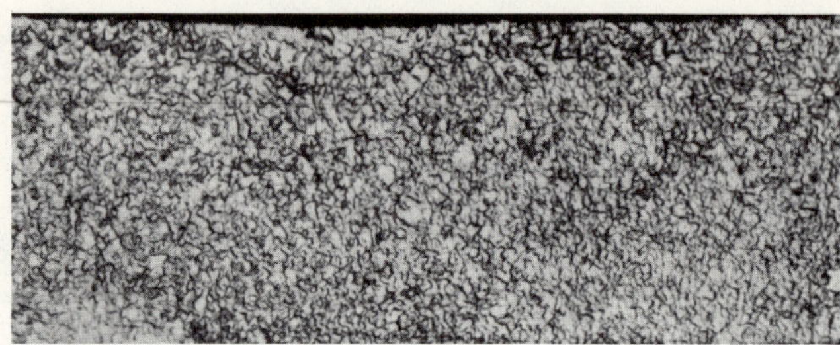

Figure 13.5. Structure of High-Strength Nickel-Iron Alloy Like That in Figure 13.3 after Annealing 15 Minutes at 400 C. 1000× (36)

agent. (76) The γ-phase consisted of equiaxial grains of about 0.5-μm diameter. Heat treatment formed a random polycrystalline texture. Similar lamellar microtextures were reported for high-strength alloys containing 25 to 50 weight percent iron deposited from a chloride sulfate solution with or without sodium 1, 3, 6-naphthalene sulfonate as a stress reducer. (36) Figures 13.3 and 13.4 show different structures for high-strength alloy deposited in a bath with a pH of 2.8 and lower-strength alloy produced with a pH of 2.1. The stress reducer was added in both cases. The annealed structure of the high-strength alloy from a bath with a pH of 2.5 (or higher) is shown in Figure 13.5.

Structures of nickel-iron alloys containing 9 to 86 percent nickel electrodeposited from sulfate-citrate solutions were examined by X-ray diffraction. (77) Alloy containing 82 and 91 percent iron crystallized with a bcc lattice. Those containing 14, 36, and 48 percent iron had an fcc lattice. The alloy containing 69 percent iron was a mixture of bcc and fcc.

Powder deposits from iron-nickel-ammonium sulfate solution containing fluoride were fcc for alloy with iron contents up to 50 percent and bcc at 87 weight percent iron. (78) Films of nickel-iron-molybdenum alloys containing up to 5 percent molybdenum electrodeposited onto evaporated films of silver on glass were solid solutions having an fcc lattice. (63)

Nickel-Phosphorus Alloys

Electrodeposited nickel-phosphorus alloys are harder and more resistant to corrosion than unalloyed nickel deposits. (79) Alloy deposits containing 9 percent phosphorus were better than unalloyed nickel coatings for protecting steel outdoors and retained a bright finish whereas alloy coatings containing 3 percent phosphorus blackened during exposure. (80) Alloy containing 3 to 15 weight percent phosphorus is suitable for electroplating rigid joints, although its resistance to fatigue is lower than that of electrodeposited chromium. (81) Thermally treated deposits of the phosphorus alloys are very resistant to abrasion in contact with steel.

The properties of the electrodeposited alloys are similar to those of the electroless nickel-phosphorus alloys, but differences in density and electrical resistivity have

TABLE 13.8

Density, Hardness, and Resistivity Data for Electrodeposited Nickel-Phosphorus Alloys (79)

Phosphorus Content, percent	Hardness, kg/sq mm			Density, g/cu cm	Resistivity, microhm-cm[a]	Solution Composition, mole/liter[b]				Current Density, amp/sq dm
	Before Heating	After Heating 400 C	After Heating 800 C			$NiSO_4 \cdot 7H_2O$	$NiCl_2 \cdot 6H_2O$	H_3PO_4	H_3PO_3	
0	260	220	—	8.92	8 (8)	0.67	0.21	0.5	—	—
2–3	600	850	240	8.75	30 (15)	0.67	0.21	0.5	0.015	5–40
6–7	630	900	450	8.42	74 (22)	0.67	0.21	0.5	0.015	5–40
13–14	750	900	750	7.92	118 (40)	0.57	0.19	0.5	0.5	10–20

[a] Figures in parentheses are resistivity values after heating at 800 C.
[b] The pH of all solutions ranged from 0.5 to 1.0. Solution temperature ranged from 75 to 95 C.

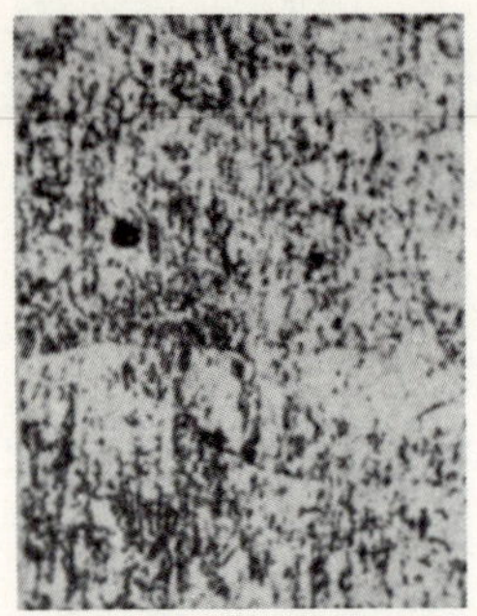

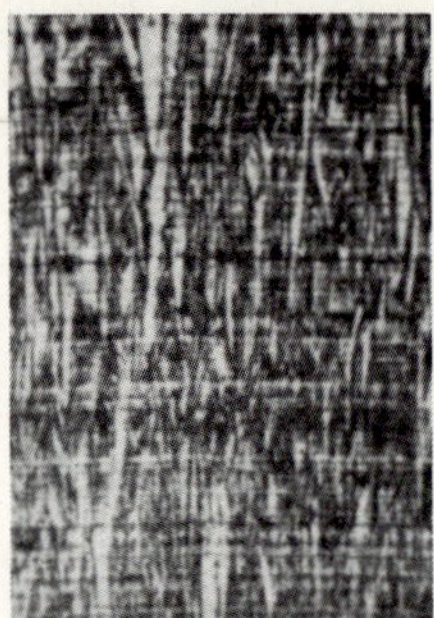

Figure 13.6. Photomicrographs of Electrodeposited Nickel Alloy Containing (left) 2 and (right) 13 Percent Phosphorus after Etching with Nital. (250×) (79)

been noted. Unlike the electroless alloys, which exhibit hardness values independent of phosphorus content, the electrodeposited alloys show increasing hardness with increasing phosphorus. The electrodeposits resisted wear better than the electroless alloys. (82)

Physical Properties

Density decreased with increasing phosphorus content. Data for electrodeposited nickel containing 2 to 13 percent phosphorus are detailed in Table 13.8. These alloys are approximately 2 percent higher in density than electroless nickel alloys with similar phosphorus contents. (See Figure 22.1 in Chapter 22, Electroless Nickel.)

Electrical resistivity data also are tabulated in Table 13.8. These data were reported for solutions with a pH of 0.5 to 1.0. (79) Lower resistivities were reported elsewhere for alloys deposited in solutions with a pH of 2.0, as follows: (83)

Phosphorus Content, percent	Specific Resistivity, microhm-cm
3	13 to 22
9	28 to 34

The above values are lower than resistivities which the same investigators report for electroless nickel, which ranged from 28 to 34 and 51 to 58 microhm-cm for electroless deposits from alkaline and acid baths, respectively.

Hardness

Table 13.8 shows hardness increasing from 600 to 750 kg/sq mm as the phosphorus content increased from 2 to 13 percent. Lower values from 500 to 550 kg/sq mm were reported elsewhere for alloy containing 10 to 11 percent phosphorus (4) or

13 percent phosphorus. (84) However, heat treatment at 400 or 500 C greatly increased hardness to the range of 850 to 900 kg/sq mm in one report, (79) or 1000 kg/sq mm in another. (84) Heat treatment at 600 C increased the hardness of a 10 to 11 percent phosphorus alloy from only 500 or 550 kg/sq mm to the range of 1200 to 1300 kg/sq mm. (83) These increases in hardness are a result of the precipitation of nickel phosphide from supersaturated solid solutions, which was noted in heat-treated deposits by X-ray analysis. (84)

Like electroless nickel-phosphorus alloys, the electrodeposited alloys retain high hardness at temperatures to about 400 C. (4) (See data for electroless nickel-phosphorus in Figure 22.4 in Chapter 22, Electroless Nickel.) However, the alloys softened appreciably when hardness was measured at 800 C, which is associated with their low melting point (about 1000 C in comparison with about 1450 C for unalloyed nickel). (79)

Ductility and Stress

Ductility of the electrodeposited alloys containing about 2 percent phosphorus was slight. (79) The deposits became brittle with increasing phosphorus content.

Stress in electrodeposited nickel-phosphorus alloys increases with increasing current density and decreasing temperature. (85) An increase in the pH of a bath containing phosphoric acid and sodium hypophosphite from 1.1 to 1.4 or 1.5 also increased stress. A saccharin addition to the bath reportedly reduced stress. (79,85)

Structural Features

Alloy deposits containing about 2 percent phosphorus exhibited a fibrous grain interrupted by bands, whereas a deposit containing 13 percent phosphorus showed only a lamellar (banded) structure after cross sectioning and etching. (79) These structures are shown in Figure 13.6. Heat treating at 400 C had little influence on

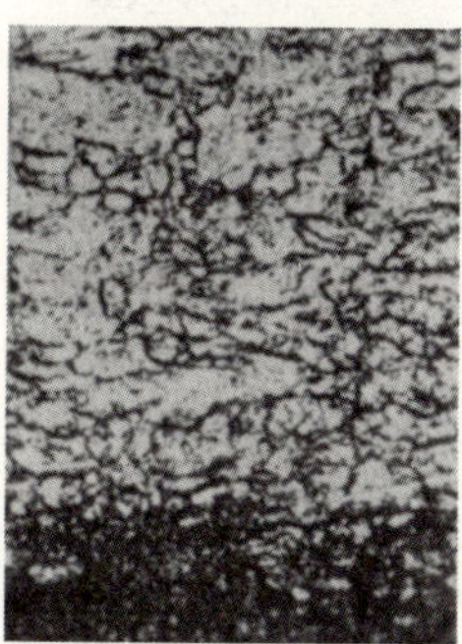

Figure 13.7. Photomicrographs of Electrodeposited Nickel Alloy Containing (left) 2 and (right) 13 Percent Phosphorus after Heat Treatment at 800 C. (250×) (79)

the grain structure, whereas heat treating at 600 or 800 C produced the structure shown in Figure 13.7.

Nickel phosphides were identified by X-ray examination after the alloy containing 13 percent phosphorus was heated at temperatures of 350 or 400 C. (84)

Nickel-Tungsten Alloys

Hardness

Tungsten codeposited with nickel increases the hardness of the deposits and improves high-temperature oxidation resistance. Usually, tungsten codeposition is achieved by adding sodium tungstate (Na_2WO_4) to a nickel sulfate bath which contains a complexing agent such as sodium citrate and is operated at a pH of 5.5 to 8.9.

Table 13.9 lists data indicating that hardness increases with tungsten content. (71,86–88) The hardness of deposits containing 25 to 40 percent tungsten usually exceeds 500 kg/sq mm and ranges to about 700 kg/sq mm.

Heat treatment at 400 to 600 C increased the hardness of alloy containing 28 to 40 percent tungsten. The increase in hardness was proportional to the tungsten content. For example, the hardness of deposits containing 28 and 40 percent tungsten was increased from about 525 to 575 and 1075 kg/sq mm, respectively, as a result of a 600 C heat treatment. (86) These alloy deposits were obtained at 10 amp/sq dm in a 60 C bath with a pH of 8.6 to 8.9. Molar concentrations of nickel and cobalt were as follows:

Tungsten Content of Deposit, percent	Concentration in Solution, M	
	Nickel	Tungsten
28	0.152	0.15
33	0.152	0.30
40	0.076	0.15

Heat treatment of 900 C reduced the hardness of the 28 and 33 percent tungsten alloy to 270 kg/sq mm. (86) The 40 percent tungsten alloy retained a hardness of 350 kg/sq mm after such a heat treatment.

Tensile Strength and Ductility

The tensile strength of an alloy deposit containing 20 percent tungsten ranged from 78 to 110 kg/sq mm (110,000 to 156,000 psi), (71) which compares with about 50 kg/sq mm (71,000 psi) for unalloyed nickel deposited in a similar solution containing no sodium tungstate. The 20 percent tungsten alloy was deposited at 3 to 6 amp/sq dm in a 90 C solution containing about 0.2 M nickel sulfate, 0.37 M so-

TABLE 13.9

Hardness of Nickel-Tungsten Alloys before and after Heat Treating

Tungsten Content, weight percent	Hardness, kg/sq mm		Heat Treating Temperature, C	Type of Bath	Current Density, amp/sq dm	Reference
	Before Heating	After Heating				
25	552 to 662	379 to 783	600 C, 1 hr	Sulfate-chloride-tungstate-citrate (90 C)	3 to 6	71
36	693	752	600 C, 1 hr	Ditto	Ditto	71
28	500 to 550[a]	850	400	Sulfate-tungstate-citrate (60 C)	10	86
33	Ditto	875	400	Ditto	Ditto	86
40	"	1030	500	"	"	86
28	"	575	600	"	"	86
33	"	875	600	"	"	86
40	"	1075	600	"	"	86
28	"	270	900	"	"	86
33	"	270	900	"	"	86
40	"	350	900	"	"	86
30	400 to 700	500 to 800	600	Sulfate-tartrate-tungstate	—	87
>40	814	—	—	Sulfate-citrate-tungstate (>70 C)[b]	—	88

[a] Value for alloys containing 16 to 40 weight percent tungsten.

[b] 10 g/l of hydrazine sulfate was added to the solution operated at a pH of 8.4.

dium tungstate, 1.0 M ammonium chloride, and 0.3 to 0.6 M sodium citrate. The pH was 5.5.

Alloy deposits containing at least 14 percent tungsten exhibited zero ductility in tensile pull tests. (71) However, heat treating at 600 C ductilized alloys containing not more than 20 percent tungsten.

A modulus of elasticity of 17,000 kg/sq mm (24,000,000 psi) was reported for an alloy containing 25 percent tungsten. (71) In comparison, the modulus for nickel ranges from 16,100 to 21,600 kg/sq mm (23,000,000 to 30,800,000 psi), depending on the type of solution adopted for depositing the nickel.

Conditions are reported for producing smooth deposits of nickel-tungsten up to 0.02 inch thick. (87) In other work, (88) the alloy deposited onto soft steel had a hardness of 247 kg/sq mm from a solution containing 20 g/l $NiSO_4 \cdot 7H_2O$, 60 g/l $Na_2WO_4 \cdot 2H_2O$, 60 g/l sodium citrate, and 10 g/l hydrazine sulfate at pH 8.0 and bath temperatures over 70 C.

Resistivity

Specific resistivities for alloys containing 30 and 33 percent tungsten were reported to be 84 and 93 microhm-cm, respectively. (71) Heat treatment decreased these values, however. Alloys containing over 20 percent tungsten are nonmagnetic.

Structure

Alloy deposits containing up to 50 percent tungsten are single-phase supersaturated solid solutions. (71,89,90) Deposits containing 15 to 45 percent tungsten, which were deposited in a pyrophosphate bath, exhibited a face-centered-cubic structure. (89) A two-phase structure was identified after heat treatment at 600 C. (71) Thus the increase in hardness observed with such a heat treatment is a result of the precipitation of the second phase in the case of alloys containing more than 30 percent tungsten, which is the limit of solubility in cast alloys.

Other Nickel Alloys

Binary alloys of nickel with aluminum, chromium, columbium (niobium), germanium, manganese, molybdenum, palladium, rhenium, silver, thallium, tin, and zinc obtained wholly or partially by electrodeposition are reported for a variety of applications. Some applications of the alloys are well established. Others are novel. Electrodeposited nickel-tin alloys, for example, are useful corrosion-resistant coatings developed originally by the Tin Research Institute. Electrical resistivity and corrosion resistance of the Nichrome series of nickel-chromium alloys are primarily useful to the electrical industry, but electroplates are attractive for decorative-finishing applications. Nickel-rhenium alloys are specifically useful as magnetic materials, with hardness surpassing those of binary alloys of nickel with other metals of the iron group.

Electrical Resistivity of Nickel-Silver Alloys

A maximum electrical resistivity of 18 microhm-cm was reported for nickel alloy containing 20 percent silver. (91,92) Figure 13.8 shows the change of electrical resistivity with silver content that suggests a change of atomic structure. The hardest deposit (360 kg/sq mm) had the maximum resistivity.

Tensile Properties: Nickel-Manganese and Nickel-Chromium Alloys

The ultimate tensile strengths of nickel alloy containing 0.1 to 0.5 percent manganese ranged from 117 to 131 kg/sq mm (167,000 to 187,000 psi). (70) Corresponding yield strengths were 99 to 114 kg/sq mm (140,000 to 163,000 psi) with elongations of 5 to 6.5 percent. At a temperature of 400 C, ultimate tensile strength was reduced to 34 to 46 kg/sq mm (48,000 to 66,000 psi) and yield strength to 22 to 34 kg/sq mm (31,000 to 49,000 psi), while elongation increased to 15 to 30 percent. At 265 C, the alloys withstood 56 kg/sq mm (80,000 psi) stress for 804 hours, and the endurance limit for 10^6 cycles was 39 kg/sq mm (55,000 psi). The density of the alloy was 8.86 g/cu cm, which compares with 8.9 g/cu cm for unalloyed nickel. The manganese alloy was deposited at 5 amp/sq dm in a 60 C solution prepared with 80 g/l nickel sulfamate, 30 g/l manganese sulfamate, 30 g/l boric acid, and a wetting agent. The pH was 3.5.

A tensile strength of 34 kg/sq mm (49,000 psi) was reported for an alloy containing 17 percent chromium, 6 percent tungsten, and 2 percent aluminum, produced by codepositing nickel-plated particles of the alloying elements with nickel, followed by homogenization during heating at 1040 C for 52 hours. (93)

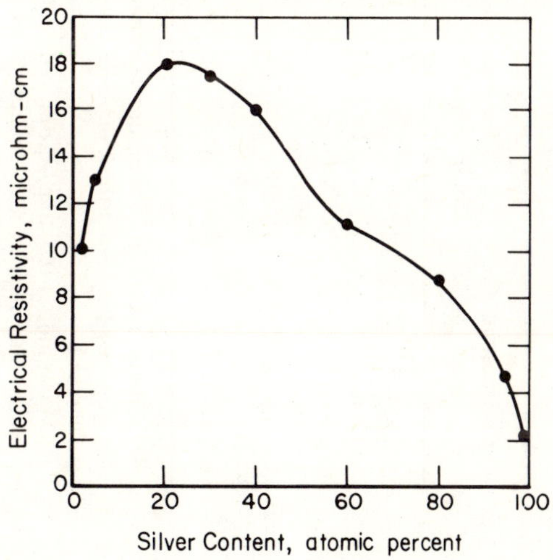

Figure 13.8. Resistivity of Electrodeposited Nickel-Silver Alloys. (91, 92)

TABLE 13.10

Microhardness of Binary Nickel Alloys

Codeposited Metal	Alloy Composition	Microhardness, kg/sq mm	Type of Bath	Current Density, amp/sq dm	Reference
Aluminum	16–30% Al	700–725	Two solutions[a]	—	94
Chromium	15–17% Cr	350–450[b]	Sulfate with glycine (40 C)	15–30	95
Germanium	≤5% Ge	800[c]	Ammoniacal oxalate	0.6	96
Rhenium	10% Re	900[d]	Sulfate-rhenate (20–70 C)	1–20	97
Ruthenium	15–20% Ru	800–1000	—	—	98
Silver	5 to 40% Ag[e]	290–460[f]	Pyrophosphate (25 C)	0.5	91, 92
Zinc	50% Zn	520	Ammoniacal sulfate (20 C)	1.0	99, 100

[a] Data for nickel and aluminum deposited in layers from two solutions and diffused by heating at 550 to 700 C.

[b] Hardness for 10-μm-thick deposits containing 2% carbon, 1.2% nitrogen, and 1.6 to 2.2% hydrogen. Thicker deposits were cracked.

[c] Maximum at 0.6 amp/sq dm, reduced to 450 kg/sq mm at 3 amp/sq dm. Value of 800 kg/sq mm was reduced to 720 kg/sq mm after heating for 1 hour at 300 C.

[d] In comparison, the hardness of unalloyed rhenium was 1100 to 1700 kg/sq mm with a maximum at a current density of 20 amp/sq dm. Alloy containing 90% rhenium exhibited a hardness of 2000 kg/sq mm.

[e] Atomic percentages.

[f] Maximum (460 kg/sq mm) at 20 atomic percent silver which decreased to 290 kg/sq mm after 1 year.

Hardness Data

Table 13.10 lists hardness data for seven binary alloys of nickel. Alloys containing 15 to 20 percent ruthenium, (98) 10 percent rhenium, (97) or about 5 percent germanium (96) were the hardest of this binary group. Their microhardness ranged from 800 to 1000 kg/sq mm, which is in the same range as that for electrodeposited chromium. The solution for plating the rhenium alloy was prepared with nickel sulfate, potassium perrhenate, and ammonium sulfate, and was adjusted with sulfuric acid to give a pH of 3.0. (97) The germanium alloy plating bath was prepared by dissolving 0.5 g/l germanium oxide and about 18 g/l nickel hydroxide in an ammoniacal solution of ammonium chloride. (96)

Alloys containing 16 to 30 percent aluminum having hardnesses from 611 to 788 kg/sq mm were obtained by electrodepositing aluminum in a 1000 C fused-salt bath and diffusing it with electrodeposited nickel. (94) By comparison, aluminum deposited in a 160 C salt bath and diffused with nickel at 550 to 600 C produced an alloy with a hardness of 700 to 725 kg/sq mm.

Nickel-chromium alloy with a hardness of 350 to 450 kg/sq mm was electrodeposited from a solution 2 N in the glycine complex of trivalent chromium, $Cr(Gl)_4{}^{3+}$, 2 N in nickel sulfate, and 0.27 N in free glycine at a pH of 2.5 to 2.7 and 30 to 40 C using a current density of 15 to 30 amp/sq dm with a current efficiency of 17 to 30 percent. (95) Anodes were graphite or lead immersed in 10 percent sulfuric acid separated from the cathode by a ceramic diaphragm. Additions of 0.05 g/l sodium lauryl sulfate reduced pitting in alloy deposits containing 15 to 17 percent chromium.

Binary alloys containing 5 to 40 percent silver with hardness ranging from 290 to 460 kg/sq mm were deposited in pyrophosphate solutions. (91,92) Deposits with a higher silver content were softer. For example, the hardness of alloys containing 60, 80, and 90 percent silver was 250, 200, and 150 kg/sq mm, respectively. The plating baths contained a total of 6 g/l nickel plus silver and 100 g/l potassium pyrophosphate, and were operated at a temperature of 25 C with a current density of 0.5 amp/sq dm.

A nickel alloy containing 50 percent zinc, which exhibited a hardness of 520 kg/sq mm, was deposited in a nickel sulfate - zinc sulfate - ammonium sulfate solution increased in pH to 9.6 to 10.4 with ammonium hydroxide. (99,100) A current density of 1 amp/sq dm and a temperature of 20 C were adopted for a current efficiency in the range of 85 to 100 percent.

Stress

Large internal stresses were reported for bright electrodeposited nickel-molybdenum alloys containing up to 25 percent molybdenum. (101) X-ray diffraction showed a lattice parameter of 3.55 angstroms for the expanded bcc structure that did not relate to the parameters for nickel or molybdenum according to Vergard's law.

TABLE 13.11

Summary of Structural Data for Nickel Alloys

Codeposited Metal	Grain Structure	Crystallographic Structure	Reference
Cobalt (14–72%)	Banded	Bcc solid solution	20
Germainum (1–5%)	Banded[a]	—	96
Iron	Banded	Single-phase fcc	77
Molybdenum (1–25%)	—	Single-phase bcc[b]	101
Silver (2–99%)	—	Supersaturated solid solutions	92
Rhenium	—	Fcc solid solutions	105
Thallium	—	Cubic[c]	106
Thallium and cobalt	—	Cubic[d]	106
Tin (56 to 66%)	—	Supersaturated solid solutions	107, 108
Tungsten (to 50%)	Banded	Single-phase, supersaturated solutions	71, 89, 90
Zinc (14%)	—	Single-phase fcc	109
Zinc (67 and 75%)	—	Single-phase cubic	109
Zinc (83 and 92%)	—	Two-phase hcp	109

[a] No change was detected after heating at 300 C for 1 hour.
[b] The lattice parameter was 3.55 angstroms.
[c] Preferred (100) and (210) orientations were identified for deposits at <3 amp/sq dm. Only the (210) orientation was observed for alloys deposited at 3 to 6 amp/sq dm.
[d] Orientation was random at current densities <4 amp/sq dm. A (100) orientation emerged at higher current densities.

Nickel alloy containing up to 3 percent niobium electrodeposited from fluoroniobate solutions were microcracked and stressed too badly for corrosion-protective applications. (102)

Some nickel-rhenium alloy deposits were reported to have high tensile stress of up to 315 kg/sq mm (450,000 psi) which can be removed by low-temperature annealing. (103) By contrast, nickel-rhenium alloys harder than rhenium, which were deposited in conventional sulfate-rhenate solutions, exhibited small stresses. (97) Nickel-zinc alloy deposited at 1 to 3 amp/sq dm in sulfate solutions that contained ammonium sulfate exhibited compressive stress. (104)

Structure

Structural data for several binary alloys of nickel are summarized in Table 13.11. The alloys of silver, tin, and tungsten were deposited as supersaturated solid solutions. All but zinc-rich nickel alloy exhibited a cubic lattice structure.

References

(1) Lainer, V. I., and Velichko, Yu. A., "Electrolytic Method for Making Forming Dies for Plastics From Nickel-Cobalt Alloy," *Vestnik Mashinostroeniya, 41* (1), 47–51 (1961).

(2) Lainer, V. I., "Electroforming of Nickel-Cobalt Mold Inserts for Pressure and Injection Molding of Plastics," *Electrodeposition of Alloys*, Edited by V. A., Averkin, Translated by the Israel Program for Scientific Translations, Jerusalem, 132–144 (1964). OTS 64–11015.

(3) Zhogina, V. M., and Kaznachei, B. Ya., "Electrodeposition of Magnetic Alloys," *Trudy Chetvertogo Soveshcheniya po Elektrokhimii, Moscow*, 506–511 (1956) (Pub. 1959).

(4) Vyacheslavov, P. M., "Electroplating With Alloys," *MCL-46411*, Aerospace Technical Intelligence Center, Wright-Patterson Air Force Base, Ohio (October 31, 1960). Translated from *Gal'vanicheskiye Pokrytiya Splavami*, Chapter III, 26–30 (1959). AD 257 132.

(5) Kaznachei, B. Ya., and Zhogina, V. M., "Effect of Conditions of Electrodeposition on the Magnetic Characteristics of Nickel-Cobalt Alloys," *Metallovedenie i Obrabotka Metallov*, 77–90 (1957).

(6) Friemel, F., "Electrodeposition of a Magnetizable Nickel-Cobalt Deposit of Low Permeability," German (East) Patent 52,869 (December 20, 1966).

(7) Kaznachei, B. Ya., and Zhogina, V. M., "Electrodeposition of Highly Coercive Nickel-Cobalt Alloys," *Tr. Vses. Nauchn.-Issled., Inst. Zvukozapisei, No. 1*, 91–93 (1957).

(8) Grant, R. W., "Electrolytic Bath for Deposition of Magnetic Films," U.S. Patent 3,463,708 (August 26, 1969). Assigned to Mohawk Data Sciences Corporation.

(9) Freitag, W. O., and Mathias, J. S., "Electrodeposited Magnetic Alloys for Computer Uses," *Electroplating and Metal Finishing, 17*, 42–47 (1964).

(10) Wolf, I. W., "Electrodeposition of Magnetic Materials," *Journal Applied Physics, Supplement, 33* (3), 1152–1159 (1962).

(11) Fedot'ev, N. P., Vyacheslavov, P. M., and Kashcheeva, E. A., "An Investigation of Some 'Structure-Sensitive' Properties of Ni-Co Galvanic Alloys," *Journal Applied Chemistry (USSR), 36* (11), 2396–2398 (1963); *Zh. Prikl. Khimii, 36* (11), 2474–2477 (1963).

(12) Brenner, A., Zentner, V., and Jennings, C. W., "Physical Properties of Electrodeposited Metals. I. Nickel. 3. The Effect of Plating Variables on the Structure and Properties of Electrodeposited Nickel," *Plating, 39*, 865–894, 899–927, 933 (1952).

(13) Endicott, D. W., and Knapp, Jr., J. R., "Electrodeposition of Nickel-Cobalt Alloy: Operating Variables and Physical Properties of the Deposits," *Plating, 53*, 43–60 (1966).

(14) Safranek, W. H., "Don't Overlook Electroforming," *Product Engineering, 32* (23), 44–49 (June 5, 1961).

(15) McFarlen, W. T., "An Evaluation of Electroformed Nickel-Cobalt Alloy Deposits," *Plating, 57*, 46–50 (1970).

(16) Belt, K. C., Crossley, J. A., and Watson, S. A., "Nickel-Cobalt Alloy Deposits From a Concentrated Sulfamate Electrolyte," *Trans. Inst. Metal Finish., 48*, 133–137 (1970).

(17) Ehara, S., Kishi, Y., Maruyama, K., and Kase, H., "An Investigation of Bright Cobalt-Nickel Alloy Plating," *Journal of the Metal Finishing Society of Japan, 9* (9), 334–338 (1958).

(18) Ledford, R. F., "Cobalt-Nickel Deposition in Electrotyping," *Plating, 36*, 560–565 (1949).

(19) Belt, K. C., Crossley, J. A., and Kendrick, R. J., "Properties of Electrodeposits From a Concentrated Nickel Sulfamate Solution," *Interfinish '68 Tagungsberichtsband*, Deutsche Gesellschaft für Galvanotechnik, e.V., Düsseldorf, Germany, 222–228 (1968).

(20) Korovin, N. V., "Alloy Deposition. III. Alloys for Wear Resistance and Low Friction Application," *Electroplating and Metal Finishing, 17*, 188–192 (1964).

(21) Sree, V., and Rama Char, T. L., "Electrodeposition of Nickel-Cobalt Alloys From the Pyrophosphate Bath," *Journal Electrochem. Soc., 108*, 64–70 (1961).

(22) Wolf, I. W., and McConnell, V. P., "Nickel-Iron Alloy Electrodeposits for Magnetic Shielding," *Proceedings Am. Electroplaters' Soc., 43*, 215–218 (1956).

(23) "Thin Films and Magnetic Memory Elements," British Patent 1,156,986 (July 2, 1969). Assigned to Japan Telegram and Telephone Corporation.

(24) Hayasaka, T., Masuzawa, K., Odani, Y., Senzaki, S., Gomi, Y., and Fujisawa, K., "Producing Magnetic Memory Elements," U.S. Patent 3,383,761 (May 21, 1968). Assigned to Japan Telegram and Telephone Corporation.

(25) Wolf, I. W., Katz, H. W., and Brain, A. E., "Fabrication and Properties of Memory Elements Using Electrodeposited Thin Magnetic Films of 82–18 Nickel-Iron," *Proceedings of the Electronics Components Conference*, 15–20 (1959).

(26) Mathias, J. S., and DiGuilio, G. V., "Electrolytic Deposition of Magnetic Films," Belgian Patent 638,864 (February 17, 1964). Assigned to Sperry Rand Corporation.

(27) Sallo, J. S., "Current Status of Plated Wire Memories," *Extended Abstracts, Abstract No. 33*, Electrochemical Society (Spring, 1970).

(28) Ginberg, A. M., Granovskii, Yu. V., Titov, Yu. E., and Klyachko, Yu, A., "Choosing Optimum Electrodeposition Conditions by the Box-Wilson Method," *Zashchita Metallov, 1* (6), 716–718 (1965).

(29) Rozhdestvenskaya, A. K., Bakhvalov, G. T., Vishkarev, A. B., and Rozenblat, M. A., "Electrodeposition of Thin Films of Soft Magnetic Nickel-Iron Alloys in a Sulfonamide Electrolyte," *Tr. Vses. Nauchn.-Issled. Inst. Magnitn. Zapisii Tekhnol. Radioveshch. i Telev., No. 4*, 172–188 (1966).

(30) Kaznachei, B. Ya,. and Rozhdestvenskaya, A. K., "Electrodeposition of a Nickel-Iron Alloy From a Sulfonamide Electrolyte," *Tr. Vses. Nauchn.-Issled. Inst. Magnitn. Zapisi i Tekhnol. Radioveshch. i Telev., No. 4*, 164–171 (1966).

(31) Toledo, E., and Semienko, P., "Effective Nickel Salts in Permalloy Plating Baths for Production of DRO Wire," *Plating, 57*, 504–507 (1970).

(32) Ohno, I., "Electrodeposition of Iron-Nickel Alloys. VI. Magnetic Properties of Electrodeposited Iron-Nickel Alloy Thin Films," *Kinzoku Hyomen Gijutsu (Journal of the Metal Finishing Society of Japan), 19* (5), 201–202 (1968).

(33) Uehara, Y., "Electrodeposited Nickel-Iron Thin Films," *Japanese Journal of Applied Physics, 2* (8), 451–462 (1963).

(34) Maeda, M., and Mukasa, K., "Internal Stresses of Electrodeposited Nickel-Iron Films," *Japanese Journal of Applied Physics, 6* (7), 895 (1967).

(35) Politycki, A., and Gotthard, H., "Effect of Manufacturing Conditions on the Magnetic Properties of Electrolytically Produced Permalloy Layers," *Z. Angew. Phys., 14*, 363–365 (1962).

(36) Levy, E. M., "Nickel-Iron Alloys Electrodeposited From Sulfate-Chloride Electrolyte: Effects on Mechanical Properties and Microstructure of Stress-Reducing Agent, Heat Treatment, pH, and Chemical Compositions," *Plating, 56*, 903–908 (1969).

(37) Wolf, I. W., "Stress-Reducing Organic Additives in the Sulfate-Chloride Bath for Nickel-Iron Deposition," *Electrochemical Technology, 1*, 164–167 (1963).

(38) Koshmanov, V. V., and Primak, N. M., "Effects of Current Density, Material, and the Pretreatment of the Base Metal on the Internal Stresses of Electrodeposited Iron-Nickel Films," *Izvestiya Vysshikh Vchebnykh Zavedeniy SSSR., Fiz., 10* (9), 40–43 (1967).

(39) Kostyuk-Kul'gavchuk, L. P., "Obtaining Magnetic Films by an Electrolytic Method," *Dokl. Akad. Nauk Beloruss. SSR, 8* (5), 309–311 (1964).

(40) Doyle, W. D., "Epitaxial Influence of the Substrates on the Anisotropy in Thin Electrodeposited Nickel-Iron Cylindrical Films," *Physica Status Solidi, 17* (1), K67–K69 (1966).

(41) Luborksy, F. E., "Influence of the Substrate and Electrolyte on the Composition and Properties of Thin Electrodeposited Nickel-Iron Films," *IEEE Transactions on Magnetics, 5* (2), 106–111 (1969).

(42) Reimer, L., "Magnetic Properties and Electron-Microscopic Structure of Electrolytically Deposited Thin Films of Nickel-Iron Alloys," *Zeitschrift für Physik, 150*, 99–105 (1958).

(43) Il'yushenko, L. F., Sheleg, M. U., and Boltushkin, A. V., "Preparation and Properties of Thin Electrolytically Precipitated Permalloy Films," *Dokl. Akad. Nauk Beloruss. SSR, 13* (8), 683–686 (1969).

(44) Wolf, I. W., "Composition and Thickness Effects on Magnetic Properties of Electrodeposited Nickel-Iron Thin Films," *Journal Electrochemical Society, 108*, 959–964 (1961).

(45) Elmanova, V. A., "Magnetic Properties of Electrochemical Films," *Izv. Akad. Nauk SSSR, Ser. Fiz., 31* (3), 445–447 (1967).

(46) Jostan, J., "Method for Electrodeposition of Thin Ferromagnetic Films," German Patent 1,258,234 (January 4, 1968). Assigned to Telefunken Patentverwerttungsgesellschaft m.b.H.

(47) Morioka, S., Takahashi, M., Sawada, Y., Shimodaira, S., and Ogawa, S., "The Preparation of 80% Ni-Permalloy Films by Electrodeposition and Their Magnetic Properties," *Nippon Kinzoku Gakkaishi, 25*, 679–683 (1961).

(48) Zentner, V., "Electrodeposited and Electroless Magnetic Alloys for Computers," *Plating, 52,* 868–872 (1967); *J. Intern. Applic. Cobalt, Brussels, Belgium,* 152 (June 1964).

(49) Stephen, J. H., "Electrodepositing Nickel-Iron Films," British Patent 859,725 (January 25, 1961). Assigned to United Kingdom Atomic Energy Authority.

(50) Kaznachei, B. Ya., Balashova, N. N., Shuvalova, M. A., and Belyaeva, L. A., "Effect of Composition of Pyrophosphate Electrolytes and Conditions of Electrodeposition on the Magnetic Properties of Nickel-Iron Alloys," *Tr. Vses. Nauchn.-Issled. Inst. Magnitn. Zapisi i Tekhnol. Radioveshch. i Televid., No. 1,* 168–179 (1964).

(51) Il'yushenko, L. F., and Kostyuk-Kul'gavchuk, L. P., "Decrease in Coercive Force and Increase in Squareness Coefficient of Hysteresis Loops of Electrolytically Deposited Iron-Nickel Films. I," *Vestsi Akad. Navuk Beloruss. SSR, Ser. Fiz.-Mat. Navuk, No. 1,* 114–119 (1967).

(52) Edigaryan, A. A., Papyan, M. V., and Pogosyan, T. A., "The Effect of Thermomagnetic Treatment on Magnetic Properties of Electroplated Ferromagnetic Films," *Izv. Akad. Nauk Arm. SSR, Fiz., 2* (4), 259–265 (1967).

(53) Stephen, J. H, "Electrodepositing Magnetic Alloy Films," British Patent 925,144 (May 1, 1963). Assigned to United Kingdom Atomic Energy Authority.

(54) Cisman, A., De Sabata, C., Muscutar, I., and Mihalca, I., "Conditions for the Formation of the Hexagonal Crystal Structure in Electrolytic Nickel-Iron Layers," *Rev. Roum. Phys., 12* (3), 273–277 (1967).

(55) Le Mehaute, C., and Rocher, E., "Method of Obtaining a Non-Magnetostrictive Magnetic Electrolytic Deposit," British Patent, 1,081,316 (August 31, 1967). Assigned to International Business Machines Corporation.

(56) Sagal, M. W., "Preparation and Properties of Electrodeposited Cylindrical Magnetic Films," *Journal Electrochem. Soc., 112,* 174–176 (1965).

(57) Bogenschuetz, A. F., and Jostan, J. L., "Electrodeposition of Nickel/Iron/Antimony and Nickel/Iron/Selenium Thin Films for Magnetic Storage Tapes," *Metalloberflaeche, 23* (4), 97–100 (1969).

(58) Freitag, W. O., DiGuilio, G., and Mathias, J. S., "Electrodeposited Ni-Fe-As Magnetic Thin Films," *Journal Electrochem. Soc., 113,* 441–445 (1966).

(59) See Reference (9).

(60) Firestone, S., "Magnetic Properties of Electrodeposited Nickel-Iron-Cobalt Films," *USAEL TR-2417* (August 1964), 28 pp. AD 609 829.

(61) Le Mehaute, C., and Rocher, E., "Electrodeposition of Stress-Insensitive Ni-Fe and Ni-Fe-Cu Magnetic Alloys," *Electroplating and Metal Finishing, 18,* 300–304, 308 (1965).

(62) Freitag, W. O., and Mathias, J. S., "Electrodeposited Nickel-Iron-Molybdenum Thin Magnetic Films," *Journal Electrochem. Soc., 112,* 64–67 (1965).

(63) Vagramyan, A. T., Ginberg, A. M., Fedotova, N. Ya., and Ginberg, T. A., "Effect of Ultrasound on Electrodeposition of Nickel-Iron-Molybdenum Alloys," *Elektrokhimiya, 2* (5), 551–556 (1966).

(64) Freitag, W. O., Mathias, J. S., and DiGuilio, G., "The Electrodeposition of Nickel-Iron-Phosphorus Thin Films for Computer Memory Use." *Journal Electrochem. Soc., 111,* 35–39 (1964).

(65) Fraenz, I., and Jostan, J. L., "Inclusion of Selenium in Electrodeposited Thin Permalloy Films and Its Influence on Their Magnetic Properties," *Metall, 23* (5), 418–424 (1969).

(66) Bogenschuetz, A. F., and Jostan, J. L., "Electrodeposition of Thin Magnetic Films of Ternary Permalloy Alloys," *Oberflaeche-Surface, 9* (12), 271–273 (1968).

(67) Medernach, J. W., Bobb, L. J., and Hunter, N. A., "Deposition of Permalloy Films on Organic-Polymer-Coated Substrates," *Journal Applied Physics, 37* (5), 2013–2015 (1966).

(68) Lloyd, J. C., and Smith, R. S., "Relation of Coercive Force to Structural Properties of Electroplated Films," *Canadian Journal Physics, 40,* 454–462 (1962).

(69) Morioka, S., Takahashi, M., Sawada, Y., and Watanabe, Y., "The Effects of Heat Treatment on the Magnetic Properties of Electrodeposited 80% Ni-Permalloy Films," *Nippon Kinzoku Gakkaishi, 28* (11), 722–727 (1964).

(70) Stephenson, Jr., W. B., "Development and Utilization of a High-Strength Alloy for Electroforming," *Plating, 53,* 183–192 (1966).

(71) Brenner, A., Burkhead, P., and Seegmiller, E., "Electrodeposition of Tungsten Alloys Containing Iron, Nickel, and Cobalt," *Research Paper No. 1834, Journal Research National Bureau of Standards, 39* (4), 351–383 (1947).

(72) Morkhov, M. I., and Atchi, L. P., "Electrodeposition of Ni in the Presence of Fe," *Korroziya i Bor'ba s Nei, 6* (5–6), 10–15 (1940).

(73) Aotani, K., "Properties of Electrodeposited Ni:Fe Alloys," *Journal of the Electrochemical Society of Japan, 20,* 31–34 (1952).

(74) Vagramyan, A. T., and Solov'eva, Z. A., *Technology of Electrodeposition,* Robert Draper, Ltd., Teddington (1961), 398 pp.

(75) Marti, J. L., "The Effect of Some Variables Upon Internal Stress of Nickel as Deposited From Sulfamate Electrolytes," *Plating, 53,* 61–71 (1966).

(76) MacInnis, R. D., and Gow, K. V., "Microstructure of Some Electrodeposited Nickel-Iron Alloys," *Plating, 57,* 626–628 (1970).

(77) Fedorova, N. S., "X-Ray Diffraction Structure Investigation of Electrodeposited Iron-Nickel Alloys," *Zh. Fiz. Khimii, 32,* 1211–1213 (1958).

(78) Pomosov, A. V., Murashova, I. B., and Artamonov, V. P., "Electrodeposition of a Powdered Nickel-Iron Alloy," *Porosh. Met., 9* (5), 1–6 (1969).

(79) Brenner, A., Couch, D. E., and Williams, E. K., "Electrodeposition of Alloys of Phosphorus and Nickel or Cobalt," *Plating, 37,* 36–42 (1950); *37,* 161–164 (1950).

(80) de Minjer, C. H., and Brenner, A., "Studies on Electroless Nickel Plating," *Plating, 44,* 1297–1305 (1957).

(81) Lasas, A., "Hard Nickel Coatings," *Mokslas ir Tech., No. 8,* 34–35 (1961).

(82) Kozarev, Kh., and Mestudzhian, I., "Electrochemical Deposition of a Nickel-Phosphorus Coating," *Mashinostroene (Sofia), 16* (6), 258–260 (1967).

(83) Ro, B., "On Properties of Electric Conductivity and Corrosion Resistance of Nickel-Phosphorus Alloy Deposits. Studies on Chemical Plating (4)," *Journal of the Metal Finishing Society of Japan, 11* (9), 331–334 (1960).

(84) Mitani, H., Shoji, K., and Kanbe, T., "Phase Changes in Electrolytic Ni-P Deposits During Heat Treatment," *Journal of the Metal Finishing Society of Japan, 17* (10), 384–388 (1966).

(85) Dolzhenkov, A. T., and Lashas, A. A., "Internal Stresses in Electrolytic Nickel-Phosphorus Coatings," *Dokl. Mosk. Inst. Inzh. Sel'skokhoz. Proizv., 1* (4), 61–66 (1964).

(86) Morioka, S., Sawada, Y., and Shimazaki, R., "Hardness of Electrodeposited Plating Film of W-Ni Alloy," *Journal of the Metal Finishing Society of Japan, 16* (11), 512–516 (1965).

(87) "Electrodeposition of Tungsten Alloys," *Corrosion and Material Protection, 5* (2), 10, 17 (1948).

(88) Ishida, T., Yoshioka, S., Mizuno, K., and Kudo, T., "Electroplating With Tungsten and Cobalt, Iron, or Nickel," Japanese Patent 5261 (June 25, 1959).

(89) Offermans, H., and Stackelberg, M. V., "Electrodeposited Tungsten-Cobalt, Tungsten-Nickel, and Tungsten-Iron Alloys," *Metalloberflaeche, 1,* 142–144 (1947).

(90) Vasu, K. I., and Rama Char, T. L., "Electrodeposition of Nickel-Tungsten Alloys From the Pyrophosphate Bath," *Metal Finishing Journal, 13* (1), 5–9, 13 (1967).

(91) Fedot'ev, N. P., and Vyacheslavov, P. M., "The Phase Structure of Binary Alloys Produced by Electrodeposition," *Plating, 57,* 700–706 (1970).

(92) Burkat, G. T., Fedot'ev, N. P., and Vyacheslavov, P. M., "Electrodeposition of Silver-Nickel Alloys," *Journal Applied Chemistry (USSR), 41* (2), 405–408 (1968).

(93) Browning, M. E., Webster, Jr., W. H., and Macklin, B. A., "Deposition Forming Processes for Aerospace Structures," *AMF Progress Report No. 7, Contract AF 33(657)-7016,* American Machine and Foundry Company (May 1, 1963).

(94) Couch, D. E., and Connor, J. H., "Nickel-Aluminum Alloy Coatings Produced by Electrodeposition and Diffusion," *Journal Electrochem. Soc., 107,* 272–276 (1960).

(95) Kudryavtsev, N. T., and Tsupak, T. E., "Conditions for Electrolytic Preparation of a Nickel-Chromium Alloy From Sulfate Solutions in the Presence of Aminoacetic Acid," *Tr. Mosk. Khim.-Tekhnol. Inst., No. 44,* 96–101 (1963).

(96) Andryushchenko, F. K., and Gavyrina, N. N., "Electrolytic Deposition of Nickel-Germanium Alloys," *Protective Metallic and Oxide Coatings, Metal Corrosion and Electrochemistry,* edited

by N. P. Fedot'ev, Akademiya Nauk SSSR. Translated by the Israel Program for Scientific Translations, Jerusalem, 159–164 (1968).

(97) Pishikin, A. M., and Popereka, M. Ya., "Electrodeposition of Bright Electrodeposits of Rhenium and Rhenium-Nickel Alloys," *Blestyashchie Komb. Metal. Pokrytiya, No. 2,* 3–12 (1967).

(98) Kadaner, L. I., Avakyan, R. B., and Kiryukhina, Z. N., "Electrochemical Deposition of Ruthenium-Cobalt, Ruthenium-Nickel, Ruthenium-Cobalt-Nickel, and Platinum-Nickel Alloys," *Elektrolit. Osazhdenie Splavov, 2,* 43–50 (1968).

(99) Batyuk, Zh. V., and Marchenko, N. A., "Electrodepositing a Zinc-Nickel Alloy From an Ammoniacal Electrolyte," *Tr. Khar'kovsk. Politekhn. Inst., 45,* 48–53 (1963).

(100) Marchenko, N. A., and Batyuk, Zh. V., "Electrolytic Deposition of Zinc-Nickel Alloys," *Zh. Prikl. Khimii, 37* (3), 595–600 (1964).

(101) Frantsevich, I. N., Frantsevich-Zabludovskaya, T. F., and Zhelvis, E. F., "Electrolytic Production of Nickel and Molybdenum Alloys," *Journal Applied Chemistry (USSR), 25* (4), 387–396, 511 (1952); *Zh. Prikl. Khimii, 25* (4), 350–361 (1952).

(102) Kalyuzhnaya, P. F., and Gavrilova, Z. P., "Deposition Conditions and the Physico-Chemical Properties of Nickel-Niobium Alloy Coatings," *Ukr. Khim. Zh., 33* (3), 327–333 (1967).

(103) Pishikin, A. M., Popereka, M. Ya., and Khazankin, V. B., "Electrodeposited Nickel-Rhenium Films. II. Internal Stresses. Ferromagnetic Properties," *Izvestiya Vysshikh Uchebnykh Zavedeniy SSSR, 12* (2), 11–14 (1969).

(104) Sushkov, Ya., P., and Suvorova, O. A., "Internal Stresses and Structure of Electrolytic Nickel-Zinc Alloys," *Izv. Akad. Nauk Kaz. SSR, Ser. Khim., 20* (6), 42–48 (1970).

(105) Pishikin, A. M., and Popereka, M. Ya., "Electrodeposition of Nickel-Rhenium Alloys," *Elektrolit. Osazhdenie Splavov, 2,* 15–21 (1968).

(106) Pobedimskii, G. R., and Zhikharev, A. I., "Texture of Electrodeposited Nickel-Thallium and Nickel-Cobalt-Thallium Alloys," *Tr. Kazansk. Khim.-Tekhnol. Inst., No. 36,* 278–280 (1967).

(107) Nawrocka, A., "Nickel-Tin Electrodeposits Obtained From Pyrophosphate Solutions," *Ochrona Przed Korozja, 9* (6), 138–144 (1966).

(108) Rooksby, H. P., "An X-Ray Study of Tin-Nickel Electrodeposits," *Journal Electrodepositors' Technical Soc., 27,* 153–160 (1951).

(109) Rama Char, T. L., and Panikkar, S. K., "Electrodeposition of Nickel-Zinc Alloys From the Pyrophosphate Bath," *Electroplating and Metal Finishing, 13,* 405–412 (1960).

Chapter 14

Nickel Composites

Composites of nickel and solid dispersed particles or fibers obtained by electrodeposition processes include mixtures with alumina, boron, carbon, silicon carbide, titanium dioxide, and tungsten. Particles with diameters of 0.01 to 5 μm, which are dispersed in the electrodeposition bath and become positively charged, are attracted to the cathode and encapsulated with the electrodeposited metal. Fiber-reinforced composites are made by winding a continuous filament around a symmetrical cathode during deposition. Filaments incorporated in nickel include boron, silicon carbide, and tungsten.

Stimuli for the development of electrodeposited composites include needs for high-strength materials with a high modulus of elasticity, especially at high temperatures. Light-weight particles such as alumina and fibers such as silicon carbide are of special interest because of prospects for high modulus to density ratios. Hard coatings with good resistance to wear offer another potentially useful application for composites of electrodeposited nickel and particles of hard oxides or carbides.

Mechanical Properties

Composites of Particles and Whiskers

Mechanical property data for composites of nickel and alumina, silicon carbide, or titanium dioxide particles are summarized in Table 14.1. Nickel with 5.6 weight percent silicon carbide was the strongest of the particle composites, but another containing 5 weight percent (10.4 volume percent) titanium dioxide was nearly as strong (85 kg/sq mm or 121,000 psi). A composite of alumina (1 percent by weight) retained higher strength after annealing at 650 C, however. The yield strength of this deposit after heat treating was 42 kg/sq mm (60,000 psi), in comparison with 33 and 13 kg/sq mm (47,000 and 18,000 psi) for the titanium dioxide composite and particle-free nickel, respectively. Supplementary data for annealed deposits are given in Table 14.2.

A composite of 1.7 weight percent alumina retained a tensile strength of 32 kg/sq mm (46,800 psi) after heat treating at 1100 C, which also reduced brittleness. (2)

TABLE 14.1
Mechanical Property Data for Nickel-Particle Composites

Dispersed Material[a]	Plating Bath	Tensile Strength		Yield Strength		Hardness, kg/sq mm	Modulus of Elasticity		Reference
		kg/sq mm	psi	kg/sq mm	psi		kg/sq mm	psi × 10^6	
Alumina, 1.0%	Sulfamate (50 C)	77[b]	110,000[b]	52.5[b]	75,000[b]	270 to 320[b]	—	—	1
Alumina, 1.7%	Watts (50 C)	45[c]	64,000[c]	—	—	—	—	—	2
Alumina, 2%	Watts (50 C)	21[d]	30,000[d]	—	—	514[e]	—	—	4
Alumina, 2.5%	Chloride (50 C)	—	—	35[f]	50,000[f]	—	—	—	3
Alumina, 4.5%	Watts (50 C)	68	97,000	45	64,000	350	—	—	5
Alumina, 6.0%	Watts (50 C)	75	107,000	56	80,000	380	22,000[g]	31[g]	5
Alumina, 9.4%	Watts (50 C)	70	100,000	45	64,000	400	22,000[g]	31[g]	5
Silicon carbide, 3.2%[h]	Sulfamate	174	249,000	—	—	—	—	—	6
Silicon carbide, 3.6%[h]	Sulfamate	202 to 229	288,000 to 327,000	—	—	—	31,500	44	6
Silicon carbide, 5.6%	Watts (50 C)	85	121,000	60	86,000	410	21,000[g]	30[g]	5
Silicon carbide, 9.6%	Watts (50 C)	74	105,000	55	79,000	450	21,000[g]	30[g]	5
Titanium dioxide, 5%	Sulfamate (50 C)	81[b]	115,000[b]	55[b]	78,000[b]	300 to 420[b]	—	—	1

[a] Weight percent.
[b] The effects of heat treating are summarized in Table 14.2.
[c] Tensile strength was 32 kg/sq mm (46,800 psi) after heat treating at 1100 C. The 100- to 500- angstrom particles formed 2500 angstrom clusters during heat treating.
[d] Tensile strength was 2.3 kg/sq mm (3,300 psi) at 1095 C. Elongation at 1095 C was 34%.
[e] Hardness decreased to 205 kg/sq mm at 260 C and 74 kg/sq mm at 425 C.
[f] Yield strength for alloy deposit after heating to 750 C. The yield strength of nickel containing no alumina, which was deposited in a similar solution, was only 8 kg/sq mm (11,400 psi) after a similar heat treatment.
[g] The modulus of the matrix nickel was 18,000 kg/sq mm (26,000,000 psi).
[h] Whiskers about 1 μm in diameter were incorporated in the deposit.

This composite was deposited at 2 amp/sq dm on a rotating cathode disc while 100 to 500-angstrom alumina particles were dispersed with air agitation in a 50 C Watts-type bath with a pH of 3 to 4. The tensile strength of particle-free nickel from this bath was only one-fourth of that of the composite after heat treating.

A composite of 2.5 weight percent (6 volume percent) alumina retained a yield strength of 35 kg/sq mm (50,000 psi) after annealing at 750 C. (3) In comparison, the yield strength of similarly annealed, particle-free nickel deposited with similar conditions was only 8 kg/sq mm (11,400 psi). The 200 to 300-angstrom alumina particles (150 g/l) were dispersed by mechanical stirring in a Watts-type bath to deposit the composite at a current density of 2 amp/sq dm. Current density shifts from 1 to 5 amp/sq dm or temperature changes from 20 to 80 C had little effect on the amount of incorporated alumina; pH changes from 2 to 5 had only a slight effect, but the amount of alumina incorporated in the deposit decreased appreciably when the pH was reduced to 1.0.

Extensive hardness data for particle composites have been reported. These data are summarized in Table 14.3. Particles of silicon carbide, tungsten carbide and zirconium carbide, and of silicon nitride increased hardness to > 500 kg/sq mm. (8) Dispersion of alumina and chromic oxide also increased hardness to > 500 kg/sq mm.

During abrasion tests, the wear of composites containing 5 to 9 weight percent alumina or silica was less than one-half that of an unalloyed Watts-type nickel plate. (5) Composites of nickel and tungsten particles were also reported to be better than nickel for resisting wear. (9) These tungsten composites were deposited at

TABLE 14.2

Effect of Heat Treating Nickel-Alumina and Nickel-Titanium Dioxide Composites (1)

Dispersed Material	Heat Treating Temperature, C	Tensile Strength, kg/sq mm (psi)	Yield Strength, kg/sq mm (psi)	Elongation, percent	Hardness, kg/sq mm
None[a]	No heating	70 (100,000)	51 (72,000)	3	200
	370 (4 hours)	60 (85,000)	39 (55,000)	4	150
	650 (4 hours)	34 (48,000)	13 (18,000)	32	100
Alumina, 1.0%[b]	No heating	77 (110,000)	53 (75,000)	2	330
	370 (4 hours)	77 (110,000)	53 (75,000)	2	340
	650 (4 hours)	53 (75,000)	42 (60,000)	2.5	290
Titanium dioxide, 5%[c]	No heating	81 (115,000)	55 (78,000)	2	350
	370 (4 hours)	79 (113,000)	56 (80,000)	2	360
	650 (4 hours)	46 (66,000)	33 (47,000)	6	230

[a] Deposited at 10 amp/sq dm in a 50 C bath prepared with 320 g/l $NiNHSO_3$, 22 to 30 g/l H_3BO_3, a proprietary bromide compound and 0.5 g/l of a wetting agent. The pH of the solution was 4.0.

[b] 50 g/l Al_2O_3 (0.07 μm diameter, average) was dispersed in the solution described in Footnote (a) by forced circulation.

[c] 100 g/l TiO_2 (0.2 μm in diameter) was dispersed in the solution described in Footnote (a) by forced circulation.

TABLE 14.3

Hardness Data for Nickel-Particle Composites

Dispersed Materials	Particle Size, μm	Amount of Dispersed Particles Encapsulated With Deposit, percent by weight	Hardness, kg/sq mm	Reference
Alumina	0.01–0.05	24	270 to 230	1
Alumina	0.5–1.0	1.8	318 to 384	7
Alumina	30	4.8	514[(a)]	4
Alumina	1–2	50	562	8
Boron nitride	1–5	25	453	7
Carborundum	0.3–0.5	7	350	7
Chromium carbide	2–3	40	420	8
Chromium oxide	1	30	533	8
Corundum	0.1–1	5	330	7
Silicon carbide	3–5	40	515	8
Silicon nitride	1–2	30	551	8
Tantalum carbide	3.5	40	518	8
Titanium carbide	3.4	40	504	8
Titanium oxide	0.2	10	300 to 420	1
Tungsten carbide	2–3	40	515	8
Zirconium carbide	3–5	40	533	8
Zirconium dioxide	0.5–1	40	453	7

(a) Hot hardness at 260 and 430 C was 205 and 74 kg/sq mm, respectively.

high current densities in a Watts-type bath at a temperature of 40 C and contained particles ranging in diameter from 0.3 to 2 μm.

A composite of nickel and silicon carbide whiskers was much stronger than the particle composites. A 3.8 weight percent (10 volume percent) composite exhibited a tensile strength of 202 to 229 kg/sq mm (288 to 327,000 psi). (6) Its modulus of elasticity was increased to 31,500 kg/sq mm (44,600,000 psi), which is about 75 percent higher than the modulus for unalloyed nickel electrodeposited in a Watts-type solution. The whiskers were dispersed in a nickel sulfamate plating bath by mechanical stirring.

Deposits with a thickness of 0.02 to 0.07 mm, which contained 25 or 60 volume percent of molybdenum disulfide particles, exhibited friction coefficients of 0.05 and 0.18, respectively. (10) These coefficients were in the range of those for other films containing molybdenum disulfide. Twenty weight percent of the particles were dispersed in a nickel sulfamate bath with a pH of 2.0. The 60 volume percent deposits were obtained at a current density of 25 amp/sq dm.

Composites of Filaments

Mechanical property data for composites of nickel and continuously wound filaments are summarized in Table 14.4. High tensile strengths approaching or exceeding 140 kg/sq mm (200,000 psi) were reported for composites containing 34 or 42 volume percent boron, 50 volume percent silicon carbide, or 35 to 50 volume percent tungsten. The modulus of elasticity of the nickel was slightly increased by the incorporaton of boron or tungsten filaments, but silicon carbide was more effective. Composites with 40 or 50 volume percent silicon carbide increased the modulus to 28,000 and 31,500 kg/sq mm (40 and 45 psi $\times 10^6$), respectively, which corresponded to a 100 and 125 percent increase over the modulus of the unalloyed nickel matrix, electrodeposited in the same sulfamate solution. (6)

The property data in Table 14.4 relate to composites produced in nickel sulfamate solutions. A typical bath contained 60 to 75 g/l nickel ions, and 22.5 to 35 g/l boric acid; the pH ranged from 3.5 to 5.0, and temperature from 50 to 55 C. A current density as high as 20 amp/sq dm was adopted for some experiments.

The notched tensile strength of nickel composites containing 6 to 7 volume percent of boron filaments was about equal to the notched tensile strength of the nickel matrix electroformed in the sulfamate bath. (13) The ratio of notched and unnotched strength was about 1.1. Increasing the boron filament content to 12 percent reduced this ratio to about 0.8. However, the ratio for tensile specimens pulled at −195 C (−320 F) was 1.0. At this temperature, notched and unnotched specimens showed a tensile strength of 91 kg/sq mm (129,000 psi), which was 43 percent higher than the strength of an unnotched composite at room temperature.

Fatigue data indicated a twofold increase in endurance for a composite containing 6 volume-percent fibers, in comparison with the endurance life of the nickel matrix. (13) Evidently the fibers acted as crack arrestors.

The bond strength of nickel to encapsulated filaments is an important criterion of the usefulness of composites of filaments. The bond strength of nickel to boron fibers and the tensile strength of the fibers are improved by etching them before incorporation in the nickel matrix. (17) Heat treatment at 500 C further improved bond strength. However, a coating of boron nitride was more effective in improving bond strength, as follows: (18)

	Average Bond Strength			
	Before Heating		After Heat Treatment at 500 C	
Coating	kg/sq mm	psi	kg/sq mm	psi
None	1.3	1830	2.3	3206
Boron nitride	2.0	2820	3.7	5310

Because boron nitride is a nonconductor, the coated filaments can be closely spaced with high loading in the nickel matrix with less danger of void formation between filaments.

Heating the boron-nickel composite at 650 C weakened the pullout strength of the fibers. (18) A significant decrease in the tensile strength of nickel-boron composites when heat treated at temperatures above 500 C was attributed to the formation of nickel boride. (11,13) Boron fibers protected with boron nitride were not affected by heat treatment at 650 C. (18) However, indications of a breakdown in the boron nitride coating occurred after heating at 900 C.

TABLE 14.4

Mechanical Property Data for Nickel-Filament Composites

Filament Material[a]	Tensile Strength kg/sq mm	Tensile Strength psi	Modulus of Elasticity kg/sq mm	Modulus of Elasticity psi × 10⁶	Reference
Boron, 15%[b]	81.5	116,000	19,500	28	6, 11
Boron, 23%[b]	87[c]	124,000[c]	23,000[d]	32.7[d]	6, 11, 12
Boron, 24%[e]	126	180,000[f]	—	—	13
Boron, 34%[e]	144	205,000[f]	—	—	13
Boron, 34%[b]	112	159,000	22,400	32.0	6
Boron, 42%[b]	132	188,000	22,800	32.6	6, 11
Boron, 51%[g]	132	188,000	23,600	33.6	11
Boron, 56%[g]	108	154,000	27,600	39.2	11
Silicon carbide, 20%[b]	70	100,000	21,000	30	6
Silicon carbide, 34%	106[h]	152,000[h]	—	—	14
Silicon carbide, 40%[b]	105	150,000	28,000	40	6
Silicon carbide, 50%[b]	130	185,000	31,500	45	6
Tungsten, 16%[i]	105	150,000	19,000	27	6
Tungsten, 20%[i]	120	170,000	17,500	25	6
Tungsten, 25%	28[j]	40,000[j]	—	—	15
Tungsten, 30%	116	166,000	21,000	30	6
Tungsten, 35%[k]	148[l]	211,800[l]	—	—	16
Tungsten, 43%	140	~200,000	—	—	17
Tungsten, 50%[i]	154	220,000	24,000	34	6

[a] Volume percent of filament encapsulated in nickel deposited in a sulfamate bath.

[b] 0.1 mm (4 mil) diameter filament.

[c] The tensile strength of the filament-free matrix was 88 kg/sq mm (125,000 psi) before heat treatment. Heat treatment of the nickel and the nickel composites changed tensile strength as follows:

Tensile Strength

Heat Treatment Temperature, C	Nickel kg/sq mm	Nickel psi	Nickel-Boron Composite kg/sq mm	Nickel-Boron Composite psi
300	64	91,000	102	145,000
500	46	65,000	44	63,000
700	—	—	22 to 41	31,000 to 59,000

Footnotes to TABLE 14.4 (Continued)

(d) The modulus of the filament-free, nickel matrix was 17,300 kg/sq mm (24.7 $\times$ 10^6 psi). Heat treatment of the nickel and nickel composite reduced the modulus as follows:

Heat Treatment Temperature, C	Modulus of Elasticity: Nickel, kg/sq mm	Nickel, psi $\times$ 10^6	Nickel-Boron Composite, kg/sq mm	Nickel-Boron Composite, psi $\times$ 10^6
300	20,000	28.6	21,000	30
500	18,400	26.3	16,000 to 22,500	23 to 32
700	—	—	12,600 to 22,500	18 to 32

(e) 0.1 mm (4 mil) diameter filament with 0.0125 mm (0.5 mil) tungsten core.

(f) The strength of the filament-free matrix was 98 kg/sq mm (140,000 psi).

(g) These composites contained 8 or 10% voids.

(h) A composite of 46 volume percent silicon carbide filaments exhibited a tensile strength of 75 kg/sq mm (106,000 psi) at 600 C.

(i) 0.05 to 0.1 mm (2 to 4 mil) diameter filaments.

(j) This low strength probably was a result of voids in the deposit. In similar composites with more than 25 volume percent tungsten, large voids were noted.

(k) 0.025 mm (1 mil) diameter filament.

(l) Tensile strength after annealing at 650 C was 116 kg/sq mm (165,000 psi).

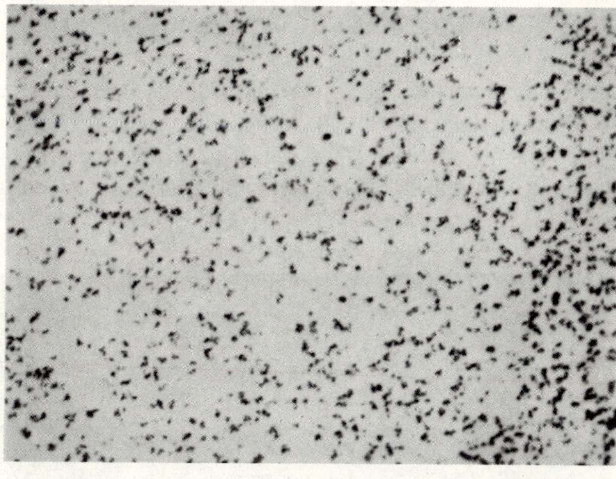

Figure 14.1. Cross Section of Nickel Composite Containing 10.4 Volume Percent of 0.2 μm Titanium Dioxide Particles. (1000$\times$) (1)

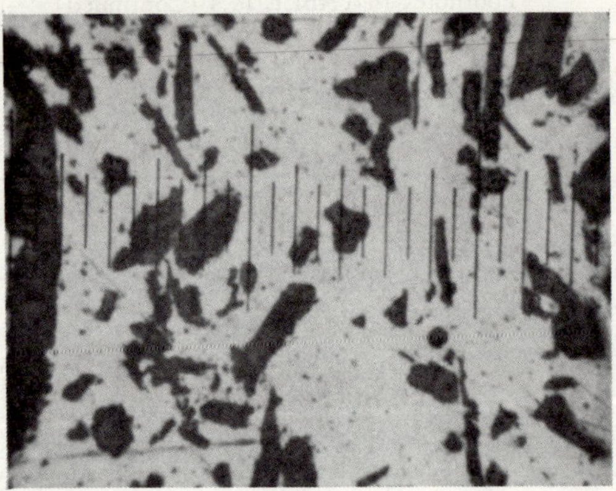

Figure 14.2. Cross Section of Nickel Composite Containing about 10 Volume Percent of Randomly Oriented, 0.025 mm Silicon Carbide Whiskers. (400×) (6)

The tensile strength of a composite of 23 or 25 volume percent boron filaments was increased from about 84 kg/sq mm (120,000 psi) to 95 kg/sq mm (135,000 psi) by heat treating at 300 C. (12) The modulus was unchanged. On the other hand, a 500 C heat treatment reduced tensile strength to about 45 kg/sq mm (64,000 psi), which was about the same as the tensile strength of a similarly annealed, unalloyed nickel electrodeposit.

Electroless cobalt-phosphorus alloy and electroless phosphorus-free nickel obtained with hydrazine was a useful matrix material for thornel yarn, subsequently compacted by hot isostatic or swaging techniques. (19) Tensile strength of a 50 volume percent composite was about 50 percent of the theoretical strength, at room temperature, but decreased excessively at 760 C, which was attributed to the break up of the fibers during fabrication. Breaking of graphite fibers also was observed when composites were produced by hot pressing preplated graphite filaments at

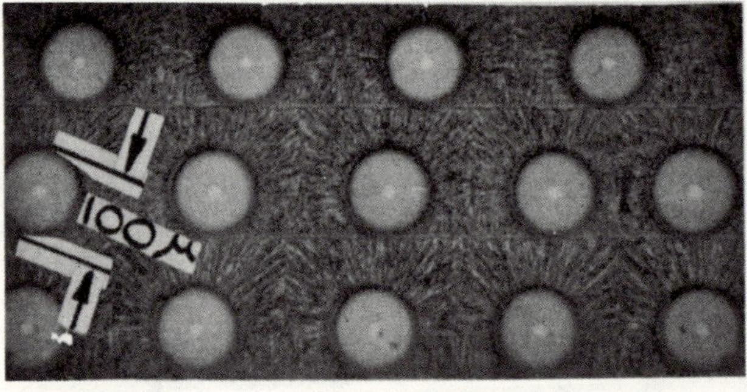

Figure 14.3. Cross Section of Nickel Composite Containing Aligned, 0.1 mm Boron Filaments. (17)

temperatures of 1050 to 1150 C with pressures of 2.1 to 3.2 kg/sq mm (3000 to 4500 psi). (20) Heat treatment above 750 C of the individual nickel-plated fibers reduced their tensile strength of > 240 kg/sq mm (>340,000 psi) to < 160 kg/sq mm (<225,000 psi), in the case of a high-modulus, or from > 175 kg/sq mm (> 250,000 psi) to < 140 kg/sq mm (< 200,000 psi) in the case of a lower modulus graphite fiber.

Structural Features

Figure 14.1 shows a dispersion of 5 percent (10.4 volume percent) titanium dioxide in nickel electrodeposited at 10 amp/sq dm in a nickel sulfate solution containing 100 g/l dispersed, 0.2 μm, titanium dioxide particles. (1) The interparticle spacing averages 0.05 μm. Figure 14.2 shows a cross section of randomly oriented silicon carbide whiskers in electrodeposited nickel. (6) The composite contained about 10 volume percent of 0.025 mm whiskers. A cross section of a composite of boron filaments and nickel is shown in Figure 14.3.

One team of investigators reported that the aluminum particles dispersed in their nickel matrix differed only slightly in size from the loose particles dispersed in the electrodeposition bath. (2) Titanium dioxide particles in the nickel were slightly smaller than those added to the bath. Agglomeration of 100- to 500- angstrom particles of alumina occurred during heat treatment at 1100 C. (1) The clusters had dimensions of approximately 2500 angstroms.

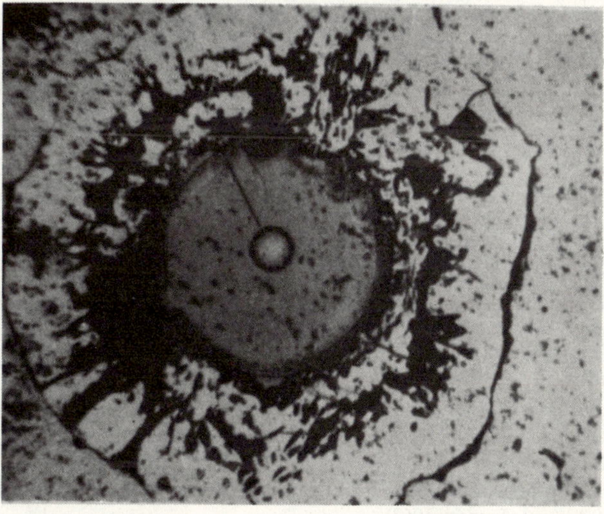

Figure 14.4. Cross Section of Nickel-Boron Filament Composite Heat Treated at 900 C for 3 Hours. (600×) (6)

Structural failure that occurred during heat treatment at 900 C of a nickel-boron composite is shown in Figure 14.4. (6) Microprobe analysis of the interfacial zone between carbon and nickel, after heat treatment at 650 C, indicated the formation of a diffusion zone. (18)

References

(1) Greco, V. P., and Baldauf, W., "Electrodeposition of Ni-Al_2O_3, Ni-TiO_2 and Cr-TiO_2 Dispersion-Hardened Alloys," *Plating, 55*, 250–257 (1968).

(2) Gilliam, E., McVie, K. M., and Phillips, M., "The Structure of Nickel Electrodeposited with Alumina Particles," *J. Inst. Metals, 94* (6), 228–229 (1966).

(3) Sautter, F. K., "Electrodeposition of Dispersion-Hardened Ni-Al_2O_3 Alloys," *Journal Electrochem. Soc., 110*, 557–560 (1963).

(4) Browning, M. E., Leavenworth, Jr., H. W., Webeter, Jr., W. H., and Dunkerley, F. J., "Deposition Forming Processes for Aerospace Structures," *Technical Documentary Report No. ML-TDR-64-26, Final Report under Contract AF 33(657)-7016*, American Machine & Foundry Company (January, 1964), 133 pp. N64–18564. AD 433 559.

(5) Broszeit, E., Heinke, G., and Wiegand, H., "On the Mechanical Properties of Electrodeposited Nickel Dispersion Coatings with Inclusions of Al_2O_3 and SiC," *Metall, 25*, 470–475 (1971).

(6) Withers, J. C., and Abrams, E. F., "The Electroforming of Composites," *Plating, 55*, 605–611 (1968).

(7) Saifullin, R. S., "Nickel Plating from an Electrolyte with Dispersed Particles," *Journal Applied Chemistry (USSR), 39*, 764–767 (1966); *Zh. Prikl. Khimii, 39*, 810–814 (1966).

(8) Kedward, E. C., and Kiernan, B., "Electrodeposited Composite Coatings for Wear Resistance," *Metal Finishing Journal, 13* (148), 116–120, 128 (1967).

(9) Saifullin, R. A., and Nadeeva, F. I., "Nickel Coatings Containing Inclusions of Simple Substances," *Elektrokhim. Protsessy Elektroosazhdenii Anodnom Rastvorenii Metal*, 28–33 (1969).

(10) Vest, C. E., and Bazzarre, D. F., "Co-Deposited Nickel-Molybdenum Disulfide," *Metal Finishing, 65* (11), 52–58 (1967).

(11) Alexander, J. A., and Stuhrke, W. F., "Effect of Temperature on Mechanical Properties of Boron-Electrodeposited Nickel Composites," *Amer. Soc. Test. Mater., Spec. Tech. Publ. No. 427*, 34–52 (1966).

(12) Alexander, J. A., "The Elevated Temperature Reactivity of Boron-Metal Matrix Composite Materials," *Summary Technical Report AFML-TR-67-101, Contract AF 33(615)-3155*, General Technologies Corp. (July 1967).

(13) Adsit, N. R., "Electroforming to Make Composite Materials," *Amer. Soc. Test. Mater., Spec. Tech. Publ. No. 427*, 27–33 (1966).

(14) Alexander, J. A., Cunningham, A. L., and Chuang, K. C., "Investigation to Produce Metal Matrix Composites with High-Modulus, Low-Density Continuous-Filament Reinforcements," *Technical Report AFML-TR-67-391, Contract AF 33(615)-2862*, General Technologies Corp. (February 1968).

(15) Baker, A. A., Harris, S. J., and Holmes, E., "Fiber Reinforcement of Metals by a Filament-Winding and Electroforming Technique," *Metals & Materials, 1* (7), 211–219 (1967).

(16) Ahmad, I., Greco, V. P., and Barranco, J. M., "Electroforming of the Composites of Nickel Reinforced with Some High-Strength Filaments," *Report No. WVT-6709*, Benet Laboratories, Watervliet Arsenal (January 1967).

(17) Greco, V. P., Wallace, W. A., and Cesaro, J. N. L., "Bond-Strength Characteristics of Electrodeposited Nickel on Boron and Silicon Carbide Filaments (Reinforced Composites)," *Plating, 56*, 262–270 (1969).

(18) Wallace, W. A., and Greco, V. P., "Electroforming High-Strength Continuous, Fiber-Reinforced Composites. Tungsten, Boron, Boron Nitride Coated Boron Filaments in Nickel Matrix," *Plating*, *57*, 342–348 (1970).

(19) Niesz, D. E., Kistler, Jr., C. W., and Fleck, J. N., "Development of Filament-Reinforced Metals," *U.S. Clearinghouse Fed. Sci. Tech. Inform* AD 649 509.

(20) Sara, R. V., "Fabrication Properties of Graphite-Fiber, Nickel-Matrix Composites," *Proceedings of the 14th National Symposium and Exhibition*, Soc. Aerospace Materials and Process Engineers (November 1968).

Chapter 15

The Platinum-Group Metals and Their Alloys

Table 15.1 compares the physical properties of the electrodeposited and wrought precious metals. Each of the platinum-group electrodeposits is significantly harder than its annealed wrought metal counterpart. Wear resistance is important for deposits used for electrical contacts. Rhodium deposits are used for electrical contacts because of high wear resistance, and palladium deposits because of lower costs. Ruthenium is of interest because of high hardness and low cost, which is only slightly higher than that of palladium. Because of high cost, thin coatings of the precious metals are desirable for providing protection of the basis metal. Thus low stress, ductility, and freedom from porosity are important properties.

Some of the precious metals find engineering applications as protective coatings in chemical environments or at high temperatures. In decorative applications, high reflectivity, tarnish and scratch resistance are important properties of precious metal deposits.

By analogy with other electrodeposited metals, it seems likely that density, strength, and ductility of the platinum metals could be improved by increasing the concentration of the platinum metals ions in solution or increasing agitation. Low concentration solutions and little or no agitation have been employed heretofore, as a rule. No data on tensile strength and only limited data on ductility has appeared in the literature.

Iridium and Iridium Alloys

A hardness of 90 kg/sq mm (DPN 10-gram load, 5 μm section) was reported for iridium deposited at 0.15 amp/sq dm in a 75 C bath prepared with hydrated iridium dioxide (5 g/l Ir) and hydrobromic acid (0.1 M HBr). (1) Deposits thicker than 1 μm showed cracks. Internal stress was on the order of 42 kg/sq mm (60,000 psi). Iridium was highly stressed and cracked when deposited in solutions prepared with 5 to 15 g/l iridium chloride tetrahydrate and 25 to 50 g/l sulfamic acid at 50 to 80 C and 0.2 to 0.6 amp/sq dm with auxiliary electrodes for a-c electrolysis (which favored a high cathode efficiency). (2) Table 15.2 shows the hardness of iridium and iridium-platinum alloys obtained from a fused sodium cyanide bath. (3) (4) The

TABLE 15.1

Comparison of the Properties of the Electrodeposited and Wrought Platinum-Group Metals

Metal	Specific Gravity, g/cu cm		Electrical Resistivity, microhm-cm		Hardness, kg/sq mm	
	Electro-deposit	Wrought Metal	Electro-deposit	Wrought Metal	Electro-deposit	Wrought Metal
Iridium	No data	22.65	No data	4.71	800–900	200–240
Osmium	No data	22.61	No data	8.12	No data	300–650
Palladium	11.9	12.02	10.7	9.93	75–400	37–40
Platinum	21.37	21.45	9.83	9.85	155–530[a]	37–42
Rhodium	No data	12.41	8.5	4.33	755–1095	120
Ruthenium	No data	12.45	No data	6.80	750–1300[a]	200–350

[a] Excludes softer deposits from a fused cyanide bath.

value of 800 kg/sq mm has been used as the maximum hardness for iridium in reviews of precious metal electrodeposits. (5)

The potential commercial applications of iridium deposits are limited, but iridium provides an interesting combination of physical and chemical properties. (2) Iridium is one of the most inert of all metals, being insoluble in aqua regia, hydrofluoric acid and all minerals acids.

Iridium is resistant to anodic corrosion in most aqueous solutions and has a lower overvoltage than platinum for some reactions. (1) It also resists attack by a number of molten salts and molten metals including the alkali metals, lead, and mercury. Some potential applications of iridium deposits include their use for inert electrodes, sliding contacts, reflectors and mirrors, vacuum tube elements, and laboratory equipment.

TABLE 15.2

Hardness of Iridium and Iridium-Platinum Alloys Deposited from a Fused Sodium Cyanide Bath at 600 C

Platinum Content, weight percent	Metal Content of Bath, weight percent	Current Density, amp/sq mm	Hardness, kg/sq mm	Reference
0	0.3 Ir	1.4 to 5.1	800	3
0	2 Ir	5	727[a]	4
35	4 Ir + 0.2 Pt	5	433 to 550[a]	4
100	0.2 Pt	5	78[a]	4

[a] Knoop hardness, 100-gram load.

Osmium

Osmium deposits are said to have exceptionally good wear properties, perhaps superior to any other metallic coating with the exception of osmium-iridium alloys. (6) For comparison with chromium in wear tests, the osmium deposits with a thickness of about 2 μm (80 microinches) were obtained from an alkaline solution containing osmium as an anionic complex formed by heating osmium tetroxide with sulfamic acid. No property data have been reported, however.

Palladium

Early references dating to 1935 indicated that the Brinell hardness of palladium deposits ranged from 190 to 435 compared to 49 for annealed wrought metal. (7) A hardness of 150 kg/sq mm (Brinell) was reported for palladium deposited from proprietary brush plating solutions. (8) The nominal stress value for palladium coatings was reported to be about 42 kg/sq mm (60,000 psi), which was lower than the stress of rhodium deposits in a correlation of stress with melting point. (9) Such a high stress is not typical of all palladium deposits, however.

In the 1960's an increased interest in palladium deposits for electrical contacts spurred the development of a number of new baths to provide thin, pore-free deposits. Many developments were concerned with bath life when using insoluble anodcs, Nuctral baths are used for plating copper and brass contacts, The acid chloride bath is used to provide thick protective coatings.

Interest in palladium is partly due to its favorable cost in comparison with the other platinum metals. For example, ruthenium is approximately 70 percent higher in cost. Rhodium and platinum are five to six times higher in cost.

Hardness and Wear Resistance

As a rule, palladium deposits are five to ten times harder than the annealed wrought metal or two to three times harder than the cold-worked metallurgical counterpart. Hardness varies from 75 to 400 kg/sq mm depending on the type of plating bath and operating conditions, as shown in Table 15.3. A hardness of 300 to 400 kg/sq mm has been reported for deposits obtained in chloride, phosphate-chloride, nitrite, and nitrate baths. On the other hand, deposits from bromide solutions exhibited hardness values ranging from 75 to 200 kg/sq mm. A relatively high palladium ion concentration (30 g/l) favored the softer deposits in this range. (21)

Additions of saccharin to a tetraamino palladous chloride bath reduced hardness significantly, as shown in Table 15.4. (22) However, the number of pores detected in 1.25 to 25 μm-thick deposits was increased. No pores were detected in 1.25 μm-thick palladium when saccharin was omitted from 20 C solution containing 30 g/l of palladium ions with a pH of 9.0. On the other hand, pores were detected in deposits obtained at a higher temperature, at a current density above or below 2.4

TABLE 15.3

Hardness and Stress of Palladium Electrodeposits

Plating Bath	pH	Palladium Concentration, g/l	Temperature, C	Current Density, amp/sq dm	Hardness, kg/sq mm	Stress[a] kg/sq mm	Stress[a] psi	References
1 Palladous chloride	<1	1–3	70	0.2	189	—	—	10
2 Palladous chloride	<1	50	50	1.6	160 to 200	4.6 to 7.0	6,600 to 10,000	11, 12
3 Palladous phosphate	—	20	18–20	—	—	70	100,000	13, 14
4 Palladous phosphate	5.6	1.2	—	0.1	—	5[b]	7,000	15
5 Palladous phosphate-chloride	—	—	—	—	400	17.5	25,000	3
6 Palladous pyrophosphate	7.5	7.0	20	0.3	105	—	—	16
7 Sodium palladous nitrite[c]	6.0	10	40	0.1–0.4	400	—	—	17
8 Sodium palladous nitrite	6.0	5	50	0.4–1.0	387–435[d]	28	40,000	18, 19
9 Tetraamino palladous bromide	9.0	15	30	1	150 to 200	15	21,500	20
10 Ditto	9.2	30	50	4	75 to 100	1.5 to 3	2,100 to 4,300	21
11 ″	—	5–30	25	1	150	9 to 10	13,000 to 14,000	21
12 Tetraamino palladous chloride	9.0	20	20	1.6	370–390	—	—	22
13 Tetraamino palladous chloride[e]	9.0	20	20–40	0.5–1.5	190–196[d]	—	—	23
14 Ditto	9.0	20	30	1.0	220–240	19	27,000	20
15 Tetraamino palladous chloride	9.0	20	25	1.5	212	56.8	81,000	24
16 Tetraamino palladous nitrate	~7	10	60–80	0.5–2.0	300–400	21 to 50	30,000 to 71,000	25, 26
17 Tetraamino palladous nitrite	9.0	3.2	50	1	~350	—	—	27
18 Ditto	9–10	4	—	—	300	2.8 to 5.6	4,200 to 8,000	17, 19
19 ″	9.0	10	50	1	300[d]	3 to 6	4,300 to 8,500	20
20 Tetraamino palladous nitrite[f]	—	15	—	—	300 to 400	17 to 21	24,000 to 30,000	26
21 Tetraamino palladous nitrite[g]	8.3	12	30	1	310	24	34,000	20

[a] Values for these deposits (about 5 μm); lower values were reported for thicker (25 μm) deposits.
[b] The addition of 7.5 g/l furfural increased stress to 10 kg/sq mm.
[c] The solution also contained 30 g/l sodium chloride.
[d] DPN with 6- to 17-gram loads.
[e] Operated with a diaphragm between the anolyte and catholyte.
[f] Proprietary bath containing organic acids.
[g] The solution contained 100 g/l ammonium sulfamate.

TABLE 15.4

Influence of Saccharin on the Hardness of Palladium Electrodeposited in a Tetraamino Chloride Bath[a]

Thickness, μm	Saccharin Concentration, g/l	Hardness, KHN_{25g}
30	0	370 to 390
30	0.5	240 to 260
40	1.0	250 to 270

[a] Tetraamino palladous chloride bath with 40 g/l $Pd(NH_3)_2Cl_2$, 10 g/l NH_4Cl, 50 ml/l NH_4OH, and 25 g/l $(NH_4)_2SO_4$, pH 9, 1.6 amp/sq dm, 20 C. (22)

amp/sq dm in solutions with a lower pH. The bath was operated with no diaphragm separating the anode and cathode zones, in contrast with other reports on tetraamino palladous chloride solutions which cited slightly lower hardness values for deposits obtained in diaphragmed tanks.

A reduction in the pH of the tetraamino nitrite bath from the range of 9–10 to 7–8 increased brightness and hardness. (17) Ordinarily, hardness ranged from 300

TABLE 15.5

Wear and Hardness of Electrodeposits (27)

Deposit	Thickness, μm	Wear Volume,[a] 10^{-6} cu cm	Microhardness at 10g load (DPN) kg/sq mm
Rhodium[b]	12.7	0.70	870
Palladium[c]	38.1	4.80	320
Gold[d]	50.7	10.20	65

[a] Wear of brush materials running on 2.5 μm of rhodium; 8.5 cm/sec speed, 10g loading; wear volume for 100 hour duration estimated from measurements made after 170 hours.

[b] Rhodium deposited at 1 amp/sq dm, 50 C, no agitation from solution containing 10 g/l rhodium and 70 ml/l sulfuric acid (conc).

[c] Palladium deposited at 1 amp/sq dm, 50 C, mechanical stirring, from electrolyte containing 3.2 g/l palladium (as "P" salt), 10 g/l ammonium nitrate, 10 g/l sodium nitrite, ammonium hydroxide (0.88 N) to pH 9.

[d] Gold deposited at 0.6 amp/sq dm, 50 C, with mechanical stirring of electrolyte containing 12 g/l potassium gold cyanide, 90 g/l potassium cyanide, 0.16 g/l potassium silver cyanide.

TABLE 15.6

Mechanical Wear of Palladium and Gold Electrodeposits[a]

Test Conditions	Track Wear		Brush Wear	
Slip Ring[b]	Max Depth μm	Wear Volume $\times 10^{-8}$ cm³	Wear Volume (60% Pd–40% Ag) $\times 10^{-8}$ cm³	
Palladium	0.25	51	16.4	
Gold	2.0	1640	51	
Semipermanent[c] (point contact)	Mean "socket" Wear, μm		"Plug" Wear, μm	
Palladium	0.25		1.54	
Gold	0.76		2.80	
Semipermanent[d] (line contact)	Mean "socket" Wear, μm		"Plug" Wear, μm Av	Max
Palladium	2.80		0.38	0.89
Gold	2.28		0.64	0.89

[a] Palladium deposits from conventional P-salt bath with 4 g/l palladium, 100 g/l ammonium nitrate, 10 g/l sodium nitrite, ammonia to pH 9–10, 50 C, 1 to 4 amp/sq dm with agitation. (19)

[b] Deposit thickness 2.5 microns. Brush diameter 0.66 mm (0.026 in.). Load 10 g. Linear speed 8.4 cm/sec. Time 100 hr.

[c] Deposit thickness: "Plug" 2.5 microns, "socket" 2.5 microns. Radius of plug contact 1.58 mm (0.062 in.). Contact load 50 g, stroke length 1.27 cm (0.5 in.), No. of operations 1000.

[d] Deposit thickness: "Plug" 1.25 microns, "socket" 5 microns. Radius of plug contact 6.1 mm (0.24 in.). Length of line contact 3.1 mm (0.12 in.). Other conditions as in Footnote (c).

to 350 kg/sq mm. The pH reduction that occurs during bath operation lowers the cathode efficiency. (26) This type of solution (commonly called the P-salt bath) is prepared by dissolving diamino palladous nitrite in a strong solution of ammonium hydroxide to form the tetraamino complex compound. Ammonium nitrate is formed by electrolysis, but is unstable and decomposes to nitrogen and water. (26)

Electroless palladium from solutions containing tetraamino chloride, ammonium hydroxide, the sodium salt of ethylene diamine tetraacetic acid and hydrazine (3 g/l) ranged in hardness from 150 to 350 kg/sq mm (measured with a 25-gram load), averaging 257 kg/sq mm. (28) Electroless palladium achieved by utilizing hypophosphite as a reducing agent contained 1.5 percent phosphorous and had a hardness of 165 kg/sq mm. (29) A palladium-phosphorous alloy deposit containing 99.9 percent palladium was reported to have a hardness of 240 kg/sq mm. (30)

The hardness and wear resistance of palladium, rhodium, and gold deposits are correlated in Table 15.5. The palladium was deposited in a tetraamino nitrite bath. Supplemental wear data for palladium and gold deposits are listed in Table 15.6. The wear characteristics of palladium and gold plug and socket contacts appear in Table 15.7.

Heat treatment at 850 C for 22 hours reduced the hardness of a palladium deposit on titanium from the range of 160–200 to 60 kg/sq mm (100-g load). (11,12) The original deposit had a thickness of 180 μm. Hardness at depths of 100 to 170 μm after heat treatment was in the range of 350 to 380 kg/sq mm, indicating the formation of a palladium-titanium diffusion alloy.

Palladium deposited on molybdenum with an initial hardness of 100 to 150 kg/sq mm was softened to 38 kg/sq mm after heat treatment for 20 hours at 1070 C. The diffusion alloy with molybdenum exhibited a hardness of 416 kg/sq mm. The diffusion rate was primarily from the substrate into palladium and the diffusivity of molybdenum was 2×10^{-11} sq cm/sec. (11,12)

Stress

The initial stress of palladium deposits varies from 1.5 to 70 kg/sq mm, depending on the plating bath formulation and conditions. As shown in Table 15.3, palladium with a stress of only 1.5 to 3 kg/sq mm (2,100 to 4,250 psi) was deposited in tetraamino bromide baths when their temperature was 50 C. (21) Such deposits were crackfree and free of porosity, even when thickness ranged from only 1 to 2.5 μm. (22) Higher stress up to 15 kg/sq mm was reported for deposits from bromide baths operated at 25 or 30 C. (20,21)

Relatively low stress ranging from 2.8 to 6 kg/sq mm (4,000 to 8,500 psi) was reported for some deposits from tetraamino palladous nitrite baths by some investi-

TABLE 15.7

Wear of Dissimilar Electrodeposits (Line Contact) (19)

Contact Finish		Mean Penetration on Socket Coating, μm	Max. Penetration on Plug Coating, μm	Distribution of Wear on Plug
Socket	Plug			
Pd (P-salt)	Au (acid)	0.51	1.22	Localized
Pd (TPN)[a]	Au (acid)	3.36	0.82	Uniform
Au (cyanide)	Pd (P-salt)	3.36	1.22	Uniform
Au (cyanide)	Pd (sulfamate)	2.72	0.20	Localized
Au (acid)	Pd (sulfamate)	2.09	0.20	Uniform

[a] TPN = tetraamino palladous nitrate.

gators. (17,20) An increase in the current density for plating in nitrite solutions evidently increases stress, judging by the cracks observed when a current density of 4 amp/sq dm was employed in place of 2 amp/sq dm. (19) This trend was noted in the case of deposits produced in 25 or 50 C baths containing 4 or 10 g/l of palladium ions.

Palladium with a relatively low stress below 7 kg/sq mm (10,000 psi) was deposited in chloride and phosphate baths maintained at 50 C. (11,12,15) On the other hand, stress in a deposit from a 20 C phosphate solution reached 70 kg/sq mm. (13,14) Deposits from sodium palladous nitrite and tetraamino nitrate solutions also were highly stressed (>21 kg/sq mm or >30,000 psi). (18,25,26) Palladium from a solution prepared with sodium chloropalladite (14 g/l), sodium chloride (40 g/l), and boric acid (10 g/l) evidently was also highly stressed, because cracking was noted. (22) The bath was adjusted to pH 6.0–6.5, and temperature to 40 C, and was operated at a current density of 0.5 amp/sq dm.

Additions of furfural and protalbinic acid as brighteners in the phosphate bath increased stress in the deposits from about 5 to 10 kg/sq mm. (15) The deposits were produced at 0.1 amp/sq dm in a bath with a pH of 5.6.

Stress in palladium has been associated with codeposited hydrogen. (14) Removal of hydrogen induced a compressive stress. Another researcher related the high stress of palladium from phosphate baths to its high overvoltage (800 to 850 millivolts), by comparison with the overvoltage of other metals. (31)

Ductility

In terms of elongating palladium-plated copper rods until cracking appeared, the best palladium deposits were obtained in a tetraamino palladous bromide bath. (20) The harder deposits from other baths were not so ductile, as detailed in Table 15.8, which compares ductility values with hardness and stress. Deposits obtained in sodium palladous nitrite solutions operated with soluble anodes were described as brittle and cracked, (18) whereas those from a tetraamino palladous nitrate solution were said to be ductile though highly stressed. (25) The nitrate bath is now considered obsolete. (26)

Maximum ductility of 20 percent was obtained for palladium deposited at 4 amp/sq dm at 50 C in the bromide bath with 30 g/l palladium, a pH of 9.2 and no agitation. (21) With an anode current density of 10 amp/sq dm, ductility could be maintained above 15 percent. The stress of the deposits was only 1.5 to 3 kg/sq mm and the hardness was 75 to 100 kg/sq mm (VHN_{30g}) under the above optimum conditions. For operation at 25 C, ductility was reduced, whereas stressed increased. With agitation, as in barrel plating, ductility was further reduced to about 4 to 5 percent at 25 or 50 C. These deposits had a reflectivity of about 45 percent (relative to polished silver). Of several metal additives at 0.1 g/l, silver increased reflectivity to 62 percent while reducing ductility to 15 percent. Additions of 0.3 g/l furfuraldehyde increased reflectivity to 64 percent but reduced ductility to 4 percent; a higher additive concentration resulted in porosity.

TABLE 15.8

Stress, Hardness and Ductility Palladium Deposits from Four Baths (20)

Bath	Tensile Stress, kg/sq mm[a] 5 μm	Tensile Stress, kg/sq mm[a] 25 μm	Hardness, kg/sq mm[b]	Relative Ductility, percent[c]
Tetraamino chloride[d]	19	—	220 to 240	2.2 to 4.2
Tetraamino nitrite (P-salt)[e]	3 to 6	—	300	0.8
P-salt/sulfamate[f]	24	5	310	1.2
Tetraamino bromide[g]	15	3	150 to 200	8 to 19

[a] The stress in deposits was measured by the "bending strip" method using beryllium-copper strips 10.76 cm × 0.48 cm × 0.25 mm thick.

[b] Hardness VHN_{30g} was measured on the surface of coatings.

[c] Relative ductility was determined by elongation of plated copper rods until the deposit cracked. Deposit thickness was 5 μm on 10.7 cm annealed copper rods of 0.318 cm in diameter. The rods were elongated in a tensometer until protection of the basis metal was lost (0.5 percent blackened area by electrographic test with cadmium sulfide paper). A zero-value would indicate that the deposit was porous as plated.

[d] Deposits obtained at 1 amp/sq dm and 30 C from a bath at pH 9.0 containing 20 g/l palladium as tetraamino palladous chloride, and 10 g/l ammonium chloride using a diaphragm.

[e] Deposits obtained at 1 amp/sq dm and 50 C from a tetraamino palladous nitrite bath at pH 9.0 containing 10 g/l palladium as P-salt, 100 g/l ammonium nitrite, and 10 g/l sodium nitrate.

[f] Deposits obtained at 1 amp/sq dm and 30 C from a tetraamino palladous nitrite bath at pH 8.3 containing 12 g/l palladium as P-salt and 100 g/l ammonium sulfamate.

[g] Deposits obtained at 1 amp/sq dm and 30 C from a bath at pH 9.0 containing 15 g/l palladium as tetraamino palladous bromide, and 30 g/l ammonium bromide using a diaphragm.

Minimum ductility in terms of resisting cracking during bend tests for palladium deposited in tetraamino chloride solution was reported for deposits at 2.4 amp/sq dm from unagitated, 20 C baths at a pH of 9. (22) A lower pH, a lower or higher current density, or a higher temperature reduced ductility. The electrolyte was prepared with 40 to 80 g/l dichloroamine palladium, 10 g/l ammonium chloride, 25 g/l ammonium sulfate and 50 ml/l ammonium hydroxide.

Specific Gravity and Resistivity

Table 15.9 compares the specific gravity and resistivity of palladium electrodeposits from a tetraamino palladous nitrate bath with the values of the metallurgical form. The specific plating conditions were not identified by the author and are

TABLE 15.9

Specific Gravity and Resistivity of Palladium Deposits Compared to Metallurgical Values

Property	Electrodeposit[(a)]	Metallurgical[(b)]
Specific gravity, g/cu cm	11.9	12.02
Specific resistance at 0 C, microhm-cm	10.7	9.93
Thermal conductivity	0.17	0.17

[(a)] Deposits at 0.5 to 2 amp/sq dm from a tetraamino palladium nitrate bath assumed to contain 10 g/l Pd, at 60 to 80 C, which had a hardness of 300 to 400 kg/sq mm. (25)

[(b)] *Metals Handbook* values; a hardness of 37–44 DPN is cited for annealed palladium and 105–110 DPN for 50 percent cold worked metal. (40)

assumed to be typical for the tetraamino palladous nitrate bath. (25) The neutral pH bath is favored for plating printed circuits. It operates at 100 percent current efficiency and is claimed to produce pore-free deposits which are bright and ductile. The deposit will remain tarnish-free in air though an oxide film forms on its surface at a temperature of about 400 C.

In comparison with the resistivity (10.7 microhm-cm) reported for palladium

TABLE 15.10

Contact Resistance of Wrought and Electrodeposited Metals (27)

	Contact Resistance at 70 g, ohms[(a)]		Contact Resistance at 0.05 g, ohms	
Metal[(a)]	Wrought	Electrodeposit[(b)]	Wrought	Electrodeposit[(c)]
Gold[(d)]	0.0004	0.0005	0.05	0.06
Palladium[(e)]	0.0021	0.0064	0.21	0.16
Rhodium[(f)]	0.0019	0.0047	0.40	0.35

[(a)] Measurements carried out at 0.5 ampere with crossed rods of 0.0275 cm diameter.

[(b)] 12.7 μm thickness.

[(c)] 0.5 to 12.7 μm thickness.

[(d)] Gold deposited at 0.6 amp/sq dm, 50 C, with mechanical stirring of electrolyte containing 12 g/l potassium gold cyanide, 90 g/l potassium cyanide, and 0.16 g/l potassium silver cyanide.

[(e)] Palladium deposited at 1 amp/sq dm, 50 C, with mechanical stirring of electrolyte containing 3.2 g/l palladium (as "P" salt), 10 g/l ammonium nitrate, 10 g/l sodium nitrite, and ammonium hydroxide (0.88 N) to pH 9.

[(f)] Rhodium deposited at 1 amp/sq dm, 50 C, 50 C, and no agitation of solution containing 10 g/l rhodium and 70 ml/l sulfuric acid (conc.).

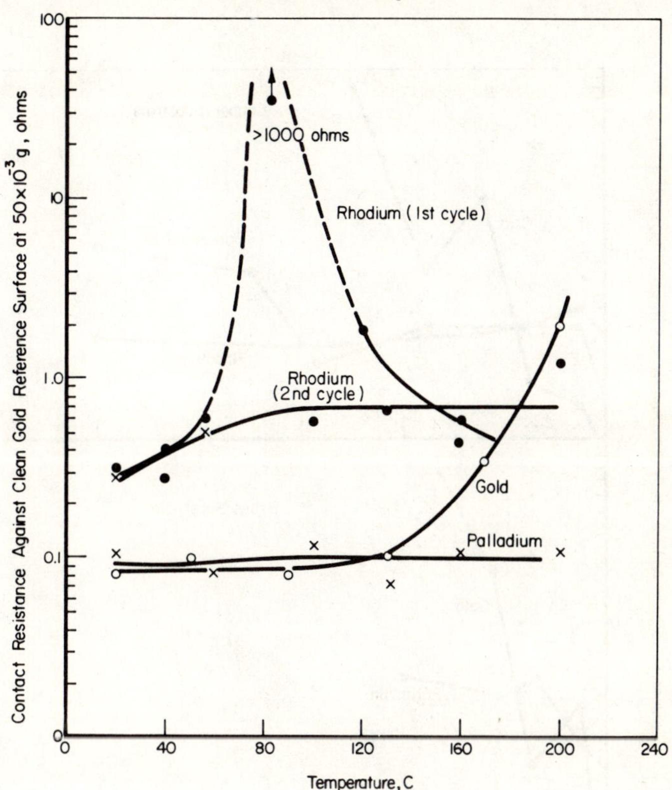

Figure 15.1. Changes in Contact Resistance with Temperature for 2.5-μm Electrodeposits on Copper during a 15-Minute Heat Treatment in Air at Progressively Higher Temperatures. (27)

Electrodeposition conditions are described in Table 15.10, footnotes.

TABLE 15.11

Effect of Sulfur Dioxide Exposure on the Contact Resistance of Copper and Brass Electroplated with Palladium, Gold or Rhodium[(a)]

Thickness of Deposit, μm	Contact Resistance at 0.5 Gram Against Clean Gold					
	Copper Substrate			Brass Substrate		
	Gold	Palladium	Rhodium	Gold	Palladium	Rhodium
0.5	(b)	0.50	2.4	(b)	(b)	(b)
1	(b)	0.77	0.99	(b)	3.5	6.5
2	(b)	0.86	0.62	(b)	0.80	1.3
3	1.20	0.30	2.5	(b)	1.25	8.7
6	1.75	0.45	(c)	0.13	0.55	(c)
12	0.20	0.47	(c)	0.07	0.35	(c)

(a) Electrodeposition conditions are summarized in the footnotes to Table 15.10. (27)

(b) Basis metal was exposed.

(c) Severe attack occurred adjacent to cracks in the electrodeposits.

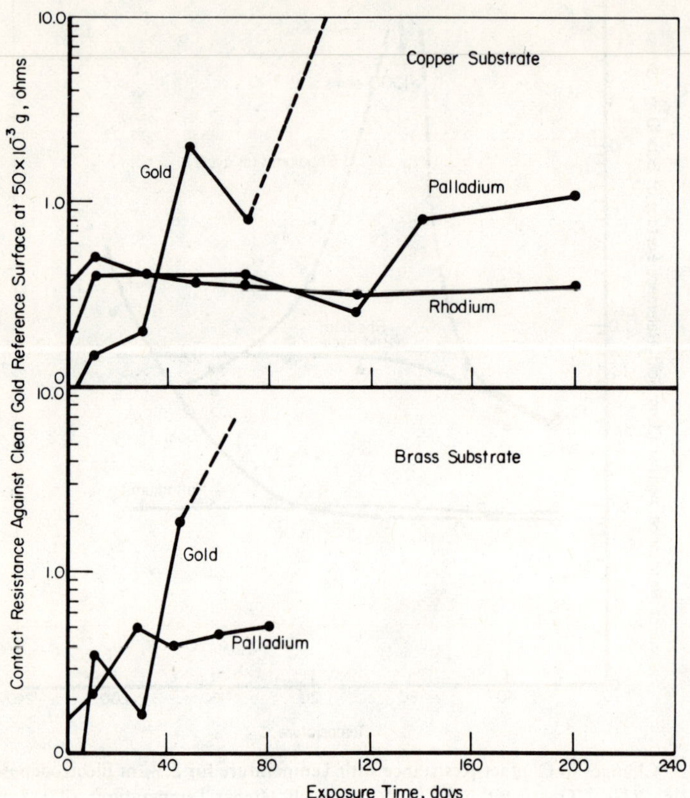

Figure 15.2. Variation of Contact Resistance with Time for Electrodeposits of Gold, Palladium, and Rhodium on Copper and Brass.

Thickness of 0.5, 1.5, 2.5, and 5 μm behaved similarly, so a single curve is shown for each coating on each basis metal. With rhodium, only 2.5-μm deposits were exposed. (27)

from a tetraamino nitrate bath, which is about 8 percent higher than that cited for the metallurgical form, a higher resistivity of 15 microhm-cm was determined for a softer deposit from a pyrophosphate solution prepared with palladous chloride and potassium pyrophosphate. (16) The pyrophosphate bath was maintained at 20 C whereas the tetraamino nitrate solution was heated to 60–80 C.

Table 15.10 compares the contact resistance of wrought and electrodeposited palladium, rhodium, and gold. Figure 15.1 shows that palladium deposits on copper maintain a lower contact resistance at higher temperatures than gold or rhodium on copper. Table 15.11 shows that palladium deposits provide better surface protection of copper and brass to sulfur dioxide exposure. After such exposure the contact resistance of palladium-plated copper was less than that for gold deposits in thicknesses up to 3 μm. Figure 15.2 shows that for thicknesses of 0.5 to 5 μm, palladium deposits maintain lower contact resistance than gold deposits for extended exposure in a laboratory atmosphere.

Palladium Alloys

Palladium-Cobalt Alloys

Table 15.12 shows hardness and electrical resistivity data for palladium-cobalt alloy deposits from a pyrophosphate bath. (16) The values increase markedly as the cobalt content of the deposit is increased, reaching a maximum of 540 kg/sq mm at 65 percent and 74 microhm-cm at 60 percent cobalt. X-ray structural analysis revealed that palladium-cobalt alloys form a continuous series of solid solutions. Pure hexagonal epsilon cobalt was observed in cobalt rich alloys (> 90 percent Co). A relative minimum in hardness of 295 kg/sq mm was obtained at 96 percent cobalt and a relative minimum in electrical resistivity of 48 microhm-cm was obtained at 94 percent cobalt.

Palladium-Nickel Alloys

Table 15.13 shows the hardness, wear resistance and internal tensile stress of palladium alloys containing up to 30 percent nickel deposited in an aminochloride electrolyte. (24) The effects of changing current density, temperature, and pH on the properties of the deposits are summarized in Table 15.14. An increase in the

TABLE 15.12

Hardness and Electrical Resistivity of Palladium-Cobalt Alloy Deposits[a]

Cobalt Content, weight percent	Microhardness, kg/sq mm	Electrical Resistivity, microhm-cm
0	105	15
10	230	23
20	315	34
30	380	49
40	430	61
50	480	69
60	525	74
70	530	71
80	490	61
90	370	51
100	330	53

[a] Deposits obtained at 0.3 amp/sq dm and 20 C from a bath containing 7 g/l of metal with palladium as $PdCl_2$ and cobalt as $K_6[Co(P_2O_7)_2]$ which was prepared by dissolving $CoSO_4$ in excess $K_4P_2O_7$, 200 g/l free $K_4P_2O_7$, 20 g/l Rochelle salt and pH 7.5. (16)

TABLE 15.13

Properties of Palladium-Nickel Alloy Deposits[a]

Nickel Content, weight percent	Hardness[b] kg/sq mm	Wear Resistance,[c] 10^3 cycles	Internal Stress,[d] kg/sq mm
0	212	20	56.8
5	230	70	55
10	248	100	47
15	270	118	39.4
20	290	163	29
25	312	241	19
30	334	350	10

[a] Alloys were deposited at 25 C and 1.5 amp/sq dm from a solution of pH 9 containing 20 g/l Pd as $Pd(NH_3)_4Cl_2$, 0 to 30 g/l Ni as $Ni(NH_3)_6Cl_2$, 20 to 25 g/l NH_4Cl. (24)

[b] Microhardness with a load of 100g and deposit thickness of 10 μm.

[c] The wear resistance was determined with a contact load of 200 g/sq mm and a bronze point making reciprocating movements over a base plate having an alloy deposit thickness of 5 μm until the base was exposed.

[d] Internal stress determined by the flexible cathode method with a deposit thickness of up to 3 μm.

TABLE 15.14

Effect of Plating Variables on Properties of Palladium-Nickel Alloy Deposits[a]

Plating Variable	Change in Property Value: Nickel Content of Deposit, weight percent	Hardness, kg/sq mm	Wear Resistance, 10^3 cycles	Internal Stress, kg/sq mm
Increase current density from 1.5 to 5 amp/sq dm	Increase from 25 to 35	Increase from 320 to 350	Increase from 280 to 390	Decrease from 28 to 20
Increase bath temperature from 25 to 55 C	Decrease from 25 to 45	Increase from 350 to 220[b]	Increase from 280 to 350	Constant 20 ± 2
Increase bath pH from 8.0 to 9.5	Decrease from 35 to 11	Decrease from 320 to 280	Decrease from 380 to 210	Increase from 22 to 40[c]

[a] The electrolyte at 25 C contained 30 g/l Pd and was operated at 1.5 amp/sq dm with a pH of 9.0 unless specified otherwise. (24)

[b] Hardness decreased sharply from 350 to 220 kg/sq mm above 45 C.

[c] Internal stress increased sharply from 27 to 40 kg/sq mm above pH 9.0.

nickel content of the deposit was associated with a decrease in current efficiency to as low as 80 percent. An increase in the ammonium chloride concentration from 10 to 70 g/l increased current efficiency to 98 percent, but had little effect on the properties of the deposit.

Palladium alloys with 10 to 70 percent nickel were claimed in a patent disclosing a similar plating bath which also contained sulfate ions and an addition agent. (32) An alloy of 40 percent nickel had a hardness of 535 kg/sq mm. The 10 μm thick deposit was unaffected by exposure to ammonia gas for 24 hours or immersion in sea water for 6 days. The alloy deposit was obtained at 1 amp/sq dm, and 30 C at a pH of 7.5 from a bath containing 20 g/l tetraamino palladous chloride (10 g/l palladium), 50 cu cm/l ammonia solution (28 percent NH_4OH), 50 g/l ammonium sulfate, 50 g/l hydrated nickel sulfate (10.4 g/l nickel), and 10 g/l sodium 1, 4, 6-naphthalene trisulfonate.

Palladium-Silver Alloys

Table 15.15 shows the electrical resistivity and microhardness of palladium-silver alloy deposits from a lithium chloride bath. (10) The maximum in electrical resistivity was obtained for a deposit with about 40 percent silver, which also is the maximum for the metallurgical alloys. However, the resistivity of the electrodeposit was about 2.5 times higher. Maximum hardness of about 320 kg/sq mm was obtained for a 15 percent silver alloy deposit. The hardness was about 240 kg/sq mm over the range of 25 to 60 percent silver. The maximum in hardness for the cast, vacuum-annealed metallurgical alloys was at 25 percent silver. All of the electrode-

TABLE 15.15

Properties of Palladium-Silver Alloy Deposits[a]

Silver Content, weight percent	Electrical Resistivity, microhm-cm	Hardness, kg/sq mm
0	48	189
20	56	301
40	110	237
60	85	231
80	48	168
100	~2	87

[a] Deposits obtained at 70 C and 0.2 amp/sq dm from a solution containing 500 g/l LiCl, 10 g/l HCl, 0.68 to 2.24 g/l Ag and 0.08 to 106 g/l Pd. (10)

posited alloys had the same grain size and the higher hardness for deposits with 5 to 25 percent silver was attributed to a difference in texture.

The alloy composition of the deposit was influenced only slightly by bath temperature over the range of 20 to 80 C. (10) Changing the current density had a greater effect, particularly below 0.3 amp/sq dm where smooth bright deposits were obtained. The ratio of palladium to silver ions in the bath below a ratio of about 3, which produced an alloy with 25 percent silver, also influenced alloy composition. The acid concentration did not affect alloy composition. The best deposits were obtained with 4 to 15 g/l hydrochloric acid. With 20 g/l hydrochloric acid or more the deposits had high internal stress and were cracked.

Platinum

Hardness

Early references dating to 1935 indicated that the Brinell hardness of platinum deposits ranged from 606 to 642 kg/sq mm compared to 47 kg/sq mm for annealed wrought metal. (7) The platinum deposits were probably obtained from the early chloroplatinic acid baths using a boiling solution containing ammonium and sodium phosphates, first developed in 1885. (23)

A hardness of 453 to 495 kg/sq mm (VHN_{10g}) was reported for platinum deposits from a stabilized platinate bath containing 11.5 g/l platinum as sodium hexahydroplatinate and 10 g/l sodium hydroxide operated at 75 C and 0.8 amp/sq dm. (23) Platinum deposited in sulfamate solutions heated to 75 C exhibited about the same hardness. (11) However, deposits from other baths were not so hard, as shown in Table 15.16. For example, the hardness of platinum from a fluoborate solution was 220 kg/sq mm. (35) Deposits from chloroplatinic acid baths ranged from 155 to 350 kg/sq mm in hardness. (34) Those in the range of 155 to 250 kg/sq mm were deposited at a rate in the range of 10 to 25 μm/hour in solutions containing at least 225 g/l hydrochloric acid. According to the results of rolling wires with 12.5 μm-thick coatings, these relatively soft deposits were characterized as ductile. The harder deposits, which were brittle and cracked, were plated at a maximum rate of 10 μm/hour. Heating the soft or hard deposits above 600 C caused blistering, due to inclusions arising from the hydrolysis of chloroplatinic acid, which occurs at a pH of 2.2.

A hardness of 84 to 93 kg/sq mm (KHN_{25g}) was reported for deposits at 0.3 to 3 amp/sq dm from a fused cyanide bath with 0.3 weight percent platinum at 600 C. (3,23) A value of only 78 kg/sq mm was reported for deposits at 1.5 amp/sq dm from the fused cyanide bath with 0.2 weight percent platinum at 600 C. (4)

Heat treatment for 22 hours at 850 C reduced hardness of deposits from a diaminonitro sulfamic acid solution from the range of 400 to 520 kg/sq mm (VHN_{100g}) to the range of 30 to 50 kg/sq mm. (11) The diffusivity of the nickel substrate into platinum was estimated as 6×10^{-10} cm^2/sec at 1000 C.

TABLE 15.16

Hardness and Stress of Platinum Deposits

Plating Bath	pH	Platinum Conc, g/l	Temp, C	Current Density, amp/sq dm	Hardness, kg/sq mm	Stress kg/sq mm	Stress psi	Reference
Alkaline platinate	(a)	11.5	75	0.8	453–495	—	—	23
Aminonitrite	>7	10	60	2	—	7[b]	10,000[b]	33
Aminonitrite	>7	10	60	2	—	25[c]	36,000[c]	33
Chloroplatinic acid	<1	20	60–70	1.0–5.8[d]	155–250			34
Chloroplatinic acid	<1	20	60–70	1.0–5.8[e]	300–350	cracked deposits		34
Fluoboric acid	—				220			35
Fused cyanide	—	0.2–0.3%	600	0.3–3.0	78–93			3, 4, 23
Sulfamate[f]		40	75	.1.6	400–530	30–>75[g]	43,000–>105,000	11

[a] The solution contained 10 g/l sodium hydroxide.

[b] Value for a solution containing 1 to 2 g/l of a 5% solution of NH_4OH and 100 to 280 g/l $NaNO_3$.

[c] Value for a solution containing 50 g/l of a 10% solution of NH_4OH, 280 g/l $NaNO_3$ and 100 g/l NH_4NO_3.

[d] Deposits obtained at >10 μm/hour in solutions containing 225 to 325 g/l hydrochloric acid.

[e] Deposits obtained at <10 μm/hour in solutions containing 150 to 300 g/l hydrochloric acid.

[f] Solution prepared with diaminodinitro platinum and sulfamic acid. The authors reported a platinum metal concentration of 40g /l, but may have intended 40 g/l $Pd(NH_3)_2(NO_2)$.

[g] Supplemental data are given in Table 15.17.

Ductility

Ductile platinum was claimed from aqueous solutions at 70 C containing 20 g/l of platinum as chloroplatinic acid and 220 to 330 g/l hydrochloric acid, and was deposited at 12 to 23 μm/hour as shown in Figure 15.3. (34) Deposits (12 μm) on wires of 0.1 cm diameter were tested by rolling each plated cathode to give a 20 percent reduction in thickness. Brittle and cracked deposits which had hardness values in the range of 300 to 350 kg/sq mm (KHN_{25g}) were obtained at lower deposition rates. Slightly softer deposits (250 to 300 kg/sq mm) cracked during the rolling test. The ductile deposits which did not crack during rolling had hardness values in the range of 155 to 250 kg/sq mm (open circles in Figure 15.3).

Cathode current densities ranged from 1.0 to 5.8 amp/sq dm. (34) However, the rate of deposition was affected by the ratio of the concentrations of chloroplatinic acid and chloroplatinous acid and by the amount of stirring. The best results were obtained without any stirring except that provided by thermal convection currents with the bath at 60 to 70 C. For ease of control, the electrolyte should contain 240 to 260 g/l hydrochloric acid; electrolyte with less than 240 g/l became more acid during use and those with more than 260 g/l lost acid steadily.

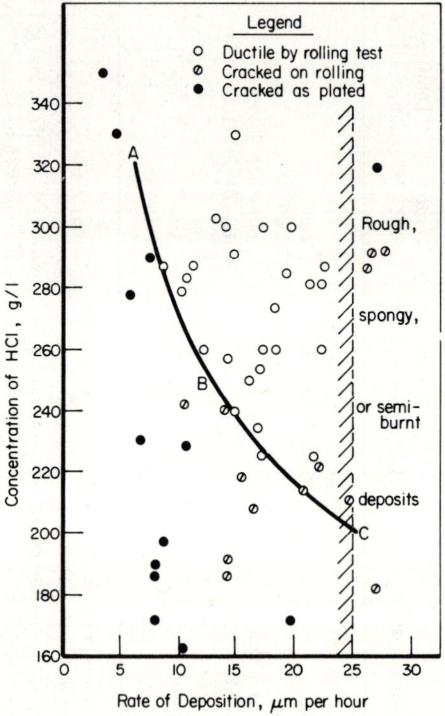

Figure 15.3. Effect of Acidity and Rate of Deposition on Quality of Plates from Chloroplatinic Acid Solution. (34)

TABLE 15.17

Stress in Platinum Deposits on Nickel and Copper[a]

Deposit Thickness, μm	Tensile Strength, kg/sq mm	
	on Nickel	on Copper
0.05	—	30
0.22	220	70
0.60	310(max)	80
1.0	270	95
4.0	170	105(max)
10.0	110	75

[a] Deposits obtained at 1.6 amp/sq dm and 75 C from a solution containing 40 g/l Pt as $Pt(NH_3)_2(NO_2)_2$ and 80 g/l sulfamic acid. (11,12)

Stress

The internal stress in platinum deposits from a diaminodinitro platinum/sulfamic acid bath on nickel was much higher than on copper as shown in Table 15.17. (11,-12) Cracking in 2.5 to 7.6 μm-thick deposits was often observed on nickel substrates, whereas cracking occurred at 17.7 μm thickness or greater on copper substrates. The maximum stress on nickel reached 362 kg/sq mm (500,000 psi) at a thickness of 0.56 to 1.8 μm and the minimum on copper was 30 kg/sq mm (43,000 psi) at 0.05 μm.

Russian studies indicated that platinum of greater than 3 μm thickness on titanium and tantalum was sometimes cracked and exfoliated when deposited in an aminonitrite electrolyte containing a high concentration of ammonium ion. (33) The internal stress was 25 kg/sq mm (36,000 psi) for deposits obtained at 60 C and 2 amp/sq dm from a solution containing 10 g/l platinum (as $H_2PtCl_6 \cdot 6H_2O$), 280 g/l sodium nitrite, 100 g/l ammonium nitrate, and 50 g/l ammonium hydroxide (10 percent solution). The stress was reduced to 7 kg/sq mm (10,000 psi) at 60 C and 2 amp/sq dm for deposits from a bath containing 10 g/l platinum (as $H_2PtCl_6 \cdot 6H_2O$), 100 to 200 g/l sodium nitrite, and 1 to 2 g/l ammonium hydroxide (5 percent solution). Stress increased to 37 kg/sq mm (53,000 psi) when the temperature of the solution was reduced to 30 C or to 70 kg/sq mm (100,000 psi) when the current density was increased to 10 amp/sq dm. Deposits from the nitrite bath containing 1 to 2 g/l ammonium hydroxide were not as porous as those from the solution containing 50 g/l ammonium hydroxide.

Adhesion

Adhesion of platinum deposits to tantalum was improved by vacuum heat treating at 500 C for 2 hours and 900 C for 1 hour. (33) Deposits of 5 to 7 μm thickness on titanium were firmly attached. A bond strength of about 25 kg/sq mm was calculated for platinum-plated titanium heated to 790 C, in comparison with a bond strength of only 0.3 to 1.4 kg/sq mm for specimens heated to 700 C, or lower.

Specific Gravity and Resistivity

The specific gravity of platinum deposits was reported to be 21.37 g/cu cm. (10) The specific resistivity was 9.83 microhm-cm at 0 C.

Platinum Alloys

Platinum-Cobalt Alloy

For magnetic applications, an alloy deposit containing 77 percent platinum and 23 percent cobalt was reported to have a coercivity of 4300 oersteds and an energy product of 9.5×10^6 gauss-oersteds. (36)

Platinum-Iridium Alloys

A hardness range of 166 to 216 kg/sq mm was reported for alloys of 4 to 30 percent iridium, as shown in Table 15.18. (37) Hardness and stress were minimized at an iridium content of about 10 percent, which was controlled by adjusting the temperature of the solution to 30 C. The scratch hardness of a platinum-iridium alloy (iridium content not specified) was reported to be 345 relative to values of 186 to 210 for platinum (Table 15.19). (35)

TABLE 15.18

Hardness and Stress of Platinum-Iridium Deposits from a Bromide Electrolyte[a]

Iridium Content, weight percent	Temperature, C	Hardness, kg/sq mm (DPN_{10g})	Stress, kg/sq mm
4	20	212	Cracked above 1 μm
10	30	166	7
30	70	216	Cracked above 1 μm

[a] Deposits of 5 μm thickness obtained at 1 amp/sq dm from a solution containing bromoplatinic acid and bromoiridic acid with 1.5 g/l of iridium metal ions and 3.5 g/l of platinum ions, pH 1.0 to 2.0, no agitation, and various temperatures to control the iridium content of the deposit. (37)

TABLE 15.19

Hardness of Platinum and Platinum Alloy Deposits from Various Electrolytes (35)

Metal or alloy	Electrolyte	Hardness, kg/sq mm		
		Stretch Hardness	Microhardness	Vickers Hardness Calculated
Platinum	Alkaline solution	210	—	—
Platinum	Fluoboric acid	186	220	120
Platinum-Rhodium	Sulfamic acid	239	—	—
Platinum-Rhodium	Fluoboric acid	374	346	230
Platinum-Iridium	—	345	—	—

Platinum-Rhodium Alloy

Platinum alloy deposits containing 20 to 75 percent rhodium from a cis-dinitrodiamine electrolyte had a hardness of 760 to 780 kg/sq mm. (38) Deposits were obtained at 1 to 12 amp/sq dm, 20 to 95 C, from an electrolyte with a pH of 4 to 11 containing 0.055 g-equiv/l platinum as $[Pt(NH_3)_4]\ (NO_2)_2$, 0.07 to 0.36 g-equiv/l rhodium as $[Rh(NH_3)_6](NO_2)_3$, and 6 to 60 g/l sodium nitrite and ammonia. The alloys were solid solutions of rhodium in platinum, as with their metallurgical counterparts. They were finely crystalline, lustrous in thin layers up to 10 μm, and dull at greater thickness. At 450× magnification, a fine network of cracks were observed at the surface. Their number and depth increased with the rhodium content of the alloy.

Thicker alloy deposits, brilliant at 100 μm thickness, were obtained from hydrochloric acid solutions at higher current efficiency. (39) The hardness of alloys containing 45 percent rhodium was 780 kg/sq mm. Deposits were obtained at 2 to 12 amp/sq dm, 20 to 70 C, from an electrolyte containing 120 g/l hydrochloric acid, 3 g/l platinum and 3 g/l rhodium. The deposits had a network of cracks visible at 450× magnification. The porosity was 7 to 10 pores/sq cm for a 10 μm thick deposit and 1 to 2 pores/sq cm for a 20 μm thick deposit.

Platinum-rhodium alloy deposits from sulfamic acid and fluoboric acid solutions had a hardness of 230 to 346 kg/sq mm (Table 15.19). (35) Table 15.19 shows comparative hardness data for platinum and platinum-alloy deposits.

Rhodium

Hardness

The hardness of rhodium deposits spans a wide range and depends on the electrolyte, deposition conditions, and method of measurement. Values in the range of

TABLE 15.20

Hardness of Rhodium Deposits from Sulfate Baths

Rhodium Concentration, g/l	Sulfuric Acid, ml/l	Temperature, C	Current Density, amp/sq dm	Thickness, μm	Hardness, kg/sq mm	Reference
2 to 4	Up to 600 g/l	35 to 40	1	Up to 127	800 to 820	44
2.1 to 10.5	13 to 130	—	—	25	775 to 820	45
10	70	50	1	12.5	870	27
No data	—	20 to 60	0.3 to 1	—	800	30
No data	—	—	—	—	850 to 890	46
8 to 11	18 to 125	43 to 52	0.8 to 2	>25	690 to 1095	47
20	150	20 to 70	3	Up to 250	819 to 1039	48

800 to 1000 kg/sq mm appear to be representative of deposits from the commonly used sulfate electrolyte. Early references dating to 1935 indicate a Brinell hardness of 600 to 800 kg/sq mm for rhodium deposits compared to 100 kg/sq mm for wrought metal. (7,41) For a 6 μm-thick rhodium coating on brass, the Vickers hardness was reported to be 800 kg/sq mm with a 15-gram load and 150 kg/sq mm with a 300-gram load. (42) However, the latter value is not considered representative because a minimum thickness of 6.3 μm for a 10-gram load was shown to be necessary for measuring rhodium hardness with a diamond indenter penetration less than 1/10 of the coating thickness. (27) Likewise, reported hardness values of 540 to 640 kg/sq mm (VHN_{20g}) are suspected to be too low. (43)

Table 15.20 shows that the hardness values reported by several sources range

TABLE 15.21

Hardness of Rhodium Deposits from Different Solutions

Electrolyte	Hardness, kg/sq mm	
Sulfate[(a)]	898 to 999	(DPN)
Phosphate-Sulfate[(b)]	770 to 930	(DPN)
Acid Chloride[(c)]	900	($Eberhard_{23g}$)
Brush Plating[(d)]	500	(Brinell)

(a) Room temperature deposits from rhodium sulfate electrolyte; deposits obtained at 70 C ranged in hardness from 819 to 1039 kg/sq mm (see Table 15.22). (48)

(b) Rhodium phosphate in free sulfuric acid (see Table 15.22). (48)

(c) H_3RhCl_6. (23)

(d) Proprietary solution at 156 amp/sq cm. (49)

TABLE 15.22

Comparison of Hardness and Stress of Rhodium Deposits from Sulfate and Phosphate-Sulfate Baths[a]

Method of Preparing Rhodium and Rhodium Salt			Hardness, kg/sq mm[d]		Stress, kg/sq mm[e]	
Rhodium[b]	Precipitation of Rhodium Hydroxide[c]	Deposition Temperature, C	Sulfate	Phosphate-Sulfate	Sulfate	Phosphate-Sulfate
A	1	20	898	770	139	139
A	2	20	910	846	108	121
A	3	20	910	846	131	173
A	3	70	1039	—	—	—
A	4	20	900	846	153	159
B	4	20	945–960	870	94	124
B	4	70	871	—	—	—
B	4	20	945	870	122	148
C	4	20	998–999	930	96	166
C	4	70	819	—	—	—
C	4	20	998	930	133	135

[a] Deposits obtained at room temperature and 70 C at 3 amp/sq dm from a solution containing 20 ml/l free sulfuric acid and 20 g/l rhodium (as rhodium sulfate) or 150 ml/l free sulfuric acid and 20 g/l rhodium (as rhodium phosphate). (48)

[b] Rhodium black prepared by: fusion of rhodium powder with lead followed by extraction of lead with 1:4 nitric acid solution (Method A) or low temperature reduction of insoluble rhodium chloride in hydrogen (Method B) or reduction of rhodium chloride solution with alkaline sodium formate solution (Method C).

[c] After extraction of rhodium black with sulfuric acid solution, rhodium hydroxide was precipitated: (1) at room temperature with a slight deficiency of ammonium hydroxide, (2) at room temperature with an excess of ammonium hydroxide, (3) at about 60 C with a slight deficiency of ammonium hydroxide, or (4) at about 60 C with an excess of ammonium hydroxide.

[d] Hardness (DPN) measured on polished sections of deposits of 150 to 250 μm thickness using a Bergsman microhardness tester with a load of 150 g.

[e] Internal stress determined by deflection of thin metal strip plated on one side only to a thickness of 2.5 to 5 μm; calculations of stress were based on linear portion of deflection-time curves.

TABLE 15.23

Effect of Selenic Acid Additions on the Hardness of Rhodium Deposits from a Sulfate Electrolyte[a]

Temperature, C	Current Density, amp/sq dm	Added Sulfuric Acid, ml/l[b]	Hardness, kg/sq mm[c] No Selenic Acid	0.1 g/l Selenic Acid	1 g/l Selenic Acid
25	1	nil	910	830	770
50	1	nil	640	790	790
75	1	nil	560	820	820
25	5	nil	870	880	680
50	5	nil	700	890	790
75	5	nil	650	560	830
25	1	20	—	—	700
25	1	50	—	—	600
25	1	100	—	—	710
50	2	nil	—	—	780
50	2	20	—	—	710
50	2	50	—	—	720
50	2	100	—	—	650
75	2	nil	—	—	820
75	2	20	—	—	660
75	2	50	—	—	720
75	2	100	—	—	820

(a) 10 g/l rhodium. (50)

(b) Refers to acid added to electrolyte in addition to that already present in the concentrated sulfate solutions from which baths were prepared (equivalent to approximately 15 ml/l in the final electrolyte).

(c) Hardness values below about 700 kg/sq mm are suspect because of inadequacy of sample for hardness testing.

from 775 to 1095 kg/sq mm for rhodium deposits from sulfate baths. Table 15.21 compares the hardness values reported for other electrolytes. (48) The hardness values for the sulfate and phosphate-sulfate electrolyte were measured on polished sections and were reproducible. In one investigation, deposits from sulfate electrolytes were consistently harder, as shown in Table 15.22. For the room temperature deposits in Table 15.22, hardness could be correlated with the method of preparing the rhodium salt for the electrolyte, and hardness values probably reflect electrolyte purity.

Table 15.23 shows the effect of selenic acid additions (for reduction of stress) on the hardness of deposits from a sulfate electrolyte. (50) Hardness values below 700 kg/sq mm are questionable because of the sample quality for hardness determination. A 0.1 g/l addition of selenic acid usually increased hardness. A maximum hardness of 930 kg/sq mm was obtained with 0.05 g/l selenic acid at a current density and temperature of 5 amp/sq dm and 50 C, respectively.

The addition of 500 mg/l copper sulfate to the sulfate bath reduced the hardness of rhodium deposits from 875 to 675 kg/sq mm. (51)

Table 15.24 shows the large effect of post-plating surface treatments on the hardness values of rhodium deposits. A true hardness of approximately 1000 kg/sq mm was obtained only on buffed or polished surfaces and on polished cross sections. (52) Thick rhodium deposits from a sulfate bath had a hardness in the range of 800 to 1000 kg/sq mm (VHN_{1000g}) compared to a value of 120 kg/sq mm for annealed metal. (53) Deposits with a thickness of 125 to 250 μm exhibited surface and cross-section hardness from 850 to 900 kg/sq mm. Similar values were obtained for 7.5 to 25 μm-thick deposits using a 30-gram load.

Wear Resistance

Relatively thin rhodium deposited in a proprietary sulfate solution with a hardness of 850 to 890 kg/sq mm showed better resistance than thicker deposits of chromium with a hardness 800 to 1300 kg/sq mm (see Table 16.2, Chapter 16, Rhenium and Rhenium Alloys). (46) Although the rhodium was supposedly crackfree, cracks developed during hardness testing. Even so, the rhodium was less brittle than the chromium; the superior wear resistance of the rhodium was correlated with this lesser brittleness. Rhenium deposits of lower hardness (520 to 560 kg/sq mm) were less brittle than rhodium and showed better wear resistance. No cracking occurred in the rhenium during hardness testing.

Data on the wear resistance and hardness of rhodium compared to palladium and gold was shown previously in Table 15.5. (27)

TABLE 15.24

Effect of Surface Treatment of Rhodium Deposits on Hardness (52)

Condition	Hardness,[a] kg/sq mm
Freshly plated	600 to 700
Scratch brushed	300 to 350
Buffed	950 to 1200
Polished with diamond paste	920 to 1200[b]
Tumbled with aluminum oxide	350 to 500

[a] Vickers hardness with a 25-g load.

[b] Approximately 1000 kg/sq mm when measured on a cross section.

TABLE 15.25

Tensile Stress of Rhodium Deposits from a Phosphate-Sulfate Electrolyte (48)

Rhodium, g/l	Sulfuric Acid, ml/l	Current Density, amp/sq dm	Temperature, C	Internal Stress, kg/sq mm
50	20	3	20	218
10	20	3	20	204
20	20	3	20	209
20	50	3	20	204
20	150	3	20	174
20	100	3	20	178
20	100	1	20	194
20	100	10	20	171
20	100	3	40	157
20	150	3	5	365[a]
20	150	3	20	178[a]
20	150	3	40	147
20	150	3	60	132

[a] Higher than normal stress resulted from overheating the solution; prior to overheating, deposits obtained at 20 C had a stress of 155 kg/sq mm.

Stress

Reports on rhodium deposits show high tensile stress values. The degree of stress depends on the method of preparing the rhodium salts, as shown in Table 15.22. Stress for rhodium deposits from a room temperature phosphate-sulfate bath ranged from 121 to 173 kg/sq mm (172,000 to 246,000 psi). (48) Slightly lower stress values of 94 to 153 kg/sq mm (134,000 to 218,000 psi) were obtained for rhodium deposits from the sulfate bath. Several other investigators reported lower stress ranging from < 50 to about 100 kg/sq m (< 70,000 to about 140,000 psi) for deposits from sulfate solutions.

Stress is reduced by an increase in sulfuric acid concentration, the current density and/or the temperature, as shown in Table 15.25 for deposits from a phosphate-sulfate bath. For a sulfate bath, the stress was approximately 200 kg/sq mm with 20 ml/l sulfuric acid and about 175 kg/sq mm with 100 ml/l sulfuric acid at a current density of 3 amp/sq dm. (48) Stress was reduced to 132 kg/sq mm (185,000 psi) by increasing the temperature from 20 to 60 C of a solution containg 20 g/l rhodium and 150 g/l sulfuric acid.

In the case of solutions containing 0.3 to 1.05 g/l rhodium and 40 to 50 g/l sulfuric acid, the effect of current density (from 0.5 to 1.5 amp/sq dm) and temperature (from 20 to 60 C) on the internal stress of the deposit was slight. (54)

Table 15.26 shows the wide variations in stress for deposits produced from sulfate baths prepared with various lots of rhodium concentrate, as a function of deposit thickness. (55) In general, stress decreased with increasing thickness but not always as shown for Lot 4, Table 15.26. Spectrographic analysis of various lots did not show any correlation of trace metallic impurities with the stress data. However, deposits of 5 to 7.5 μm with a relatively low stress of 14 to 28 kg/sq mm were rough and dark; those with a stress of 28 to 46 kg/sq mm were somewhat rough and dark; those with a stress of 46 to 70 kg/sq mm were smooth and white or bright. Electrolyzing new solutions improved the appearance of deposits but increased stress. For example, a bath producing dark deposits of 28 kg/sq mm stress, produced white deposits of 42 to 46 kg/sq mm stress after operation for 3 amp-hr/l.

Organic contaminants can increase the stress of rhodium deposits, as shown in Tables 15.27 and 15.28. (56) For example, stress cracking has been observed to originate at the boundary of the rhodium and a masking paint. Studies of contamination by the paint components showed that the resins and plasticisers had no effect on the stress. However, commercial-grade solvent naphtha containing sulfur compounds (essentially thiophen) can increase stress appreciably, particularly at high solution operating temperature, as shown in Table 15.27. A masking paint formulated with a low sulfur content solvent resulted in a small increase in stress. To avoid influencing stress measurements, the masking paint used to mask one side of strips or spirals for stress measurement should not contaminate the electrolyte.

TABLE 15.26

Tensile Stress of Rhodium Deposits from a Sulfate Bath for Various Lots of Rhodium-Concentrate[(a)]

Thickness, μm	Internal Stress, kg/sq mm[(b)]				
	Lot 4	Lot 5	Lot 8	Lot 11	Lot 12
1.25	81	91	34	—	—
2.5	81.5	83	24	45.5	71
3.75	82	78	20	45	63
5.0	82.5	74	15	44.5	56
6.75	83	71	13	44	49
7.5	—	—	—	43.5	43
8.75	—	—	—	43	38
10	—	—	—	42.5	33
11.75	—	—	—	42	29

(a) Stress measurements made with a Brenner-Senderoff Spiral Contractometer and copper helices for deposits at 1.5 amp/sq dm and 49 C ± 1 C in a solution containing 10 g/l rhodium and 25 ml/l sulfuric acid. (55)

(b) Differences in internal stress for various lots of rhodium concentrate used to prepare plating bath is probably related to impurities, but no correlation was obtained of stress level with spectrographic analytical data.

TABLE 15.27

Effect of Contamination from Masking Paint on the Stress of Rhodium Deposits from Sulfate Baths[a]

	Stress, kg/sq mm[b]		
Temperature, C	Uncontaminated Solution	Masking Paint A (low sulfur content)[c]	Masking Paint B[c][d]
30	70[e]	72[e]	91[e]
40	87	89	114
50	80	82	92
60	69	71	91
70	59	61	100

[a] Deposits were obtained at 0.5 amp/sq dm for 30 minutes from a solution containing 8 g/l rhodium and 30 g/l sulfuric acid. (56)

[b] Stress calculated from deflection of beryllium-copper strip (3.5 × 0.375 × 0.005 inch) plated on one side in a special fixture.

[c] A copper panel (1 × 2 inches) was coated with paint and air dried for 24 hours prior to immersion in 200 ml of plating solution at temperature under consideration, and then removed.

[d] The sulfur contamination from commercial naphtha solvent in the paint was identified as thiophen (see Table 15.28).

[e] Deposits at 30 C were cracked; all other deposits were crackfree.

When a rhodium sulfate electrolyte is operated at a pH of 5.5, the deposit consists primarily of yellow rhodium hydroxide. (50) The addition of sulfuric acid results in a deposit that is primarily metallic but severely cracked. Some nonmetallic matter is occluded in the typical deposit since it loses 2 to 3 percent of its weight on heating to 1000 C in hydrogen. Increasing the concentration of sulfuric acid to about 100 ml/l gives a typically smooth gray deposit with a progressive reduction in stress to a constant value. Stress was not further reduced by increasing the amount of acid addition.

A normal value for the internal stress of rhodium deposits from a sulfate electrolyte containing 5 g/l rhodium and 20 ml/l sulfuric acid at 50 C and 2 amp/sq dm is reported to be 56 kg/sq mm (80,000 psi). (50) Deposits are uncracked up to a thickness of 5 μm, which is typical for many important industrial applications. Cracking may occur in thicker deposits, but with proper plating techniques to secure adhesion, flaking of deposits does not occur. One author (50) claims that there is no evidence that fine cracks in rhodium on suitable undercoats has any deleterious effect on performance primarily as electrical contact surfaces. Such a claim ought to be limited to exclude cracks that extend completely through the deposit, however.

TABLE 15.28

Effect of Sulfur Compound Contamination of Sulfate Bath on Stress of Rhodium Deposits[a]

Contaminant Concentration			
Sulfur Content of Solvent, g/l[b]	Thiophen, molarity[c]	Stress, kg/sq mm	Type of Deposit According to Electrographic Printing
0.0005	—	78.9	Crackfree
0.0009	—	80.2	Crackfree
0.0027	—	84.5	Crackfree
0.0175	—	94.3	Crackfree
0.3650	—	52.7	Cracked
—	0.0001	79.5	Crackfree
—	0.0003	96.5	Crackfree
—	0.0010	43.6	Cracked

[a] Deposits obtained at 0.5 amp/sq dm and 40 C in a solution containing 8 g/l rhodium and 30 g/l sulfuric acid. (56)

[b] Commercial grade solvent naphtha used in masking paint; 1 gram added to 200 ml of plating solution at 40 C. The solution was stirred for 30 minutes and filtered before the stress measurement.

[c] Thiophen (sulfur compound identified in naphtha solvent) added to bath as in (b).

TABLE 15.29

Effect of Substrate Material on Stress and Porosity of Rhodium Deposits from a Sulfate Bath (48,53)

Substrate			Rhodium Deposit		
				Nonporous	Porous
Metal	Hardness, kg/sq mm	Stress, kg/sq mm	Stress, kg/sq mm	Thickness, μm	Thickness, μm[a]
Silver	50	nil	105.5	2.5 to 7.5	12.5 to 25
Gold	50	nil	95.5	2.5 to 12.5	18 to 25
Palladium	160	45.0	101.2	2.5 to 25	—
Nickel	380	17.4	102.5	—	12.5

[a] Porosity was associated with cracks.

TABLE 15.30

Effect of Aluminum and Magnesium Additions on the Stress of Rhodium Deposits[a]

Rhodium Concentration, g/l	Stress, kg/sq mm		
	No Addition	6 g/l Magnesium	6 g/l Aluminum
4	55	59.5	65.1
5	64.5	68.1	81.0

[a] Crackfree deposits of 12.5 μm thickness produced at 1 amp/sq dm in a solution at 50 C with 50 ml/l sulfuric acid. (50)

The substrate material has little effect on the stress of rhodium deposits from a sulfate bath, as shown in Table 15.29. However, the number of cracks in the deposit was increased when applied on a stressed basis metal. (48,53) X-ray and electron diffraction studies of residual stress in thin deposits of rhodium, rhenium, ruthenium, chromium, and gold on polycrystalline and single crystal substrates indicated that coating stress could not be predicted from simple lattice mismatch of coating and substrate. (57)

Additions of 10 to 100 mg/l copper darken the deposit; mirror bright deposits are obtained with copper additions of over 500 mg/l; with 800 to 1000 mg/l copper the deposits were cracked and crazed, which indicated increasing stress with increasing copper [see Footnote (a), Table 15.26, for plating conditions]. (55) Additions of lead, cobalt, nickel, silver, zinc, tin, and mercury were either ineffective or detrimental to rhodium plating. Phenol sulfonic acid was an effective brightener with best results at 1 to 1.5 mg/g rhodium. For 7 μm-thick deposits, stress was increased from 49 to 56 kg/sq mm after an addition of 150 mg/l phenol sulfonic acid.

For crackfree deposits from sulfate baths, the temperature should be 50 C and the rhodium and sulfuric acid concentrations should be at least 5 g/l and 15-to-30 g/l, respectively. (51) The addition of 500 mg/l copper sulfate increased the internal stress from about 40 to 70 kg/sq mm. At thicknesses of 1 to 3 μm, the deposits showed an internal stress of only 26 kg/sq mm (37,000 psi), but stress increased with increasing thickness of the deposit. Saccharin can be used in place of magnesium sulfate to reduce stress. The crackfree deposits showed a high resistance to corrosion in an atmosphere containing hydrogen sulfide.

Cracks in rhodium deposits up to 12.5 μm thick were eliminated by the addition of 20 g/l aluminum as aluminum sulfate to a solution containing 10 g/l rhodium and 50 ml/l sulfuric acid. (50) However, aluminum additions above 5 g/l tended to produce rough deposits of unacceptable finish. Table 15.30 shows the increase in stress resulting from aluminum and magnesium additions to solutions with a lower rhodium content (4 or 5 g/l). Crackfree deposits were obtained up to a thickness of 12.5 μm with additions of 6 g/l aluminum or magnesium to a bath containing 6 g/l rhodium. Foils isolated by dissolving away the brass metal showed no tendency to curl.

In the case of solutions containing only 0.35 to 1.05 g/l rhodium and 40 to 50 g/l sulfuric acid, the stress increased with temperature when magnesium sulfate or aluminum sulfate was present in the solution. (54) The addition of 50 g/l magnesium sulfate or 50 to 100 g/l aluminum sulfate lowered the stress. A larger stress reduction was obtained by adding 1 g/l selenic acid, however.

In a subsequent study by the same authors, it is reported that internal stress can be reduced by decreasing the concentration of sulfuric acid to 16 g/l with a rhodium concentration of 2.7 g/l and heating the deposits in a vacuum at 400 to 500 C for 20 to 30 minutes. (58) During heating, the hydrogen content decreased from 0.06 to 0.015 percent.

Deposits of rhodium of 12.5 to 25.4 μm thickness that are free of macrocracks can be obtained with additions of 0.1 to 1.0 g/l selenic acid to sulfate solutions containing a minimum of 10 g/l rhodium and 15 to 200 ml/l sulfuric acid at 50 to 75 C and 1 to 2 amp/sq dm. (59,50) The stress with a selenic acid addition is slightly higher than that without the additive for deposits up to about 1 μm in thickness when stress relief attributed to microcracking becomes apparent; residual stress then decreases steadily to very low values as deposit thickness increases. (60) In contrast with this trend, rhodium deposited in solutions containing aluminum or magnesium salts increases in stress and develops macrocracks with increasing thickness. Table 15.31 shows that a low stress was obtained for 2.5 to 5 μm-thick deposits with selenic acid additions. (50) The deposits obtained with selenic acid contain microcracks that can be developed by etching. Cracks formed initially tend to heal with increasing thickness until a deposit is produced which may be regarded as crackfree from a corrosion protection viewpoint.

When the standard beryllium-copper strip for measuring stress was replaced by silver, the absolute stress of thin (5 μm) deposits was reduced from 106 to 57 kg/sq

TABLE 15.31

Relative Stress of Rhodium Deposits from a Sulfate Electrolyte with Various Addition Agents[a]

Rhodium, g/l	Sulfuric Acid, ml/l	Selenic Acid, ml/l	Magnesium, g/l	Copper, g/l	Deflection of Cathode, mm			
					1.3 μm	2.5 μm	3.8 μm	5.0 μm
10	nil	nil	nil	nil	2.79	6.65	10.5	14.0
10	nil	1.0	nil	nil	4.23	5.85	6.29	5.92
10	50	1.0	nil	nil	3.63	2.54	2.01	1.69
10	50	1.0	nil	1.0	2.78	1.69	0.84	0.29
5	50	nil	6	nil	5.55	11.50	15.2	(b)
5	50	nil	6	1.0	7.75	13.65	15.95	16.70

[a] Deposits at 1 amp/sq dm and 50 C. (50)

[b] An absolute stress of 81 kg/sq mm was obtained for deposits of up to 12.5 μm (see Table 15.30).

TABLE 15.32

Apparent Correlation of Stress in Deposited Metal with Melting Point (9)

Metal	Melting Point, C	Stress in Electrodeposit,[a] kg/sq mm
Cadmium	320	−0.35 to −2.1
Zinc	420	−0.7 to −1.4
Aluminum	760	0
Silver	960	±1.4
Gold	1062	+1.05
Copper	1083	+1.4
Nickel	1450	+14
Cobalt	1495	+14
Iron	1535	+28
Palladium	1552	+42
Chromium	1875	+42
Rhodium	1966	+70

[a] Approximate literature values for deposits of 2.5 to 25 μm.

mm. (50) The lattice constant of rhodium is 4.07 which is closer to that for silver (4.0774) than for copper (3.6080).

Whereas a 5 μm-thick rhodium deposit with a stress of 70 kg/sq mm was fragmented when the silver basis metal was dissolved in nitric acid, the rhodium deposit remained intact when deposited from a solution containing a proprietary additive to reduce stress to 14 to 21 kg/sq mm. (9)

The high stress in rhodium deposits has been correlated with the high melting point of rhodium, as shown by the data in Table 15.32.

High stress and cracking also have been correlated with the cathode efficiency of depositing rhodium in phosphate, sulfate, fluoborate, and sulfamate baths. (51)

Resistivity

An electrical resistivity of 8.5 microhm-cm was reported for rhodium deposited at 0.3 to 1 amp/sq dm from a sulfate bath at 20 to 60 C; (30) the contact resistance (50 grams) was 1.7 milliohms. The resistivity of the electrodeposit was about twice the resistivity of wrought rhodium, which is 4.33 microhm-cm at 0 C. Another source says that the resistivity of plated rhodium is about the same as that for cast metal. (61)

In the section on palladium, data on the contact resistance of rhodium, palladium, and gold were compared for both electrodeposits and wrought metals in Table 15.10; the effect of sulfur dioxide exposure on the contact resistance of the deposits was compared in Table 15.11. (27,62) The effect of high temperature on contact resistance of rhodium deposits was compared to palladium and gold deposits in Figure 15.1 and the effects of extended exposure in a laboratory atmosphere were compared in Figure 15.2.

The contact resistance of commercial rhodium-, silver-, and copper-plated and stored contacts varied over wide limits from 5 to 490 milliohms. (52) Freshly plated contacts or freshly abraded contracts showed resistances of 0.5 to 10 milliohms when measurements were made with 40- to 50-g loads between two gold-plated electrodes.

Ruthenium

Hardness

Ruthenium deposits have the highest hardness of all electro-deposited precious metals. Table 15.33 shows a correlation of atomic number with the maximum hardness values of electrodeposited and wrought, annealed metals. The maximum hardness of about 1300 kg/sq mm for ruthenium in Table 15.33 (5) refers to deposits from nitrosyl sulfamate baths with aluminum sulfate and magnesium sulfate additions as shown in Table 15.34. Deposits from the nitrosyl sulfamate bath without additives have a hardness of about 800 kg/sq mm measured on cross sections. (63) Baths prepared from the N-bridged complex ruthenium salt, introduced in the past few years, have a higher hardness of 950 to 1150 kg/sq mm. (65)

The lowest hardness for ruthenium from an aqueous solution (a phosphate electrolyte) was 650 kg/sq mm. (63) Thick deposits obtained from fused cyanide ranged in hardness from 610 to 937 kg/sq mm. (3,23)

Table 15.35 shows the effect of heat treatment for 30 minutes at temperatures from 100 to 800 C on the hardness of deposits from electrolytes based on the N-

TABLE 15.33

Maximum Hardness of Precious Metal Electrodeposits Related to Atomic Number (5)

Atomic Number	Metal	Hardness, kg/sq mm	
		Electrodeposit	Wrought Annealed
44	Ruthenium	1300	200–350
45	Rhodium	1000	120–140
46	Palladium	400	37–44
47	Silver	60	25–30
76	Osmium	—	400–500
77	Iridium	800	200–240
78	Platinum	500	37–42
79	Gold	60	25–27

TABLE 15.34

Hardness and Stress of Ruthenium Deposits from Various Electrolytes

Electrolyte	Hardness, kg/sq mm	Stress, kg/sq mm	Reference
Fused cyanide	785–937	—	23
Fused cyanide	611–892[a]	—	23
Fused cyanide[b]	610–935	—	3
N-bridge complex[c]	750–850	70–85	64
N-bridge complex[d]	750–850[e]	60–85	65
N-bridge complex[d]	900–1100[f]	60–85	65
Nitrosyl chloride	950–1150	—	23
Nitrosyl sulfamate[g]	800[h]	24–57	63
Nitrosyl sulfamate (plus Al)[i]	840–1280	28	63
Nitrosyl sulfamate (plus Mg)[j]	800–1370	—	63
Nitrosyl sulfate[k]	950–1150	—	63
Phosphate[l]	650	42–56	63
Sulfamate[m]	750–800	42	63
Sulfate[n]	750–800	42–56	63
Tetranitro-nitrosyl complex[o]	—	36	63

[a] Lower maximum hardness than for other fused cyanide deposits may have been the result of using periodic reverse current.

[b] Deposits obtained at 0.1 to 1.8 amp/sq dm in a fused sodium and potassium cyanide bath at 560 C containing 0.5 percent ruthenium for obtaining thick deposits of 39 μm (0.0015 inch).

[c] Electrolyte based on N-bridged ruthenium complex, $K_3[Ru_2NCl_8(H_2O)_2]$, prepared by reaction of $K_2Ru(NO)Cl_6$ with stannous chloride in hydrochloric acid to produce deep red crystalline product. The electrolyte was prepared by dissolving the complex salt in water to provide 10 g/l Ru, adjusting the pH to 1.3 and adding 10 g/l ammonium formate. Deposits were produced at 0.75 amp/sq dm and 70 C.

[d] Electrolyte based on N-bridged ruthenium complex, $(NH_4)_3[Ru_2NCl_8(H_2O)_2]$, prepared by refluxing ruthenium chloride in hydrochloric acid with sulfamic acid to produce a dark red crystalline product. The electrolyte was prepared by dissolving the complex salt in water to provide 12 g/l Ru, adding 12 g/l sulfamic acid (and 1 to 2 grams per gram of ruthenium deposited) and adjusting the pH to 1.0–1.5 with ammonium hydroxide. Deposits were produced at 1 amp/sq dm and 70 C.

[e] Microhardness (VHN_{20g}) measured on surfaces.

[f] Microhardness (VHN_{10g}) measured on sections.

[g] Deposits obtained at 2 to 8 amp/sq dm and 50 C from a solution containing 10 to 50 g/l sulfamic acid and 5 g/l ruthenium prepared by dissolving ruthenium nitrosyl trihydroxide in sulfamic acid.

[h] Hardness (DPN) measured on section; lower values obtained on surface of deposit which was rough at suitable thickness for measurement.

[i] Hardness increased with addition of aluminum (as sulfate) with a maximum at 6 g/l; at higher concentrations up to 25 g/l the hardness fell below the reference value of 840 kg/sq mm, and occluded material was as much as 12 percent by weight of deposit.

[j] Hardness was increased with addition of magnesium (as sulfate) with a maximum at 10 g/l.

Footnotes to TABLE 15.34 (Continued)

(k) Refers to a thick, gray, nodular and cracked deposit (39 μm or 0.0015 inch) produced during the electrolytic recovery of radioactive ruthenium from a dilute solution (5×10^{-4} mole/l).

(l) Same as for Footnote (k) except that phosphoric acid was used in place of sulfuric acid.

(m) Crackfree deposits up to a thickness of 5 μm were obtained at 2 to 8 amp/sq dm and 50 to 70 C with a ruthenium concentration of 5 g/l using nitrosyl-free ruthenium chloride to prepare ruthenium trihydroxide for reaction with sulfamic acid.

(n) Deposits obtained at 8 amp/sq dm and 70 C from solution containing 40 g/l sulfuric acid and 5 g/l ruthenium prepared by reacting nitrosyl-free ruthenium chloride with sulfuric acid.

(o) Deposits obtained at 8 amp/sq dm and 70 C in an electrolyte containing 5 g/l ruthenium prepared by the reaction of sodium tetranitro-nitrosyl ruthenate with sulfamic acid at 100 C.

bridged complex. The decrease in hardness with increased temperature is similar to that obtained for hard and highly stressed deposits of other metals. (65)

Wear Resistance

The wear resistance of a 0.6 μm ruthenium deposit from a nitrosyl sulfamate bath was better than that of a 0.6 μm rhodium deposit as shown in Table 15.36. (60)

TABLE 15.35

Effect of Heat Treatment on Hardness of Ruthenium Deposits from a Solution Based on N-Bridged Complex[a]

Heat Treatment Temperature, C[b]	Hardness, kg/sq mm	
	Surface (20-g load)	Section (10-g load)
None	750–850	900–1100
100	840	1080
200	840	1070
300	830	1020
400	760	920
500	530	760
600	500	700
700	500	680
800	500	690

(a) Deposits of 5 μm thickness obtained at 70 C and 1 amp/sq dm from a solution at pH 1 to 1.5 containing 12 g/l ruthenium as the N-bridged complex, $(NH_4)_3[Ru_2NCl_8(H_2O)_2]$. See Footnote (d), Table 15.34. (65)

(b) 30-minute heat treatment.

TABLE 15.36

Wear of Ruthenium and Rhodium Deposits[(a)]

Coating Thickness, μm	Maximum Depth of Wear, μm		
	Ruthenium[(b)]	Rhodium[(c)]	Rhodium[(d)]
0.63	0.3	0.63	—
1.25	0.15	0.15	1.0
2.5	0.15	0.15	0.3
5.0	—	—	0.15

(a) Wear of plated slip rings (brass cylinders) rotating under spring loaded probes of 60% palladium–40% silver as brushes at a speed of 3 to 4 inches/second with a load of 10 grams for 100 hours. (60)

(b) Ruthenium deposited from nitrosyl sulfamate electrolyte containing 5 g/l ruthenium at 70 C and 4 amp/sq dm to give crackfree coatings up to 2.5 μm.

(c) Rhodium deposited from sulfate electrolyte.

(d) Low stress rhodium deposited from sulfate electrolyte with a selenic acid addition.

Thicker deposits of ruthenium and rhodium were equal in wear resistance. The hardness and general characteristics of the ruthenium coatings were similar to those of rhodium from a sulfate bath.

Stress

The tensile stress of ruthenium deposits ranges from 24 to 85 kg/sq mm (34,000 to 121,000 psi) depending on the electrolyte and deposition conditions, as shown in Table 15.34. The sulfamate electrolyte [Footnote (m), Table 15.34] provided crackfree deposits up to a thickness of 5 μm. The stress, which was on the order of 42 kg/sq mm (60,000 psi), was slightly higher than that for comparable deposits from the nitrosyl sulfamate bath. (63)

Deposits from the nitrosyl sulfamate bath prepared from nitrosyl trihydroxide [Footnote (g), Table 15.34] were bright to a thickness of 5 μm and crackfree to a thickness of 2.5 μm with slight cracking at 5 μm. Cracking increased beyond 5 μm thickness and the deposits became frosted in appearance. (63) The minimum stress at 50 C was obtained at 4 amp/sq dm as shown in Table 15.37. For deposits at 70 C and 8 amp/sq dm with ruthenium maintained at the nominal level of 5 g/l, the internal stress was consistently in the order of 28 to 42 kg/sq mm with a current efficiency of 5 to 7 percent. There was considerable variation of stress from 24 to 57 kg/sq mm, among nine different batches of electrolyte. No correlation between stress and current efficiency from 1.3 to 10.8 percent, or free acid concentration from 0.1 to 0.5 normal was established.

TABLE 15.37

Variation of Internal Stress with Current Density for Ruthenium Deposits from a Nitrosyl Sulfamate Electrolyte[(a)]

Current Density, amp/sq dm	Tensile Stress kg/sq mm
2	46
4	34
6	42
8	55

[(a)] Deposits at 50 C from a solution containing 5 g/l ruthenium prepared from ruthenium nitrosyl trihydroxide dissolved in 5 to 50 g/l sulfamic acid. (63)

Deposits from an electrolyte prepared from the sodium tetranitro-nitrosyl ruthenate complex and sulfamic acid provided consistent results for three separate batches of solution. (63) A stress of 37 kg/sq mm, and a current efficiency of 7 percent were noted at 70 C and 8 amp/sq dm. However, the electrolyte was less stable in operation than the nitrosyl sulfamate electrolyte.

With the addition of aluminum (as sulfate) above 2.5 g/l to the nitrosyl sulfamate electrolyte, there was a significant reduction in cracking with some loss of brightness. (63) Although current efficiency was improved by adding aluminum, little effect was seen on stress, which was of the order of 28 kg/sq mm (40,000 psi) with aluminum contents from 2.5 to 25 g/l. The effect of additions of magnesium

TABLE 15.38

Effect of Current Density on Stress of Ruthenium Deposits from a Solution Based on N-Bridged Complex[(a)]

Current Density, amp/sq dm	Stress, kg/sq mm	
	50 C	70 C
0.5	61	66
1	64	83[(b)]
2	69	75
4	76	70
6	—	82

[(a)] See Footnote (d) Table 15.34 for electrolyte preparation. (65)

[(b)] Stress was reduced to 80 kg/sq mm with a decrease in the concentration of ruthenium from 12 g/l to 6 g/l.

was less well defined. All deposits showed some cracking with a minimum incidence at 15 g/l magnesium (as sulfate).

Table 15.38 shows the stress of deposits from electrolytes based on the N-bridged complex as a function of current density and temperature. (65) The range of 60 to 85 kg/sq mm is higher than the stress of 30 to 45 kg/sq mm reported for deposits from the nitrosyl sulfamate electrolyte. (63) Cracking was seen in the deposits from the N-bridged complex at a thickness of 1.5 μm in comparison with a transition thickness of 2.5 μm for cracking in deposits from the nitrosyl sulfamate bath.

A tendency for edge cracking to develop some days after the plating of coatings that were initially crackfree has been noted. (60) A post-treatment in warm 50 percent ammonia solution has been recommended as an effective method of stabilizing stressed deposits.

The use of periodic reverse current with auxiliary electrodes is reported to reduce the stress of deposits from a nitrosyl sulfamate bath, but no quantitative data were given. (66)

References

(1) Tyrell, C. J., "The Electrodeposition of Iridium," *Trans. Inst. Metal Finish., 43* (4), 161–168 (1965).

(2) Conn, G. A., "Iridium Plating," *Plating, 52* (12), 1258–1261 (1965).

(3) Reid, F. H., "Electrodeposition of the Platinum-Group Metals," *Metallurgical Reviews, 8* (30), 167–211 (1963).

(4) Andrews, R. L., Kenahan, C., B., and Schlain, D., "Electrodeposition of Iridium from Fused Sodium Cyanide and Aqueous Electrolytes," *Reports of Investigation No. 7023*, United States Bureau of Mines (1967).

(5) Vines, R. F., "Precious Metal Electrodeposits for Electrical Contact Service," *Plating, 54* (8), 923–925 (1967).

(6) Greenspan, L., "Electrodeposition of Osmium," *Englehard Industries Technical Bulletin, 10* (2), 48–49 (1969).

(7) Such, T. E., "The Physical Properties of Electrodeposited Metals," *Metallurgia, 56* (334), 61–66 (1957).

(8) "Improvements in Plating Processes: The Dalic Process," *Product Finishing (London), 6* (9), 77–79 (1953).

(9) Kushner, J. B., "Stress in Electroplated Metals," *Metal Progress, 81* (2), 88–93 (1962).

(10) Andreeva, G. P., Fedot'ev, N. P., and Vyacheslavov, P. M., "Electrodeposition and Properties of Silver-Palladium Alloys," *Protective Metallic and Oxide Coatings, Metal Corrosion and Electrochemistry*, edited by N. P. Fedot'ev, Akademiya Nauk SSSR. Translated by the Israel Program for Scientific Translations, Jerusalem, 226–231 (1968).

(11) Cramer, S. D., and Schlain, D., "Electrodeposition of Palladium and Platinum from Aqueous Electrolytes," *Plating, 56* (5), 516–522 (1969).

(12) Cramer, S., Kenahan, C., Andrews, R., and Schlain, D., "Electrodeposition of Thick Coatings of Platinum and Palladium on Refractory Metals from Aqueous Electrolytes," *Report of Investigations No. 7016,* United States Bureau of Mines (1967).

(13) Vagramyan, A. T., and Solov'eva, Z. A., *Technology of Electro-Deposition*. English Translation, Robert Draper Ltd., Teddington (1961), 398 pp.

(14) Ostroumov, V. V., "Mechanical Stresses in Palladium Electrodeposits," *Zh. Fiz. Khimii, 31* (8), 1812–1819 (1957).

(15) Ostroumov, V. V., "Influence of Certain Organic Substances on the Palladium-Plating Process," *Journal Applied Chemistry (USSR), 31* (3), 389–394 (1958); *Zh. Prikl. Khimii, 31* (3), 402–408 (1958).

(16) Vyacheslavov, P. M., Fedot'ev, N. P., Burkat, G. K., and Semenova, N. L., "Electrodeposition of Palladium-Cobalt Alloys," *Zashchita Metallov, 5* (3), 349–354 (1969).

(17) Wise, E. M., "Palladium Plating," *Palladium, Recovery, Properties and Uses*, Chapter 6, Academic Press (1968).

(18) Atkinson, R. H., and Roper, A. R., *Journal Electrodepositors Soc., 8* (10), (1933).

(19) Reid, F. H., "Palladium Plating—Processes and Applications in the United Kingdom," *Plating, 52* (6), 531–539 (1965).

(20) Angus, H. C., "Physical Properties of Palladium Electrodeposits," *Trans. Inst. Metal Finish., 44* (2), 41–49 (1966).

(21) Stevens, J. M., "Electrodeposition of Ductile Palladium," *Trans. Inst. Metal Finish., 46* (1), 26–31 (1968).

(22) Fatzer, G. D., "Electrodeposited Palladium as a Low-Energy Circuit Contact Coating," *Plating, 50* (11), 1000–1005 (1963).

(23) Atkinson, R. H., "Platinum Metals," *Modern Electroplating*, 2nd Ed., Chapter 13 edited by F. A., Lowenheim, John Wiley & Sons (1963).

(24) Vinogradov, S. N., and Kudryavstev, N. T., "Electrodeposition of Palladium-Nickel Alloy," *Protection of Metals, 4* (2), 128–133 (1968); *Zashchita Metallov, 4* (2), 145–151 (1968).

(25) Philpot, J. E., "Palladium Plating," *Electroplating and Metal Finishing, 15* (1), 20 (1962).

(26) Heathcote, L. A., "Bright Palladium Plating," *Platinum Metals Review, 9* (3), 80–82 (1965).

(27) Angus, H. C., "Properties and Behavior of Precious-Metal Electrodeposits for Electrical Contract," *Trans. Inst. Metal Finish., 39* (1), 20–28 (1962).

(28) Rhoda, R. N., "Electroless Palladium Plating," *Trans. Inst. Metal Finish., 36* (3), 82–85 (1959).

(29) Pearlstein, F., and Weightman, R. F., "Electroless Palladium Deposition," *Plating, 56* (10), 1158–1161 (1969).

(30) Loebich, O., and Zilske, W., "Comparative Evaluation of Electroplated Precious Metal Coating in the Electrical Industry," *Metalloberflaeche, 22* (2), 34–38 (1968).

(31) Vagramyan, A. T., and Petrova, Yu. S., *The Mechanical Properties of Electrolytic Deposits*, USSR Academy of Sciences Press for Institute of Physical Chemistry, Moscow (1960). English translation by Consultants Bureau (1962), 108 pp.

(32) Kabushiki Kaisha, S. S., and Nisshin Kasei, K. K., "Palladium-Nickel Alloy Plating Bath," British Patent 1,143,178 (February 1969).

(33) Kudryavtseva, L. V., Kharlamova, K. N., and Moskhov, M. I., "Platinum Plating of Titanium and Tantalum Electrodes in Ammonia-Nitrite Electrolytes," *Zashchita Metallov, 1* (5), 500–504 (1965).

(34) Atkinson, R. H., "Electrodeposition of Platinum from Chloroplatinic Acid," *Trans. Inst. Metal Finish., 36* (1), 7–16 (1958).

(35) Hänsel, G., "The Electrodeposition of Platinum and Platinum Alloys with Respect to Increase in Hardness," *Metalloberflaeche, 21* (8), 238–242 (1967).

(36) Morral, F. R., "Survey of Recent Developments in Electroplating Cobalt Alloys," *Plating, 52* (9), 879–888 (1965).

(37) Tyrrell, C. J., "The Electrodeposition of Platinum-Iridium Alloys," *Trans. Inst. Metal Finish., 45*, 53–57 (1967).

(38) Koslova, G. B., and Bakhvalov, G. T., "Deposition of Platinum-Rhodium Alloy from a Cis-dinitrodiamine Electrolyte," *Zashchita Metallov, 1* (4), 370–373 (1965).

(39) Koslova, G. V., and Lainer, V. I., "Electrodeposition of Platinum-Rhodium Alloy from Hydrochloric Acid Electrolyte," *Zashchita Metallov, 1* (5), 511–514 (1965).

(40) *Metals Handbook*, 8th Ed., *Vol. 1, Properties and Selection of Metals*, American Society for Metals, Metals Park, Ohio (1961).

(41) Buss, G., "Properties and Corrosion Protection of Electrodeposited Coatings and Anodized Aluminum," *Metall, 17* (6), 567–571 (1963).

(42) Keil, A., and Merkle, E., "Hardness Measurements on Electrodeposited Coatings. II," *Metalloberflaeche, 8A* (9), 129–131 (1954).

(43) Hosdowich, J. M., "Electroplated Coatings," *Surface Protection Against Wear and Corrosion*, American Society for Metals, Cleveland, 41–55 (1954).

(44) Laister, E. H., and Benham, R. R., "Rhodium Plating and Its Modern Applications," *Trans. Inst. Metal Finish., 29*, 1–25 (1953).

(45) Weisberg, A. M., "Heavy Rhodium Electroplates Now Possible," *Materials & Methods, 37* (3), 85–87 (1953).

(46) Georges, J. M., and Rabinowicz, E., "Effect of Film Thickness on the Wear of Hard Electrodeposits," *Wear, 14* (3), 171–180 (1969).

(47) Weisner, H. J., "Some Experiences in Heavy Rhodium Plating," *Proc. Am. Electroplaters' Soc., 39*, 79–89 (1952).

(48) Reid, F. H., "Some Experimental and Practical Aspects of Heavy Rhodium Plating," *Trans. Inst. Metal Finish., 33*, 105–140 (1956).

(49) Machu, W., "Brush Plating for Localized Metal Deposition (Dalic Process)," *Metallwaren-Industrieand Galvanotechnik, 49* (11), 467–481 (1958). Cf. Hughes, H. D., *Trans. Inst. Metal Finish., 33*, 424–439 (1956).

(50) Reid, F. H., "Some Experimental Observations on the Effect of Addition Agents on Stress and Cracking in Rhodium Deposits," *Trans. Inst. Metal Finish., 36* (3), 74–81 (1959).

(51) Hänsel, G., "Electrodeposition of Crack-free Rhodium Coatings of Increased Thickness," *Metalloberflaeche, 20* (2), 67–70 (1966).

(52) Staub, H., "High Electrical Resistance on Rhodium Plated Contacts," *Galvanotech. Oberflaechenschutz, 7* (7), 167–173 (1966).

(53) Rhodes, E. C., "The Modern Practice of Rhodium Plating," *Chimie et Industrie, 75* (6), 1272–1289 (1956).

(54) Kadaner, L. I., and Naumenko, K. M., "Method for Decreasing Internal Strain in Rhodium Coatings," *Priborostroenie*, (9), 19–20 (1966).

(55) Wiesner, H. J., and Meers, H. A., "Further Studies in Heavy Rhodium Plating," *Plating, 43* (3), 347–355 (1956).

(56) Hill, J., "Stress in Electrodeposited Rhodium. Effects of Contamination by Organic Compounds," *Plating, 52* (5), 417–419 (1965); *Platinum Metals Review, 8* (2), 55–59 (1964).

(57) Feaster, R. V., and Block, R. J., "Residual Stress Measurements on Thin Electrodeposited Films," *United States Atomic Energy Commission ORO-3401-10* (1969); *Nuclear Science Abstracts, 23* (11), 20501 (1969).

(58) Kadaner, L. I., and Medyanik, V. N., "Relation Between Plasticity of Electrolytic Deposits and Their Hydrogen Content," *Navodorozhivanie Metal. Bor'ba Vodorodn. Khrupkost'yu*, 182–183 (1968); Ref. *Zh. Khim. Abstr. No. 19L257* (1968).

(59) Reid, F. H., "Crack-Free Rhodium Deposits," *Metalloberflaeche, 15* (2), 33–37 (1961).

(60) Reid, F. H., "Platinum Metal Plating—A Process and Applicational Survey," *Trans. Inst. Metal Finish., 48* (3), 115–123 (1970).

(61) Fluhmann, W., "Electroplating of Precious Metals in Electrical Engineering," *Schweizer Elektrotechnischer Verein Bulletin, 55* (18), 883–887 (1964).

(62) "Science for Electroplaters. 110. Rhodium," *Metal Finishing, 64* (9), 84–88 (1966).

(63) Reid, F. H., and Blake, J. C. "Electrodeposition of Ruthenium," *Trans. Inst. Metal Finish., 38* (2), 45–51 (1961).

(64) Bradford, C. W., and Cleare, M. J., "New Ruthenium Plating Bath," *Platinum Metals Review, 13* (3), 90–92 (1969).

(65) Reddy, G. S., and Taimsalu, P., "Electrodeposition of Ruthenium," *Trans. Inst. Metal Finish., 47* (5), 187–193 (1969).

(66) Conn, G. A., "Ruthenium Plating," *Plating, 56* (9), 1038–1040 (1969).

Chapter 16

Rhenium and Rhenium Alloys

Rhenium

Early plating baths for rhenium contained potassium or ammonium perrhenate and sulfuric acid with a pH of about 1.0, operated at 10 amp/sq dm and room temperature to produce bright deposits which were hard and brittle. (1) It was postulated that rhenium was deposited as rhenium hydride, which was unstable in air and moisture. Heating the deposit to 950 and 1000 C in hydrogen for about 15 minutes produced tarnish-resistant metal. This procedure could be used to build up successive 10 μm layers of rhenium to a thickness of 75 μm. Specimens heated in hydrogen remained bright after long exposure in air whereas unheated specimens oxidized in a short time during exposure to laboratory atmospheres.

The maximum thickness of good quality rhenium coatings was reported to be 2 μm when deposited from a solution containing 12 g/l potassium perrhenate, 30 g/l sulfuric acid and 50 g/l ammonium sulfate at room temperature and 20 amp/sq dm. Deposits thicker than 2 μm were cracked and had poor adhesion. (2) Although relatively thick rhenium coatings were obtained by successive deposition of 2 μm layers with intermediate heat treatment in hydrogen at temperatures above 800 C, an alloy was formed with the basis metal by diffusion. Diffused rhenium coatings have been shown to increase the ductility of tungsten wire. (11) Studies of the physical properties of niobium and molybdenum with diffused rhenium coatings have also been reported. (3)

The characteristics of rhenium deposits from sulfamic acid solutions were improved by the addition of magnesium sulfamate and aluminum compounds. (4) Magnesium sulfamate reduced internal stress permitting the deposition of coatings thicker than 1 μm at higher current density. Deposits were obtained at 60 to 80 C, 10 to 15 amp/sq dm, with moderate agitation in a solution containing 10 g/l potassium perrhenate, 10 g/l sulfuric acid (to lower pH to 1 or 1.5 for brighter and clearer deposits with increased efficiency), 25 g/l ammonium sulfate (to increase conductivity and efficiency, particularly at low current density), and 30 g/l magnesium sulfamate to reduce stress.

A study of the sulfuric acid, citric acid, citrate and phosphate baths indicated differences in characteristics of the rhenium deposits. (5) After a water rinse, all of the deposits had a mirror-like luster, but deposits from the sulfuric acid bath turned

TABLE 16.1

Hardness and Stress of Rhenium Deposits

Plating Bath	pH	Rhenium Conc, g/l	Temp, C	Current Density, amp/sq dm	Hardness, kg/sq mm	Stress kg/sq mm	Stress psi	Reference
Phosphate[a]	5.7	7.0	60	13.5	300	—		5
Phosphate[b]	—	—	—	—	520–560	—		6
Phosphate	—	—	—	30	650	10.5	15,000	7
Phosphate[a]	5.7	7.0	60	40	800	—		5
Phosphate[a]	5.7	7.0	60	54	1000	—		5
Sulfate[c]	3.0	8.0	20–70	1–20	1100–1700	—		8
Sulfate[d]	<1.0	8.4	18–20	80–100	287–294[e]	—		9
Sulfate[d][f]	<1.0	8.4	18–20	140[f]	322–330[e][f]	(f)		9

[a] The solution contained 10 g/l NH_4ReO_4, 200 g/l $NH_4H_2PO_4$ and 80 g/l ml/l NH_4OH (25%, by weight).

[b] Proprietary phosphate solution (ACR Renplate).

[c] The solution contained 11.2 g/l $KReO_4$, 200 g/l $(NH_4)_2SO_4$ and H_2SO_4.

[d] The solution contained 12 g/l $KReO_4$ and 35 g/l H_2SO_4.

[e] Relative hardness values of a special alloy formed by fusion of rhenium deposits on copper foil.

[f] Values obtained with periodically reversed current reduced stress 30 percent compared to direct current.

dark within a few hours or a few days. Those from the citrate and citric acid baths remained bright and metallic for four to seven months, and deposits from a slightly acid phosphate bath maintained their luster for more than a year.

Hardness

Hardness data for rhenium, summarized in Table 16.1, include values ranging from 1100 to 1700 kg/sq mm, which were obtained in a sulfate bath. (8) Deposits from a phosphate bath ranged in hardness from 300 to 1000 kg/sq mm, depending on the cathode current density. (5) Increasing the current density from 13.5 to 54 amp/sq dm increased hardness from 300 to 1000 kg/sq mm, as shown in Table 16.1. Rhenium with a hardness >1050 kg/sq mm was obtained from the phosphate bath at higher current densities and higher rates (60 μm/hr), but these deposits were not structurally sound.

Stress

The stress of a medium-hard rhenium deposit (650 kg/sq mm) from a phosphate solution was 10.5 kg/sq mm (15,000 psi). (7) Harder deposits from a sulfate bath tend to crack because of higher stress when thickness exceeds 1 or 2 μm. (2) An addition of magnesium sulfamate and aluminum compounds reportedly reduced the stress of rhenium deposited in sulfate baths, but stress data were not quantified. (4)

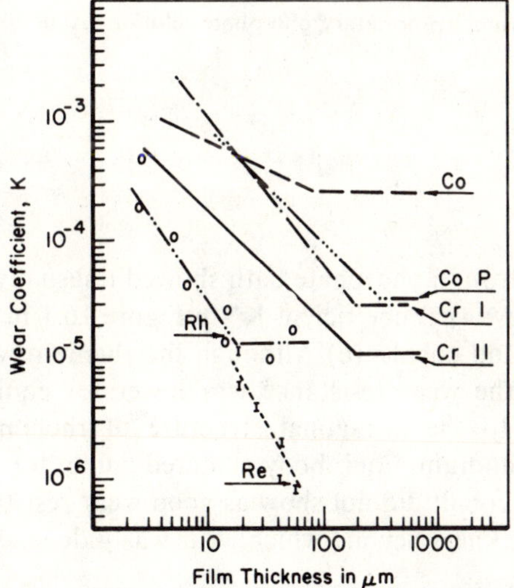

Figure 16.1. Wear Coefficient K as a Function of the Thickness of Different Electrodeposits. (6)

Plated specimens run against 52100 bearing steel at a hardness level of about 800 kg/sq mm at a speed of 6 cm/sec, no lubricant, applied load, and 1 kg/sq mm mean stress. See Table 16.2 for properties of deposits and plating conditions.

TABLE 16.2

Properties of Rhenium and Other Electrodeposits (6)

Metallic Deposit	Substrate	Distance Between Cracks, μm	Hardness of Deposit, kg/sq mm	Brittleness of Deposit	Transition Thickness, μm[(a)]
Cobalt[(b)]	Copper	No cracks	200–225	Low	100
Cobalt-phosphorus alloy[(c)]	Copper	No cracks	600–625	Low	330
Ditto	Nickel	No cracks	600–625	Low	330
Chromium I[(d)]	Copper	250	800–900	High	200
Chromium II[(e)]	Copper	30	1100–1300	High	200
Rhodium[(f)]	Nickel	No cracks	850–950	Low	20
Rhenium[(g)]	Nickel	30	520–560	Intermediate	>6
Rhenium[(g)]	Copper	30	520–560	Intermediate	>6

(a) The transition film thickness is the deposit thickness above which wear coefficient was independent of thickness.

(b) Cobalt deposits obtained at 4 amp/sq dm and 30–40 C from a solution containing 350 g/l $CoSO_4 \cdot 7H_2O$, 35 g/l H_3BO_3, 45 g/l $CoCl_2 \cdot 5H_2O$, 25 g/l NaCl, pH 3 to 4.

(c) Cobalt phosphorus deposits obtained at 40 amp/sq dm and 75 to 80 C from a solution containing 180 g/l $CoCl_2 \cdot 6H_2O$, 50 g/l H_3PO_4, 40 g/l H_3PO_3, pH 1.0.

(d) Chromium I deposits obtained at 50 amp/sq dm and 74 to 77 C from a solution containing 250 g/l CrO_3, 2.5 g/l H_2SO_4, 5 g/l HF and 10 g/l Na_2SiF_6.

(e) Chromium II deposits obtained at 55 to 58 C; other conditions as in Footnote (d).

(f) Rhodium deposits from a proprietary sulfate electrolyte (No. 221 Englehard, Inc.).

(g) Rhenium deposits from a proprietary phosphate solution (ACR Renplate).

Wear Characteristics

Rhenium deposits from a phosphate bath showed unusually good wear resistance, as evidenced by the low wear coefficient K* in Figure 16.1 in comparative tests with rhodium, chromium and cobalt. (6) Although the rhenium was not as hard as rhodium or chromium, the wear resistance was lower for equivalent thickness. This might be attributed to the hexagonal structure of rhenium, compared to face-centered-cubic for rhodium, and body-centered-cubic for chromium. However, hexagonal structured cobalt did not show as good wear resistance. Table 16.2 shows the transition film thickness beyond which wear was independent of thickness.

* Wear coefficient K was computed from the following equation:

$$K = \frac{3p\Delta V}{L\Delta x}$$

where p = hardness of softer material
L = load
Δx = sliding distance.
ΔV = volume of wear.

TABLE 16.3

Crystal Lattice Constants for Rhenium and Rhenium-Chromium Alloy Deposits on Molybdenum (2)

Deposit	Lattice Constant[a] a	c	V
Annealed rhenium metal	—	—	28.857
Rhenium[b]			
as deposited	2.761	4.542	29.984
after 6 hr at 250 C in air	2.766	4.496	28.879
after 15 min at 1000 C in hydrogen	2.782	4.543	30.490
Rhenium-chromium alloy[c]	2.759	4.456	29.378

[a] $V = 0.865\ a^2c$ where V is the volume of the cell; a and c are the lattice constants.

[b] Deposits obtained at room temperature and 20 amp/sq dm from a solution containing 12 g/l $KReO_4$, 80 g/l H_2SO_4, and 50 g/l $(NH_4)_2SO_4$.

[c] Alloy deposits obtained at room temperature and 100 amp/sq dm from a solution containing 30 g/l $HReO_4$, 15 g/l CrO_3, 80 g/l H_2SO_4, and 50 g/l $(NH_4)_2SO_4$.

Rhenium Alloys

Rhenium-cobalt, rhenium-iron, and rhenium-nickel alloys deposited from citric acid electrolytes had better corrosion and tarnish-resistance than pure rhenium. (10) Rhenium deposits from citric acid electrolyte became dark in ten days of exposure to laboratory air, and in three months the entire plate was covered with dark, nonadherent corrosion products. Rhenium-iron and rhenium-nickel alloy deposits remained untarnished for three-months; rhenium-cobalt became somewhat dull and gray.

TABLE 16.4

Hardness of Rhenium and Rhenium-Nickel Alloys[a]

Nickel Content, percent[b]	Hardness, kg/sq mm
0	1100 to 1700
10	2000
90	900

[a] Deposits with relatively small internal stress were obtained at 1 to 20 amp/sq dm and 20 to 70 C in a solution containing 200 g/l $(NH_4)_2SO_4$, H_2SO_4 to pH 3, nickel as $NiSO_4 \cdot 7H_2O$ and rhenium as $KReO_4$ in metal mole ratios of 1:5 to 5:1 with a total metal concentration of 0.06 mole/l. (8)

[b] Possibly atomic percent consistent with Russian practice in reporting.

Rhenium-Chromium Alloy

Deposits of rhenium-chromium alloy up to 10 μm in thickness can be obtained with no cracks. (2) Dense deposits of the alloy were firmly attached only to copper and nickel and did not adhere to molybdenum, chromium or tungsten substrates. According to X-ray studies, alloying chromium decreased the lattice constants and cell volume as shown in Table 16.3, indicative of substitution solid solution. The decrease of lattice constants and volume for rhenium heated to 250 C is associated with loss of hydrogen, whereas the increase when heated to 1000 C is associated with rhenium-molybdenum alloying by diffusion.

Rhenium-Nickel Alloy

Table 16.4 shows the hardness of rhenium-nickel alloys deposited in a sulfate bath. (8) Hardness reached a maximum of 2000 kg/sq mm at a nickel content of 10 percent. Internal stresses in the alloy and pure rhenium were relatively small. Deposits with about 10 percent nickel also had the greatest luster. The rhenium-nickel alloys did not show any signs of corrosion after 32 months of exposure in an industrial city atmosphere at ordinary temperatures. Homogeneous supersaturated solid solutions were obtained for low concentrations of either component. (11) At equal contents of the components, two-phase systems consisting of a finely dispersed mixture of solid solutions of rhenium and nickel were observed.

References

(1) Root, G. S., and Beach, J. G., "Electroplating of Rhenium," *Rhenium,* edited by B. W. Gonsor, Elsevier, 181–188 (1962).

(2) Kvokova, I. M., and Lainer, V. I., "Electrodeposition of Thick Dense Rhenium Coatings," *Zashchita Metallov, 1,* 515–520 (1965).

(3) Korobkova, G. A., Solomko, Yu. V., and Mikhaylov, S. K., "Rhenium Coatings on Refractory Metals," *New Materials in the Machining Industry,* edited by L. Ya. Popilov, Mashinostroyeniye, Leningrad, 96–109 (1967).

(4) Meyer, A. R., "The Electrodeposition of Rhenium," *Trans. Inst. Metal Finish., 46,* 209–211 (1968).

(5) Camp, E. K., "Recent Developments in the Application and Electroplating of Rhenium," *Plating, 52,* 413–416 (1965).

(6) Georges, J. M., and Rabinowicz, E., "Effect of Film Thickness on the Wear of Hard Electrodeposits," *Wear, 14,* 171–180 (1969).

(7) Camp, E. K., "Method and Bath for Electroplating Rhenium," U.S. Patent 3,285,839 (November, 1966).

(8) Pishikin, A. M., and Popereka, M. Ya., "Electrodeposition of Bright Electrodeposits of Rhenium and Rhenium-Nickel Alloys," *Blestyashchie Komb. Metal. Pokrytiya, 2,* 3–12 (1967); *Zh. Khim. Abstr. No. 5L383* (1969).

(9) Sklyarenko, S. I., Lavrov, I. I., and Shamagin, Yu. P., "Electrolytic Production of Rhenium Coatings from Aqueous Solutions with the Use of Current or Alternating Polarity," *Journal Applied Chemistry (USSR), 34,* 1219–1223 (1961); *Zh. Prikl. Khimii, 34,* 1281–1286 (1961).

(10) Netherton, L. E., and Holt, M. L., "Electrodeposition of Rhenium-Cobalt and Rhenium-Iron Alloys," *Journal Electrochem. Soc., 99,* 44–47 (1952).

(11) Pishikin, A. M., Popereka, M. Ya., and Khazankin, V. B., "Electrodeposited Nickel-Rhenium Films," *Izvestiya Vysshikh Vchebnykh Zavedeniy SSSR, Fiz., 12,* 7–10 (1969).

Chapter 17

Silver and Silver Alloys

The decoration and protection of flatware, hollow ware, jewelry and art objects were the major uses for silver plating to 1940. Industrial and engineering uses for silver plating have subsequently been increasing. The unique electrical, mechanical and chemical properties of silver, which include high electrical and thermal conductivity, low-contact resistance, high-load carrying capacity, nonwelding characteristics, chemical resistance and nontoxic characteristics, are useful attributes for many applications. Expanding uses of silver plating for engineering applications, (1,2) especially in the electrical and electronic industries, has resulted in increased attention to process and product innovations intended to improve physical properties, performance and reliability of plated components, and to meet specific functional requirements.

Silver is usually electrodeposited in alkaline cyanide - silver cyanide electrolytes. Proprietary processes for bright, fine-grained silver (3 6) are based on cyanide baths containing additions of organic compounds and/or metallic inorganic salts. Noncyanide, mildly-acid, silver nitrate electrolytes containing phosphates and/or organic acids as bath additives have been disclosed. (7–10) Mirror-bright ductile silver deposits are claimed, using ammonium-sodium phosphates as brightener additives in a silver nitrate bath. (7)

Silver, copper, and gold are used extensively as contact materials. The cost of silver is about 2 percent of the cost of gold, on a volume basis, but about 20 times that of copper. Both silver and copper tarnish quickly in moist sulfide atmospheres. However, the electrical conductivity of silver sulfide and silver oxide is similar to that of silver metal. Silver sulfide films have low shear strength and silver oxide is rather soluble in silver metal, which are advantages for the use of silver as a contact material.

Gold does not tarnish but has only about 75 percent the conductivity of silver. A cost saving with plated gold versus solid gold for electrical contacts is obvious. However, any advantages for electrodeposited silver or silver alloys versus solid silver generally must be related to factors other than metal cost, per se. Increased hardness, improved tarnish resistance, wear resistance and thermal stability have been emphasized in research efforts on electrodeposited silver and silver alloys.

Significant changes in metallurgical structure, hardness, wear resistance, electrical properties and other characteristics of electrodeposited silver and silver

TABLE 17.1

Physical and Mechanical Properties of Unalloyed Electrodeposited and Wrought Silver

	Unannealed Electrodeposits	Annealed Electrodeposits Temperature,			Wrought Silver (13)	
		Annealed at 480 C	Annealed at 550 C	Annealed at 880 C	Annealed	Cold Worked (50%)
Tensile Strength,						
kg/sq mm	24 to 34[a]	19	17	10.5	16 to 19	31 to 38
10^3 psi	34 to 48[a]	27	24	15	23 to 27	44 to 54
Elongation,						
Percent in 2 in.	12 to 19 (16)	50	—	25	50 to 55	3 to 5
Density, g/cu cm	9.2 to 10.5	—	—	—	10.49	—
Hardness (Aged)						
VHN	40 to 185	—	40	—	25	85
Electrical Resistivity,						
microhm-cm	1.6 to 1.9	—	—	—	1.59	—
Stress						
kg/sq mm	1.5 to 2.5[b]				—	—
psi (tensile)	2100 to 3500[b]				—	—

[a] A tensile strength of 34 kg/sq mm (48,000 psi) was reported for silver deposited in a solution containing 105 g/l AgCN, 112.5 g/l KCN, 30 g/l KOH, 15 g/l K_2CO_3, 0.075 g/l $Na_2S_2O_3$ and 0.75 ml/l H_2PO_2. Temperature and current density were 49 C and 6 amp/sq dm (60 amp/sq ft). (11)

[b] Stress in silver deposited at 1.8 amp/sq dm (18 amp/sq ft) in a 21 C solution containing 37.5 g/l Ag, 37.5 g/l free KCN, 18.8 g/l KOH and 37.5 g/l K_2CO_3. PR current reduced stress to 1.05 kg/sq mm (1,500 psi). (14)

alloys are affected by changes in solution composition, additives, and/or plating conditions. Improved silver electroplates have been developed, but much of the published property data were not related to specific operating parameters that can affect specific property values.

Mechanical Properties

Tensile Strength and Ductility

Property data for electrodeposited and wrought silver are compared in Table 17.1. Silver with a tensile strength as high as 34 kg/sq mm (48,000 psi) has been electrodeposited in a cyanide bath. (11) This strength level is about equal to that for cold worked, wrought silver. The lowest value reported for electrodeposited silver—24 kg/sq mm (12)—is higher than the tensile strength of annealed, wrought silver. The electrodeposits are appreciably more ductile than cold worked, wrought silver but not as ductile as the annealed wrought metal.

The tensile strength of a silver alloy containing 2.1 percent thallium was 33.5 kg/sq mm. (47,700 psi). (15) Elongation was 25 percent. After heating 2 hours at 500 C, tensile strength and elongation were changed to 20.5 kg/sq mm (29,400 psi) and 46 percent, respectively. Even after heat treating, the electrodeposited alloy was stronger than homogenized castings of the same composition. The plating bath contained 35 g/l silver cyanide, 19 g/l potassium cyanide, 38 g/l potassium carbonate and 0.5 g/l thallous cyanide. The current density was approximately 0.5 amp/sq dm (5 amp/sq ft).

Hardness

Hardness values for unalloyed electrodeposited silver range from 40 to 185 VHN. The softest deposits are nearly two times as hard as annealed wrought silver, which probably is a result of inclusions and lattice strains, which induce a fine grain size. Silver deposited in a 40 C cyanide solution at 0.5 amp/sq dm (5 amp/sq ft) increased in hardness from 50 to 78 VHN as the free potassium cyanide concentration was increased from 5 to 120 g/l. (16)* With a free cyanide concentration of 60 g/l, hardness decreased from 78 to 50 VHN when the plating bath temperature was increased from 40 to 60 C. Harder deposits, from 130 to 140 VHN, were obtained in another, 25 C bath, by comparison with the 80 to 90 VHN silver deposited at 40 C in the same solution. (17) A decrease in the silver concentration from 50 to 17 g/l reduced hardness from 137 to 94 kg/sq mm for

* The deposition potential was increased from −323 to −650 millivolts as the free cyanide was increased from 5 to 120 g/l.

TABLE 17.2

Hardness of Electrodeposited Silver[a] from Cyanide Electrolytes as Related to Current Density and Storage Time Versus Bath Additive (19)

Bath[b] Additive	Deposit Hardness, VHN							
	Current Density, amp/sq dm, and Storage Time							
	0.8	0.6	0.4	0.2				
	Initial	Initial	Initial	Initial	½ Month	1 Month	3 Months	6 Months
None	110	120	94	110	105	102	100	90
CS_2, 1-½ ml/l	140	155	170	148[c]	—	135	130[c]	130
Na_2SeO_4, 0.8 g/l	120	120	125	125	124	107	65	66
K_2SO_4, 50 g/l	110	120	110	140	132	105	95	85
Cobalt, 1.7 g/l	100	110	125	108	98	87	84	88

(a) Electrical resistivity was 1.5 to 1.8 microhm-cm.

(b) The solutions contained 26 to 30 g/l silver, 20 to 42 g/l free KCN and 30 to 34 g/l K_2CO_3.

(c) Grain size increased from 2.36 to 3.22 $\times$ 10^{-6} cm during 3 months storage.

silver deposited in a 20 C solution containing 20 g/l free potassium cyanide. (18) Thus, low operating temperatures, a low silver concentration and a high free cyanide concentration favor high hardness values.

Solution agitation reduced hardness slightly. (16) Changes in current density also influenced hardness, but the influence was relatively small in solutions containing no addition agent. (16,19) Periodically reversed current increased hardness from the range of 76–78 to 100–105 VHN for silver deposited at 1 amp/sq dm (10 amp/sq ft) in a bath at 20 C containing 35 g/l silver cyanide and 54 g/l potassium cyanide. (20) PR current also refined grain size and increased density.

Additions of sodium selenate and potassium sulfate increased hardness slightly. (19) A carbon disulfide addition had a larger effect, as shown in Table 17.2. Additions of citrate, tartrate or aspartic ions increased hardness to values similar to those obtained with a carbon disulfide addition. (21)

A significant decrease in hardness usually occurs with time at ambient temperatures, after plating, as shown in Table 17.2. (19) This decrease in hardness has been associated with recrystallization and grain growth. Grain size of silver deposited at 0.2 amp/sq dm ($\sim$2 amp/sq ft) in a cyanide bath containing carbon disulfide increased from 2.36 to 3.22×10^{-6} cm during a three-month storage period, and grain growth was associated with a corresponding decrease in hardness from 148 to 130 VHN. Silver deposited in solutions containing other addition agents also decreased in hardness during 3 to 6 months of aging at room temperature. Significantly, however, carbon disulfide was the only addition agent that was effective for inducing high hardness (>90 VHN) in silver deposits aged for 6 months.

Silver deposited in cyanide solutions containing 60 to 120 g/l free potassium cyanide ranged from 70 to 90 VHN in hardness and was annealed by heating at 300 C, or above. (16) Hardness was generally reduced to about 40 VHN after 2 hours of heating at 400 to 800 C. This reduction in hardness has been associated with recrystallization and grain growth.

TABLE 17.3

Hardness of Silver and Silver-Lead Alloy Electrodeposits[a]

Deposit	VHN Hardness at Room Temperature: Deposited at 25 C	Deposited at 40 C	Annealed[b]	Hot Hardness at 250 C
Ag	130–140	80–90	—	25
Ag–4% Pb	170–190	130–140	70	40

[a] Data for deposits obtained in a cyanide-tartrate solution. (18)
[b] 2 hours at 250 C.

TABLE 17.4

Hardness and Electrical Resistivity of Silver Alloy Electrodeposits

Alloy Constituents	Hardness,	Electrical Resistivity, microhm-cm	Reference
Antimony[a]			
0.7%	100	1.9	22
9.6%	164	11.6	22
Bismuth[b]			
1 to 2.6%	90 to 180	8 to 10.4	23
Cadmium			
2%	90	—	24
20%	140	—	24
50%	290	—	24
Cobalt			
20%	120	4.0	25
32%	150	7.0	25
50%	275	19.0	25
Copper[c]			
20%	240	7.5	25, 26
60%	240	12	25, 26
85%	340[d]	22	25, 26
Lead[e]			
4%	180	10.5	18, 27
10.2%	—	22.5	27
Nickel			
8%	170	45	25
20%	188	85	25
32%	200	110	25
80%	430	170	25
80%[f]	470	19	28
Palladium[g]			
12%	180	—	29
60%	250	10	29
90%	320	—	29
Thallium[h]			
9.5%	90	—	30

[a] Deposited in cyanide solutions containing 24 g/l Ag, 60 g/l Rochelle salts, 25 g/l Na_2CO_3 and 3–5 g/l NaOH. To codeposit 0.7 Sb, 5 g/l of Sb was added to the bath. (22)

[b] Deposited in solutions at 25 C containing 25–50 g/l Ag, 25 g/l Bi, 35 g/l $K_2C_4O_4H_2$, 25 g/l KOH and 20–150 g/l KCN. (23)

[c] Deposited at 5 amp/sq dm in solutions at 20–25 C containing 100 g/l $K_4P_2O_7$ and 25 g/l Ag + Cu (as silver nitrate + copper sulfate); the pH was 9 to 11. (26)

[d] Hardness was reduced to 185 VHN after 6 months' aging at room temperature. (26)

[e] Deposited at 4 amp/sq dm in solutions containing 0.22 M/l AgCN, 0.3 M/l NaCN, 0.015 M/l Pb acetate, 0.018 M/l NaOH and 0.21 M/l tartrate ions. (18)

Although there has been little commercial use of silver nitrate plating solutions, hardness and resistivity data for silver deposits produced in such solutions were reported in 1967. (16) The initial hardness of silver deposited in 0.5 N silver nitrate solutions ranged from 55 to 185 VHN, depending on the addition agent used for grain refinement. The hardest deposits, 170 to 185 VHN, were obtained in baths containing citric or tartaric acid (0.7 to 5.8 g/l) or asparagine (2.7 g/l). Hardness after annealing at 600 C was decreased to a value as low as 40 VHN.

Alloying 4 percent lead with silver deposited in a cyanide-tartrate bath at 25 C increased hardness appreciably, from about 135 VHN to the range of 170 to 190 VHN. (18) Hardness measured at 250 C also was greater for the silver-lead alloy, as shown in Table 17.3.

Hardness values for silver alloys containing antimony, bismuth, copper and other metals are compared in Table 17.4. With 20 and 85 percent copper, hardness reached 240 and 340 VHN, respectively. (25,26) These hard deposits were obtained in a silver nitrate - copper sulfate - potassium pyrophosphate bath. The 85 percent copper alloy decreased in hardness from 340 to 185 VHN during 6 months at room temperature.

A 12 percent thallium alloy with a hardness of 180 VHN was a solid solution with an fcc structure. (30) It was deposited in an acid chloride electrolyte.

Composites of silver and alumina (31) and silver and Carborundum (32) showed hardness values similar to the range for unalloyed silver deposits. The amount of alumina (7 or 10%) codeposited with the silver (in an alkaline-cyanide type bath with added potassium nitrate at 30 C) was related to the amount (90 or 230 g/l) dispersed in the silver plating bath and independent of the current density (1 to 3.5 amp/sq dm). The initial as-deposited hardness value of 110 VHN was reduced by annealing at 200, 400 and 600 C to 38 VHN, about equal to the hardness of annealed high-purity silver deposits. Appreciable improvement in wear resistance of silver containing 0.2 to 3 percent codeposited Carborundum was noted. (32)

Wear Resistance

Improved wear resistance for electrodeposited silver has not been correlated with increased hardness, but appears to be related to another, undefined factor. Service tests for 1 year on silver plated cutlery showed that hard (165 KHN) and soft (75

Footnotes to TABLE 17.4 (Continued)

(f) Deposited at 0.5 amp/sq dm in 25 C solutions prepared with $Ni_2P_2O_7$, $KAg(CN)_2$ and $K_4P_2O_7$. (28)

(g) The solution contained 500 g/l $LiCl_2$, 97 g/l HCl, 2.5–3.7 g/l Ag and 0.9–2.7 g/l Pd. (29)

(h) Deposited in a solution at 20 C containing 32 g/l AgCN, 25 g/l KCN, 30 g/l K_2CO_3 and 6 g/l Tl_2SO_4. (30)

VHN) deposits were about equal in wear, based on loss of weight, which was linear with time. (33) Scratching and loss of polish was more evident for the softer silver coating during the early stages of the service test, but became less noticeable with time.

Silver deposited in four different plating solutions with hardness values up to about 160 VHN, showed no differences in wear. (34) With a hardness between 160 and 170 VHN, abrasion resistance decreased notably, and with silver hardnesses at 190 or 200 VHN the abrasive wear reached values several orders higher than noted for normal silver deposits. Not only hardness but also embrittlement of the harder silver deposits by occluded hardener material were considered important in the reduction in resistance to abrasive wear. For silver plated tableware the hardness of the coating should not exceed 170 VHN for longest life. Ductile, hard silver deposits, > 170 VHN, are not yet available.

Seizure Resistance as Bearing Overlays

Electrodeposited and annealed silver exhibited better resistance to seizure than double-refined, remelted wrought silver. (35) Most impurities are detrimental to seizure resistance. But lead had no such harmful effect. Thus deposits of silver-lead alloy serve as strong antifriction bearing material. Performance of this material for bearings, as compared with that of other material for bearings is as follows:

Bearing Material	Shaft[a] Speed to Seizure, rpm	Bearing Pressure at Seizure, kg/sq mm
Electrodeposited silver with 3.4% Pb[b]	3700	4.65
Electrodeposited silver with 3.6% Pb[c]	3400	3.94
Bronze: 72% Cu, 25% Pb, 3% Sn	3000–3200	3.06–3.48
Silver with 0.5% Cu	2650	2.39

[a] Finely ground carburized SAE 2512 steel hardened to 57 Rockwell C.
[b] 62 μm overlay on a silver bearing.
[c] 150 μm overlay on a silver bearing.

Electrical Resistivity and Density

Electrical resistivity values for electrodeposited silver are inversely related to density of the silver. This relationship, for silver deposited in cyanide solutions, is as follows:

Density, g/cu cm	Specific Resistivity, microhm-cm
9.18	1.83
9.84	1.75
10.5	1.59

TABLE 17.5

Electrical Resistivity of Silver Deposited in a Cyanide Solution, as a Function of Current Density and Thickness (37)

Current Density, amp/sq dm	Resistivity, microhm-cm			
	0.25 mil	0.5 mil	1 mil	1.5 mil
0.15	1.75	1.72	1.71	1.69
0.3	1.80	1.75	1.73	1.71
0.6	1.85	1.78	1.74	1.73

The low value in the above compilation was reported for silver deposited at about 1.2 amp/sq dm in a solution containing 40.5 g/l silver cyanide, 40 g/l potassium cyanide and 20 g/l potassium carbonate. (36) Reversing the current for 4 seconds after each 9 seconds of direct plating increased the density of the deposit to 9.84 g/cu cm and reduced resistivity to 1.75 microhm-cm. The conditions for depositing silver with a density of 10.5 g/cu cm and a specific resistivity of 1.59 microhm-cm were not detailed. (16)

TABLE 17.6

Properties of Electrolytic Silver Deposited in Silver Nitrate Solutions with Organic and Inorganic Additives[(a)]

Additive	Amount in Deposit, percent, by wt	Density, g/cu cm	Specific Resistance, microhm-cm	Hardness VHN
Asparagine	0.7	—	17	—
Ditto	2.7	8.966	—	178
Boric Acid	0.6	9.135	—	—
Ditto	5.18	7.958	—	—
Citric Acid	2.4	9.214	—	—
Ditto	3.5	—	1300	157–185
Glycol	0.02	—	1.9	55
Metaphosphoric Acid	0.87	10.260	300	—
Ditto	2.65	9.825	960	—
Tartaric Acid	0.7	—	63	143
Ditto	1.2	—	270	170
None, Pure Silver from Cyanide Bath	—	10.502	1.59	43–100

[(a)] Refers to 0.5 N silver nitrate solutions. (16)

TABLE 17.7

Effects of Lead Content and Annealing Temperature on the Resistivity of Silver-Lead Deposits from Cyanide-Tartrate Baths (27)

	Resistivity, microhm-cm		
		Annealed (2 hours)	
Deposit	Unannealed	At 200 C	At 500 C
Ag	1.7	1.7	2.0
Ag–2% Pb	6.7	3.3	6.7
Ag–4% Pb	10.5	3.7	8.3
Ag–6.7% Pb	15.0	3.7	8.8
Ag–8.2% Pb	18.3	3.8	8.8
Ag–10.2% Pb	22.5	5.0	10.0

The alloys appeared to be supersaturated solid solutions of lead in silver.

Silver deposits with a thickness of 1.5 mils exhibited a lower resistivity than thinner electroplates, as shown in Table 17.5. (37) A low current density of 0.15 amp/sq mm (1.5 amp/sq ft) favored a low resistivity (1.69 microhm-cm).

Organic and inorganic additions to the silver nitrate (0.5 N) bath increased resistivity. Density values also were reduced with these additions. Examples are listed in Table 17.6. (16) A resistivity as high as 1300 microhm-cm was reported for silver deposited in a solution containing citric acid. A boric acid addition reduced density to 7.96 g/cu cm, in comparison with 10.50 g/cu cm for pure silver deposited in a cyanide solution.

As a rule, resistivity of the silver deposited in the silver nitrate solutions containing the organic or inorganic acids was reduced appreciably, to only 2 or 3 microhm-cm by heating to 300 C. (16) Gas evolved when the hard, impure silver deposits were heated at 200 to 600 C. A silver deposit with borate inclusions expanded 24 mm/m of length during heating to 480 C and then contracted > 50 mm/m in length.

The electrical resistivities of silver-lead alloy deposits as functions of lead content and annealing temperature are given in Table 17.7. (27) Resistivity of the silver-lead alloys increased with lead content. Annealing for 2 hours at 200 C reduced resistivity appreciably, to a greater degree than annealing for 2 hours at 500 C. Grain growth was generally associated with reductions in hardness and resistivity. Precipitation of an insoluble silver-lead compound from the supersaturated, solid solutions of lead in silver occurred during heating at 500 C.

Resistivity data for silver alloys of antimony, bismuth, copper and other metals in Table 17.4 show values from 1.9 to 170 microhm-cm, depending on the

Figure 17.1. Columnar Structure of a Field-Oriented, High-Purity Silver Deposit. (150×) (16)

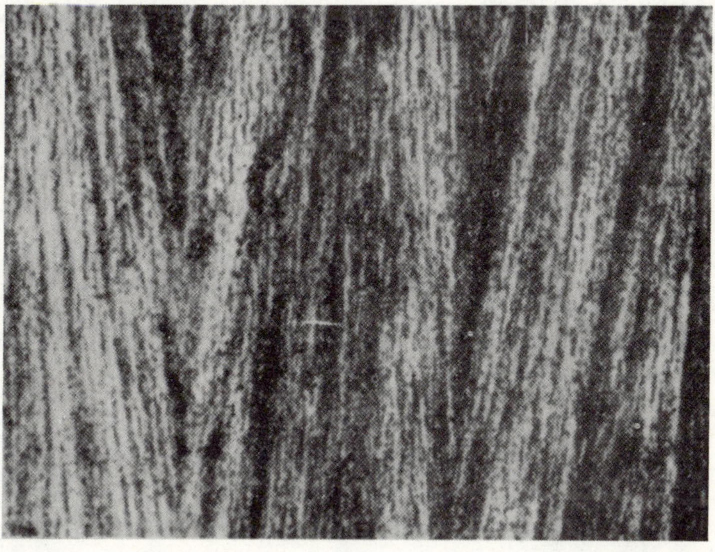

Figure 17.2. Fibrous Structure of Field-Oriented Silver Deposited in a Nitrate Bath Containing Citrate Ions. (400×) (16)

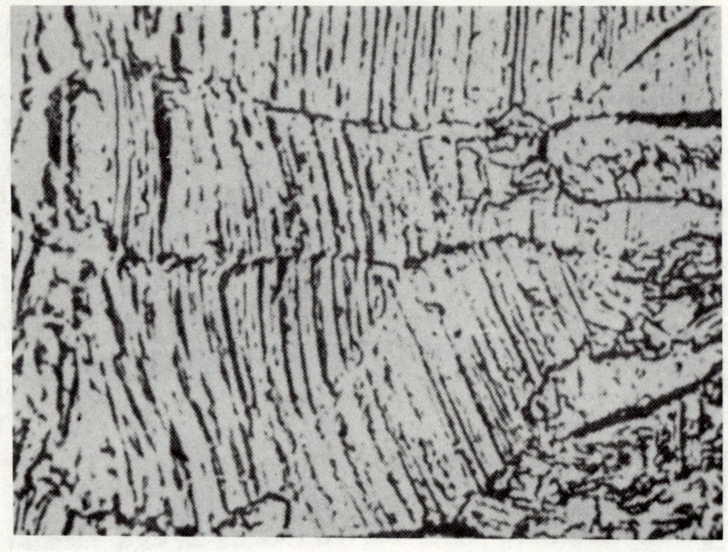

Figure 17.3. Bright Silver with a Coarse Lamellar Structure Deposited in a Cyanide Bath. (1000×) (16)

concentration of the codeposited metal. The copper alloys exhibited very high resistivities, several times greater than the resistivity of the wrought alloys, which is < 2 microhm-cm.

Proprietary addition agents containing antimony or arsenic increased resistivity of silver deposited in cyanide solutions, but selenium and sulfur had little effect, as follows: (38)

Agent	Resistivity, microhm-cm × 10^6
None	1.66 to 1.78
Antimony	16.6
Arsenic	6.1 to 33
Selenium	1.75 to 1.77
Sulfur	1.46 to 1.70

Stress

Low values of stress (customarily tensile) have been reported for silver electrodeposits. Relatively hard silver deposited at a current density of 1.8 amp/sq dm in a cyanide bath at 21 C exhibited a tensile stress of only 1.5 to 2.5 kg/sq mm (2100 to 3500 psi). (14) The solution contained 37.5 g/l silver, 37.5 g/l free potassium cyanide, 18.8 g/l potassium hydroxide and 37.5 g/l potassium carbonate. With periodic current reversal, stress was reduced to 1.05 kg/sq mm (1500 psi).

Structure

A columnar structure is generally observed for relatively soft, high-purity silver deposited in cyanide solutions containing no brightening agent. Figure 17.1 shows an example. Such a structure has been classified as field oriented, in contrast to finer-grained structures classified as basis oriented, unoriented dispersion, or intermediate types. (16)

Brighteners and impurities, which are occluded in the deposit and increase its hardness, frequently induce a fibrous structure like that in Figure 17.2 or a mixed structure of coarse lamellae and fibers like that in Figure 17.3. A finer lamellar

Figure 17.4. Structure of a Bright Silver-40 Percent Cadium Electrodeposit. (350×) (16)

Figure 17.5. Structure of Silver-Lead Alloy Deposit Containing 9 Percent Lead. (500×) (16)

structure has been observed for silver alloy electrodeposits. Figures 17.4 and 17.5 show the structures of bright silver containing 40 percent cadmium and a silver- 9 percent lead alloy, respectively. (16) Such structures are characteristic of electrodeposited solid solutions.

Bismuth and other metals are frequently codeposited in a supersaturated solid solution. With lead, the quantity incorporated is more than twice the maximum saturation cited by phase diagrams at 650 C (which is 5.2 percent). (16) It is possible to codeposit larger quantities of these metals in solid solutions with no more than the normal lattice expansion corresponding to that for normal solid solution formation. Figure 17.6 summarizes the data reported for the constants of several silver alloys as a function of the amount of codeposited metal. A solid solution was not obtained by codepositing zinc, but solid solutions with up to 50 percent cadmium were reported. Differences in the phases identified in the zinc and cadmium alloy electrodeposits and recrystallized alloys appear in Figure 17.7.

Electrodeposits of silver containing as much as 20 percent copper were supersaturated solid solutions exhibiting only the silver phase after heat treatment at 300 C.

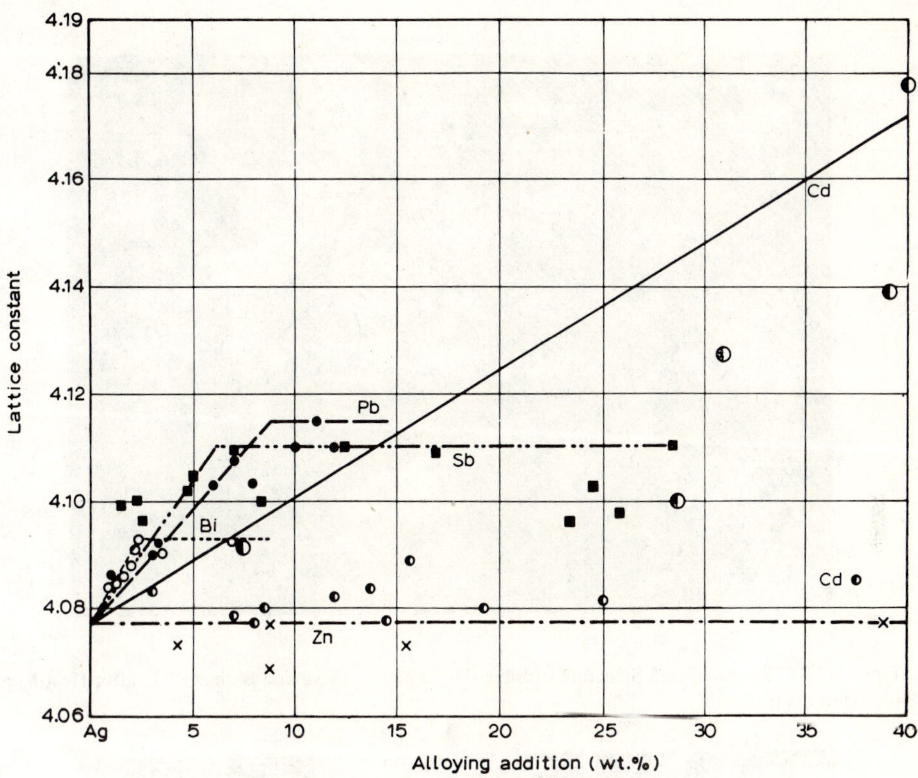

Figure 17.6. Lattice Constants of Silver-Lead Alloy Electrodeposits. (16)

(25) The high resistivity of the 20 percent copper alloy (7.5 microhm-cm) is a consequence of the stability of the supersaturated solid solution.

Heat treating recrystallizes the structure of electrodeposited silver. Figures 17.8 and 17.9 show the recrystallized structures of a soft and a slightly harder deposit, both obtained at a current density of 0.5 amp/sq dm, in a cyanide solution containing 60 g/l potassium cyanide and maintained at 40 and 20 C, respectively.

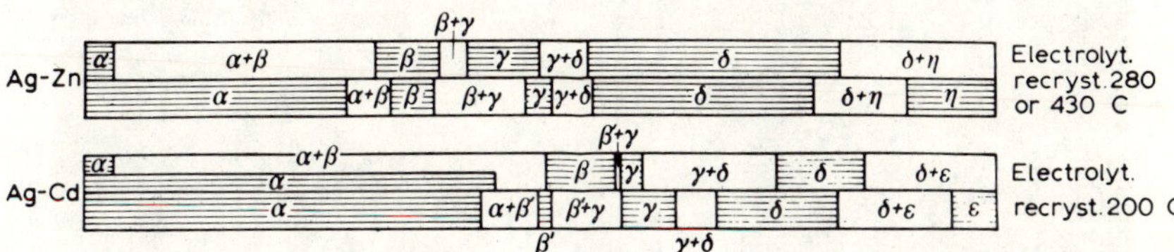

Figure 17.7. Phase Ranges of Electrodeposited and Recrystallized Silver-Zinc and Silver-Cadmium Alloys. (16)

Figure 17.8. Recrystallized Structure of Silver Deposited in a Cyanide Bath at 40 C after Heating at 700 C. (150×) (16)

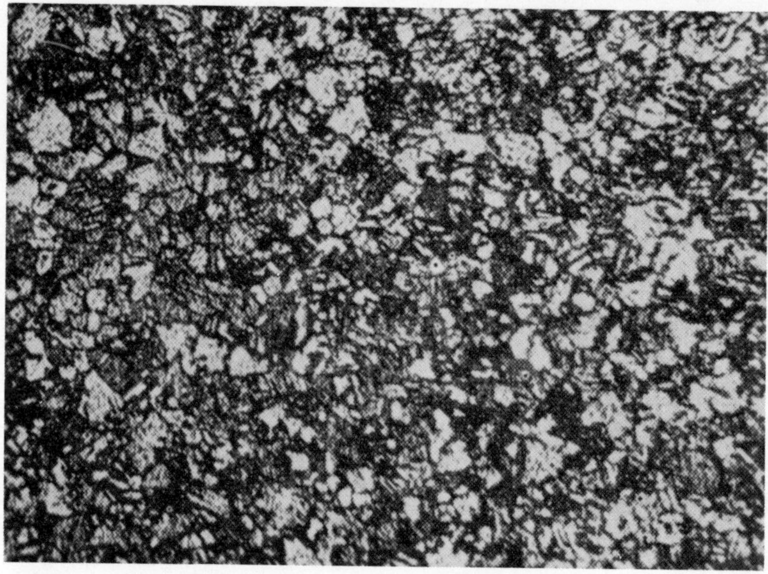

Figure 17.9. Recrystallized Structure of Silver Deposited in a Cyanide Bath at 20 C after Heating at 700 C. (150×) (16)

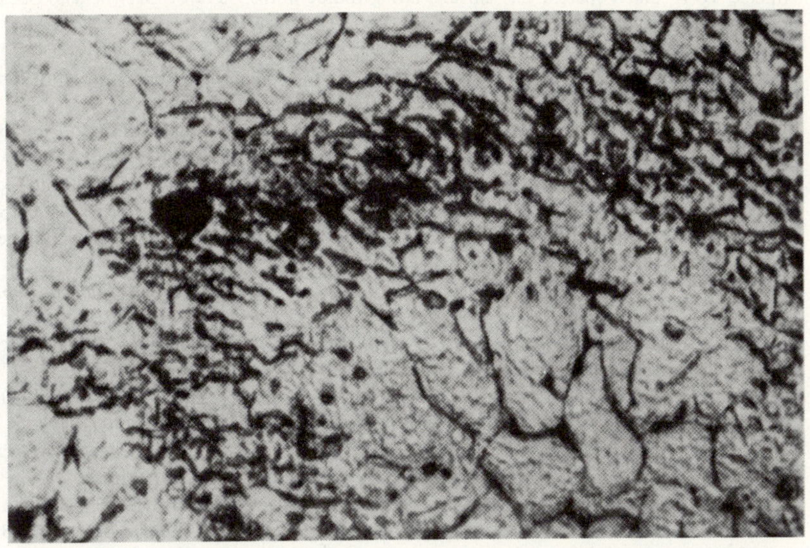

Figure 17.10. Lead Segregation from a Silver-Lead Alloy Deposit Subjected to a High Temperature and Cooled. (250×) (17)

(16) The softer deposit was transformed into larger grains by heating for 0.5 hour at 700 C.

Lead segregation in a silver-lead alloy bearing overlay subjected to conditions resulting in a high temperature is illustrated in Figure 17.10. (16) Lead initially codeposited in a supersaturated solid solution was precipitated during cooling.

References

(1) Foulke, D. G., "Silver," *Modern Electroplating,* 2nd ed., John Wiley and Sons Publishing Co., 326–341 (1963).

(2) Greenspan, L., "Modern Silver Plating Practice," *Metal Finishing, 56* (6) 353–359 (1958).

(3) Kardos, O., "Bright Silver Plating," U.S. Patent 2,666,738 (1954). Assigned to Hanson-Van Winkle-Munning Company.

(4) Greenspan, L., "Mirror Bright Silver Plating," U.S. Patent 2,735,808 (Feb 21, 1956). Assigned to The American Platinum Works.

(5) Ostrow, B. D., "Bath for Electroplating Silver," U.S. Patent 2,777,810 (Jan 15, 1957). Assigned to Elechem Corporation.

(6) Offermanns, H. and Skaliks, W., "Electrodeposition of Silver Antimony Alloys," U.S. Patent 3,425,917 (Feb 4, 1969). Assigned to Schering, A. G.

(7) Grebe, K. R. and Powers, J. V., "Noncyanide Silver Plating Bath," U.S. Patent 3,406,107 (Oct 15, 1968). Assigned to International Business Machines Corp.

(8) Sanigar, E. B., "Electrodeposition of Silver from Sulfate, Nitrate, Fluoborate, and Fluoride Baths," *Trans. Electrochem. Soc., 49,* 307 (1931).

(9) Alperse, D. K. and Toperek, S., "Silver Plating Iodide Baths," *Trans. Electrochem. Soc., 74,* 31 (1948).

(10) Fisher, J., German Patent 1,062,515 (July 30, 1959).

(11) Hart, J. S. and Heussner, C. E., "Silver Plating Aircraft Engine Bearings," *Monthly Rev., Am. Electroplaters' Soc., 33,* 142–149 (1946).

(12) Lamb, V. A., and Metzger, Jr., W. H., "Electroforming--A Method for Producing Intricate Shapes," *The Tool Engineer, 33* (2), 55–62 (1954).

(13) Butts, A. and Coxe, C. D., *Silver Economics, Metallurgy and Use,* D. Van Nostrand Company (1967), 488 pp.

(14) Walker, P. M., Bentley, N. E., and Hall, L. E., "Electroforming in Electronic Engineering," *Trans. Inst. Metal Finish., 32* (11), 349–363 (1955).

(15) Hensel, F. R., "Silver-Thallium Antifriction Alloys," *Metals Technology, Technical Publication No. 1930, AIME* (October, 1945).

(16) Raub, E., and Müller, K., *Fundamentals of Metal Deposition,* Elsevier, New York (1967).

(17) Fedot'ev, N. P., Vyacheslavov, P. M., and Luzan, M. D., "Electrodeposition of Silver Having Increased Hardness," *Trudy Leningrad, Tekhnol. Inst. im Lensoveta, No. 53,* 54–63 (1959).

(18) Raub, E., "Silver-Lead Alloys for Bearing Materials," *Z. Metall., 40,* 167–170 (1949).

(19) Fedot'ev, N. P., Vyacheslavov, P. M., and Gribel', V. I., "Change in Hardness of Electrolytic Silver with Time," *Journal Applied Chemistry (USSR), 37* (6), 1368–1376 (1964).

(20) Marchenko, N. A., Lekhovitskii, I. N., and Buyanova, A. N., "Electrolytic Deposition of Silver with Periodic Reversal of Direct Current," *Journal Applied Chemistry (USSR), 31* (10), 1496–1503 (1958); *Zh. Prikl. Khimii, 31* (10), 1511–1520 (1958).

(21) Raub, E., "Bright Electrodeposits," *Metalloberflaeche, 5B,* (2), 17–25 (1953).

(22) Fedot'ev, N. P., Vyacheslavov, P. M., Kruglova, E. G., and Burkat, G. K., "Electrolytic Deposition of Silver-Antimony Alloy," *Journal Applied Chemistry (USSR), 37* (3), 590–593 (1964); *Zh. Prikl. Khimii, 37* (3), 586–590 (1964).

(23) Raub, E., and Engel, A., "The Structure of Electrodeposited Alloys. III. Copper-Lead and Silver-Bismuth Alloys," *Z. Metall., 41* (12), 485–491 (1950).

(24) Raub, E., and Wullhorst, B., "The Structure of Galvanic Alloy Deposits. IV. The Silver-Cadmium Alloys," *Metallforschung, 2,* 33–41 (1947).

(25) Fedot'ev, N. P., Vyacheslavov, P. M., "The Class Structure of Binary Alloys Produced by Electrodeposition," *Plating, 57,* 700–706 (1970).

(26) Burkat, G. K., Fedot'ev, N. P., Vyacheslavov, P. M., and Muntsiyan, A. I., "Electrodeposition of Silver-Copper Alloy from a Pyrosphosphate Electrolyte," *Journal Applied Chemistry (USSR), 40* (11), 2391–2394 (1967); *Zh. Prikl. Khimii, 40* (11), 2497–2500 (1967).

(27) Raub, E., and Engel, A., "The Structure of the Galvanic-Precipitation Alloys. I. Ag-Pb Alloys," *Z. Elektrochem., 49,* 89–97 (1943).

(28) Burkat, G. T., Fedot'ev, N. P., and Vyacheslavov, P. M., "Electrodeposition of Silver-Nickel Alloys," *Journal Applied Chemistry (USSR), 41* (2), 405–408 (1968).

(29) Andreeva, G. P., Fedot'ev, N. P., and Vyacheslavov, P. M., "Electrodeposition and Properties of Silver-Palladium Alloys," *Protective Metallic and Oxide Coatings, Metal Corrosion and Electrochemistry,* edited by N.P. Fedot'ev, Akademiya Nauk SSSR. Translated by the Israel Program for Scientific Translations, Jerusalem, 226–231 (1968).

(30) Raub, E., and Sautter, F., "The Structure of Electrodeposited Alloys. IX. Silver-Thallium Alloys," *Metalloberflaeche, 9B,* (10), 145–147 (1955).

(31) Hoashi, J., "Electrodeposited Copper-Aluminum Oxide and Silver-Aluminum Oxide Coatings," *Journal of the Metal Finishing Society of Japan, 15* (7), 258–262 (1964).

(32) Saifullin, R. S., and Zentsova, E. P., "Composite Silver-Corundum Electrodeposits," *Protection of Metals, 3* (5), 510–514 (1967); *Zashchita Metallov, 3* (5), 594–598 (1967).

(33) Heilmann, G., "Service Tests of the Wear-Resistance of Bright-Silver-Plated Cutlery," *Metall, 11* (6), 517–518 (1957).

(34) Schlegel, H., "The Relationships Between Hardness and Abrasion Resistance of Silver Plated Tableware," *Mitt. Forschungsgesell. Blechverarb., No. 5/6,* 121–125 (1965).

(35) Dayton, R. W., and Faust, C. L., "Bearing Properties of Electrodeposited Silver and Silver-Lead Alloys," *Silver in Industry,* edited by Lawrence Addicks, Reinhold Publishing Company, 234 (1940).
(36) Colmer, W., *Nat'l Bureau of Standards Circ. 529,* 43 (1953).
(37) Balmer, J. R., and Bailey, T. W. A., "Electroforming Wave Guides," *Electroplating and Metal Finishing, 12* (7), 267–270 (1959).
(38) Krusentjern, A. V., Keil, A., and Wellner, P., "The Electrical Conductivity of Electroplated Silver Coatings," *Metalloberflaeche, 20* (1), 10–13 (1966).

Chapter 18

Tin and Tin Alloys

Tin combines corrosion protection with good solderability and is one of the easiest metals to electrodeposit. However, the mechanical properties of electrodeposited tin appear to have little interest for its known uses. No data appear in the literature on ultimate tensile or yield strength or modulus of elasticity.

Tin plating solutions (especially sodium stannate baths) have high throwing power. In contrast to hot-dipped coatings, electrodeposited tin can be adjusted in thickness to meet individual requirements for specific applications. For example, up to 0.05 mm (0.002 inch) is deposited for abrasion and corrosion resistance or lubricity, whereas only 0.0075 mm (0.0003 inch) can provide solderability. (1)

Deposits from alkaline stannate baths have a white, mat surface. By contrast, bright tin for decorative applications is deposited in acid sulfate plating solutions with proprietary addition agents. Steel strip is plated with 0.4 to 0.8 μm mat tin that is subsequently flow-brightened by heating above the melting point, followed by rapid quenching. Thus the properties of finished electrotin would be expected to be similar to metallurgical tin.

Tin alloy containing 20 to 25 percent zinc is electroplated as a protective coating on steel. More than 12 percent zinc is required to codeposit an alloy that provides anodic protection to steel. Like unalloyed zinc, tin alloy containing > 30 percent zinc tends to form bulky white corrosion products during exposure to humid atmospheres.

Tin alloys with about 35 percent nickel are obtained as bright deposits with a faint rose-pink tinge for decorative applications. The deposits are hard and combine tarnish resistance, wear resistance, and chemical resistance with corrosion protection. Solution compositions can vary appreciably in metal-ion concentrations with relatively little change in alloy composition.

Tin-copper alloy deposits include bronze with a 10 to 20 percent tin content and speculum, corresponding in composition to the cast alloy, with about 42 percent tin. Commercial speculum plating solutions produce a bright deposit. Electrodeposited speculum is white and much harder and less ductile than electroplated bronze. Unlike bronze, which is customarily lacquered to prevent tarnish, speculum exhibits good tarnish resistance.

Tin-lead alloy electrodeposits have been adopted as protective coatings which can be readily soldered. Tin is codeposited with copper and/or lead to produce overlays on steel and aluminum bearings.

Published property data for tin and its alloys of copper, lead, iron, nickel, and zinc include information on hardness and electrical resistivity. Property and structural data for some of these alloys are unique because comparable properties and structure are not obtained with cast or cold-rolled alloys.

Electrodeposited Tin

Structure and Orientation

Electrodeposited tin always has a crystal structure similar to the common metallurgical white tin (β form, body-centered tetragonal) with a density of 7.29 g/cu cm and a specific resistance of about 11 microhm-cm. (1) Gray tin (α form, diamond-type-cubic structure), which has a density of 5.76 g/cu cm and a specific resistance of about 300 microhm-cm, (1) was not observed even when tin was deposited at only 5 C. (2) An allotropic transformation from β to α can occur at temperatures below about 13 C, but the transformation rate is very slow at temperatures above −40 C. (1) It is influenced by many factors and might only be of concern for plated parts used at cryogenic temperatures. One rare instance of transformation of a plated coating at −212 C was reported for bright tin. (3)

Tin deposited from alkaline stannate solutions has a preferred orientation with (100) crystal planes parallel to the surface. (4) In deposits from acid sulfate solutions, the (110) planes were parallel to the deposit surface. Tin deposits from alkaline stannate solutions were not as smooth as those from acid sulfate solutions, according to relative measurements with a surface-roughness gage or microscopic examination of magnified cross-section profiles. Microscopic techniques show greater roughness than a diamond-tracer instrument, because tin is indented by the tracer. Tin coatings deposited from an acid bath had better low-friction properties than deposits from an alkaline stannate bath. (4)

Bright tin deposits consist of microcrystals with a grain size of 0.5 to 0.8 μm and higher orientation than dull deposits. (2) The change in surface state and crystal structure by annealing bright tin at 150 to 200 C was not observed for dull deposits.

Hardness

The hardnesses of electrodeposited and metallurgical tin are compared in Table 18.1. Relative to values of 4 to 5 and 7.2 kg/sq mm for high-purity, cast, and Straits tin, respectively, the hardness of dull electrodeposits ranges from 8 to 12 kg/sq mm, as a rule. Bright deposits have a hardness range of 15 to 35 kg/sq mm.

The comparative hardness of tin deposited from an alkaline stannate bath and an acid sulfate bath was 2.0 and 1.9, respectively, on the Mohs scale. (13) Spectroscopic examination of deposits from alkaline baths showed small amounts of sodium and magnesium; deposits from acid baths contained traces of copper. Compared to deposits from the alkaline bath, the deposits from the acid bath contained three times as much hydrogen (85 and 30 cu cm H_2 per 100 g tin, respectively). The steel substrate absorbed 20 and 0 cu cm hydrogen per 100 g base metal during plating in the acid and alkaline baths, respectively.

A relative hardness of 1 BHN for tin (compared to about 20 BHN for lead and 70 BHN for zinc) was reported for deposits obtained by brush plating at 150 amp/sq dm to an average thickness of 60 μm using proprietary solutions. (14,15)

Electrical Resistivity

The specific resistance of pure metallurgical tin at 0 C is 11 microhm-cm and the temperature coefficient is 0.0047 microhm-cm per degree C over the range from 0 to 100 C. (1) Thus values measured at 20 to 30 C would be about 11.1 microhm-cm.

TABLE 18.1

Hardness Data for Electrodeposited and Wrought Tin

Form	Plating Bath	Microhardness, kg/sq mm	Reference
Cast, high purity	—	3.9	1
Cast, high purity	—	4 to 5	5
Cast, high purity	—	5.3	6
Annealed Straits tin[a]	—	7.2	6
Cold-rolled Straits tin[b]	—	8.4	6
Dull electrotin	Sodium stannate[c]	8 to 9	5
Dull electrotin	Sodium stannate[c]	8	7
Dull electrotin	Sodium stannate[c]	10 to 12	3
Dull electrotin	Cyanide[d]	10	8
Dull electrotin	Cyanide[e]	12	9, 10
Dull electrotin	Stannous fluoborate[c]	12	9
Dull electrotin	Stannous fluoborate[f]	9 to 12	11
Bright electrotin	Stannous sulfate[c]	15 to 30	7
Bright electrotin	Stannous sulfate[g]	30 to 35	3
Bright electrotin	Stannous sulfate[h]	20	12

[a] Tin with 99.8 percent purity, annealed 1 hr at 190 C.

[b] Cold rolled to 30 percent reduction.

[c] Plating conditions not identified.

[d] Cyanide solution for tin-zinc alloy [see Footnote (a), Table 18.7].

[e] Cyanide solution for tin-copper alloy [see Footnote (a), Table 18.2].

[f] Fluoborate solution for tin-lead alloy; plating conditions for unalloyed tin were not identified.

[g] Bright tin deposited at 18 to 29 C and 4.3 amp/sq dm from a solution containing 30 g/l stannous sulfate, 105 ml/l sulfuric acid, and a proprietary addition agent.

[h] Bright tin deposited at room temperature and 1.3 to 2.6 amp/sq dm from a solution containing: 60 g/l $SnSO_4$, 50 g/l H_2SO_4, 60 g/l creosolsulfonic acid, 40 ml/l sodium n-octyl sulfate solution (27%), 10 ml/l formaldehyde, 20 ml brightener, and 20% by volume of isopropanol.

Russian workers report a variety of values for electrodeposited tin, with no reference to the temperature of measurement, which might be assumed as 20 C. In a report on electroplated tin-zinc (8) and tin-lead (9) alloys, the value of tin was given as 18 and 12 microhm-cm, respectively. In the study of electroplated tin-zinc alloys, the value for tin was reported to be 18 microhm-cm, in comparison with 11.8 microhm-cm for metallurgical tin, showing a 50 percent increment in specific resistance of the electroplated tin. Because the alkaline solutions used for the tin-zinc alloy plating contained cyanide for codeposition of zinc, this high value of 18 microhm-cm may not be representative of pure tin deposited from conventional stannate or stannous tin solutions which contain no cyanide ions. A resistivity of 11.4 microhm-cm was reported for tin deposited in a fluoborate solution. (37)

The contact resistance (at 410 ma) of bright tin was reported to be slightly less than that of gold (34.1 μv, in comparison with 36.8 μv for gold and 41.0 μv for solder alloy plate). (3)

Tin-Copper Alloys

Structure, Hardness, and Electrical Resistivity

Table 18.2 shows the increase in specific resistance and microhardness of tin-copper alloys with increasing copper content to a maximum at 43 percent copper. (9,10) This maximum corresponds to the eta (η) phase (Cu_6Sn_5). A second maximum occurred at 74 percent copper, which corresponds to the delta (δ) phase ($Cu_{31}Sn_8$).

The delta phase occurs in metallurgical alloys above 350 C and is typically found in cast bronze. Tin alloy with > 40 percent copper should be in the form of the epsilon (ϵ) phase (Cu_3Sn), according to the equilibrium diagram. However, the transformation of the metastable delta phase to the epsilon plus alpha phases is negligible at room temperature. Supplemental data on tin-copper alloy electrodeposits are discussed in Chapter 7, Copper Alloys.

Speculum Plating

Commercial speculum plating covers tin-copper alloys in the range of 55 to 60 percent copper which have the combination of high reflectivity and good tarnish resistance. At greater than 61 percent copper, the alloy develops a brown tarnish, and at less than 45 percent copper, the deposit is cloudy. The preferred composition is 57 or 58 percent copper for low specific resistance and high microhardness, as shown by the data in Table 18.2. In speculum plating, the concentrations of tin and copper in the solution are kept constant; the copper content of the deposit can be increased by decreasing the free cyanide concentration and/or increasing the sodium hydroxide concentration. A 1 g/l change in the latter changes the alloy composition about 1 percent.

TABLE 18.2

Properties of Tin-Copper Alloys Deposited in Cyanide Solutions[a]

Copper, wt %	Specific Resistance, microhm-cm	Microhardness, kg/sq mm	Phase Structure
0	16	26	tin
5	20	57	eta + tin
10	26	63	eta + tin
20	42	106	eta + tin
30	60	218	eta + tin
40	88	374	eta
43	100[b]	436[b]	eta
48	76	280[c]	eta
57	30[c]	425	eta + delta
74	78[b]	665[b]	delta
100	2	120	copper

[a] The alloys were deposited at 2 amp/sq dm and 65 C in an alkaline cyanide solution containing 0.02–0.41 N Sn, 0.01–0.40 N Cu, 0.3 N KCN (free), 0.2 N NaOH (free). The ratio of copper and tin concentrations in the solution varied at a constant total metal concentration of 0.42 N; constant concentration of other components. (9, 10)

[b] This value corresponds to a maximum.

[c] This value corresponds to the minimum between two maxima.

Tin-Nickel Alloys

Structure and Orientation

Deposits from commercial plating solutions usually contain about 35 percent nickel. Electrodeposited tin alloys containing 31 to 43 percent nickel have a unique single-phase structure. (16–18) There is no corresponding metallurgical phase. The electrodeposit has a crystal structure similar to the hexagonal nickel arsenide type of the gamma Ni_3Sn_2 phase. Recent studies have shown that the single-phase structure depends only on the nickel content of the deposit being within the range of 31 to 43 percent nickel; the structure is independent of the electrodeposition conditions or the type of solution adopted (chloride or fluoride solutions). (16) The single-phase deposit is a metastable compound, which recrystallizes and decomposes at temperatures above 325 C into two phases: Ni_3Sn_3 and Ni_3Sn_2, (17) as expected from the equilibrium diagram.

X-ray data have shown that the lattice constants for the single-phase alloy differ from those reported for the gamma phase, even when the electrodeposited alloy has the same composition (42.6 percent nickel for Ni_3Sn_2). (16) The axial ratio of the

TABLE 18.3

Relation of Hardness to Wear for Tin-Nickel Alloy Coatings Compared to Nickel Coatings (28)

Coating	Vickers Hardness, kg/sq mm	Weight Loss From Wear,[a] mg
Tin-nickel	548	32.6
	565	33.6
	602	40.4
Bright nickel	560	7.6
Dull nickel	286	10.2

[a] The quantity of deposit abraded for a path length of 1000 m on a rotating disk running under a load of 50 g in an emery-oil suspension.

hexagonal cell constants of the single-phase deposit is relatively constant (from 31 to 43 percent nickel) and smaller than for the gamma phase of cast metallurgical alloy (c/a = 1.23 and 1.26, respectively). Electrodeposits with less than about 27 percent nickel contain two phases, Sn and Ni_3Sn_4 (27 percent nickel).

Tin-nickel alloy deposits did not form compounds with copper, brass, or steel substrates during heat treatments that normally cause compound formation when tin deposits on these basis metals are heated. (18)

The average crystal size of both bright and dull deposits of tin-nickel alloys is between 0.01 and 0.05 μm. (16) Brightness of bright deposits is attributed to a strongly preferred orientation in which the (110) planes of the hexagonal crystallites tend to lie parallel to the basis metal. (17) More recent studies (19) on crystal orientation of the single-phase alloy have shown that at low values of nickel content, the deposit was oriented along the (1120) axis. On increasing the nickel content, the predominant axis changed by way of (1010), $(10\bar{1}0)$ and (1122) to (2021).

Density

The density of tin-nickel alloy deposits increases linearly with nickel content; values of 8.736, 8.828, and 8.954 g/cu cm were reported for alloys with nickel contents of 31.2, 35.0, and 38.7 percent, respectively. (16)

Hardness and Wear Resistance

For producing a 35 percent tin-nickel alloy, the plating solutions in the early 1950's contained sodium fluoride, and hardness was reported as 650 kg/sq mm for stress-free deposits. (20-22) For deposits from a fresh bath, a value of 710 kg/sq mm was reported elsewhere and was independent of indenter load and basis metal, provided the deposit thickness was 16 times the depth of penetration. (23) Russian workers reported values from 500 to 800 kg/sq mm. (24) Deposits from the plating solutions used since 1960, which contain no sodium ion, were reported to have a hardness of 650 to 700 kg/sq mm; (25,26) hardness values of 500 to 650 kg/sq mm are reported in Russian work. (27)

Comparison of the wear of tin-nickel and other electrodeposited coatings shows that hardness is not necessarily a reliable criterion for the wear behavior of a coating. (28) Pieces of the deposit may separate from the basis metal, which accounts for the excessive weight loss during abrasion of tin-nickel alloys as compared with pure nickel shown in Table 18.3. Brittleness and/or internal stress in the deposit could influence wear as well as hardness.

Effect of Heat Treatment on Structure and Hardness

Although the structure of the typical tin-nickel alloy electrodeposited with 35 percent nickel is metastable, there is no tendency toward recrystallization at room temperature, as confirmed by X-ray data after 1 year. (1) However, recrystallization occurs if the alloy is heated above about 250 C for extended periods in excess of 150 hours. (16) For shorter heating periods, higher temperatures are required to detect a structural change. No structural change occurred on heating for 1 hour at 550 C, although slight cracking was observed. A second phase (Ni_3Sn_4) appeared after heating for 1 hour at 700 C. (18) Alloy stressed in tension exhibited the same thermal-stability characteristics as compressively stressed alloy.

The effect of heat treatment on increasing the hardness of the alloy is correlated with structural change in Table 18.4. The precipitation hardening appears to depend

TABLE 18.4

Effect of Heat Treating Tin-Nickel Alloy[a] on Hardness and Structure (29)

Heat Treating Temperature, C	Microhardness,[b] kg/sq mm	Structure
None	520 to 550	Electrodeposited phase
200	640 to 650	—
400	670 to 680	Electrodeposited phase + delta (Ni_3Sn_4)
600	720 to 740	—
700	760 to 770	Gamma (Ni_3Sn_2) + delta (Ni_3Sn_4)

[a] Alloy containing 35 weight percent nickel deposited at 1 to 5 amp/sq dm at 75 C and pH 2.5 from a solution containing 50 g/l $SnCl_2 \cdot 2H_2O$, 300 g/l $NiCl_2 \cdot 6H_2O$, and 60 g/l NH_4HF_2.

[b] Vickers with a 25-g load.

TABLE 18.5

Internal Stress in Relation to Nickel Content of Tin-Nickel Alloy Deposits (16)

Nickel Content, weight percent	Stress kg/sq mm	Stress psi
34	10.5	15,000
35	2.0	2,800
36	−2.7	−3,800
37.5	−8.0	−11,300
39	−11.3	−16,000

on the amount of delta phase (Ni_3Sn_4) precipitated at specific temperature levels. (29)

Internal Stress

The internal stress of tin-nickel alloy deposits ranges from 28 kg/sq mm (40,000 psi) in tension to −28 kg/sq mm (−40,000 psi) in compression, depending on the plating conditions. (25) Because cracking tends to occur in deposits with a tensile stress above 5 kg/sq mm (7,000 psi), commercial plating baths are usually operated under conditions which minimize stress. Deposits with zero stress or compressive stress can be obtained by control of plating conditions.

Recent studies (16) indicate that changes in plating conditions which favor compressive stress in the deposit result in an increase in the nickel content of the deposit, as shown in Table 18.5. Thus a stress-free deposit is obtained with a nickel content only slightly above the nominal 35 percent nickel that is normally obtained from commercial plating solutions.

Table 18.6 shows values of internal stress in relation to plating conditions. Control of the plating solution temperature is a convenient method of controlling the type and amount of stress in the deposit. Over the normal operating range of 60 to 75 C, an increase in temperature reduces tensile stress or increases compressive stress. High compressive stress is reported for deposits obtained at 95 C. The solution temperature at which stress-free deposits were obtained provides a convenient basis for comparison of other plating conditions. Relative to the temperature for a stress-free deposit listed in Table 18.6, deposition at a higher temperature produces a compressive stress, and deposition at a lower temperature produces a tensile stress.

The relationship of internal stress and temperature obtained with the stannous fluoride bath [Footnote (c), Table 18.6] when no chloride was added was linear at both 1.3 and 2.6 amp/sq dm. However, the temperature-stress relationship was

TABLE 18.6

Internal Stress of Tin-Nickel Alloy Deposits in Relation to Plating Conditions (25)

Type of Plating Bath	Total Chloride Ion, g/l	Total Stannic Ion, g/l[a]	Current Density, amp/sq dm	Stress, kg/sq mm,[b] for Bath Temperature 60 C	75 C	95 C	Temperature for Stress-Free Deposit, C
Stannous fluoride[c]	0	0	1.3	16	0	−21	75
Ditto	66[d]	0	1.3	11	−1	−19	74.5
"	132[d]	0	1.3	3	−3	−17	66
"	0	0	2.6	25	9	−12	83
"	66[d]	0	2.6	6.3	−4.9	−11	67
"	132[d]	0	2.6	2.9	−8.7	−18	63
Stannous chloride-ammonium fluoride-sodium fluoride[e]	105	0	1.3	5.5	−6.5	−19	65
Ditto	105	0	2.6	3.9	−4.8	−9.1	64.5
"	205[d]	0	1.3	4	−13	−18	62.5
"	205[d]	0	2.6	1	−16	−19	61
Stannous chloride-ammonium fluoride[f]	89	0	1.3	7.2	−4.9	−29	70
Ditto	89	0	2.6	4.2	−7	−31	68
"	89	30	2.6	12	—	—	73
"	89	65	2.6	15	—	—	78.5
"	89	100	2.6	16	2.7	−22	76
"	89	100	1.3	11	0	−24	75
"	157[d]	100	1.3	−4.2	−17	−25	57
"	157[d]	100	2.6	−6.2	−21	−32	56

(a) Stannic ion added as stannic fluoride to simulate bath aging in which stannous tin is oxidized to stannic tin.

(b) Stress in kg/sq mm times 1.42 equals stress in 10^3 psi; positive values denote tensile stress and negative values designate compressive stress.

(c) 33 g/l SnF_2; 74–82 g/l NiF_2; pH adjusted to 2.5 with hydrofluoric acid or nickel carbonate.

(d) Ammonium chloride added to increase total chloride above that provided by $SnCl_2 \cdot 2H_2O$ and $NiCl_2 \cdot 6H_2O$.

(e) 50 g/l $SnCl_2 \cdot 2H_2O$; 300 g/l $NiCl_2 \cdot 6H_2O$; 28 g/l NaF; 35 g/l $NH_4F \cdot HF$; 4–12 g/l free HF; pH 2.5.

(f) 50 g/l $SnCl_2 \cdot 2H_2O$; 250 g/l $NiCl_2 \cdot 6H_2O$; 40 g/l $NH_4F \cdot HF$; 35 ml/l NH_4OH solution (sp. gr. 0.880); pH 2.5.

nonlinear when chloride was added. Chloride-ion additions reduced tensile stress or induced compressive stress.

The absolute stress of the deposit depends on many factors, such as current density, pH, and relative concentration of ions such as Na^+, NH_4^+, F^-, Ni^{+2}, Sn^{+4}, and various addition agents. (25,27,30) The first commercial tin-nickel alloy plating bath (referred to as Parkinson's bath) was distinguished by the inclusion of sodium fluoride in the formulation [Footnote (e), Table 18.6]. As the bath aged, stannic tin was formed by air oxidation and insoluble sodium and stannic tin compounds precipitated. To avoid problems with handling such precipitates, sodium ion is omitted in the formulation of present-day commercial plating solutions [Footnote (f), Table 18.6]. Since the metal-ion content of commercial plating solutions is provided by stannous chloride and nickel chloride, there is sufficient chloride-ion concentration to produce deposits with little or no tensile stress. Although aging of the bath by air oxidation of stannous to stannic ions tends to increase tensile stress, the regular additions of stannous chloride to maintain the stannous tin concentration provides supplemental chloride ions which counteract the effect of stannic tin. For example, about 68 g/l chloride ion is added during the time required for the accumulation of 100 g/l stannic tin ions, which increases the total chloride-ion concentration to 157 g/l.

For the commercial plating bath, compressive stress in the deposit would be favored by increase in the following variables, which also tends to increase the nickel content of the deposit: (25)

(1) Increasing the bath temperature
(2) Increasing the chloride-ion concentration
(3) Increasing the current density
(4) Increasing the ratio of fluoride ion to stannous ion
(5) Increasing the ratio of nickelous ion to stannous ion.

Russian studies have confirmed the effect of ammonium chloride additions on reducing the tensile stress of deposits from chloride-fluoride baths (50 g/l $SnCl_2 \cdot 2H_2O$, 300 g/l $NiCl_2 \cdot 6H_2O$, 60 g/l NH_4HF_2, pH 2.5 and 4.5, 50 to 70 C, 2 amp/sq dm). (27) The studies indicated that increasing the pH from 2.5 to 4.5 reduced internal stress at 55 C. Compressive stress was obtained at 70 C. Additions of phenol (10 g/l) and gelatin (1 g/l) increased tensile stress, but additions of ammonium chloride or treatment of the bath with activated carbon restored compressive stress. Additions of stannic ion (as $SnCl_4$) increased tensile stress and decreased the nickel content of the alloy. A linear decrease in nickel from 41 to 22 percent was reported as stannic-ion concentration increased to 1 N. However, with additions of fluoride ion (as ammonium fluoride) in the ratio of $Sn^{+4}:F^- = 1:1.5$, there was no change in alloy composition or stress with stannic-ion concentrations up to 2 N.

Unstressed deposits and deposits with compressive stress were less brittle than those with internal tensile stresses. (27) The compressively stressed plate showed no cracks when bent over a rod 3 mm in diameter, while those stressed in tension became cracked even when bent over a rod 18 mm in diameter. After heating for 1 hour at 140 C, the brittleness of deposits with tensile stresses decreased and became the same as for deposits with compressive stress. The outward appearance of the coating hardly changed at 140 C, but became slightly darker if heated at 290 C.

Tin-Zinc Alloys

Structure

X-ray studies showed that tin-zinc alloy deposits from alkaline stannate/cyanide solution were mixtures of nearly pure tin and zinc phases like their metallurgical counterparts. (8) Similar results were reported for deposits from a pyrophosphate-type bath. (31) Alloy deposits containing 20 percent zinc are much finer in grain size than electrodeposited tin. (33) Alloy containing 50 percent zinc was finer grained than one containing 22 percent zinc. (34)

In contrast with tin plate, which transforms rapidly to the gray, alpha phase at subzero temperatures when inoculated with gray tin, tin alloy containing 20 percent zinc showed no evidence of such a transformation after inoculation and exposure to dry-ice temperatures for 6 months. (32)

Hardness

Tin-zinc alloys are harder than tin and have good antifriction properties. Table 18.7 shows the increase in microhardness for tin-zinc alloy deposits with increasing zinc content. The microhardness data plotted as a function of atomic percent zinc showed a linear increase from 10 to 56 kg/sq mm. Within the range of 0 to 100

TABLE 18.7

Electrical Resistance and Hardness of Electrodeposited Tin-Zinc Alloys[a]

Zinc Content of Alloy, weight percent	Specific Resistance, microhm-cm		Microhardness, kg/sq mm	
	Electrodeposit	Metallurgical Alloy	Electrodeposit	Metallurgical Alloy
0	18.3	11.8	10	3.5
10	16.3	10.9	18	6.5
20	14.7	10.1	24	9.0
30	13.2	9.4	30	11.0
40	11.9	8.7	35	13.0
50	10.7	8.2	39.5	14.5
100	6.6	6.2	56	21

[a] Alloy deposition at 2 amp/sq dm and 70 C from a solution of constant total metal concentration with tin (as Na_2SnO_3) 1.0 to 5.0 N and Zn (as K_2ZnO_2 and $K_2Zn[CN]_4$) 0.02 to 0.5 N; constant total cyanide of 0.4 to 0.45 N (as KCN) and constant free hydroxide of 0.2 N (as NaOH). (8)

percent zinc, the electrodeposited alloys were 2.7 times as hard as the corresponding metallurgical alloys. (8)

A hardness value of 37 kg/sq mm was reported for an alloy with 20 weight percent zinc deposited at 70 C and 1.1 to 1.6 amp/sq dm from a sodium stannate-zinc cyanide bath containing 30 g/l tin (0.5 N), 2.5 g/l zinc (0.04 N), 4 to 6 g/l sodium hydroxide (0.1 to 0.15 N), and 25 to 28 g/l sodium cyanide (0.51 to 0.57 N). (33) Another report on tin-zinc alloys (20 to 50 percent zinc) pointed out that immersion in oxidizing solution hardened the deposit, bringing the hardness up to 37 kg/sq mm, and eliminated finger marking. (34) Still another reference gives the hardness of the 20 percent zinc alloy as about 37 kg/sq mm. (35) The hardness of a cast alloy of the same composition is 9 kg/sq mm. The greater hardness of the electrodeposits is probably associated with their fine grain size.

Electrical Resistivity

Specific resistance data as a function of atomic percent zinc showed a linear decrease from 18.3 to 6.6 microhm-cm for electrodeposited tin alloys with 0 to 100 percent zinc. (8) Compared to the metallurgical counterpart, the specific resistance of the electrodeposit was 50 percent higher at the tin-rich end of the composition range and 5 percent higher at the zinc-rich end. The electrodeposition conditions for the alloys are given in the footnote to Table 18.7.

TABLE 18.8

Electrical Resistance of Tin-Lead Alloy Deposits[a]

Lead Content, weight percent	Specific Resistance, microhm-cm
0	11.4
0.05	12.7
0.10	14.1
0.50	13.9
5.0	13.5
10.0	13.2
20.0	13.4
40	14.5
60	16.7
80	20.8
96.5	24.0
100	22.9

[a] Measurements at 25 C for deposits from a fluoborate solution. Other conditions not given. (37)

Other Tin Alloys

Tin-Lead Alloys

Tin-lead alloys deposited at different current densities and temperatures showed no fundamental differences in crystal structure, hardness, or specific resistance in comparison with their metallurgical counterparts. (36)

An increase in the lead content from 0 to 70 percent for alloys deposited from fluosilicic acid solutions produced no change in hardness compared to pure tin (12 kg/sq mm) and a relatively small linear increase in specific resistance, from 12 to 17 microhm-cm. (9)

Another source reported a minimum resistivity of 13 microhm-cm for alloy deposits containing 10 percent lead and 11.4 microhm-cm for unalloyed tin. (37) Data for other alloys deposited in fluoborate solutions are given in Table 18.8. A maximum resistivity of 14.1 microhm-cm for alloy containing 0.1 percent lead indicates a solubility limit of the same order. Supplemental information on the properties of the tin-lead alloys can be found in Chapter 10, Lead and Lead Alloys.

Tin-Iron Alloys

The microhardness of an electrodeposited tin-iron alloy containing 20 percent iron was 35 to 40 kg/sq mm, as compared to 8 to 10 kg/sq mm for pure tin. (38) Alloys containing less than 48 percent iron were deposited on steel or brass at a current density of 0.5 to 1.0 amp/sq dm in an electrolyte containing 27 g/l ferric chloride ($FeCl_3 \cdot 6H_2O$), 6.8 g/l stannous chloride ($SnCl_2 \cdot 2H_2O$), and 180 to 190 g/l sodium pyrophosphate ($Na_4P_2O_7 \cdot 10H_2O$) at 55 C, pH 8.8.

References

(1) *Tin and Its Alloys*, edited by E. S. Hedges, Edward Arnold (Publishers) Ltd., London (1960).

(2) Dohi, N., "Crystal Structure of Bright Tin Electrodeposit. Studies on Bright Tin Plating (Part 5)," *Journal of the Metal Finishing Society of Japan, 16* (7), 293–299 (1965).

(3) Bellinger, K. P., and Steeg, H. J., "Practical Bright Acid Tin Plating," *Plating, 54*, 1139–1147 (1967).

(4) Smith, J. F., "Surface Texture of Electrodeposited Tin," *Trans. Inst. Metal Finish., 46*, 199–200 (1968).

(5) MacNaughtan and Hothersall, "The Determination of the Structure of Electrodeposits by Metallurgical Methods," *Transactions of the Faraday Society, 31*, 1168 (1935).

(6) *Metals Handbook*, 8th Ed., *Vol. 1, Properties and Selection of Metals*, American Society of Metals (1961).

(7) Roper, M. E., "The Development of Tin Plating," *Metal Finishing Journal, 12* (140), 323–326, 329; (141), 361–365 (1966).

(8) Fedot'ev, N. P., Vyacheslavov, P. M., and Andreeva, G. P., "Structure and Properties of Electrodeposited Zinc-Tin Alloy," *Journal Applied Chemistry* (USSR), *36* (3), 640–641 (1963); *Zh. Prikl. Khimii, 36* (3), 671–673 (1963).

(9) Fedot'ev, N. P., and Vyacheslavov, P. M., "The Phase Structure of Binary Alloys Produced by Electrodeposition," *Plating, 57*, 700–706 (1970).

(10) Burkat, G. K., Fedot'ev, N. P., Vyacheslavov, P. M., and Smorodina, T. P., "Structure of Electrolytic Copper-Tin Alloy," *Journal Applied Chemistry (USSR), 39* (3), 658–661 (1966); *Zh. Prikl. Khimii, 39* (3), 702–705 (1966).

(11) Raub, E., and Blum, W., "The Electrodeposition of Lead-Tin Alloys," *Metalloberflaeche, 9A* (4), 54–57 (1955). Brutcher Translation No. 3631.

(12) Clarke, M., and Britton, S. C., "A New Addition Agent for Bright Tin Plating," *Trans. Inst. Metal Finish., 39*, 5–13 (1962).

(13) Gratsianski, N. N., and Kalyuzhnaya, P. F., "Chemical Resistance of Tin Coatings," *Zh. Prikl. Khimii, 21*, 341–346 (1948).

(14) Machu, W., "Brush Plating for Localized Metal Deposition (Dalic Process)," *Metallwaren-Industrie und Galvanotechnik, 49* (11), 467–481 (1958). Cf. Hughes, H. D., *Trans. Inst. Metal Finish., 33*, 424–439 (1956).

(15) "Improvements in Plating Processes: The Dalic Process," *Product Finishing (London), 6* (9), 77–79 (1953).

(16) Dutta, P. K., and Clarke, M., "Structure and Thermal Stability of Tin-Nickel Alloys Electrodeposited From Acid Baths," *Trans. Inst. Metal Finish., 46*, 20–25 (1968).

(17) Rooksby, H. P., "An X-Ray Study of Tin-Nickel Electrodeposits," *J. Electrodepositors' Soc., 27*, 153–160 (1951).

(18) Smart, R. F., and Robins, D. A., "The Structural Stability of Tin-Nickel Electrodeposits," *Trans. Inst. Metal Finish., 37*, 108–109 (1960).

(19) Kochergin, S. M., Pobedimsky, G. R., and Zhikharev, A. I., "Crystal Orientation of an Electrolytic Tin-Nickel Alloy," *Elektrokhimiya, 2* (8), 958–959 (1966).

(20) Cuthbertson, J. W., "Electrodeposition of Tin Alloys," *Industrial Finishing (London), 7* (76), 176–183, 191 (1954).

(21) Gore, R. T., and Lowenheim, F. A., "Is Tin-Nickel the New Plating Finish You Need?" *Iron Age, 177* (22), 59–61 (1956).

(22) Gore, R. T., "Tin-Nickel Alloy Plated Coatings," *Materials & Methods, 42* (4), 102–105 (1955).

(23) Ramanathan, V. R., "Hardness of Electrodeposited Tin-Nickel Alloy on Brass," *Trans. Inst. Metal Finish., 34*, 4–7 (1957).

(24) Kudryavtsev, N. T., and Tyutina, K. M., "Electrolytic Deposition of Tin-Nickel Alloy From Chloride-Fluoride Solutions," *Zashchitno - Dekorativ i Spetsial. Pokrytiya Metal., Nauch. -Tekh. Obshchestvo Mashinostroitel. Prom.*, 5–13 (1959).

(25) Clarke, M., "Internal Stress in Electrodeposited Tin-Nickel Alloy," *Trans. Inst. Metal Finish., 38*, 186–194 (1961).

(26) Couch, R. W., and Bikales, R. G., "Plating of Printed Circuits With Pyrophosphate Copper and Tin-Nickel," *Proceedings Am. Electroplaters' Soc., 48*, 176–181 (1961).

(27) Kudryavtsev, N. T., Tyutina, K. M., Kosmodamianskaya, L. V., and Yarlykov, M. M., "Study of Internal Stresses in Tin-Nickel Alloy Deposits," *Protection of Metals, 3* (2), 156–161 (1967); *Zashchita Metallov, 3* (2), 198–204 (1967).

(28) Raub, E., and Müller, K., *Fundamentals of Metal Deposition*, Elsevier Publishing Co., Ltd., Barking, Essex (1967), 268 pp.

(29) Matsumoto, S., "Effect of Heat Treatment of Electrodeposited Ni-Sn Alloy on Its Hardness and Structure," *Journal of the Metal Finishing Society of Japan, 14* (7), 253–260 (1963).

(30) Kudryavtsev, N. T., and Tyutina, K. M., "Electrodeposition of Tin-Nickel Alloys," *Electrodeposition of Alloys*, edited by V. A. Averkin. Translated by the Israel Program for Scientific Translations, Jerusalem, 70–86 (1964). OTS 64-11015. Cf: Tyutina, K. M., and Kudryavtsev, N. T., "Electrolytic Deposition of Tin-Nickel Alloy From Chloride-Fluoride Solutions," *Zh. Prikl. Khimii. 31*, 723 (1958).

(31) Vaid, J., and Rama Char, T. L., "Tin-Zinc Alloy Plating from the Pyrophosphate Bath," *Symposium on Electrodeposition and Metal Finishing, Karaikudi, 1957*, 61–62 (1960).

(32) Lowenheim, F. A., "Tin Alloy Plating: American Experience," *Trans. Inst. Metal Finish., 31*, 386–397 (1954).

(33) Cuthbertson, J. W., and Angles, R. M., "The Electrodepositon and Properties of Tin-Zinc Alloys," *Trans. Electrochem. Soc., 94*, 73–98 (1948).

(34) Bertorelle, E., and Fogliani, F., "Studies on the Electrodeposition of Tin-Zinc Alloy," *La Chimica e l'Industria, 34*, 639–645 (1952).

(35) Whittaker, A., "Production Tin-Zinc Alloy Plating," *Machinery (London), 84* (2158), 639–642 (1954).

(36) Riedel, W., "Structure and Properties of Electrolytically Deposited Tin-Lead Alloys," *Metalloberflaeche, 23* (2), 42–44 (1969).

(37) Mohler, J. B., and Sedusky, H. J., "Alloys by Electrodeposition," *Metal Finishing, 45* (12), 65–70 (1947).

(38) Izmailov, A. V., and Kudryavtseva, V. A., "Conditions for the Electrolytic Preparation of Tin-Iron Alloy," *Izvestiya Vysshikh Uchebnykh Zavedeniy SSSR, Khimiya i Khimicheskaya Tekhnologiya, 10* (9), 1031–1034 (1967).

Chapter 19

Zinc

Electrodeposited zinc is used primarily for the protection of steel from corrosion. Thickness usually ranges from 4 to > 25 μm (0.15 mil to > 1 mil) depending upon the anticipated applications and environments. Small bare spots on the steel surface are protected by galvanic action.

The mechanical properties of zinc coatings are not usually critical, and they affect the mechanical properties of the steel only when the steel and zinc are of about equal thickness. However, certain properties such as fatigue limit may sometimes be affected, and hydrogen embrittlement may result, from the plating process.

No consistent relationship exists between hardness, strength, and ductility of electrodeposited zinc. These properties are in the same general range as those for wrought zinc. Although electrodeposited zinc is usually less ductile than wrought zinc, elongations up to 51 percent have been reported, but the material recrystallized during the tests. This value is comparable to that for recrystallized wrought zinc but falls short of the moderately superplastic 200 percent elongation exhibited by highly worked zinc foil.

Stress in zinc deposits is frequently compressive, which is beneficial for maximizing the endurance limit of zinc-coated steel. The magnitude of stress can be controlled by the plating bath composition and conditions. Solution composition and current density are critical factors affecting stress. Any tensile stress initially exhibited by a zinc deposit is often converted to compressive stress as the deposit ages. The only serious consequence of stress in zinc is that adjacent areas of high stress and low stress may cause undesirable pitting corrosion.

Although the data on physical, mechanical, and electrical properties of electrodeposited zinc are limited and incomplete, available data appear reliable.

Physical and Mechanical Properties

Density

The density of electrodeposited zinc on copper appeared to be greater than the bulk density of metallurgical zinc (7.13 g/cu cm), according to calculations from stress measurements of zinc deposited in 20 C, zinc sulfate - sodium acetate solu-

TABLE 19.1

Mechanical Properties of Zinc Deposited from Low-Acid Baths (2)[a]

Material	Hardness, Knoop hardness number (400-gram load)	Tensile Strength		Ductility, percent elongation
		kg/sq mm	kpsi	
Electrodeposited zinc (8 specimens)	35–67	4.9–11.2	7–16	1–2
Electrodeposited zinc (4 specimens)	36–42	6.3–10.5	9–15	12–51
Electrodeposited zinc (1 specimen)	35	10.5	15	51
Zinc sheet—rolled	55	21.8	31	10
Zinc sheet—annealed	47	9.1	13	32

[a] Electrodeposited zinc specimens were about 0.75 mm (30 mils) thick. Low-acid baths contained (a) 240 g/l $ZnSO_4 \cdot 7H_2O$, 15 g/l $NaC_2H_3O_2 \cdot 3H_2O$, 30 g/l $Al_2(SO_4)_3 \cdot 18H_2O$, and 1 g/l licorice, or (b) 280 g/l $ZnSO_4 \cdot 7H_2O$, 35 g/l $Al_2(SO_4)_2 \cdot 18H_2O$, 18 g/l NH_4Cl and 1 g/l licorice. The baths were purified with zinc dust, ranged in pH from 2.5 to 4.0, and were operated at 25 to 40 C. Current density was 1.5 amp/sq dm. The properties of specimens from the two baths were indistinguishable. Ductility greater than about 2 percent elongation occurred only after the bath had been treated with potassium permanganate.

tions with pH of 5.4. (1) Calculated values for density decreased from 7.6 to 7.2 g/cu cm as the current density was increased from 1 to 4 amp/sq dm. With 3 g/l aminobenzoic acid added to the plating bath, the calculated density reached 7.8 g/cu cm at a current density of 1 amp/sq dm.

Strength and Ductility

Tensile strength for 0.75-mm-thick zinc electrodeposited in acid solutions ranged from 4.9 to 11.2 kg/sq mm (7,000 to 16,000 psi). (2) One group of specimens deposited in a solution treated with potassium permanganate exhibited a relatively high ductility (12 to 51 percent elongation) in comparison with another group plated in a solution treated with only zinc dust. The tensile strength and ductility data for these groups are summarized in Table 19.1. The ductile deposits, associated with high purity, were recrystallized during tensile testing.

The electrodeposited specimen showing 51 percent elongation (Figure 19.1) corresponded to the typical columnar structure frequently encountered in electro-deposited metals. Figure 19.2 shows recrystallization near the fracture of the tensile specimen. In this area, zinc recrystallized and grain growth occurred during deformation. Deformation was equivalent to hot working, and recrystallization was possible because zinc has a low melting point (420 C). The electrodeposited specimen with 50 percent elongation corresponds to recrystallized rolled sheet.

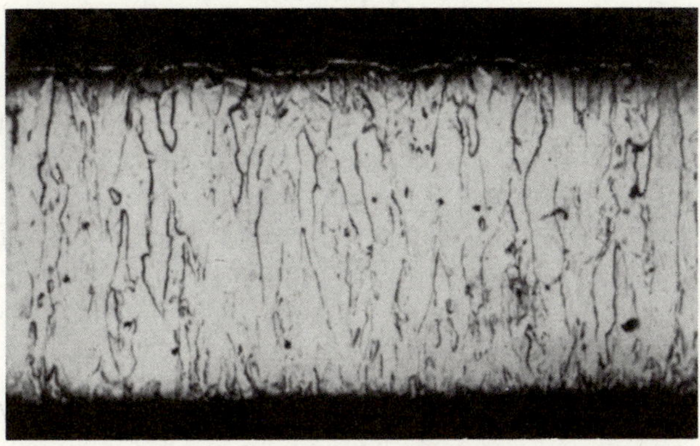

Figure 19.1. Columnar Structure of Zinc Electrodeposit Exhibiting 51 Percent Elongation. (2)

Shifts in the pH of the zinc sulfate solution from 2.5 to 4.0 or a change in temperature from 25 to 40 C had no consistent effect on strength or ductility of zinc deposited in solutions treated with zinc dust. (2) These solutions were not treated with potassium permanganate.

Strength and ductility data for thin zinc coatings (30 and 93 μm or 1.2 and 3.7 mils) on steel, determined by hydraulic-bulge testing, are given in Table 19.2. Tensile strength was higher than that for the 0.75-mm deposits. In Table 19.2, ductility is expressed as significant strain at fracture, but is equivalent to more than 3 percent elongation.

Several variables were found to have little effect. These included thickness in the range of 30 to 93 μm, temperature changes from 40 C to 50 C, and impurities in the solution such as 20 mg/l lead, 100 mg/l iron, or 5 g/l manganese. However, the ad-

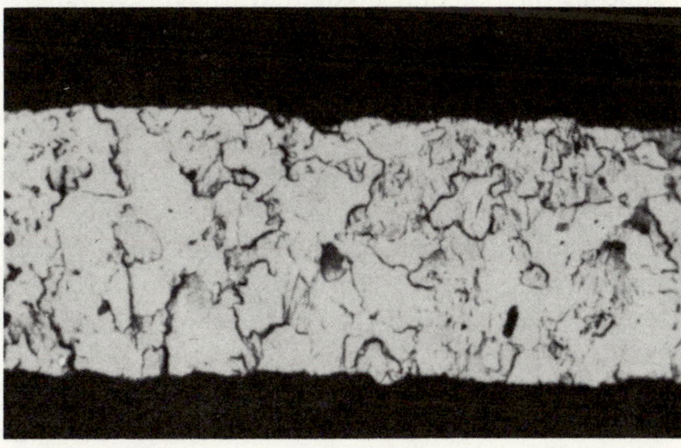

Figure 19.2. Structure of Ductile Zinc Electrodeposit near the Fracture of a Tensile-Tested Specimen Showing Recrystallization. (2)

dition of 100 mg/l cadmium increased the tensile strength 139 percent and reduced the ductility of 30-μm zinc deposited at a temperature of 50 C by 68 percent. In similar deposits made at 40 C, cadmium increased the strength 83 percent and reduced ductility by 45 percent. Thicker deposits (93 μm) had smaller increases in strength of 57 percent (40 C) and 59 percent (50 C), but similar losses in ductility (69 percent for 40 C deposits and 79 percent for 50 C deposits).

Table 19.1 compares the strength and ductility of electrodeposited and rolled zinc. The ductility of the rolled, wrought zinc is indicative of lightly worked sheet. Unannealed, heavily worked zinc (90 percent reduction in thickness) displays weakly superplastic properties and an elongation of 200 percent.

Superplasticity with 200 percent or more elongation occurs when extremely small grains (1–2 microns) are deformed rapidly to prevent recrystallization and grain growth. Extremely fine grains occur in many electrodeposited metals, but no reference was encountered where high strain rates were used to look for high ductility. It seems probable that stable fine grains could be induced with some addition agents used in zinc plating baths.

The ductility of single-crystal zinc depends upon the orientation of the basal plane of the zinc with respect to the deformation axis. If the basal plane is between 40 and 60 degrees of the deformation axis, the zinc is ductile. At lower angles, zinc deforms by twinning. At greater angles, the zinc crystal cleaves. Therefore the ductility of electrodeposited zinc will depend upon the predominate orientation of the deposited zinc crystals.

Some recent evidence shows that the orientation of zinc deposits depends upon the current density. (3) Within the range of 0.4 to 10 amp/sq dm, deposits at 10 amp/sq dm were dendritic with the stem and side branches of the dendrites growing in the (1210) direction. At 0.4 amp/sq dm, mossy growths containing large numbers of small zinc crystals were obtained. The dendritic deposit probably would deform by twinning and show brittle behavior. The mossy deposits of small crystals could have some ductility. These zinc deposits were made on oriented single-crystal substrates in an aqueous solution of potassium hydroxide saturated with zinc oxide.

Hardness and Stress

Hardness is frequently related to stress in electrodeposited zinc and other metals, but should not be used as a measure of strength or ductility because there is no consistent relationship between hardness and strength or ductility. (2) In general, stress in compression increases as the hardness increases in electrodeposited zinc. This relationship appears in Table 19.3, which summarizes the data reported for hardness and stress of zinc deposited in acid solutions.

As a rule, the hardness of electrodeposited zinc ranges from about 35 to 125 kg/sq mm. Differences in the type of indentor and in the load, which were not always reported, could account for this broad range in hardness values. Acid sulfate baths containing no grain-refining agents give values of 35 to 80 kg/sq mm. Increasing the current density increased hardness. (9) Changes in pH had little effect. (4) Gelatin additions hardened zinc deposits slightly and dextrin hardened them considerably, as follows: (7)

Dextrin, g/l	Hardness, kg/sq mm	Compressive Stress, kg/sq mm
None	59	1.0
1	112	4.0
2	115	4.5
5	115	4.8
10	126	5.3

Aging for 25 min after electrodeposition increased the hardness of zinc deposited at 2.5 amp/sq dm in a zinc sulfate bath from about 80 to 140 kg/sq mm. (9) This solution contained 150 g/l zinc sulfate, 50 g/l sodium sulfate, 20 g/l sodium chloride, and 10 g/l boric acid.

Dull zinc deposited in cyanide solutions is similar in hardness to dull zinc deposited in acid baths. A hardness of 86 kg/sq mm was reported for the cyanide zinc deposits. (13) Ultrasonic agitation in place of air agitation increased hardness of zinc deposited at 8 amp/sq dm in a cyanide bath from 50 to 103 kg/sq mm. (14)

Bright zinc deposited in cyanide solutions containing commercial brighteners exhibited hardness values of 132 to 148 kg/sq mm. (13,15) In wear tests, dull zinc with a hardness of 86 kg/sq mm exhibited 3.3 times the weight loss measured for bright zinc with a hardness of 132 kg/sq mm. (13)

Bright zinc deposited in an alkaline zinc sulfate bath containing polyethylene polyamine and ammonium chloride exhibited hardness values in the range of 180 to 200 kg/sq mm. (16) These deposits were produced at 1 to 4 amp/sq dm in a 20 C solution with a pH of 8 to 9.5.

TABLE 19.2

Tensile Strength and Ductility of Thin Zinc Coatings Deposited on Steel (2)

Thickness,		Plating	Ultimate Tensile Strength				Ductility, Significant Strain at Fracture	
			Zinc[a]		Zinc-Cadmium[b]			
μm	mils	Temp, C	kg/sq mm	psi	kg/sq mm	psi	Zn	Zn-Cd
30	1.2	40	11.5	16,500	21.0	30,200	0.144	0.079
93	3.7	40	13.0	18,400	21.0	29,900	0.162	0.051
30	1.2	50	12.5	17,200	29.0	41,100	0.152	0.049
93	3.7	50	12.5	18,000	20.0	28,700	0.138	0.029

(a) The electrolyte was prepared by mixing 62.5 g/l zinc sulfate and 5 g/l sulfuric acid and enough potassium permanganate to make the solution pink. The excess acid was neutralized with ZnO. The neutral solution was analyzed for Zn and sulfuric acid added to give a zinc concentration of 75 g/l and a sulfuric acid concentration of 235 g/l.

(b) The plating solution also contained 100 mg/l cadmium.

TABLE 19.3

Hardness and Stress Data for Zinc Electrodeposited in Acid Baths

Stress[a]		Hardness, kg/sq mm	Plating Bath	Current Density, amp/sq dm	Temperature, C	Remarks
kg/sq mm	psi					
−0.1 to −0.95	−140 to −1,400	49 to 52	240 g/l $ZnSO_4 \cdot 7H_2O$[b]	1	18	Compressive stress decreased with increasing pH
<−0.7	<−1,000	No data	Acid with glucose[c]	—	—	
−0.4 to −0.5	−600 to −700	Ditto	Sulfate[d]	1 to 4	19–22	Adding 5 g/l aminobenzoic acid increased compressive stress to 1.7 kg/sq mm
−0.02 to −1.2	−300 to 1,700	″	450 g/l $ZnSO_4 \cdot 7H_2O$[e]	1 to 10	20–30	
−0.3 to −1.0	−400 to −1,400	51 to 59	240 g/l $ZnSO_4 \cdot 7H_2O$[f]	0.25–2.0	20	
−2.6	−3,700	No data	143 g/l $ZnSO_4 \cdot 7H_2O$ with gum arabic[g]	—	—	
−2.8 to −5.0	−4,000 to −7,000	Ditto	Sulfate[h]	—	—	Stress decreased with increase in thickness
−2.5 to −3.5	−3,500 to −5,000	65 to 88	240 g/l $ZnSO_4 \cdot 7H_2O$ with 0.2 to 1.2 g/l gelatin[i]	1	20	Gelatin increased hardness and compressive stress
−4.0 to −4.8	−5,700 to −6,900	112 to 125	240 g/l $ZnSO_4 \cdot 7H_2O$ with 1 to 10 g/l dextrin[i]	1	20	Dextrin increased hardness and compressive stress
2.0 to 8.0[j]	2,900 to 11,400[j]	50 to 80	150 g/l $ZnSO_4 \cdot 7H_2O$ with 10 g/l H_3BO_3[k]	1.25 to 7.5	20	Hardness and stress increased as current density was increased
2.2 to 9.0[j]	3,100 to 12,900[j]	No data	$ZnSO_4$[l]	1.5	20	Stress increased as grain size decreased with increasing zinc sulfate concentration
15[j]	21,000[j]	Ditto	200 g/l $ZnSO_4$ with 120 g/l $Al_2(SO_4)_3$[m]	1.5	20	Aluminum sulfate increased stress

0.25 to 1.2[n]	350 to 1,700[n]	"	450 g/l $ZnSO_4 \cdot 7H_2O$[c]	1 to 10	20–30	Increasing current density increased stress
0.25 to 8.0[n]	350 to 11,400[n]	"	215 g/l $ZnSO_4 \cdot 7H_2O$ and 1.25 g/l thiourea[o]	1 to 10	20–30	
0.8 to 11.4[n]	1,100 to 16,300[n]	"	150 g/l $ZnSO_4$[p]	0.5 to 2.5	20	Stress varied with substrate texture

[a] A negative sign designates compressive stress.

[b] The solution contained 240 g/l $ZnSO_4 \cdot 7H_2O$ and 30 g/l $Al(SO_4)_3 \cdot 18H_2O$, 50 g/l $KAl(SO_4)_2$ and 120 g/l $Na_2SO_4 \cdot 10H_2O$. (4)

[c] Details were not disclosed. (5)

[d] The solution contained 150 g/l zinc sulfate and 40 g/l sodium acetate and was adjusted to a pH of 5.4. (1)

[e] The bath contained 450 g/l $ZnSO_4 \cdot 7H_2O$ and 30 g/l $Al_2(SO_4)_3 \cdot 18H_2O$. (6)

[f] Zinc with thickness of 25–30 μm deposited in solutions containing 240 g/l $ZnSO_4 \cdot 7H_2O$, 30 g/l $Al_2(SO_4)_3 \cdot 18H_2O$, 50 g/l $KAl(SO_4)_2$, and 120 g/l $Na_2SO_4 \cdot H_2O$ (pH = 3.5). (7)

[g] The bath contained 143 g/l $ZnSO_4 \cdot 7H_2O$ and 0.25 g/l $NaCOOCH_3$ and 1 g/l gum arabic. (6)

[h] Zinc deposited in zinc sulfate-sodium acetate-ammonium chloride solution at 0.15 to 0.75-mil thickness. (8)

[i] Bath same as solution in Footnote (f) with additions of 0.2 to 1.2 g/l gelatin or 1, 2, or 5 g/l dextrin. (7)

[j] Stress shifted from tensile to compressive within an hour after plating.

[k] Solution of 150 g/l $ZnSO_4$, 50 g/l Na_2SO_4, 20 g/l NaCl and 10 g/l H_3BO_3 (pH = 4). (9)

[l] $ZnSO_4$ concentration varied from 50, 100, 200, 300, to 400 g/l to give 10 μm deposits. (10)

[m] Solution of 200 g/l $ZnSO_4 \cdot 7H_2O$ with 120 g/l $Al_2(SO_4)_3 \cdot 18H_2O$. (10)

[n] Inexplicit with respect to sign (tensile or compressive).

[o] The solution contained $ZnSO_4 \cdot 7H_2O$, 30 g/l $Al_2(SO_4)_3 \cdot 18H_2O$ or $KAl(SO_4)_2 \cdot 12H_2O$. (11)

[p] Thin zinc deposits were made in solutions of pH = 3.6 containing 150 g/l $ZnSO_4 \cdot 7H_2O$, 100 g/l $Al_2(SO_4)_3 \cdot 18H_2O$, with different substrate of Zn, Cd, Cu, Ni, Sb, Bi, or Pb. (12)

Inclusions of electrolytic nickel powder in electrodeposited zinc had little effect on hardness or wear resistance. (17) The use of ultrasonic agitation during deposition of a zinc-iron alloy decreased its coefficient of friction and increased wear resistance. (18)

Stress in zinc deposited in acid baths is usually compressive. Typical values range from −0.02 to −5.0 kg/sq mm (−300 to −7,000 psi) for sulfate solutions containing no grain-refining agents other than glucose or gum arabic. Additions of gelatin and dextrin increased compressive stress values appreciably for 30-μm zinc deposited at 2 amp/sq dm, as follows: (7)

Addition Agent	Compressive Stress	
	kg/sq mm	psi
None	1.0	1,400
Gelatin, 1.2 g/l	3.5	5,000
Nitroso β naphthol,	2.0	3,000
Dextrin, 1 g/l	4.0	5,700
Dextrin, 5 g/l	4.8	6,900

A large increase in tensile stress from 2.2 to 9.0 kg/sq mm (3,100 to 12,900) was reported for zinc deposited in concentrated solutions, in comparison with zinc deposited in dilute baths. (10) Stress was related to grain size, as follows:

$Zn \cdot SO_4$ Concentration, g/l	Grain Size, μm	Stress	
		kg/sq mm	psi
50	4.2	2.2	3,100
100	5.5	4.2	6,000
200	1.5	8.0	11,300
300	1.5	9.0	12,900

The above tensile values for stress shifted to compressive, however, within 50 minutes after electrolysis. (10) A similar shift in stress was reported after 5 days of aging for zinc deposited in a bath containing glucose. (19)

Increasing the current density for zinc deposited in a bath containing sodium chloride and boric acid increased tensile stress appreciably, as follows: (9)

Current Density, amp/sq dm	Stress		Hardness, kg/sq mm
	kg/sq mm	psi	
1.25	2	2,890	50
3.75	4	5,600	80
5.0	5	7,000	—
7.5	8	11,400	—

Stress in these deposits also shifted from tensile to compressive within 15 to 60 minutes after plating. (9) Zinc deposited in a bath containing 1.25 g/l thiourea also exhibited a high stress. (11) Microcracks were noted in electron photomicrographs of these deposits.

Brittleness of zinc is not associated with high stress, but brittleness of zinc-plated steel is attributed to hydrogen absorption during plating. (4) A ductile zinc alloy containing 6.5 to 9.5 percent nickel was reported to have a tensile stress of 0.35 to 1.0 kg/sq mm (500 to 1,400 psi). (20)

Specific Resistivity and Hardness

The specific resistivity of zinc deposited in a cyanide bath was approximately 7.5 microhm-cm. (21) Resistivity increased gradually to 9.5 microhm-cm as the amount of codeposited cadmium was increased to about 88 percent. Microhardness increased from 90 to 110 kg/sq mm as 10 percent of cadmium was codeposited with the zinc and then decreased to about 35 kg/sq mm when the cadmium content reached 88 percent.

The resistivity of the zinc electrodeposit is about 35 percent higher than handbook values for wrought, high-purity zinc, which probably is a result of salts occluded at grain boundaries.

Effect of Electrodeposited Zinc on the Properties of the Substrate

Compressively stressed zinc (and other metals) improves the fatigue limit of steel, whereas electroplate stressed in tension reduces it. (22) Since zinc is ordinarily stressed in compression or changes to a compressive stress after short aging, (9,10,19) conventional deposits should not harm the fatigue limit of steel. However, zinc deposits produced in cyanide solutions reduced the fatigue strength of the substrate about 5 to 10 percent. (23) These zinc deposits may have been stressed in tension.

The effects of hydrogen pickup during zinc plating on the embrittlement of steel substrates have been known for many years. The cracking of case-hardened, wheel-hub bolts during zinc plating (24) is just one example of such embrittlement.

Structure

The microscopic structure of a typical zinc deposit produced in an acid sulfate bath was shown in Figure 19.1. A similar columnar or fibrous structure was reported for zinc deposited in cyanide baths with varying concentrations of sodium cyanide and sodium hydroxide. (25) Increasing the free cyanide concentration refined the grain structure. Recrystallized structures like that in Figure 19.2 would be expected for zinc subjected to a high stress or the low baking temperatures adopted for removing hydrogen from the steel substrate.

During the deposition of single crystals of zinc on single crystals of iron, the basal (0001) plane of the hexagonal-close-packed zinc aligned in a parallel fashion with the (110) plane of the iron. (26) The lattice constants of hcp zinc deposited in a cyanide electrolyte were: a = 2.676 and c = 4.954, (21) in comparison with handbook values of 2.6649 and 4.970 angstroms. On a polycrystalline substrate, β- and

γ-phase formation in zinc-nickel alloys was favored by agitating the bath, in comparison with a mixture of α and β phases for zinc deposited with no agitation. An increase in the temperature of the bath to 40 C promoted the formation of α phase. (27)

References

(1) Sadek, Hussein, and Rizk, Antwan S., "Stress During the Deposition of Zinc on Copper Substrate," *J. Chem. U.A.R., 8* (3) 239–50 (1965).

(2) Read, H. J., Smith, W. H., and Joseph, W. B., "The Tensile Strength and Ductility of Zinc Electrodeposited from Acid Baths," *Proceedings Am. Electroplaters' Soc., 51*, 61–65 (1964).

(3) Naybour, R. D., "Morphologies of Zinc Electrodeposited from Zinc-Saturated Aqueous Alkaline Solution," *Electrochim. Acta, 13* (4), 763–769 (1968).

(4) Fedot'ev, N. P., "Physical and Mechanical Properties of Electrodeposited Metals," *Plating, 53* (3), 309–317 (1966).

(5) Newell, I. L., "Stress in Electrodeposited Metals," *Metal Finishing, 58* (10), 56–61, (1960).

(6) Vagramyan, A. T., and Solov'eva, Z. A., *Technology of Electrodeposition*. English Translation, Robert Draper Ltd., Teddington (1961), 398 pp.

(7) Fedot'ev, N. P., and Khonikevich, A. A., "Internal Stresses in Electrolytic Deposits of Zinc, Obtained from Sulfate Electrolytes," *Journal Applied Chemistry (USSR), 33* (2), 352–357 (1960); *Zh. Prikl. Khimii, 33* (2), 355–362 (1960).

(8) Kushner, J. B., "A New Instrument for Measuring Stress in Electrodeposits," *Proceedings Am. Electroplaters' Soc., 41*, 188–195 (1954).

(9) Popereka, M. Ya., "Secondary Internal Stresses in Zinc Deposits," *Journal Applied Chemistry (USSR), 37* (3), 553–558 (1964); *Zh. Prikl. Khimii, 37* (3) 547–553 (1964).

(10) Popereka, M. Ya., and Koshmanov, V. V., "Influence of the Electrolyte Composition on the Structure and Internal Stresses of Zinc Deposits," *Journal Applied Chemistry (USSR), 38* (9), 1968–1973 (1965); *Zh. Prikl. Khimii, 38* (9), 2011–2017 (1965).

(11) Gorbunova, K. M., and Popova, O. S., "Internal Stresses in Electrolytic Deposits. I. Stresses in Zinc Deposits," *Zh. Fiz. Khimii, 30* (2), 269–276 (1956).

(12) Popereka, M. Ya., and Koshmanov, V. V., "Influence of the Material of the Cathode Surface on Internal Stresses in Zinc Deposits," *Journal Applied Chemistry (USSR), 38* (6), 1275–1279 (1965); *Zh. Prikl. Khimii, 38* (6), 1296–1300 (1965).

(13) Weiner, R., "Mechanical Properties of Several Bright Silver Electrodeposits," *Galvanotechnik, 57* (1), 15–20 (1966).

(14) Lanyi, R. J., Lane, D. H., Forbes, C. A., and Ricks, H. E., "The Application of Ultrasonic Energy to Metal Processing," *SAE Trans., 71*, 520–540, 562 (1963).

(15) Weiner, R., and Klein, G., "Hardness and Wear-Resistance of Electrodeposits," *Metalloberflaeche, 7B* (1), 1–7 (1955).

(16) Ryabchenkov, A. V., Kokorek, N. R., (USSR), "Bright Zinc Plating in Polyethylene-polyamine electrolyte," *Blestyashchie Komb. Metal. Pokrytiya, (2)*, 83–87 (1967); Ref. *Zh. Khim.* (1969), Abstr. No. 5L366.

(17) Saifullin, R. S., and Dryazgova, E. A., "Lead Plating from Electrolytes Containing Dispersed Particles," *Tr. Kazan. Khim.-Tekhnol. Inst. Cmeni S. M. Kirova (34)*, 160–165 (1965).

(18) Karkhanin, V. P. and Sushkevich, M. V., "Antifriction Properties of Electroplated Alloys of Zinc with Iron Prepared in an Ultrasonic Field," *Tr. Stavropol. Sel'skokhoz. Inst. (USSR) (25)*, 85–89 (1967).

(19) Pinner, R., "Stress in Electrodeposits," *Electroplating and Metal Finishing, 9* (12), 391–396 (1956); *Ibid. 10*, (1), 7–11 (1957).

(20) "Zinc-Nickel Alloy Deposition," U.S. Patent 3,420,754 (July 1, 1969). Assigned to Pittsburgh Steel Co., Pittsburgh, Pa.

(21) Fedot'ev, N. P., and Vyacheslavov, P. M., "The Phase Structure of Binary Alloys Produced by Electrodeposition," *Plating, 57*, 700–706 (1970).

(22) Such, T. E., "The Physical Properties of Electrodeposited Metals," *Metallurgia, 56* (334), 61–66 (1957).

(23) Vagramyan, A. T., and Petrova, Yu. S., *The Mechanical Properties of Electrolytic Deposits*, USSR Academy of Sciences Press for Institute of Physical Chemistry, Moscow (1960). English translation by Consultants Bureau (1962), 108 pp.

(24) Weigand, H., "The Interaction of Material and Component With Electroplating Techniques," *Technica, 13* (7), 435–438 (1964).

(25) Ferenc, H., (Hung.) "Galvanic Zinc Deposition. Effect of the Composition of Cyanide Zinc Electrolytes on the Coating Structure," *Galvanotechnik, 60* (6), 433–437 (1969).

(26) Evans, D. J., and Hopkins, M. R., "An Electron-Diffraction Investigation of the Structure of Electrodeposited Coatings on Iron Single Crystals," *Journal Electrodepositors' Technical Soc., 28* (8) (1952). Advance Copy, 10 pp.

(27) Bagautdinova, S. G., "Structure and Mechanism of Formation of Bright Zinc-Nickel Alloy Electroplates," *Nekotorye Vopr. Teorii i Praktiki Ispol'z. v Galvanotekhn. Neyadovit. Elektrolitov., Kazan*, 76–77 (1964).

Chapter 20

Uncommon Metals

Antimony deposits have been examined as protective coatings for steel and zinc, but the brittleness and weakness of the metal has deterred any commercial applications of antimony plating. (1,2) Some attention has been given to bismuth electrodeposits because they are more ductile than antimony. Even so, no commercial applications have been developed. Some interest has been expressed in indium, but chiefly as a minor alloying constituent for lead alloy bearings. Despite the limited interest in these electrodeposits, some property data have been reported and are summarized in this chapter.

In contrast with antimony, bismuth, and indium, which are deposited in aqueous solutions, molybdenum, tungsten, and other refractory metals have received more attention. Potential electroforming applications and potential uses as protective coatings at high temperatures have spurred research and development on the electrodeposition of the refractory metals, which must be deposited in fused-salt solutions at high temperatures. Complex shapes with good thickness uniformity have been electroformed in fluoride salt mixtures at about 750 C, which have better throwing power than the aqueous solutions of most metals. (3) Molybdenum, niobium, tantalum, and tungsten, which range in melting points from 2468 to 3410 C, have been fabricated into shapes by electroforming from the fused-fluoride baths.

Fused-salt baths have also been adopted for electrodepositing beryllium, silicon, thorium, uranium, and zirconium, and limited property data have been reported for some of these uncommon electrodeposited metals. The literature on electrodeposition in fused-salt baths also includes reports on the formation of compounds such as tungsten carbide (4) and zirconium diboride. (3)

Antimony and Bismuth

Antimony deposits obtained in acid solutions of antimony fluoride were brittle and usually contained cracks. (1) Composites of copper, lead, antimony, and chromium resisted corrosion in salt-spray tests, but were unsatisfactory in an industrial environment. A bumper plated with successive layers of nickel, lead, antimony, and chromium corroded badly during the second winter of service in Detroit. The poor

TABLE 20.1

Hardness Data for Electrodeposits of the Uncommon Metals

Metal	Type of Bath	Solution Temperature, C	Current Density, amp/sq dm	Microhardness, kg/sq mm	Reference
Antimony	Aqueous, cyanide	No data	No data	200	5
Bismuth	Aqueous, $BiCl_3$ and HCl	14 or 30	0.2–0.5	17–23 (20 g)[a]	6
Bismuth	Ditto	No data	0.3	28	7
Bismuth	Ditto plus citric acid and glycerol	No data	0.3	60	7
Indium	Aqueous brush plating bath	—	100	<1	8
Molybdenum	Fused mixture of $Na_4P_2O_7$, NaCl, $Na_2B_4O_7$, and MoO_2	—	—	186	9
Molybdenum	Fused mixture of $NaBO_2$, $LiBO_2$, Na_2MoO_4, Li_2MoO_4, and MoO_3	900	3.1–6.2	250–270 (100 g)	10
Molybdenum	Fused mixture of MoF_4, LiF, KF, and NaF	~750	3.0	187	3
Niobium	Fused mixture of NbF, LiF, KF, and NaF	700–800	2.5	70–100	3, 11
Silicon nitride[b]	Fused fluorides of Na, K, Li, and Si	450–700	2.0	476[b]	12
Thallium	Fused chlorides of Th and Na	—	—	78	13
Titanium boride	Fused mixture of $NaBO_2$, $LiBO_2$, Na^2TiO_3, Li_2TiO_3, and TiO_2	900	6.2	5000	14
Tungsten	Fused fluorides of W, Li, K, and Na	~750	~2.5	360	3
Tungsten carbide	Fused mixture of Na_2WO_4, $NaBO_2$, and NaOH	1000–1025	37	3480–3970	4
Uranium	Fused fluorides of U, Li, Na, and K	750	2.5	230	15
Vanadium crystals	Fused chlorides of V and Na	800	—	90–116[c]	16

[a] Aging at room temperature for 100 hours softened all deposits to the range of 14 to 16 kg/sq mm.

[b] The silicon coating obtained by electrodeposition was converted to SiN_4 by heating at 1000 to 1250 C in a nitrogen atmosphere.

[c] Derived by conversion from Rockwell B measurements.

ductility and cohesive strength of antimony appear to be major drawbacks to its use for corrosion protection.

The hardness of antimony deposited in a cyanide solution was 200 kg/sq mm, which compares with a hardness of only 17 to 60 kg/sq mm for bismuth, electrodeposited in baths described in Table 20.1. A tensile stress of 7.0 kg/sq mm (10,000 psi) was reported for antimony deposited at 2.2 amp/sq dm in an acid solution (pH 4.0) prepared with antimony oxide, citric acid, gluconic acid, and potassium citrate. (17) Stress was independent of temperature from 49 to 71 C. The addition of hydrogen peroxide to the bath prevented cracking of the deposits but did not change the stress. Because the stress of such crack-free plate was 7 kg/sq mm, the investigators reasoned that the tensile strength of the antimony deposit was >7 kg/sq mm (10,000 psi). Elongation was estimated at about 0.09 percent.

The electrical resistivity of electrodeposited antimony was reported to be 32 microhm-cm. (5) Resistivity and hardness data for antimony-gold alloy electrodeposits are given in Table 20.2.

Columnar structures were reported for dull antimony deposits with a random orientation or a slight (100) and (110) orientation, based on X-ray examinations. (1) A high degree of (100) and/or (110) preferred orientation and a fine-grained structure were characteristic of bright antimony deposits. A (100) preferred orientation with (100) planes parallel with the substrate was also noted for antimony deposited in a perchlorate bath. (18)

Changing the solution temperature from 18 to 30 C had only a slight effect on the hardness of bismuth deposited in solutions prepared with 48 g/l bismuth chloride and 117 g/l hydrochloric acid. (6) Increasing the current density from 0.18 to 0.48 amp/sq dm slightly reduced the hardness of bismuth deposited at each temperature. All deposits were slightly softer, only 14 to 16 kg/sq mm, after 70 to 100 hours of standing at room temperature. The addition of glycerol and citric acid to the solution increased hardness to 60 kg/sq mm. (7) The grain size was reduced from 2.1 to 1.2 μm by a 2-g/l addition of gelatin. (19)

TABLE 20.2

Resistivity and Hardness Data for Antimony-Gold Alloy Electrodeposits[a]

Gold Content, atomic percent	Resistivity, microhm-cm	Hardness, kg/sq mm	Phase Structure From X-Ray Data
0	32	200	—
15	50	230	$AuSb_2 + Sb$
35	70	320	$AuSb_2$
55	55	300	$AuSb_2 + Au$
88	22	230	$AuSb_2 + Au$

[a] Deposited in cyanide solutions. (5)

Bismuth deposited at 0.25 amp/sq dm in solutions prepared with 48 g/l bismuth chloride and 100 ml/l hydrochloric acid was stressed in compression. (20) Stress decreased from 6 to 3 kg/sq mm as plate thickness increased and further decreased after plating, which was attributed to recrystallization.

The Refractory Metals

Table 20.1 lists hardness data for electrodeposited molybdenum, niobium, and tungsten. These deposits from fused mixtures of fluoride salts at about 750 C exhibited a coarse, columnar structure, whereas tantalum and tungsten deposits from fused-fluoride salts were finer grained. (3) Molybdenum deposited in a 900 C bath containing sodium and lithium borates, sodium and lithium molybdates and molybdenum dioxide also exhibited a columnar structure. (10) However, the deposits from this solution were slightly harder, ranging from 250 to 270 kg/sq mm, by comparison with 187 kg/sq mm for the molybdenum deposited in the fused-fluoride bath.

A tensile strength of 30 to 33 kg/sq mm (42,000 to 47,000 psi) was reported for niobium deposits from a fused mixture of niobium, lithium, sodium, and potassium fluorides. (11) The current density and temperature were about 3.0 amp/sq dm and 750 C, respectively. The elongation of the deposits ranged from 25 to 30 percent, and the density was equal to the value reported for wrought niobium (8.57 g/cu cm). A density of 6.4 g/cu cm was reported for zirconium deposited in a fused-fluoride bath, (21) which compares with 6.49 g/cu cm for close-packed-hexagonal, metallurgical zirconium with a low hafnium content. (22)

Vanadium, titanium, and vanadium-titanium alloys were deposited adherently on an iron substrate by using a 900 C fused bath containing sodium and potassium chlorides and vanadium and/or titanium fluorides. (23) Only loose crystals were deposited on a titanium substrate. Over the iron substrate, titanium diffused with the basis metal to form solid solutions of titanium in iron and solid solutions of iron in titanium.

Fine-grained, nonporous coatings of zirconium with a thickness of about 125 μm (5 mils) were deposited on uranium in a fused-salt mixture of zirconium, potassium, and lithium fluorides. (24) Current density and temperature ranged from 2.0 to 4.0 amp/sq dm and 625 to 675 C, respectively.

Refractory Compounds

Some high-temperature compounds have been electrodeposited in fused-salt baths. For example, tungsten carbide with a hardness of 3480 to 3970 kg/sq mm was deposited in a fused mixture at 1000 C of sodium tungstate, sodium borate, and sodium hydroxide. (4) The carbon introduced into the bath by using graphite anodes reached a level of 5.4 percent in the deposit, which compares with 6.1 percent for pure tungsten carbide. Dense titanium boride deposits with a thickness of 75 to 150 μm (3 to 6 mils) were deposited on Inconel at 6.2 amp/sq dm in a fused bath at 900 C containing sodium and lithium titanates, sodium and lithium borates, and titanium dioxide. (14) The compound had a melting point of 2600 C and a surface hardness of 5000 kg/sq mm.

References

(1) Du Rose, A. H., "Some Notes on the Electrodeposition of Antimony," *Proceedings Am. Electroplaters' Soc., 43*, 151–156 (1956).

(2) Fedot'ev, N. P., Grilikhes, S. Ya., and Naryshkina, I. B., "Electrodeposited Coating of Antimony and Its Properties," *Journal Applied Chemistry (USSR), 32*, 2876–2877 (1959); *Zh. Prikl. Khimii, 32*, 2798–2799 (1959).

(3) Mellors, G. W., and Senderoff, S., "The Electroforming of Refractory Metals," *Plating, 51*, 972–975 (1964).

(4) Bentz, G., Provost, G., and Urlan, C., "Preparation of Tungsten Carbide by Electrodeposition," *Bureau of Mines, BM-RI-7247*, 13 (1969).

(5) Fedot'ev, N. P., and Vyacheslavov, P. M., "The Phase Structure of Binary Alloys Produced by Electrodeposition," *Plating, 57*, 700–706 (1970).

(6) Popereka, M. Ya., and Avramenko, O. I. "Electrodeposition of Bismuth and Certain Properties of Bismuth Deposits," *Journal Applied Chemistry (USSR), 38*, 1744–1750 (1965); *Zh. Prikl. Khimii, 38*, 1783–1789 (1965).

(7) Avramenko, O. I., "Microhardness of Electrodeposited Bismuth," *Elektroosazhd. Metal. i Ul'truzvuk. Mikrodefektoskopiya Kristallov, Sb., Novosibirsk.*, 78–81 (1965).

(8) Machu, W., "Brush Plating for Localized Metal Deposition (Dalic Process)," *Metallwaren-Industrie and Galvanotechnik, 49* (11), 467–481 (1958). Cf. Hughes, H. D., *Trans. Inst. Metal Finish., 33*, 424–439 (1956).

(9) Zadra, J. B., Heinen, H. J., and Gomes, J. M., "Electrolytic Molybdenum From a Molten Bath," U.S. Patent 3,071,523 (Jan. 1, 1963). Assigned to U.S. Dept. of the Interior.

(10) McCawley, F. X., Wyche, C., and Schlain, D., "Electrodeposition of Molybdenum Coatings," *Journal Electrochem. Soc., 116*, 1028–1032 (1969).

(11) Mellors, G. W., and Senderoff, S., "Electrodeposition of Coherent Deposits of Refractory Metals. I. Niobium," *Journal Electrochem. Soc, 112*, 266–272 (1965).

(12) Lang, F. S., "Corrosion-Resistant Silicon-Nitride Coated Steel Surfaces," Ger. Offen. 1,935,701 (Sept. 10, 1970). Assigned to American Standard Inc.

(13) Raynes, B. C., Bleiweiss, J. C., Sibert, M. E., and Steinberg, M. A., "Investigations for the Production of Thorium Metal," *Final Report for May 1, 1952 to April 30, 1954, Contract AT(30-1)-1335*, Horizons, Inc. (November 1, 1954, Decl. February 28, 1957), 165 pp.

(14) Schlain, D., McCawley, F. X., and Wyche, C., "Electrodeposition of Titanium Diboride Coatings," *Journal Electrochem. Soc., 116*, 1227–1228 (1969).

(15) Binard, M., Boisde, G., Broc, M., Chauvin, G., and Coriou, H., "Electrolytic Production of Compact Layers of Beryllium, Uranium, and of Their Alloys," French Patent 1,521,522 (April 19, 1968). Assigned to Commissariat a l'Energie Atomique.

(16) Baker, Jr., D. H., and Ramsdell, J. D., "Electrolytic Vanadium and Its Properties," *Journal Electrochem. Soc., 107*, 985–989 (1960).

(17) Soderberg, K. G., and Pinkerton, H. L., "Antimony Plating," *Plating, 37*, 254–259 (1950).

(18) Lazzari, M., Rivolta, B., and Rossi, G. A., "Structural Investigation of Bismuth Electrodeposits on Single Crystal Cathodes," *Electrochim. Metal., 3* (1), 15–22 (1968).

(19) Popereka, M. Ya., and Avramenko, O. I., "Effect of Organic Additions on the Structure and Internal Stresses in Bismuth Electrolytic Deposits," *Izvestiya Vysshikh Uchebynkh Zavedeniy SSSR, Khimiya i Khimicheskaya Tekhnologiya, 10* (2), 198–202 (1967).

(20) Popereka, M. Ya., and Avramenko, O. I., "Bismuth Electrodeposits With High Internal

Stresses," *Russian Journalof Physical Chemistry, 39*, 300–305 (1965); *Zh. Fiz. Khimii, 39*, 561–568 (1965).

(21) Mellors, G. W., and Senderoff, S., "Electrodeposition of Zirconium, Tantalum, Niobium, Chromium, Hafnium, Tungsten, Molybdenum, Vanadium, and Their Alloys," U.S. Patent 3,444,058 (May 13, 1969). Assigned to Union Carbide Corp.

(22) *Metals Handbook, Vol. 1*, 1228 (1961).

(23) Kolobov, G. A., Andreev, Yu. Ya., and Lysov, B. S., "Preparation of Coatings From Titanium, Vanadium, and Their Alloys by the Electrolysis of Fused Salts," *Sb., Mosk. Inst. Stali Splavov, No. 49*, 181–195 (1968).

(24) Nissen, D. A., and Stromatt, R. W., "Zirconium Electroplating on Uranium From Molten Alkali Fluoride Salts," *U.S. At. Energy Comm., BNWL-825* (1968), 33 pp.

Chapter 21

Chemical or Electroless Cobalt and Cobalt Alloys

The term "electroless process" is used to describe methods of plating by means of chemical reduction. Electroless plating differs in one significant respect from all the other chemical plating procedures in that it is the only chemical plating process which does not depend on the presence of a couple between galvanically dissimilar metals.

Interest in cobalt coatings by chemical reduction was first aroused with the reports of Brenner and Ridell, (1) who used sodium hypophosphite as a reducing agent. Brenner and Ridell concluded that deposition from acid solutions was impractical and recommended a cobalt chloride-sodium hypophosphite-sodium citrate-ammonium chloride bath operated at a pH of 9 to 10, controlled by additions of ammonium hydroxide. Their cobalt-phosphorus alloy was harder than nickel deposits. The presence of ions such as Cd^{2+}, Zn^{2+}, Mg^{2+}, Fe^{2+} and PO^{3-} decreased the rate of deposition, while with CN^- and SCN^- there was no deposition. Cu^{2+} and Al^{2+} did not affect the normal rate of deposition.

Table 21.1 summarizes the reducing agents proposed and patented for reducing cobalt. Results vary appreciably. For instance, bright and dull coatings were obtained with hydrazine hydrate at concentrations of 1.2 and 1.8 M, respectively. (13) Deposition rates were 2.9 and 4.5 μm/hr, respectively.

All commercial cobalt deposits are made in alkaline baths, although there is a fairly recent patent for an acid bath. (19) This solution contains a sulfur-containing compound such as thiourea, thioacetamide, phenylthiourea, carbon disulfide, thiobarbituric acid, thiosulfates, and thiocyanates, in addition to sodium pyrophosphate.

Chemical reduction, autocatalytic, or electroless deposition of cobalt has been known at least 25 years. Applications suggested for electroless cobalt coatings involve its magnetic properties, wear resistance, electrical resistivity, and thermal conductivity. Prospects exist for coating disks for digital and video applications and other products related to computer applications.

Magnetic Properties

The range of magnetic properties that can be developed, and their potential usefulness in the electronics industry, have resulted from extensive research and manu-

TABLE 21.1

Reducing Agents Used in Chemical Plating of Cobalt and Cobalt Alloys

Agent	References
Amine Boranes	2, 3
Borohydrides	4, 5, 6
Cobalt Boranes	7, 8, 9, 10
Dimethylborane (derived from diborane)	11, 12
Hydrazine	13, 14, 15, 16, 17
2-Oxazolidinone	18
Sodium Hypophosphite	1

facturing development, not only for the cobalt-phosphorus alloys, but for alloys that also contain iron, nickel, tungsten, and/or zinc. Methods for measuring magnetic properties have been compiled in a single volume. (20)

The coercivity of a magnetic material is one of the most important properties. The coercivity of electrodeposited and electroless alloys may vary between 1 and 2,000 oersteds. This range can be obtained by controlling the alloy composition, as indicated in Figure 21.1, and the deposition conditions. High-coercivity materials might be used in random-access, high-speed memory devices. A fairly extensive review of chemically deposited magnetic alloys has appeared in the Russian literature. (22) Properties required for various computer components have been discussed. (23) The thickness of the film plays an important role. The magnetic properties of the film depend on the size of magnetic crystallites or elements isolated from each other by nonmagnetic material. Magnetic properties and recording behavior of chemically deposited films have been correlated. (24)

Magnetic properties such as coercivity (H_c) depend on crystallite size, orientation, phosphorus content of the alloys, thickness of deposit, and residual stresses. Heat treatment usually improves the magnetic properties. Crystallite size and orientation are influenced by the substrate, which may be conductive or nonconductive, and on the procedure selected for surface preparation. Important factors affecting the properties are summarized in Table 21.2.

The magnetic properties of a material are based on its hysteresis loop (Figure 21.2) wherein the field (H) applied to magnetic elements is plotted along the horizontal axis. The output or resulting flux density or induction change (B) is plotted along the vertical axis. Remanence (B_r) is defined as the flux remaining in the material when the magnetic field is reduced to zero. The magnetic element in its rest state is always in a position of $+B_r$ or $-B_r$. This property, coupled with the ability to sense whether the magnetic element is in the + or − state, enables the element to perform as a binary memory. Coercive force (H_c) is defined as the applied field which is necessary to maintain the magnetic element at zero flux density. The

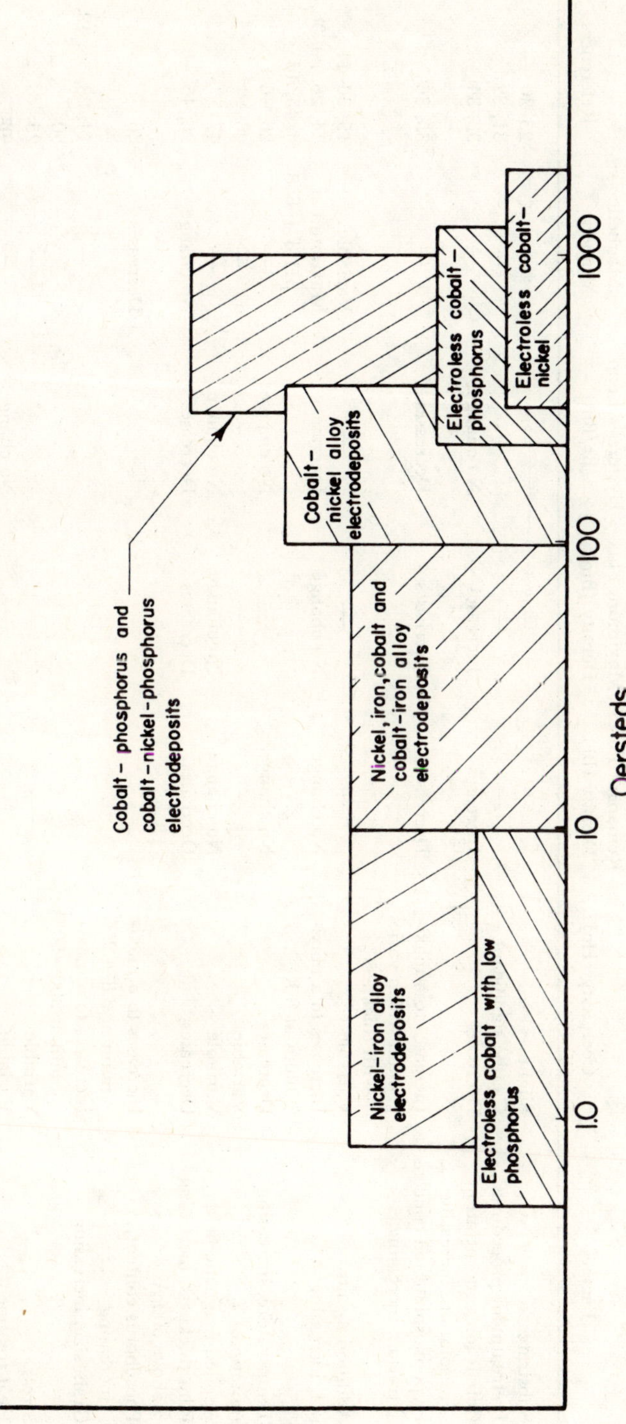

Figure 21.1. Coercivity Ranges of Electroless and Electrodeposited Metals and Alloys. (21)

TABLE 21.2

Factors Affecting Magnetic Properties

Factor	Influence on Magnetic Property					References
	Coercivity (Hc)	Remanent Flux Density (Br)	Maximum Flux Density (Bm)	Squareness of Hysteresis Loop (Br/Bm)	Magnetic Moment	
Substrate	Slight	—	—	—	—	25–30
Pretreatment procedures	Can be significant	—	—	—	—	31, 32
Cobalt ion concentration in solution, increasing	Decreases	Increases	Increases	No change	—	33, 34
Hypophosphite ion concentration, increasing	Increases to a maximum which varies with thickness	Decreases	Decreases	Decreases	—	35, 36
Addition agents	Variable	—	—	—	Variable	15, 36–44
pH, increasing	Increases to a maximum at 8.1	No change	No change	Increases	Maximum at pH 8.1–8.8	14, 26, 29, 36, 45, 46
Temperature, increasing	Decreases	No change	—	No change	—	26, 36
Agitation	Variable	—	—	—	—	26
Applied magnetic field	Variable	No change	No change	No change	—	47
Film thickness, increasing above 500 Å	Decreases	Decreases	Decreases	Decreases	No change	26, 45, 48
Phosphorus content, increasing	Increases to a maximum at 3.8–4.2%				Decreases	26
Grain size, increasing	Increases to a maximum at 300–450 Å				—	14, 28, 36, 49, 50
Stress	Variable	—	—		—	25
Heat treatment	Variable	—	—	No change	—	47

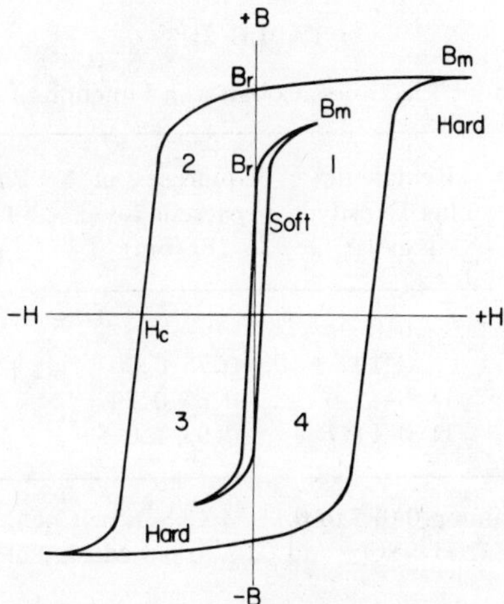

Figure 21.2. Magnetization Curves for Hard and Soft Magnetic Materials.
Changes in flux density or induction (usually expressed in gauss) are indicated on the vertical axis and the magnetic field (expressed in oersteds) is indicated on the horizontal axis.

coercive force is essentially a measure of the amount of energy which must be applied to the magnetic material in order to change its state from $+B_r$ to $-B_r$, or from $-B_r$ to $+B_r$. The maximum value along the B axis which can be obtained by applying large fields is B_m, which is a measure of the total amount of flux density or induction available in the magnetic element for the maximum field applied. The squareness of the hysteresis loop is a property of some importance in magnetic memory applications and is defined as B_r/B_m. Occasionally, workers in this field have determined the saturation magnetization (M_s), which has been measured between 0 and 160 emu/g. From the second quadrant of the hysteresis loop (Figure 21.2), the maximum energy product, $(BH)_{max}$, of the material and other characteristics are determined.

Requirements for high-density data storage are: (a) a nearly rectangular B-H loop, (b) a coercive force generally higher than 200 oersteds, and (c) a relatively low ratio of remanent magnetization (B_r) to coercive force (H_c). Substantial increases in the packing density are obtained in a magnetic recording system by the use of high-coercivity metal films. Of several techniques considered, electroless deposition was the most successful in producing films with these magnetic properties. (26)

Property Data for Cobalt-Phosphorus Alloys

Coercive force increases, in general, with increasing concentrations of hypophosphite and hydrogen ions, and decreasing temperature and thickness. (36) A low coercive force, < 12 oersteds, was reported for films obtained at a high temperature (92 C), which tends to limit the phosphorus content of the cobalt alloy to only 1.0 or 1.5 percent. (47) Table 21.3 shows coercivity increasing to as much 1200 oersteds

TABLE 21.3

Magnetic Property Data for Electroless Cobalt as a Function of Solution Temperature[(a)]

Temp, C	Coercivity, oersteds	Remanent Flux Density, k gauss	Squareness of Hysteresis Loop (Br/Bm)	Phosphorus Content, percent	Film Thickness, Å
50	1000–1200	0.8	—	5.0–5.8	600–800
75	640–800	7–10	0.75–0.90	2.0–2.4	9,000–30,000
80[(b)]	115–344	11.6–14.6	0.65–0.79	About 2	4,600–5,400
92	1–11[(c)]	11.0–14.7	0.93–1.0	1.0–1.5	1,200–6,500

[(a)] Data for solutions containing 0.075 to 0.11 M $CoSO_4$ or $CoCl_2$, 0.085 to 0.33 M $Na_3H_2PO_2 \cdot H_2O$, 0.22 to 0.85 NH_4Cl or $(NH_4)_2SO_4$, and 0.06 to 0.3 M $Na_3C_6H_5O_7$ with a pH of 8.0 to 8.7. (40, 47, 51, 52)

[(b)] A saturation magnetic moment of 100 emu/g was reported for a film with a coercivity of 250 oersteds, which was deposited in a solution heated to 88 C.

[(c)] Higher coercivities, up to 350 oersteds, were reported for films deposited in solutions having a high pH of 9.5. (47)

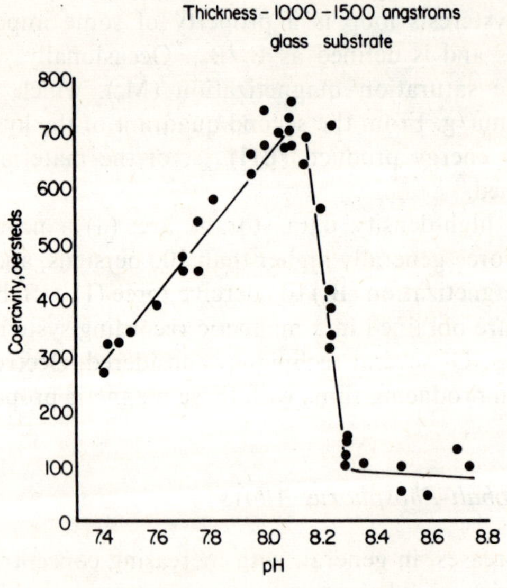

Figure 21.3. Variation of Coercivity with pH. (14)

Data for films with a thickness of 1000 to 1500 Å which were deposited on glass substrates in a cobalt chloride solution containing boric acid.

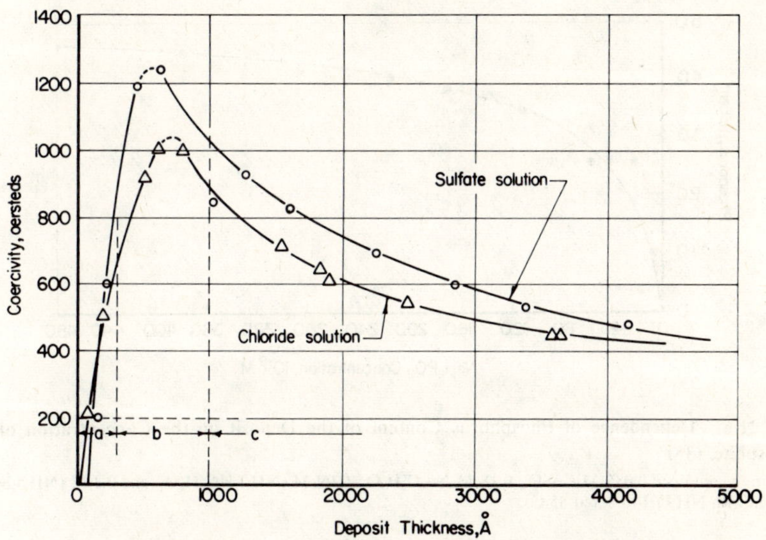

Figure 21.4. Coercivity as a Function of Film Thickness. (51)

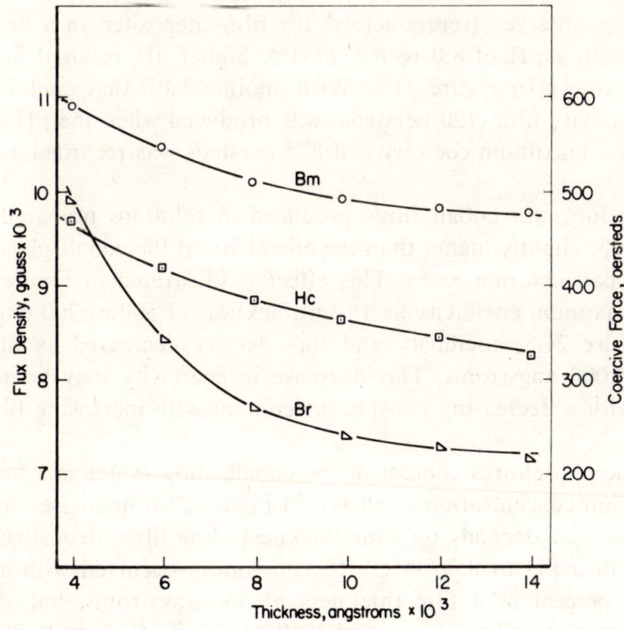

Figure 21.5. Coercive Force (H_c), Remanence (B_r), and Maximum (B_m) Flux Densities of Cobalt Deposits as a Function of Thickness. (55)

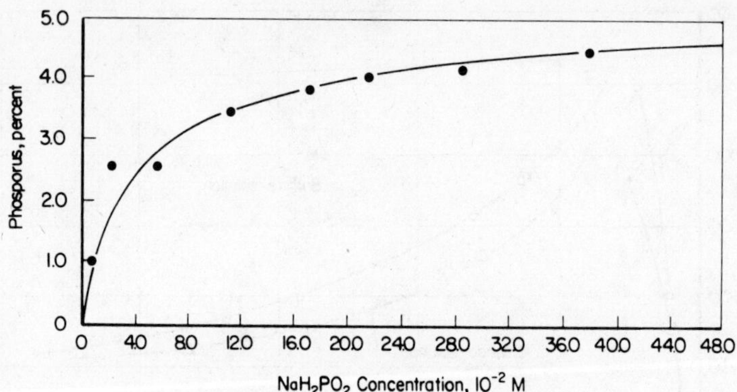

Figure 21.6. Dependence of Phosphorus Content of the Deposit on the Concentration of Sodium Hypophosphite. (35)

The solution contained 0.085 M $CoSO_4$, 0.18 M $Na_3C_6H_5O_7$, 0.09 M $(NH_4)_2HC_6H_5O_7$, and 0.3 M $(NH_4)_2SO_4$. Its pH was adjusted with NH_4OH to 8.3 at 85 C.

and remanence decreasing to 0.8 kilogauss for 600- to 800-angstrom films deposited in solutions operated at lower temperatures.

An increasing coercive force was associated with an increase in the pH of electroless plating solutions from 7.4 to 8.2. (14,54) A maximum coercivity of about 750 oersteds was observed (reproducibly) for films deposited in a boric acid buffered solution with a pH of 8.0 to 8.2. (14) A higher pH resulted in much lower coercivities, as shown in Figure 21.3. With another bath that contained no boric acid, a low-coercivity film (120 oersteds) was produced when the pH was adjusted to 7.7, whereas a maximum coercivity of 875 oersteds was recorded as the pH was raised to 9.4. (52)

The coercive force for cobalt films produced in solutions prepared with cobalt sulfate tends to be slightly higher than the coercivity of the cobalt-phosphorus alloy deposited in cobalt chloride baths. This effect is illustrated in Figure 21.4, which also shows a maximum coercivity at a film thickness of 500 to 700 angstroms. According to Figure 21.5, coercivity and flux density decreased as film thickness increased to 14,000 angstroms. This decrease in coercivity may be associated, at least in part, with a decreasing phosphorus content with increasing film thickness. (35,36)

Increasing the phosphorus content of the cobalt alloy, which is a function of the hypophosphite ion concentration as shown in Figure 21.6, increases coercivity to a maximum value that depends on film thickness. For films deposited in a cobalt sulfate bath with a pH of 8.3 (85 C), the maximum occurred with a phosphorus content of 3.5 percent at a film thickness of 560 angstroms, but shifted to 4.2 percent phosphorus at a film thickness of 2250 angstroms (Figure 21.7).

Some differences in the magnetic properties of electroless cobalt can be attributed to the selection of the complexing agent. (56) Supplementing sodium citrate, which is frequently added to electroless cobalt solutions, with sodium borate reduced coercivity. (37)

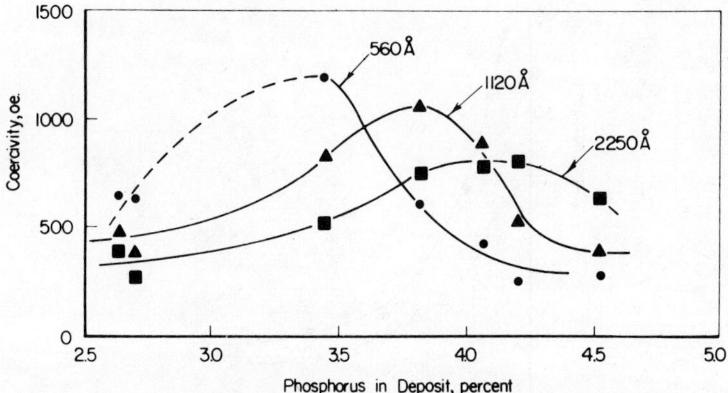

Figure 21.7. Dependence of Coercivity (H) on the Phosphorus Content of Three Different Thicknesses of Cobalt-Phosphorus Films. (35)

Addition agents containing a C=S group such as dimethylthiourea and thioacetamide decreased the coercive force of the films, whereas those containing a C=O group such as urea and acetamide increased or had little effect on the coercive force of similar films. (40) Agents with the carbon-sulfur double bond produced a significant effect on grain size, crystal orientation, and coercive force. The magnitude of the change in coercivity was dependent on the concentration of the addition agent. The solution employed for these studies consisted of 0.075 M cobalt chloride, 0.33 M sodium hypophosphite, 0.22 M ammonium chloride, and 0.06 M citrate ions and was maintained at 80 C.

Some films with a coercivity of < 100 oersteds exhibited a small crystallite size (< 250 angstroms), whereas higher coercivity films (> 300 oersteds) exhibited a larger crystallite size of 300 to 800 angstroms. (14) On the other hand, another investigator reported decreasing coercivity with an increase in grain size, as shown in Table 21.4. (55) In films containing > 3 percent phosphorus, 200- to 1000-angstrom particles were identified with an electron microprobe. (57)

Stress in electroless cobalt films deposited on single-crystal silicon substrates decreases rapidly with increasing film thickness from about 500 to 3000 angstroms.

TABLE 21.4

Grain Size and Coercive Force of Thin Films of Chemically Reduced Cobalt (55)

Thickness, Å	Grain Size, Å	Coercive Force, oersteds
700	1,000	>650
1,200	2,000	600
2,300	4,500	500

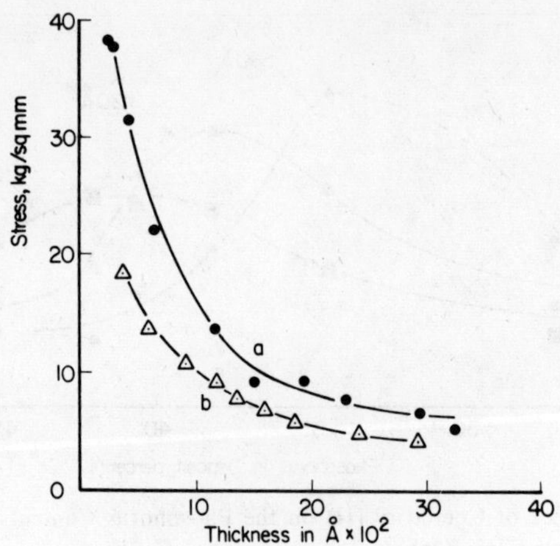

Figure 21.8. Average Stress vs Thickness for Electroless Cobalt.
(a) Deposits from concentrated solution; (b) from dilute solution. (25)

(25) As indicated in Figure 21.8, average stress was reduced to 5 or 7 kg/sq mm (7,000 or 10,000 psi) at 3000 angstroms (0.3 μm).

The application of an applied magnetic field during deposition induced a uniaxial anisotropy in films exhibiting a very low coercive force. (47)

The magnetic properties of typical films were unchanged when held at 100 C for 1100 hours, (58) but annealing at temperatures in the range of 300 to 500 C

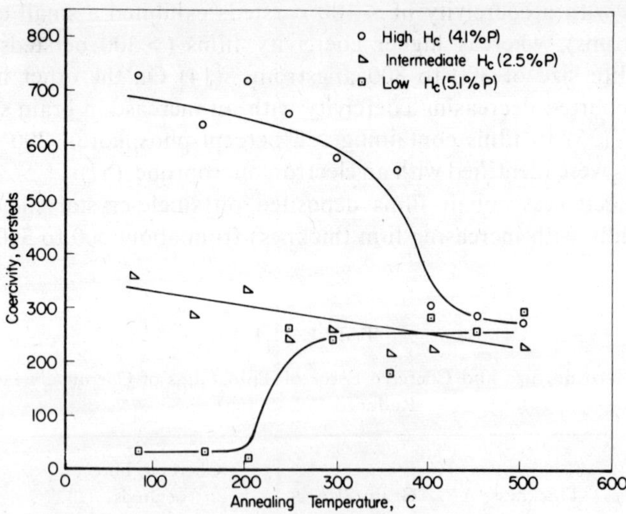

Figure 21.9. Coercivity as a Function of Annealing Temperature for Films Containing 2.5, 4.1, or 5.1 Percent Phosphorus. (59)

TABLE 21.5

Saturation Magnetic Moments For Various P Contents (35)

Percent P	σ emu/gram
2.67	156.0
2.62	147.5
3.45	114.0
3.81	118.0
4.06	120.5
4.22	127.5
4.52	122.5

increased the coercivity of low-coercivity films and decreased the coercivity of high-coercivity films, as shown in Figure 21.9. Coercive force after annealing was approximately 300 oersteds for both classes. A transition from a supersaturated solution to a two-phase structure including Co_2P probably was responsible for the shift in the coercivity of the low-phosphorus films. This transition occurs at about 320 C. (60)

The saturation magnetization of cobalt-phosphorus films, usually > 100 emu/gram, decreases with increasing phosphorus content, as shown in Table 21.5.

Data for Ternary Alloys

Cobalt-nickel alloy films with a thickness of about 2000 angstroms and a phosphorus content of 1 to 2 percent exhibited a minimum coercivity of 5 oersteds when

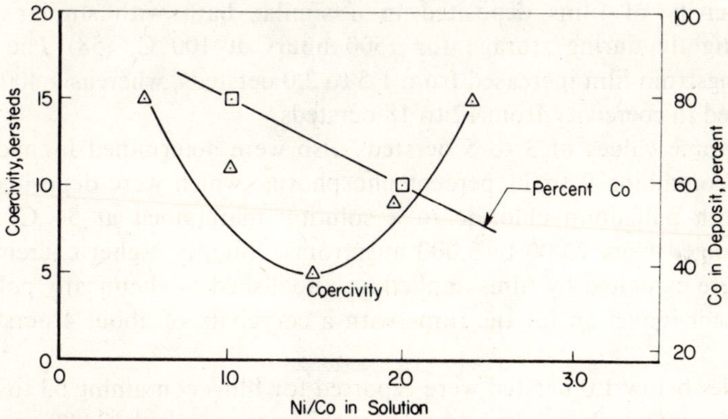

Figure 21.10. Effect of the Nickel-Cobalt Ion Ratio in Solution on the Cobalt Content and Coercive Force of Nickel-Cobalt-Phosphorus Alloy Films. (61)

TABLE 21.6

Effect of Annealing on Magnetic Properties of Cobalt-Nickel-Phosphorus Films (65)

Annealing Temperature, C	Magnetization, gausses	Coercivity, oersteds
Not annealed	32.7	*
200	115	6.5
300	131	5.4
400	150	12.7

* Very low—not measurable.

the cobalt content was about 40 percent. (61) Coercive force increased with a higher or lower cobalt content of films deposited in sulfate solutions containing 0.16 M nickel plus cobalt, 0.28 M hypophosphorous acid, 0.3 M ammonium sulfate, and 0.27 M sodium citrate, as shown in Figure 21.10. The temperature was 90 C. Dilution to a total salt concentration of about 0.3 M increased coercivity to 23 oersteds, but variatons in pH from 7.2 to 8.6 or increases in the concentrations of hypophosphite, citrate, or ammonium ions had little or no effect on coercivity. The omission of citrate ions precipitated an insoluble nickel compound. The omission of ammonium sulfate increased coercivity to 15 oersteds. Coercivity increased when the solution temperature was reduced to < 70 C. The minimum coercivity for the cobalt-nickel-phosphorus alloy over phosphor-bronze alloy was reduced to (a) 2.0 oersteds by interposing 300 to 600 angstroms of nickel and (b) 1.2 oersteds by using an electropolished, nickel-plated copper substrate. Increasing the film thickness on phosphor-bronze alloy to 7,000 angstroms reduced coercivity to 1.5 oersteds.

The coercivity of films deposited in a similar bath with similar conditions increased slightly during storage for 1500 hours at 100 C. (58) The coercivity of a 5000-angstrom film increased from 1.5 to 2.0 oersteds, whereas a 400-angstrom film increased in coercivity from 12 to 18 oersteds.

Coercive-force values of 3 to 5 oersteds also were determined for nickel-cobalt alloy films containing 9 to 11 percent phosphorus, which were deposited on glass activated with palladium chloride in a solution maintained at 54 C. (62) Film thickness ranged from 2,000 to 3,000 angstroms. Slightly higher coercivities (to 6 oersteds) were exhibited by films applied on unpolished or chemically polished copper. Remanent induction for the films with a coercivity of about 4 oersteds was 7 kilogausses.

Coercivities below 1.0 oersted were reported for films containing 33 to 45 percent cobalt and 5 percent phosphorus when film thickness reached 10,000 angstroms on copper-coated glass. (63) The solution composition and conditions were similar to those used for depositing the alloy containing 1 to 2 percent phosphorus, except for the ratio of hypophosphite and metal ions, which was increased about seven times.

A coercivity of only 0.14 oersted was reported for 20-μm-thick alloy film (200,000 angstroms) containing 23 percent cobalt and 6.9 percent phosphorus, which was deposited on a copper rod. (64) The residual induction density (B_r) and saturation induction density (B_m) were 3,000 and 6,000 gausses, respectively. With a shift in the cobalt and phosphorus contents to 37.5 and 5.5 percent, coercivity, remanence, and saturation induction values were 4.5 oersteds, 300 gausses, and 3,000 gausses, respectively.

Low-coercivity, cobalt-phosphorus-nickel films deposited in solutions containing 0.065 M cobalt sulfate, 0.095 M nickel sulfate, 0.25 M phosphorous acid, 0.3 M ammonium sulfate, and 0.27 M sodium citrate exhibited good stability at temperatures of 25 and 65 C during 1500-hour periods. (58) A 50-percent increase in coercivity was noted for films with a thickness of 400 to 3600 angstroms during heating for 1500 hours at 100 C, but values for thicker films changed very slightly at 100 C.

Annealing at 300 C reduced coercivity and increased magnetization for 16- to 18-μm-thick films deposited in a solution containing 0.04 M cobalt sulfate, 0.08 M nickel sulfate, 0.075 M sodium hypophosphite, 0.11 M ammonium sulfate, and 0.125 M sodium acetate (87 C). (65) An increase in the annealing temperature to 400 C increased coercivity, as shown in Table 21.6.

TABLE 21.7

Magnetic Property Data for Ternary Alloy Deposits

Ternary Metal, percent	Phosphorus Content, percent	Coercive Force (Hc), oersteds	Saturation Magnetization, emu/g	Film Thickness, angstroms	References
Nickel, 15–40	2 to 12	500–2000	90–110	500–2500	62, 66, 67
Nickel, 50–75[a]	1 to 11	0.1–15	40–80	2000–80,000	61, 62, 63, 64, 66, 67
Iron, 1–5	About 4	600–1480	110	500–10,000	68
Iron, 6–40	0.3–3.7	50–150	125–189	500–10,000	68
Rhenium, 30	2	0–290[b]	—	—	69
Tungsten, 9	2	250–400	—	—	69
Zinc, 3.5–4[c]	—	680–900	—	1000–5000	70

[a] Remanent flux density shifted from 0.3 to 7.0 kilogausses with a shift in the cobalt and phosphorus concentrations from 3.75 and 5.5 percent, respectively, to 23 and 6.9 percent, respectively. Saturation flux density changed from 3.0 to 6.0 kilogausses with this change in composition.

[b] Coercivity increased from 0 to 290 oersteds as film thickness was increased.

[c] For films with a thickness >300 angstroms, Br/Bm ratios (squareness) were in the range of 0.70 to 0.75.

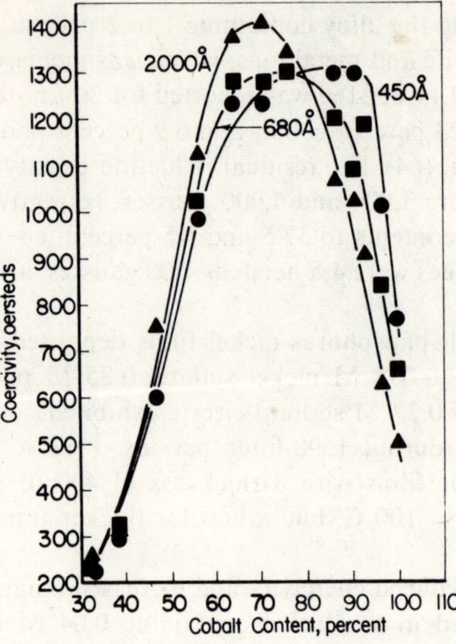

Figure 21.11. Coercivity as a Function of the Cobalt Content of Cobalt-Nickel Films Deposited in Solutions Containing Tartrate Ions. (66, 67)

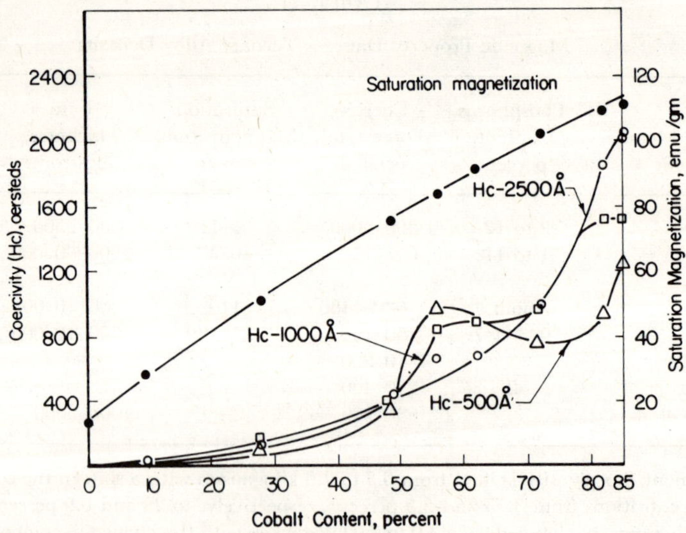

Figure 21.12. Coercivity and Saturation Magnetization as a Function of the Cobalt Content of Cobalt-Nickel Films Deposited in Solutions Containing Citrate Ions. (67)

Alloy films containing >60 percent cobalt exhibit high coercivity, as a rule, as shown in Table 21.7 and Figures 21.11 and 21.12. Coercivity was as high as 2000 oersteds for a 500-angstrom film containing 85 percent cobalt, which was deposited in a solution at 80 C containing 0.125 M nickel plus cobalt, 0.19 M sodium hypophosphite, 0.5 M ammonium sulfate, and 0.1 M ammonium citrate. (67) The pH of the solution was 8.0. Substituting malonic acid (0.12 M) for the ammonium citrate appreciably reduced the coercivity of the alloy film containing 85 percent cobalt. These films contained only a trace of phosphorus, however, whereas the high-coercivity alloy contained 2 to 2.5 percent phosphorus.

Iron-alloy films with a high cobalt content (>95 percent) also exhibited a high coercivity (600 to 1400 oersteds), whereas films with an iron content of 6 to 40 percent exhibited an appreciably lower coercivity (50 to 150 oersteds). (68) The higher values in each range corresponded to 500-angstrom films and the lower values to 2000-angstrom films (Figure 21.13). The phosphorus content of the alloy decreased rapidly from about 4 to only 0.3 percent as the iron content increased from 6 to 25 percent. These films were deposited in 80 C solutions with a pH of 8 and containing 0.09 M cobalt sulfate, 0.38 M sodium hypophosphite, 0.1 M sodium citrate, 0.3 M ammonium sulfate, and up to 0.07 M ferrous sulfate.

Coercivity data for cobalt-phosphorus alloys containing rhenium, tungsten, or zinc are tabulated in Table 21.7. The thin rhenium-containing alloy films exhibited low coercivities, whereas the films containing zinc reached high coercivities (as high as 900 oersteds). Differences in the solution temperature and the concentrations of cobalt salts and other compounds may have been largely responsible for the di-

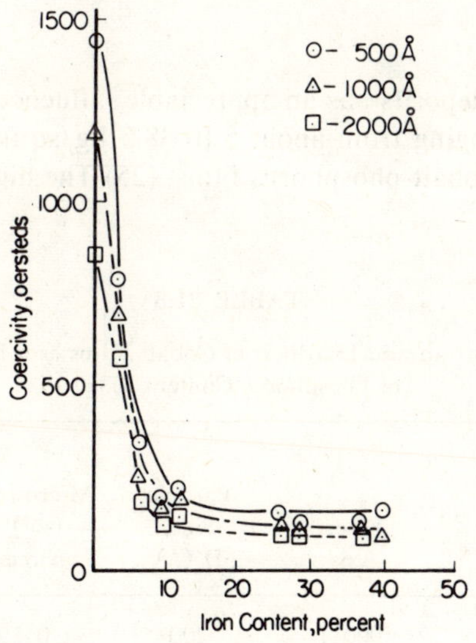

Figure 21.13. Coercivity as a Function of the Iron Content of Cobalt Alloy Films. (68)

vergence in coercivity. Solution compositions were as follows:

Salt	Co-P-Re and Co-P-W Films (69)	Co-P-Zn Films (70)
Cobalt chloride, M	0.012	0.075
Sodium hypophosphite, M	0.19	0.33
Ammonium chloride, M	0.095	0.15
Citric acid, M	—	0.07
Potassium perrhenate, M	0.03	—
Sodium tungstate, M	0.03	—
Zinc chloride, M	—	0.007
pH	8.8–8.9	8.2
Temperature, C	95	80
Phosphorus Content, percent	2	No data
Tertiary metal content of films	30 Re or 9 W	35 to 40 Zn

High coercivities, above 1000 oersteds, were reported for cobalt-phosphorus films containing aluminum. (71)

High-coercivity alloy films produced with sodium hypophosphite or phosphorous acid have been adopted for high-density recording, (40,70,72) whereas the low-coercivity alloys were suggested for high-speed computers. (23,47,52,73) Magnetic data for electroless cobalt films produced with other reducing agents have not been recorded in the literature.

Mechanical Properties

Stress

Stress in electroless deposits has an appreciable influence on magnetic properties. (25,74) High stress, ranging from about 5 to 38.5 kg/sq mm (7,000 to 55,000 psi), has been reported for cobalt-phosphorus films. (25) The higher values corresponded

TABLE 21.8

Microstrain and Hardness of Cobalt Films as a Function of Phosphorus Content (53)

Approximate Phosphorus Content, percent	Fiber Axis	Particle Size, D (Å)	Microstrain, $(\epsilon\omega^2)^{1/2}$ percent	Hardness, kg/sq mm (1-gram load)
1	(0001)	200	0.15	120
2	(1010)	100	0.11	160
3	(1010)	100	0.15	320
5	(1010)	100	0.22	380

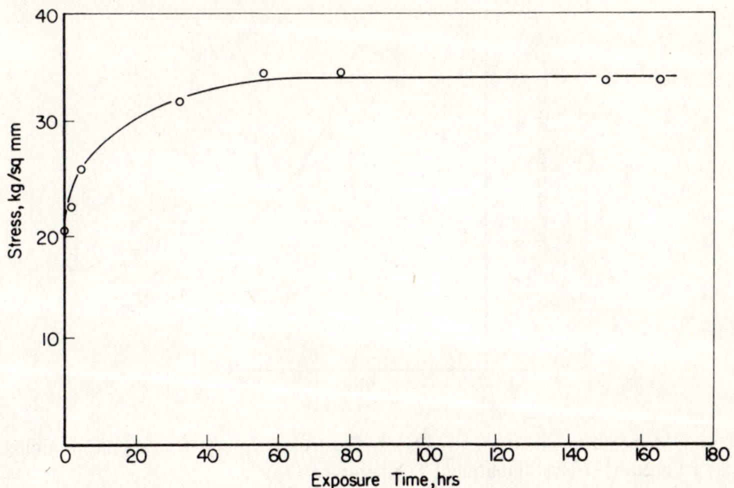

Figure 21.14. Change in Stress during Exposure to a Warm (43 C), Humid (80% RH) Atmosphere. (51)

to thin, 500-angstrom films, whereas 1000- to 2000-angstrom films exhibited a stress of 5 to 10 kg/sq mm (7,000 to 14,000 psi). Concentrated solutions produced films with high stress, in comparison with dilute solutions. The influence of film thickness and solution concentration is shown in Figure 21.8.

Exposure to a warm, humid atmosphere (80% RH) increased stress in 600- to 800-angstrom films deposited in solutions containing 0.11 M cobalt chloride or 0.13 M cobalt sulfate, 0.18 or 0.19 M sodium hypophosphite, 0.5 M ammonium sulfate and 0.12 M sodium citrate. (51) Solution pH and temperature were 8.7 and 50 C, respectively. The phosphorus content ranged from 5.0 to 5.8 percent. Figure 21.14 shows the change in stress for a typical film as a function of exposure time in the humid atmosphere. Magnetic moment increased as a result of exposure to warm, humid atmospheres. (51) The increase was attributed to an increase in surface area caused by corrosion. An overlay of rhodium is sometimes adopted to prevent such corrosion. (72)

Microstrain and hardness of electroless cobalt films increased as their phosphorus contents increased above 2 percent, as shown in Table 21.8 (53) Microstrain was determined by X-ray analysis. The pH of the 88 C solution ranged from 7.4 to 8.2, to vary the phosphorus content. The bath contained 0.085 M cobalt sulfate, 0.19 M sodium hypophosphite, 0.3 M ammonium sulfate, and 0.27 M sodium citrate.

Hardness

The hardness of electroless cobalt ranges from 120 to about 750 kg/sq mm, depending on the phosphorus content and other factors. An increase in the hypophos-

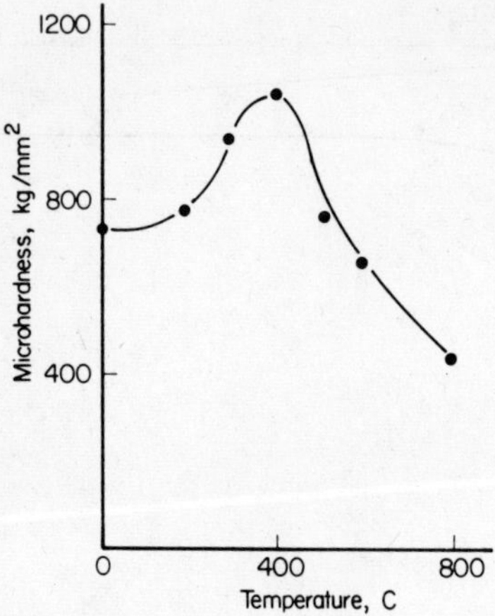

Figure 21.15. Microhardness for 30- or 35-μm Electroless Cobalt Coatings Containing 5 Percent Phosphorus as a Function of Heat-Treatment Temperature. (75)
Microhardness was measured with a 50-gram load.

phite concentration in the plating bath and an increase in the phosphorus content of the alloy increased hardness to the higher end of this range. (33,53) Heat treating for 1 or 2 hours at 300 to 500 C increased hardness to 1050 kg/sq mm for alloy containing about 5 percent phosphorus, as indicated in Figure 21.15. Data in Figure 21.15 relate to 30- to 35-μm coatings deposited on steel in 90 to 92 C solutions containing 0.16 M cobalt chloride, 0.19 M sodium hypophosphite, 0.095 M ammonium chloride, and 0.33 M sodium citrate with a pH of 9 to 10.

Cobalt-nickel alloy deposits with a phosophorus content of 3.6 percent and a hardness of 450 to 480 kg/sq mm exhibited better resistance to wear than electroless nickel or cobalt coatings, when thickness was adjusted to 13 to 14 μm. (76)

TABLE 21.9

Loss in Weight of Electroless Cobalt, Nickel and Cobalt-Nickel Alloy Coatings during Wear (76)

	Weight Loss, mg[(a)]	
Alloy[(a)]	13–14 μm Coatings	41–44 μm Coatings
Cobalt-phosphorus	8.4	15.8
Nickel-phosphorus	11.2	12.8
Cobalt-nickel-phosphorus	7.1	97.7

(a) Loss during 1,000 cycles with a 1000-gram load on a Taber Abraser CS17F wheel.

However, both the nickel and cobalt coatings showed less wear than the alloy when thickness was in the range of 41 to 44 μm, as shown in Table 21.9.

Cobalt-nickel-phosphorus alloy with an initial hardness of about 500 kg/sq mm was increased in hardness to 1000 kg/sq mm by heating in nitrogen at 500 C. (77)

Structure

Cobalt-Phosphorus Alloys

Examination by electron diffraction revealed a hexagonal-close-packed crystalline form for cobalt in many cobalt coatings. With increasing phosphorus content in the film, the face-centered-cubic structure becomes dominant in the mixture of hexagonal-close-packed and face-centered-cubic cobalt. (60)

X-ray examination has shown a fine-grained structure for high-coercive-force films. The high coercivity of such films is related to their semiparticulate nature and depends primarily on particle size and secondarily on particle interaction and their crystallographic interactions. (59)

Cobalt-phosphorus films nucleate as discrete cobalt particles with a circular area. The crystallite size of continuous deposits was in the range of 200 to 1000 angstroms. (35) The crystallites so obtained showed no correlation with the thickness, phosphorus content, or magnetic properties. On the other hand, another investigator reported a crystallite size of 1000 angstroms for 700-angstrom films and larger crystals for thicker films. (55)

Additions of N,N-dimethylurea, thioacetamide, urea, and acetamides to electroless cobalt-phosphorus baths refined the grain size and modified the orientation of cobalt films. An increase in the number of particles which were too small to diffract coherently occurred with these additions. (40)

Films may be either cloudy and white, or have a bright metallic luster. Brightness increases with pH and decreases with increase in thickness. (26) Films containing <2 percent phosphorus exhibited a crevice-type surface texture, whereas a non-crevice-type texture was noted for films with > 2 percent phosphorus. (36) Another investigator noted defects consisting of large inclusions in films with > 3 percent phosphorus. (57) The inclusions contained a Co/P ratio compatible with the compound Co_2P. Their size varied from 200 to 1000 angstroms.

The phosphorus content of the deposit correlates with crystallographic orientation. Below about 2 percent phosphorus, a random form was noted by one investigator. (78) At phosphorus contents in the range of 2 to 5 percent, an acicular structure with a (1010) fiber axis was reported. (53) The particle size appeared unchanged, however.

X-ray and electron-beam diffraction structural analysis reveals a random distribution of hexagonal crystallites with their longitudinal axis parallel with the plane of the substrate. (26,35,50,68,79) These high-coercivity films had a stacking fault frequency of about 0.13 with crystallite sizes of 200 to 700 angstroms. (80) In lower coercivity films (about 300 oersteds), orientation of the longitudinal axis was predominantly perpendicular to the film plane. Films with intermediate values of coercivity (300 to 600 oersteds) showed a random distribution. (49,80)

Magnetization decreased with increasing temperature to about 20 C, increased abruptly at 320 C, then remained constant. (60) Differential thermal analysis and X-ray diffraction analysis indicated that the transition at 320 C was accompanied by a breakdown of the solid solution and formation of Co_2P. Cobalt-phosphorus films contained hcp and fcc cobalt and Co_2P after heating to 500 C. (81) Hexagonal-close-packed crystals were not present above 550 C. At 700 C, the Co_2P decomposed. Coercivity increased sharply at 350 to 450 C.

Metallographic examination of cobalt-phosphorus deposits containing 2 to 6 percent phosphorus reveals a microstructure consisting of laminations oriented parallel to the supporting metal. This layer-like structure has been attributed to the nonuniform distribution of phosphorus. (82) Heat treating at 800 C recrystallized the metal and obliterated the layered structure. (75)

Ternary Alloy Deposits

Cobalt films containing about 4 percent iron had only the hcp phase with random crystallite orientation. (68) Films with a higher iron content showed an increasing alignment of the hexagonal longitudinal axis of the crystallites perpendicular to the plane of the film. With 20 to 50 percent iron contents, mixtures of hcp and fcc phases were found. This indicates a supersaturated solution of iron in cobalt, since under equilibrium conditions the composition should be below these values. Crystallite sizes were estimated at between 300 and 1000 angstroms for most of these films, except for the alloys with a 40–45 percent iron content, which contained very small crystallites (< 100 Å) and a large proportion of crystallites too small for resolution by X-ray analysis.

As nickel was introduced into cobalt-phosphorus alloy, the fcc phase appeared and there was an increasing orientation of the hexagonal (002) and cubic (111) axis perpendicular to the substrate. (66,67) For high nickel contents, $CoNi_3$ appeared, again with the hexagonal axis (002) perpendicular to the film plane. (66) Crystallite size varied directly with the rate of deposition. It ranged from < 200 to 700 Å for both cobalt and the high nickel-cobalt alloy films, but exceeded 1000 Å for a 73Co-25Ni-2P composition. Cobalt-nickel films with 9 percent phosphorus and a thickness of about 500 Å appeared amorphous and exhibited a maximum coercivity of 200 oersteds. (83)

Both Co_2P and Ni_2P were identified in cobalt-nickel films with a thickness of 24,000 angstroms. (84) Annealing in the range of 100–225 C increased the concentration of Co_2P. Recrystallization occurred at a temperature above 200 C. Annealing the films at 300 C caused a decrease in coercivity, caused by the destruction of the preferred orientation of the defects. (22) Increasing the temperature to

400 C caused an increase in coercivity, which may relate to coagulation of phosphides.

Cobalt-zinc-phosphorus films were discontinuous, with a particle diameter of 64 angstroms after 10 seconds of plating and 105 angstroms after 20 seconds. (70)

References

(1) Brenner, A., and Riddell, G. E., "Nickel Plating on Steel. Good Quality Deposits by Chemical Reaction," *RP 1725, Journal Research National Bureau of Standards, 37* (1), 31 (1946); "Deposition of Nickel and Cobalt by Chemical Reduction," *RP 1835, Ibid., 39* (5), 385–395 (1947). Brenner, A., Couch, D. E., and Williams, E. K., "Electrodeposition of Alloys of Phosphorus with Nickel or Cobalt," *RP 2061, Ibid., 44* (1), 109 (1950).

(2) Weisenberger, L. M., "Chemical Plating of Nickel, Cobalt, or Copper on Other Metals from Their Salt Solutions by Reduction with Amineboranes," U.S. Patent 3,431,120 (March 4, 1969). Assigned to Allied Research Products, Inc.

(3) Berzins, T., "Chemical Reduction Plating Process and Bath," U.S. Patent 3,338,726 (August 29, 1967). Assigned to E. I. du Pont de Nemours & Company.

(4) Cupr, V., "Electroless Nickel Plating from Borohydride Baths," *Povrchove Upravy, 7* (5–6), 167–170 (1967).

(5) Lang, K., and Klein, H.-G., "Chemical Plating," U.S. Patent 3,373,054 (March 12, 1968). Assigned to Farbenfabriken Bayer AG.

(6) Lang, K., and Klein, H.-G., "Method of Chemical Plating," Belgian Patent 650,755 (January 20,1965). Assigned to Farbenfabriken Bayer AG.

(7) "Chemical Plating of Nickel-Boron or Cobalt-Boron Alloy from Solution on Catalytic Surfaces," British Patent 836,480 (June 1, 1960). Assigned to E. I. du Pont de Nemours & Company.

(8) Zirngiebl, E., and Klein, H.-G., "Bath Compositions for Chemical Plating of Metals Containing Boron Nitrogen Compounds and an Organic Solubilizing Compound," U.S. Patent 3,140,188 (July 7, 1964). Assigned to Farbenfabriken Bayer AG.

(9) Hoke, R. M., "Chemical Plating of Metal-Boron Alloys," U.S. Patent 3,150,994 (September 29, 1964). Assigned to Callery Chemical Company.

(10) Mathias, J. S., and McGee, J. J., "Magnetic Alloys and Method and Compositions Useful for Preparing Same," U.S. Patent 3,416,932 (December 17, 1968). Assigned to Sperry Rand Corporation.

(11) Hoke, R. M., "Chemical Plating of Metal-Boron Alloys," U.S. Patent 2,990,296 (June 27, 1961). Assigned to Callery Chemical Company.

(12) Mochel, J. M., "Chemical Nickel Plating on Ceramic Material," U.S. Patent 2,968,578 (January 17, 1961). Assigned to Corning Glass Works.

(13) Kozlova, N. I., and Korovin, N. V., "Chemical Deposition of Cobalt and Cobalt-Nickel Alloys with Hydrazine Hydrate," *Zh. Prikl. Khimii, 40* (2), 446–447 (1967); *Journal Applied Chemistry, 40* (2), 426–427 (1967).

(14) Aspland, M., Jones, G. A., and Middleton, B. K., "Properties of Electroless Cobalt Films," *IEEE Transactions on Magnetics, MAG-5* (3), 314 (1969).

(15) Takano, O., Shigeta, T., and Ishibashi, S., "Electroless Cobalt Plating with Hydrazine as Reducing Agent," *Journal of the Metal Finishing Society of Japan, 18* (8), 7 (1967).

(16) Takano, O., Shigeta, T., and Ishibashi, S., "Electroless Plating of Cobalt and its Alloys. VI. Electroless Cobalt Plating Using Hydrazine as a Reducing Agent," *Journal of the Metal Finishing Society of Japan, 18* (8), 299 (1967).

(17) Makowski, M. P., "Electroless Cobalt Plating Bath," U.S. Patent 3,416,955 (December 17, 1968). Assigned to Clevite Corporation.

(18) Prueter, E. D., and Walles, W. E., "Electroless Coating of Cobalt and Nickel," U.S. Patent 3,472,665 (October 14, 1969). Assigned to The Dow Chemical Company.

(19) Holmen, J. O., "Electroless Coating from Acid Cobalt Baths," French Patent 1,528,173 (June 7, 1968). Assigned to Honeywell Inc.

(20) Chopra, K. L., *Thin Film Phenomena*, McGraw-Hill, New York (1969), 844 pp.

(21) Zentner, V., "Magnetic Properties of Electroplated and Chemically Deposited Cobalt Alloys," *Journeés Intern. Appl. Cobalt, Brussels, Belgium*, 152 (1965).

(22) Bondar, V. V., "Chemically Deposited Magnetic Alloys," *Itogi Nauki, Elektrokhim.*, 56 (1966) (Pub. 1968).

(23) Sallo, J. S., "Review of Deposited Metallic Films for Computer Application," *Plating, 54* (3), 257 (1967).

(24) Speliotis, D. E., Morrison, J. R., and Judge, J. S., "Correlation Between Magnetic Properties and Recording Behavior in Metallic Chemically Deposited Surfaces," *IEEE Transactions on Magnetics, MAG-1* (4), 348 (1965).

(25) Austen, H. E., and Fisher, R. D., "Internal Stress of Electroless Metal Films on Single-Crystal Silicon," *Journal Electrochem. Soc., 116* (2), 185 (1969).

(26) Blackie, I., Truman, N., and Walker, P. A., "The Preparation and Properties of Hard Magnetic Films for Use in a Magnetic Recording System," *Magnetic Materials and Their Applications*, The Institution of Electrical Engineers, *Conference Publication No. 33*, 207 (1967).

(27) Bate, G., "A Brief Review of High-Coercivity Thin Films for Digital Magnetic Recording," *Journal Applied Physics 37* (3), 1164 (1966).

(28) Fisher, R. D., and Taylor, S. D., "Observaton on the Growth Process of Cobalt Films," *Journal Applied Physics., 37* (6), 2512 (1966).

(29) Kostenich, I. F., "Chemical Cobalt Plating," *Mater. Nauch. Konf. Sovnarkhoz, Nizhne-volzhsk. Ekon. Raiona, Volgograd. Politekh. Inst., Volgograd., 2*, 207 (1965).

(30) Morton, V., and Fisher, R. D., "Effect of Electroless Nickel Substrate on Coercive Force and Microstructure of Certain Co-P Films," *Journal Electrochem. Soc., 116* (2), 188 (1969).

(31) Ruscior, C., Suciu, M., and Calugaru, Gh., "Magnetic Properties of Chemically Deposited Thin Cobalt Layers," *An. Stiint. Univ. "Al. I. Cuza," Is ai, Sect. Ib, 12*, 51 (1966).

(32) Levy, J. P., "Activation of Surfaces for Improved Chemical Plating of Nickel or Cobalt on Them," British Patent 1,003,575 (September 8, 1965). Assigned to Sperry Gyroscope Company Ltd.

(33) Cadorna, L., Cavallotti, P., and Salvago, G., "Electroless Plating. II. Electroless Cobalt from Alkaline Sulfamate Bath," *Electrochimica Metallorum, 1* (2), 177 (1966).

(34) Takano, O., Deguchi, K., and Ishibashi, S., "Electroless Plating of Cobalt and Its Alloys. III. Electroless Plating of Cobalt-Nickel Alloy from Caustic Alkaline Citrate Baths," *Journal of the Metal Finishing Society of Japan, 18* (2), 46 (1967).

(35) Judge J. S., Morrison, J. R., and Speliotis, D. E., "The Effect of the Concentration of Hypophosphite Ion on the Magnetic Properties of Chemically Deposited Co-P Films," *Journal Electrochem. Soc., 113* (6), 547 (1966).

(36) Miksic, M. G., Travieso, R., and Arcus, A., and Wright, R. H., "The Relationship Between Coercivity and the Structure and Composition of Electroless Cobalt-Phosphorus Films," *Journal Electrochem, Soc., 113* (4), 360, (1966).

(37) Alberts, G. S., Wright, R. H., and Parker, C. C., "Effect of NH_3 on Deposition from Alkaline Electroless Nickel and Cobalt Plating Baths," *Journal Electrochem. Soc., 113* (7), 687 (1966).

(38) Takano, O., Inoue, M., Aoki, K., and Ishibashi, S., "Electroless Plating of Cobalt and Its Alloys. II. Properties of Plated Films of Electroless Cobalt," *Journal of the Metal Finishing Society of Japan, 18* (1), 2 (1967).

(39) Takano, O., Yoshida, S., and Ishibashi, S., "Electroless Plating of Cobalt and Its Alloys. IV. Effects of Boric Acid in Electroless Plating Baths," *Journal of the Metal Finishing Society of Japan, 18* (3), 99 (1967).

(40) Fisher, R. D., and Chilton, W. H., "The Influence of Certain Addition Agents on the Deposition

Rate and Hard Magnetic Properties of Electroless Cobalt," *Plating, 54* (5), 537 (1967). Discussion by Weil, *Ibid. 55*, 263 (1968).

(41) Saranov, E. E., Bulatov, N. K., and Mokrushin, S. G., "Formation of Thin Cobalt and Nickel-Cobalt Alloy Films on a Glass Surface," *Izvestiya Vysshikh Uchebnykh Zavedeniy SSSR, Khimiya i Khimicheskaya Tekhnologiya, 10* (4), 403 (1967).

(42) Takano, O., Masai, S., and Ishibashi, S., "Electroless Plating of Cobalt and Its Alloys. V. Effects of Plating Condition on the Reduction Efficiencies of Electroless Cobalt Plating Baths," *Journal of the Metal Finishing Society of Japan, 18* (5), 180 (1967).

(43) Pendleton, W. W., "High Temperature Magnet Wire," U.S. Patent 3,446,660 (May 27, 1969). Assigned to Anaconda Wire and Cable Company.

(44) Schneble, Jr., F. W., Zeblisky, R. J., McCormack, J. F., and Williamson, J. D., "Process for Stabilizing Autocatalytic Metal Plating Solutions," U.S. Patent 3,403,035 (September 24, 1968). Assigned to Process Research Company.

(45) Bursuc, I. D., and Calugaru, Gh., "Magnetic Properties of Chemically Deposited Cobalt-Phosphorus Films," *Thin Solid Films, 3* (1), R5 (1969).

(46) Takano, O., and Ishibashi, S., "Electroless Plating of Cobalt and Its Alloys. VIII. Effects of Plating Conditions on Magnetic Properties of Electroless Deposited Cobalt Films in Caustic Alkaline Baths," *Journal of the Metal Finishing Society of Japan, 19* (4), 139 (1968).

(47) Ransom, L. D., and Zentner, V., "Magnetic Films of Electroless Cobalt-Phosphorus with Uniaxial Anisotropy," *Journal Electrochem. Soc., 111* (12), 1423 (1964).

(48) Takano, O., and Ishibashi, S., "Electroless Plating of Cobalt and Its Alloys. I. Precipitation Velocity of Electroless Cobalt Plating," *Journal of the Metal Finishing Society of Japan, 17* (8), 299 (1966).

(49) Jones, G. A., and Middleton, B. K., "The Structure of Electroless Cobalt Films as Revealed by Electron Microscopy," *J. Materials Science, 3* (5), 519 (1968).

(50) Aoki, K., Iseki, S., Takano, O., and Ishibashi, S., "Electroless Plating of Cobalt and Its Alloys. IX. Structure of Electroless Cobalt Plating Films," *Journal of the Metal Finishing Society of Japan, 19* (8), 301 (1968).

(51) Judge, J. S., Morrison, J. R., Speliotis, D. E., and Bate, G., "Magnetic Properties and Corrosion Behavior of Thin Electroless Co-P Deposits," *Journal Electrochem. Soc., 112* (7), 681 (1965).

(52) Moradzadeh, Y., "Chemically Deposited Cobalt-Phosphorus Films for Magnetic Recording," *Journal Electrochem. Soc., 112* (9), 891 (1965).

(53) Frieze, A. S., Sard, R., and Weil, R., "Some Properties of Electroless Cobalt," *Journal Electrochem. Soc., 115* (6), 586–591 (1968).

(54) Calugaru, Gh., "Uniaxial Anisotropy Induced by Quenching Chemically Deposited Cobalt-Phosphorus Films in a Magnetic Field," *Thin Solid Films, 3* (5), R27 (1969).

(55) Fisher, R. D., and Chilton, W. H., "Preparation and Magnetic Characteristics of Chemically Deposited Cobalt for High-Density Storage," *Journal Electrochem. Soc., 109* (6), 485 (1962).

(56) Koretzky, H., "Electroless Deposition of Thin Ferromagnetic Cobalt Films," paper presented at the Buffalo Meeting of the Electrochemical Society (October 10–14, 1965).

(57) Morrison, J. R., and Speliotis, D. E., "Electron Probe Analysis of Recording Surfaces," *IEEE Transactions on Magnetics, MAG-5* (3), 325 (1969).

(58) Lawless, G. W., "Stability of Chemically Deposited Magnetic Thin Films with Time and Temperature," *Journal Electrochem. Soc., 115* (6), 620 (1968).

(59) Speliotis, D. E., Judge, J. S., and Morrison, J. R., "Origin of High Coercivity in Chemically Deposited Cobalt-Phosphorus Films," *Journal Applied Physics, 37* (3), 1158 (1966).

(60) Kanbe, T., and Kanematsu, K., "Magnetic Properties of Solid Solution of Cobalt with Phosphorus," *Journal Physical Society of Japan, 24* (6), 1396 (1968).

(61) Lawless. G. W., and Fisher, R. D., "Electroless Plating Variables and Coercive Force of Nickel-Cobalt-Phosphorus Films," *Plating, 54* (6), 709 (1967).

(62) Hendy, J. C., Richards, H. D., and Simpson, A. W., "Some Observations on Thin Magnetic Films Made by Continuously Monitoring the Coercive Force," *J. Materials Science, 1*, 127 (1966).

(63) Holmen, J. O., and Sallo, J. S., "Magnetic Properties of Co-Ni-P Thin Films by Electroless Deposition," *Proceedings of the Intermag Conference, Washington, D.C.*, (April 6–8, 1964).

(64) Gorbunova, K. M., and Nikiforova, A. A., "The Conditions Under Which Cobalt and Nickel-Cobalt Coatings Are Formed," *Physicochemical Principles of Nickel Plating*, USSR Academy of Sciences, Moscow (1960). Translated by the Israel Program for Scientific Translations, Jerusalem, 147–153 (1963). OTS 63-11003; TS 690.

(65) Korenev, N. A., "Chemical Deposition of Nickel-Cobalt Films," *Izv. Akad. Nauk SSSR, SER. Fiz., 29* (4), 650–654 (1965); *Bulletin Acad. Sci. USSR, Physical Series, 29* (4), 651–655 (1965).

(66) Judge, J. S., Morrison, J. R., and Speliotis, D. E., "Very High Coercivity Chemically Deposited Co-Ni Films," *Journal Applied Physics, 36*, (3), Part 2, 948–9, (1965).

(67) Speliotis, D. E., Judge, J. S., and Morrison J. R., "Magnetic Properties of Chemically Deposited Films of Cobalt-Nickel-Phosphorus," *Plating in the Electronics Industry*, Second Symposium, American Electroplaters' Society, Boston, Massachusetts, 301–320 (February 6–7, 1969).

(68) DePew, J. R. and Speliotis, D. E., "Magnetic Properties of Chemically Deposited Cobalt-Iron Films," *Plating, 54* (6), 705–708 (June, 1967).

(69) Pearlstein, F., and Weightman, R. F., "Magnetic Properties of Electroless Cobalt Based Alloys," *Plating, 54* (6), 714–716 (1967).

(70) Fisher, R. D., "Preparation and Hard Magnetic Properties of Co-Zn-P Films for High-Density Recording," *IEEE Transactions on Magnetics, MAG-2* (4), 681–686 (1966).

(71) Mathias, J. S., and McGee, J. J., "Chemically Deposited Magnetic Films of Cobalt-Aluminum-Phosphorus, and Cobalt-Nickel-Aluminum-Phosphorus," *Plating in the Electronics Industry*, Symposium, American Electroplaters' Society, Newark, New Jersey (December 8–9, 1966).

(72) Morrison, J. R., and Gahr, P. E., *Plating in the Electronics Industry*, Second Symposium, Boston, Massachusetts, 279–287 (February 6–7, 1969).

(73) Haines, R. S., Judge, J. S., Morrison, J. R., and Vranka, J. S., *IBM Tech. Discl. Bull., 8*, 104 (1965).

(74) Zentner, V., "Electrodeposited and Electroless Magnetic Alloys for Computers," *Plating, 52*, 868–872 (1965).

(75) Shatsova, S. A., Andreeva, S. A., and Koroleva, N. V., "The Structure of Chemically Deposited Cobalt," *Zashchita Metallov, 3*, (6), 744–747 (1967).

(76) Unpublished data from research sponsored by the Cobalt Information Center, Brussels (1960).

(77) Takano, O., and Ishibashi, S., "Electroless Plating of Cobalt and Its Alloys. XII. Hardness of Chemically Deposited Plating Films of Cobalt-Nickel and Cobalt-Nickel-Phosphorus Alloys," *Journal of the Metal Finishing Society of Japan, 19*, 498–503 (1968).

(78) Amendola, A., Brock, G. W., Travieso, R., "Structure and Wear Properties of Thin Films of Electroless Plated Cobalt-Phosphorus on Flexible Substrates," *Lubrication Engineering, 26* (1), 22–5 (1970).

(79) Bagrowski, J. and Lauriente, M., "Electroless Ni-Co-P Films with Uniaxial Anisotropy," *Journal Electrochem. Soc., 109* (10), 987–9 (October, 1962).

(80) Bate, G., "Thin Metallic Films for High-Density Digital Recording," *IEEE Transactions on Magnetics, MAG-1* (3), 193–205 (1965).

(81) Moiseev, V. P., Sadakov, G. A., Gorbunova, K. M., "Structure Phase Transformations, and Properties of Chemically Reduced Cobalt," *Zh. Fiz. Khimii, 42* (11), 2751–2756 (1968).

(82) Kharitonyuk, A. M., "Chemical Nickel and Cobalt Plating," *Metal Science and Heat Treatment, (3)*, 234–236, (1966); *Metallovedeniye i Obrabotka Metallov., (3)*, 57–59, (1966).

(83) Richards, H. D., "Domain Wall Coercive Force of Thin Ferromagnetic Films with Uniaxial Anisotropy," *British Journal of Applied Physics, 17*, 879–884 (1966).

(84) Schmidt, E. H., and Holmen, J. O., "Diodeless Magnetic-Film-Shift-Register Designs Utilizing Shifted Hysteresis Loops," *Journal Applied Physics, 37* (3), 1378–1379 (1966).

Chapter 22

Electroless Nickel

Electroless nickel is produced by reducing nickel ions from solutions which contain phosphite, boron or hydrazine compounds.

Alloys of nickel and phosphorus are obtained with phosphite ions. The phosphorus content can be varied, normally between 2 and 12 percent, to control strength, ductility, stress, magnetic properties and structure. High phosphorus contents obtained in solutions with a low pH and a high ratio of phosphite to nickel ions favor high strength and low stress. The alloys are supersaturated solid solutions of nickel phosphide in nickel.

Nickel phosphide can be precipitated by heat treatment, which hardens the deposit about twofold. This increases stress toward the tensile direction for deposits on steel and other substrates with low expansion coefficients. The strength and ductility of alloys containing less than 7 percent phosphorus are increased by heat treatment, but higher phosphorus-nickel alloys are reduced in strength and ductility.

Hardness is not affected by the phosphorus content in the range of 2 to 11 percent, either before or after heat treatment at 400 C. Only the high-phosphorus alloys retain high hardness after heat treating at 600 or 800 C. Treatment at 400 C maximizes hardness. To minimize stress, a phosphorus content of 11 percent appears best. Heat treatment at 700 to 800 C is usually employed to avoid or minimize a loss in the fatigue strength of high-strength steel plated with electroless nickel. After heat treatment at high temperature, alloys with at least 9 percent phosphorus consist of nickel dispersed in a matrix of nickel phosphide. Alloys with less than 7 percent phosphorus are a dispersion of nickel phosphide in nickel. High-temperature heat treatment improves the corrosion resistance of electroless nickel-plated steel.

Applications

Numerous applications have been developed for electroless nickel deposits in the past. The future looks bright for continued growth where economic advantage can be made of the resistance of the coating to abrasion, wear, friction and/or corrosion.

Electroless nickel deposits have been used for tank car interiors to handle a variety of liquids which would be contaminated by contact with steel, or which are corrosive to steel. Electroless nickel deposits also find wide application in rocketry and missiles, jet engines, oil refineries, handling of petroleum products, handling of nuclear fuels and heat transfer fluids. Piping and condensing towers for solvent extraction can be coated with electroless nickel to inhibit corrosion and erosion. A coating of electroless nickel can be utilized to inhibit stress corrosion cracking of stainless steel and titanium alloys.

Characterized by high hardness, electroless nickel deposits find use in increasing the life of pumps, compressors and similar equipment. Dies used for extruding and molding plastics or casting zinc alloys can be coated with electroless nickel to inhibit erosion. The low thermal conductivity of electroless nickel (relative to that for die steels) may be an advantage in die casting, for inhibiting premature freezing of the zinc alloy onto the die cavity surfaces. In cast-iron molds for making glass, electroless nickel deposits can prevent sticking and ease release of the glassware from the mold.

Automotive applications of electroless nickel deposits on steel include ball studs, transmission thrust washers and differential pinion cross shafts. In these applications, electroless nickel is used primarily to reduce galling-type wear.

Solderability of aluminum products is enhanced when they are coated with electroless nickel.

The low coercivity of electroless nickel-phosphorus and nickel-cobalt-phosphorus alloys, and the relationship of their remanent and saturation magnetic forces, can be utilized for fast-switching memory devices in computers. The metallization of non-conducting surfaces, which has a vital role in the plating of plastics, is another widespread use for electroless nickel deposits.

Physical Properties

Brenner initiated the measurement of properties of electroless nickel deposits in 1947. He reported hardness values of 425 Knoop for a nickel-phosphorus alloy deposited in an alkaline citrate bath, 500 Knoop for an alloy produced in an acid citrate solution and 530 Knoop for an alloy deposited in an ammoniacal bath. (1) The increase in hardness resulting from heat treating was ascribed to the precipitation of phosphides.

Sodium hypophosphite is the most common agent for reducing nickel ions to metal, but several reports refer to the use of hydrazine or sodium borohydride. More property data have been reported for the nickel-phosphorus alloys deposited in solutions containing sodium hypophosphite than for electroless nickel produced with other reducing agents. The phosphorus contents of the deposits from solutions containing sodium hypophosphite usually range from 2 to 12 percent, but sometimes reach 20 percent. A high molar ratio of hypophosphite and nickel ions (>2.75) and a low pH (<4.5) favor a high phosphorus content of >10 percent. (2)

Density

The density of electroless nickel is proportional to the phosphorus content, as shown in Figure 22.1. The density of *electroless* nickel is slightly lower than the calculated value for mixtures of nickel and Ni_2P, whereas the density of *electrodeposited* nickel-phosphorus alloys is slightly higher than the calculated value. (3) Densities of 7.9 to 8.1 g/cu cm were reported for alloys containing 8 to 9 percent phosphorus, as shown in Table 22.1. Alloys containing 5 to 6 percent phosphorus have a density of 8.25 g/cu cm. The density of electroless nickel containing about 3 percent phosphorus is 8.5 g/cu cm.

The amount of occluded compounds in electroless nickel must be fairly small, as indicated by analytical data for carbon, oxygen and hydrogen which totaled < 0.05 percent. (6) One investigator reported a slight increase in density as a result of heat treatment. (10) The density of electroless nickel reduced with hydrazine was lowered from 8.5 to 7.9 g/cu cm upon heat treatment. (11) It contained an appreciable amount of oxides (0.09 to 0.46 percent oxygen).

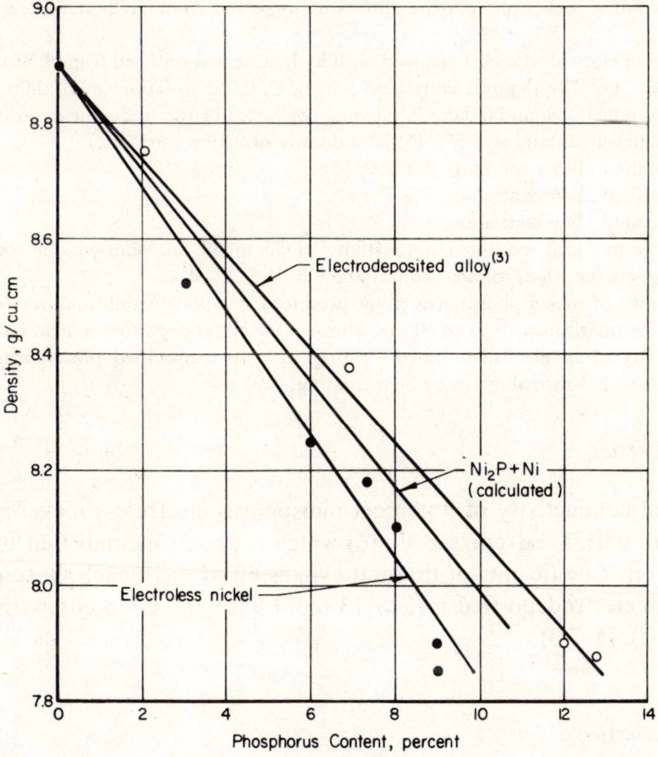

Figure 22.1. Density of Nickel-Phosphorus Alloys as a Function of the Phosphorus Content.

TABLE 22.1

Physical Properties of Electroless Nickel

Property	Property Value as a Function of Phosphorus Content			Supplemental Data
	3% P	5–6% P	8–9% P	
Density, g/cu cm	8.52(4)	8.25(4)	7.85 (5,6), 7.9(7), 8.1(4)	(a), (b), (c), (d)
Electrical resistivity, microhm-cm	30[e]	72[f]	30 to 55(8), 60 (5, 7), 110[g]	(h), (i), (j)
Thermal conductivity, cal/cm/sec/C	No data	No data	0.0105 to 0.0135 (6)	
Coefficient of thermal expansion, μm/m/C	No data	No data	13 (5, 7), 13.5 to 14.5 (9)	10.45 μm/m/C was reported for one deposit; no P content was given (10)

[a] Impurities in electroless nickel containing 8% P with a density of 7.85 g/cm^3 included 0.04% C, 0.0016% H_2, 0.002% O_2, and 0.0005% N_2. (6)

[b] The density of a nickel-phosphorus alloy was increased from 7.72 to 7.82 g/cm^3 by heat treating. (10)

[c] The density of electroless nickel reduced with hydrazine was reduced from 8.5 to 7.9 g/cm^3 by treating at 760 C. (11) The deposits contained 0.04% C, 0.003 to 0.007% H_2, 0.09 to 0.46% O_2, 0.24 to 0.29% N_2, 0.005% S, and 0.03% P. Heating converted brittle to ductile deposits.

[d] Electroless nickel containing 7.5% P had a density of 8.19 g/cm^3. (12)

[e] Electrodeposited alloy containing 2.2% P. (3)

[f] Electrodeposited alloy containing 7% P. (3)

[g] Electrodeposited alloy containing 13% P. (3)

[h] Thin films (1 to 2 μm) exhibited a resistivity of 55 to 68 microhm-cm, in comparison with 30 to 55 microhm-cm for 10μm coatings containing 9 to 10% P. (8)

[i] The resistivity of nickel-phosphorus alloy produced in alkaline solutions was only 28 to 34 microhm-cm, in comparison with 51 to 58 microhm-cm for alloys deposited in acid baths. (13)

[j] The resistivity of an electroless nickel coating with an unspecified phosphorus content was reduced from 190 to 135 microhm-cm by heat treating. (10)

Thermal Properties

The thermal conductivity of 9 percent phosphorus electroless nickel alloy ranges from 0.0105 to 0.0135 cal/cm/sec/C, (6) which is about one-half that of pure metallurgical nickel. Coefficients of thermal expansion of the nickel-phosphorus alloys are lower than electrodeposited nickel (13 to 14.5 μm/m/C, in comparison with 14 to 17 μm/m/C). (5–7,9)

Electrical Properties

The conductivity of electroless nickel ranges form 1.6 to 5.7 percent IACS, depending on the phosphorus content. The electrical resistivities (Table 22.1) of alloys containing 8 to 9 percent phosphorus are nearly 10 times the resistivity of pure metallurgical nickel, which is 9.5 microhm-cm. A reduction of resistivity occurs after heat treating. (10) The resistivity of an electrodeposited alloy containing 7 percent phosphorus is reduced from 72 to about 20 microhm-cm by heat treating at 600 C. (3) A resistivity of only 13 to 15 microhm-cm was reported for electroless nickel containing 1.3 or 4.7 percent boron that had been reduced with dimethylamine borane. (14) The resistivity of a nickel alloy containing 0.6 percent boron was only 5.3 microhm-cm, which is lower than handbook values for high-purity nickel. The boron-doped nickel corresponds to n-type, whereas phosphorus nickel is p-type. (15)

Magnetic Properties

Electroless nickel deposits produced in alkaline solutions and containing 7 to 18 percent phosphorus exhibited a coercivity of 2 oersteds or less, as shown in Table 22.2. Coercivity values of 10 to 30 oersteds were reported for some deposits containing 3 to 6 percent phosphorus. Others ranged in coercivity from 50 to 80 oersteds. Heat treating at 350 or 400 C increased coercivity considerably, for example, from 2 to 110 oersteds or from < 1 to 150 oersteds. (19)

By contrast with electroless nickel films, which exhibit a low coercivity regardless of their phosphorus content, electroless cobalt films with more than 1 to 1.5 percent phosphorus were higher in coercivity (> 115 oersteds). Alloys containing 60 percent cobalt, balance nickel, also are high in coercivity (> 500 oersteds, as a rule) but alloys containing 15 to 50 percent cobalt have a low coercivity (< 5 oersteds).

A patent cited a coercivity of 2 to 3 oersteds for electroless nickel deposited in an acid bath with a pH of about 4.0. (26) A slightly acid bath with a pH of about 6.0 was used to deposit nickel alloys containing 15 to 33 percent cobalt and 2 percent phosphorus. These alloys exhibited a coercivity of < 3 oersteds. (22) Increasing the pH to 6.4 increased the coercive force. A high ratio of remanent to saturation magnetization also was reported. Low-coercivity nickel-cobalt alloys containing 5 to 11 percent phosphorus also are included in Table 22.2.

Relatively thick films 0.03 mil (7,600 angstroms) of nickel-cobalt-phosphorus alloys were unchanged in coercivity during exposure to humid air (50 to 70 percent relative humidity) at 100 C, whereas thinner films 0.00015 to 0.015 mils (400 to 3,800 angstroms) were not. (23) Coercivity of the thicker film was stable for 1,400 hours.

Unlike electroless nickel, which exhibited an increased coercivity after heat treatment, electroless nickel-cobalt-phosphorus alloys were reduced in coercivity when they were heat treated. (27) Heat treating at 300 C minimized coercivity, which was ascribed to the "destruction of defects." These alloy films were deposited in 87 C solutions containing 0.11 M nickel sulfate, 0.12 M sodium acetate, 0.13 M ammonium sulfate and 0.16 M sodium hypophosphite.

Although several reports state that electroless nickel becomes nonmagnetic when its phosphorus content reaches 9 percent, a field of 6,500 gausses showed slight magnetization for alloys containing < 15 percent phosphorus. (28) Magnetization ap-

TABLE 22.2

Coercivity of Electroless Nickel and Nickel-Cobalt Alloy

Phosphorus Content, percent	Coercivity, oersteds		Squareness of Hysteresis Loop (Br/Bm)	Solution Composition		Operating Conditions		Reference
	No Heat Treatment	Heat Treated at 350 C		$NiSO_4$, M	NaH_2PO_2, M	pH	Temperature C	
I Nickel								
<0.5	14 to 22	No data	0.92[b]	0.2[a]	None[a]	7–11	92	16, 17
3.0 to 6.0	10 to 20	No data	No data	0.12[c]	0.19	8.0	71	18
3.0 to 6.0	50 to 80	No data	No data	0.12[c]	0.19	9.1	71	18
3.5	30	60	<0.1	0.13[d]	0.095	8.5	90	19
7.0	2	110	0.7	0.09[e]	0.19	8.5	90	19
8.6	1.4	125	0.5	0.06[e]	0.19	8.5	90	19
Ca 9	No data	325	No data	0.13[f]	0.095	8–9	71	20
9.6	<0.1	150	<0.4	0.04[e]	0.19	8.5	90	19
10.0 to 17.0	0	No data	<0.4	0.12[g]	0.19	8–9	71	18
15.0 to 18.0	0	No data	<0.4	0.033[h]	0.095	8–9	82	18

II Nickel-Cobalt Alloy								
1.0 to 2.0	2 to 8	No data	No data	0.09[i]	(i)	8.0	75–95	21
2.0	<3.0	No data	No data	0.07[j]	0.14	6.0	90	22
<5.0	1 to 4[k]	No data	—	0.09[l]	(l)	8.0	80	23
5.0	2[m]	No data	No data	0.027[n]	(n)	8.5	90	24
5.5	4.5	No data	0.10	0.06[o]	0.19	9.5	—	25
6.9	0.14	No data	0.50	0.13[p]	0.19	9.5	—	25

(a) This solution contained 0.2 M $NiCl_2$, 0.2 M $NH_2C_4H_4O_6$, and 1.0 M hydrazine.

(b) Saturation magnetization was 98% of that of bulk, wrought nickel. The thickness of these films ranged from 1,000 to 2,000 angstroms.

(c) This solution contained 0.12 M $NiSO_4$, 0.19 or 0.75 M $(NH_4)_2SO_4$, 0.12 M $Na_3C_6H_5O_7$, and 0.19 M $NaH_2P_2O_7$.

(d) This solution was prepared with $NiCl_2$ and also contained 0.48 M NH_4Cl and 0.17 M $Na_3C_6H_5O_7$.

(e) This solution was prepared with $NiCl_2$ and also contained 0.48 M NH_4Cl and 0.5 M $Na_3C_6H_5O_7$.

(f) This solution contained 0.13 M $NiCl_2$, 0.95 M NH_4Cl, 0.34 M $Na_3C_6H_5O_7$, and 0.095 M NaH_2PO_2.

(g) This solution contained 0.12 M $NiSO_4$, 0.12 M $Na_3C_6H_5O_7$, and 0.19 M NaH_2PO_2.

(h) This solution contained 0.033 M $NiSO_4$, 0.5 M $Na_3C_6H_5O_7$, 0.095 M NaH_2PO_2, and 0.4 M $Na_2B_4O_7$.

(i) This solution contained 0.09 M $NiSO_4$, 0.063 M $CoSO_4$, 0.28 M $Na_3C_6H_5O_7$, 0.3 M $(NH_4)_2SO_4$, and 0.28 M H_3PO_2. The deposits contained 40 to 50% cobalt.

(j) This solution contained about 0.07 M $NiCl_2$, 0.06 M $CoCl_2$, 0.17 M $Na_3C_6H_5O_7$, 0.75 M NH_4Cl, and 0.14 M NaH_2PO_2. The deposits contained 15 to 33% cobalt.

(k) Coercivity decreased with thickness from 4 oersteds at 2,100 Å to 1 oersted at 7,600 angstroms.

(l) This solution contained 0.095 M $NiSO_4$, 0.065 M $CoSO_4$, 0.3 M $(NH_4)_2SO_4$, 0.27 M $Na_3C_6H_5O_7$, and 0.25 M H_3PO_2. The cobalt content was about 50%.

(m) Coercivity increased to 12 oersteds as film thickness increased from 500 to 3,000 angstroms. The alloy deposits contained about 40% cobalt.

(n) This solution contained 0.027 M $NiSO_4$, 0.018 M $CoSO_4$, 0.07 M $Na_3C_6H_5O_7$, 0.16 M $(NH_4)_2SO_4$, and 4 ml/l H_3PO_2.

(o) This solution contained 0.06 M $NiCl_2$, 0.34 M $Na_3C_6H_5O_7$, 0.95 M NH_4Cl, and 0.19 M NaH_2PO_2. The cobalt content of the deposit was 37.5%.

(p) This solution contained 0.13 M $NiCl_2$, 0.34 M $Na_3C_6H_5O_7$, 0.95 M NH_4Cl, and 0.19 M NaH_2PO_2. The cobalt content of the deposit was 23%.

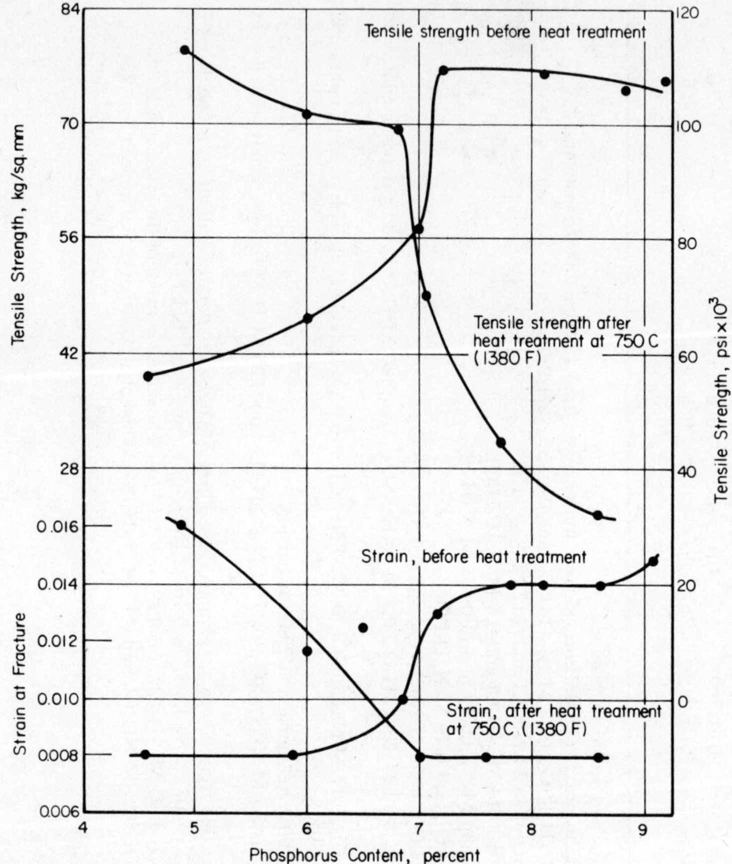

Figure 22.2. Influence of Phosphorus Content on the Strength and Ductility of Electroless Nickel. (29)

proached zero at 15 percent phosphorus, which corresponds to the compound Ni_3P. At 9 percent phosphorus, magnetic susceptibility was 4 percent, in comparison with 37.3 percent for electrodeposited nickel. (5) Saturation magnetization of electroless nickel reduced with hydrazine was nearly equal to the saturation magnetization for wrought nickel (484 gausses/cu cm). (17)

Mechanical Properties

Strength and Ductility

A tensile strength of 66,000 psi (46 kg/sq mm) was reported for electroless nickel containing about 6 percent phosphorus. (29) This is slightly stronger than nickel electrodeposited from a Watts-type solution or annealed, rolled nickel. The tensile strength of 8.2 or 9.4 percent phosphorus alloys deposited in an alkaline bath was much higher, but not as high as electrodeposited nickel containing > 0.05 percent sulfur. The tensile strength of these high-phosphorus alloys was 106,000 psi (75 kg/sq mm). (29) They were deposited in a sodium citrate-buffered nickel chloride solution. In comparison, electroless nickel containing about 8 percent phosphorus, which was deposited in an unbuffered nickel sulfate bath, exhibited a tensile

TABLE 22.3

Mechanical Properties of Electroless Nickel

Property	Property Value as a Function of Phosphorus Content: 3% P	5–6%	8–9%	Supplemental Data
Tensile strength				
psi	No data	56,000 to 68,000(29)	65,000(30), 106,000(29), 111,000 to 122,000 (31)	
kg/sq mm	No data	39 to 49(29)	45 (30), 75 (29), 78.6 to 86:1 (31)	(a) and (b)
Strain at fracture[c]	No data	0.008(29)	0.007 to 0.008 (31) 0.014 to 0.015 (29)	(d)
Elongation, percent	No data	2.0 to 2.1(32)	0.3(29), 3 to 6 (5–7)	(e)
Modulus of elasticity	No data			
psi $\times 10^6$			17 (30), 28.5 (31)	
kg/sq mm			12,000 (30), 20,000 (31)	
Hardness, kg/sq mm	500 to 800	500 to 625	500 to 800	See Table 22.4
Internal Stress				
psi	No data	9,500 to 17,000	−15,000 to 40,000	
kg/sq mm	No data	6.7 to 12.0	−10.8 to 28	See Table 22.5

[a] Heat treatment at 400 C reduced the tensile strength of 6% P and 8–9% P alloys to only 10,000 psi (7 kg/sq mm). (29) Heat treatment at 750 C increased the tensile strength of the 5–6% P alloys to 101,000 to 112,000 psi (71 to 79 kg/sq mm), but reduced the tensile strength of an 8% P alloy to 44,000 psi (31 kg/sq mm). Most of these deposits were made in 90 C bath adjusted to a pH of 8.2 to 8.4, containing 0.19 M $NiCl_2$, 0.28 M $Na_3C_6H_5O_7$, 0.95 M NH_4Cl, and about 0.19 M NaH_2PO_2; a few values for electroless nickel deposited in acid baths (pH of 4.2 to 5.7) agreed with data for the nickel deposited in the alkaline solutions.

[b] The electroless nickel with a tensile strength of 45 kg/sq mm was deposited in a bath at 80 to 100 C with a pH of 5.2 to 5.6, containing 0.057 M $NiSO_4$, 0.14 M $HC_3O_2H_6$, 0.15 g/l MoO_3, 0.06 g/l hydroxylamine sulfate, and 0.17 M NaH_2PO_2. (30) Heat treatment at 400 C and 600 C reduced tensile strength to 31,500 psi (22 kg/sq mm) and 48,000 psi (34 kg/sq mm), respectively.

[c] Determined by bulge testing. (33) A strain at fracture of 0.015 is equivalent to an elongation of >1%.

[d] Heat treatment at 400 or 600 C reduced the ductility of 5 to 9% P alloys. (29) Heat treatment at 750 C improved the ductility of the 5 or 6% P alloys, but reduced the ductility of the 8 or 9% P alloys.

[e] The ductility of the 6% P alloy was higher than the ductility of 4 or 7% P alloys. (32) Heat treating the 6% P alloy reduced elongation to only 0.2%.

strength of only 65,000 psi (45 kg/sq mm). (30) Table 22.3 includes supplemental tensile strength data.

Heat treating the phosphorus alloys at 400 to 600 C reduces the tensile strength appreciably. (29,30) For example, the tensile strength of alloys containing 5.6 to 9.4 percent phosphorus was reduced to only 10,000 psi (7.0 kg/sq mm) after 60 or 120 minutes heating at 400 C. (29) The strength of an alloy containing 8.0 to 9.4 percent phosphorus was reduced as much as 80 percent to only 18,000 psi (12.5 kg/sq mm) by heat treating at 600 C. Heat treating at 750 C reduced the strength of an 8 or 9 percent phosphorus alloy about 70 percent, but increased the strength of 6 percent phosphorus alloy about 65 percent. (29) Figure 22.2 shows these relationships. The changes in strength (and ductility) shown in Figure 22.2 are related to structural changes that occur when alloys are heat treated at 750 C. An alloy containing 5 or 6 percent phosphorus is transformed to a dispersion of nickel phosphide in nickel, whereas an alloy containing 8 or 9 percent phosphorus becomes a dispersion of nickel in nickel phosphide. (29)

Heat treatment at 400 to 600 C reduces the ductility of the phosphorus alloys. Ductility, as determined by bulge tests, was reduced about 75 percent by heat treating at 400 C or 40 to 70 percent, depending on the phosphorus content, by heat treating at 600 C. (29) Heat treatment at 750 C reduced the ductility of the 8 or 9 percent phosphorus alloys about 40 percent. The reduction in ductility and tensile strength probably was at least partly due to microcracking, which several investigators have observed microscopically in heat-treated nickel-phosphorus alloy coatings.

The ductility of electroless nickel was maximized by adjusting the phosphorus content to 5 percent and selecting a heat-treatment temperature of 750 C. (29) Even so, such a heat-treated alloy exhibited only slight ductility (a strain at fracture of 0.016, approximately equal to 1 percent elongation). A 5-hour treatment at 750 C or higher followed by slow cooling to (200 C) or below in an inert atmosphere was suggested to maximize the ductility of an alloy containing about 9 percent phosphorus, (6) but no quantitative information for such a heat-treated alloy was reported.

Hardness

Hardness values of electroless nickel deposits are a measure of the resistance of the deposits to indentation by an indenter of fixed geometry under a static load. Pyramid diamond indenters of either Vickers or Knoop shape are usually used, with the results reported as VHN (DPH) or KHN respectively. Foreign investigators report hardness values in terms of kg/sq mm. Knoop Hardness Numbers (KHN) can be considered equivalent to kg/sq mm readings with only a slight error. Vickers Hardness Numbers (VHN) differ from kg/sq mm readings by an error of 10% or less. Accordingly, all hardness values in this text are shown as kg/sq mm, either as reported in the original work or by direct unit transposition from the KHN or VHN values reported by the investigator.

Hardness values for electroless nickel are detailed in Table 22.4 The data show no significant difference in hardness for alloys containing 2 to 11 percent phosphorus. However, the alloys containing 13 or 19 percent phosphorus were softer (only 340 to 375 kg/sq mm, in comparison with at least 500 kg/sq mm for the 2 to 11 percent phosphorus alloys). Neither the pH of the solution nor the Ni^{2+}/PO_2^{3-} ratio influenced hardness. Variations in hardness (in Table 22.4) probably are due in large part to differences in the indentation load applied for measuring hardness. An alkaline electroless nickel plating bath was preferred to acid baths by one team of investigators for producing 25-μm thick coatings with a homogeneous phosphorus content, which are desirable for the measurement of hardness and other properties. (29)

An electroless nickel containing only 0.03 percent phosphorus, which was produced with hydrazine as the reducing agent, (11) was similar in hardness to de-

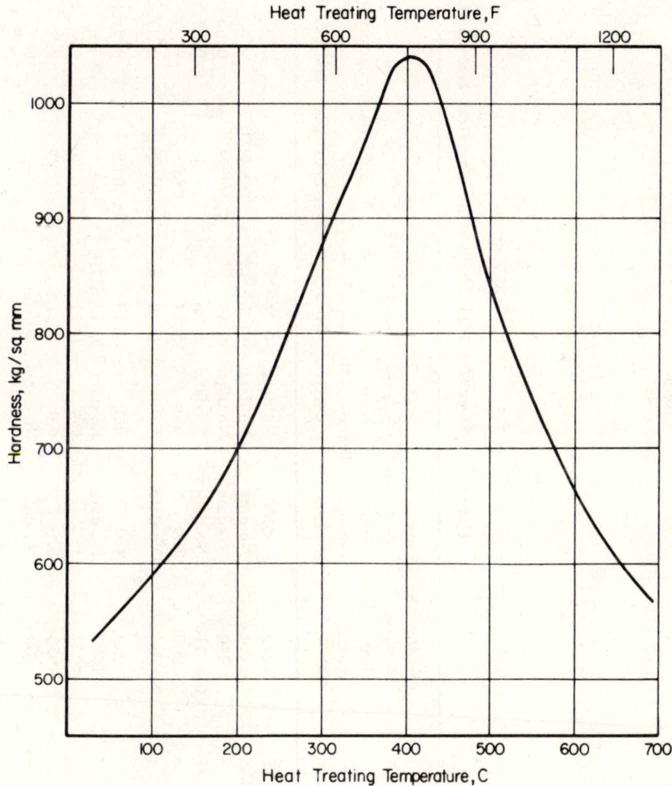

Figure 22.3. Change in Hardness of Electroless Nickel Containing about 9 Percent Phosphorus as a Function of Heat-Treating Temperature. (6)

Five-mil-thick deposits were heat treated for an hour and slowly cooled to 200 C before measurements at room temperature.

TABLE 22.4

Hardness of Electroless Nickel before and after Heat Treatment

Phosphorus Content, percent	Hardness, kg/sq mm				Solution Composition				Operating Conditions		
	Before Heat Treatment	Heat Treated, 400 C	Heat Treated, 600 C	Heat Treated, 750 or 800 C	$NiCl_2$, M	$Na_3C_6H_5O_7$, M	NH_4Cl, M	NaH_2PO_2, M	pH	Temperature C	Reference
0.03	400 to 500	275	<100	No data	0.34[a]	—	—	(a)	10.9	90	11
2 to 3	750 to 850	No data	No data	No data	0.20[b]	0.44[b]	—	0.20	9.5	70–80	34
3.6	600	850	No data	No data	0.13	0.34	0.55	0.095	11.0	—	35
3.7	500 to 600	1000 to 1100	No data	No data	0.28	0.17	0.75	0.19	9.0	80–85	36
4.0	625	900	700	300	0.13	(c)	(c)	0.03	5.0	94	37
4.7	560	No data	No data	330	0.19	0.28	0.95	0.19	8.3	90	29
5.8	550	930	440	360	0.10	0.28	0.95	0.19	8.3	90	29
6.0	500	>1000	No data	No data	0.12(SO_4)	—	—	0.19	—	—	32
6.5	780	—	—	355	0.13	0.50	—	0.095	Ca 5	90	38
7.1	540	950	520	420	0.19	0.28	0.95	0.19	8.3	90	29
7.5	415	600	No data	No data	No data	—	—	—	—	83	12
8.0	600	1000 to 1500[d]	No data	355	0.11(SO_4)	0.05	—	0.095	4 to 5	80	39
8.0	600	900	700	500	0.13	(c)	(c)	0.05	4.5	94	37
8.2	510	960	610	450	0.19	0.28	0.95	0.19	8.3	90	29
Ca 8.5	600	1000 to 1100	600 to 700	250 to 300	0.057(SO_4)	(e)	—	0.17	5.4	90	30
Ca 8.5	550	1050	650	—	(f)			(f)	Ca 5	>90	6
8.7	500	1140	No data	No data	0.27[g]	—	—	0.12[g]	—	95	40

Col 1	Col 2	Col 3	Col 4	Col 5	Col 6	Col 7	Col 8	Col 9	Col 10	Col 11	Col 12
9.0	500	1195	No data	No data	No data	—	—	—	—	—	5
9.0	500	950[h]	No data	No data	No data	—	—	—	—	—	41
9.0	570	978	No data	No data	0.09	(i)	—	0.23	—	—	42
9.4	510	—	No data	510	0.19	0.28	0.95	0.19	8.3	90	29
11.0	700 to 800	—	1300	—	0.075(SO_4)	(j)	—	0.23	4.0	92	43
11.0	600	925	800	600	0.13	(c)	(c)	0.05–0.1	4.0	94	37
12.8	375	375	No data	No data	0.13	0.34	0.55	0.095	7.0	—	35
12.9	550	1000	No data	No data	0.13	0.34	0.55	0.095	9.0	—	35
13.2	350	350	No data	No data	0.13	0.34	0.55	0.095	5.0	—	35
18 to 20	340	430	No data	No data	0.13	0.33	0.89	0.095	8–10	50–92	44

(a) This solution contained 0.34 M $Ni(C_2H_3O_2)_2$, 0.8 M $H_2C_2H_2O_3$, 0.06 M Na_4 EDTA and 0.67 M hydrazine.

(b) This solution contained 0.2 M nickel sulfamate, 0.44 M $(NH_4)_3C_6H_5O_7$, and 0.2 M NaH_2PO_2.

(c) This solution contained 0.03 M $NiCl_2$, 0.395 M $H_2C_2O_3H_2$, 0.085 M $H_2C_4O_4H_4$, and 0.072 M NaF.

(d) Heating in N_2 induced a hardness of 1000 kg/sq mm; heating in a vacuum resulted in a hardness of 1200 kg/sq mm and heating in hydrogen induced the maximum hardness of 1500 kg/sq mm.

(e) This solution contained 0.057 M $NiSO_4$, 0.14 M $HC_3H_6O_2$, 0.015 g/l MoO_3, 0.06 g/l hydroxylamine sulfate, and 0.17 M NaH_2PO_2.

(f) Proprietary Kanigen process.

(g) This solution contained 0.27 M nickel lactate and 0.12 M Na_3PO_2.

(h) Heated at 280 C for 15 to 20 hours.

(i) The solution contained 0.09 M $NiCl_2$, 0.12 M $NaC_2H_3O_2$, and 0.23 M NaH_2PO_2.

(j) The solution contained 0.075 M $NiSO_4$, 0.12 M $NaC_2H_3O_2$, 0.015 M maleic acid, and 0.23 M NaH_2PO_2.

posits containing 3 or 7.5 percent phosphorus, but the 0.03 percent phosphorus alloy softened as a result of heat treatment. For example, a hardness of 275 kg/sq mm or less was reported after heating at 420 C and a hardness of less than 100 kg/sq mm was measured after heating at 550 C. By contrast, the 4 to 11 percent phosphorus alloys increase in hardness to about 1000 kg/sq mm when heated at 400 C, which is a result of the precipitation of nickel phosphide from supersaturated solid solutions of phosphorus.

For alloys containing 4 to 11 percent phosphorus, maximum hardness is developed by heat treating in the range of 400 to 450 C. After one or two hours of heating in this temperature range, hardness usually reaches 900 to 1100 kg/sq mm, depending on the indentation load. Figure 22.3 shows hardness data as a function of heat-treating temperature for an alloy containing about 9 percent phosphorus. Two hours of heat treating at 300 C hardened electroless nickel to about 950 kg/sq mm, whereas a temperature 200 C did not increase the hardness of an alloy containing about 9 percent phosphorus. (45) On the other hand, the hardness of electroless nickel deposits containing 6 or 7 percent phosphorus was increased from about 550 to nearly 800 and 700 kg/sq mm respectively, by four hours of heat treatment at 200 C. (29) Low-temperature hardening treatments, which minimize oxidation when electroless nickel is heated in air, have been reported, as follows: (41)

Temperature C	Time, hours	Hardness After Heat Treating, kg/sq mm
232	28	825
260	25	800
288	18-½	950

Hardness after heat treating at 400 C does not appear to vary appreciably with the phosphorus content, according to the data in Table 22.4. A similar observation was made for electrodeposited nickel-phosphorus alloy containing 2 to 14 percent phosphorus, also heat treated at 400 C. (3) However, an alloy containing 9 percent phosphorus was reported to be slightly harder (500 kg/sq mm) than alloys containing 5 or 6 percent phosphorus (about 350 kg/sq mm), after heat treatment at 400 C. (29) Some electroless nickel deposits containing 13 or 19 percent phosphorus were much softer (350 to 430 kg/sq mm, in comparison with 900 to 1100 kg/sq mm for most of the 4 to 11 percent phosphorus electroless alloys) after a similar treatment. The accuracy of the hardness measurements for the alloys containing 13 or 19 percent phosphorus can be questioned because their brittleness would tend to promote cracking and crushing during indentation.

After heating at 600 C, the hardness of electroless nickel containing 5 to 8.5 percent phosphorus was reduced to the range of 450 to 700 kg/sq mm, as a rule, which reproduces the behavior of electrodeposited nickel containing 2 to 7 percent phosphorus. (3) However, electroless nickel with 11 percent phosphorus retained high hardness (800 or 1300 kg/sq mm) after heat treatment at 600 C. (37,43) This high-phosphorus alloy also retained a higher hardness (600 kg/sq mm) than low-phosphorus nickel after heating at 750 or 800 C. (37)

After heat treatment at about 400 C, the hardness of electroless nickel was about equal to the hardness of chromium electroplate. Such a heat treatment also improves adhesion to steel. A three- to six-fold increase in bond strength was reported after a one-hour treatment at 400 C, in comparison with a twofold increase after heat treatment at 300 C. (46)

Electroless nickel containing 0.6 or 4.7 percent boron, a product of the oxidation of the dimethylamine borane used for reducing nickel ions, exhibited, before heating, a hardness of 700 to 725 kg/sq mm, (14) which is slightly higher than the hardness of most of the nickel-phosphorus alloys. This high level of hardness was retained after heat treatment at 400 C. However, the nickel-boron alloys (like the nickel alloys containing 5 to 9 percent phosphorus) softened appreciably (to only 325 to 400 kg/sq mm) as a result of heat treatment at 800 C.

A maximum hardness of 1100 kg/sq mm for electroless nickel reduced with sodium borohydride was reached by heat treatment at about 400 C, which was attributed to the precipitation of Ni_3B, identified from X-ray diffraction patterns. (47) The solution contained 0.13 M nickel chloride, 1.0 M sodium hydroxide, 1.0 M ethylenediamine, 0.07 M sodium fluoride, 0.013 M sodium borohydride and 0.0002 M lead chloride.

Hardness data for electroless nickel-phosphorus alloys at elevated temperatures are shown in Figure 22.4. Specimens that were heat treated in advance of the hot-

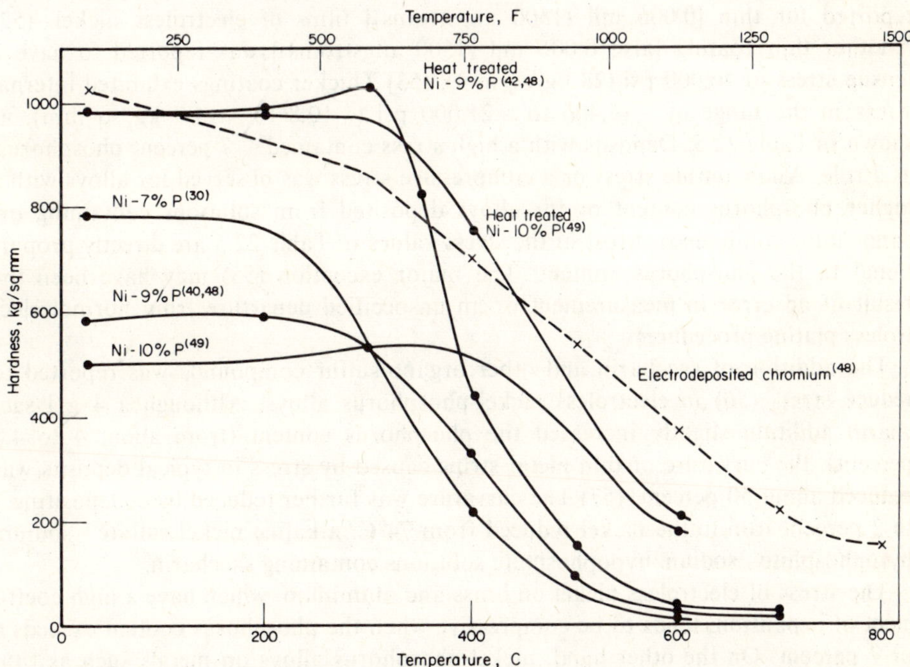

Figure 22.4. Hardness of Electroless Nickel-Phosphorus Alloys at Elevated Temperatures.

hardness measurements were much harder from 20 to 400 C than their counterparts that were not preheated. The heat-treatment temperature was 400 C. Electroless nickel-phosphorus alloys are similar in hardness to electrodeposited chromium in the temperature range of 20 to 400 C, but softer than chromium electroplate above 500 C.

Resistance to Wear and Friction

The frictional characteristics of nickel-phosphorus alloy heat treated at 650 to 900 C are similar to those of electroplated chromium. (50) Yet the coefficient of friction changed very little when the nickel-phosphorus alloy was heat treated. (30, 36,37) Resistance to wear and abrasion is improved by heat treating. (5,37,51) The wear resistance of alloys containing more than 7 percent phosphorus is maximized by heat treating at 400 to 600 C. (37) An alloy containing 11.5 percent phosphorus exhibited better wear and galling resistance than alloys with a lower phosphorus content.

Internal Stress

High values of internal tensile stress up to 50,000 psi (35 kg/sq mm) were reported for thin [0.006 mil (1500 angstroms)] films of electroless nickel. (52) Another thin coating [also 0.006 mil (1500 angstroms)] was reported to have a tensile stress of 40,000 psi (28 kg/sq mm). (53) Thicker coatings exhibited internal stress in the range of −15,400 to +21,000 psi (−10.8 to +14.7 kg/sq mm), as shown in Table 22.5. Deposits with a high stress contained < 7 percent phosphorus, as a rule. A low tensile stress or a compressive stress was observed for alloys with a higher phosphorus content or for alloys deposited from solutions containing organo-sulfur compounds. Most of the stress values in Table 22.5 are directly proportional to the phosphorus content. The major exception (55) may have been the result of an error in measurement or an unspecified departure from normal electroless plating procedures.

The addition of saccharin and other organo-sulfur compounds was reported to reduce stress (56) in electroless nickel-phosphorus alloys. Although a 4 g/l saccharin addition slightly increased the phosphorus content (from about 4 to 4.5 percent), the curvature of thin metal strips caused by stress in typical deposits was reduced about 50 percent. (57) The curvature was further reduced by codepositing 1 to 2 percent iron in the nickel reduced from 74 C, alkaline nickel sulfate - sodium pyrophosphite - sodium hypophosphite solutions containing saccharin.

The stress of electroless nickel on brass and aluminum, which have a high coefficient of expansion, tends to be compressive when the phosphorus content exceeds 6 or 7 percent. On the other hand, nickel-phosphorus alloys on metals such as titanium and beryllium, which have low coefficients of expansion, are stressed in tension when the phosphorus content is less than about 11 percent. (58) Figure 22.5 compares stress values cited as a function of the phosphorus content for several substrate metals. Heat treating (at 190 C) increased the tensile stress of deposits on beryllium (54) or titanium. (58) However, the tensile stress in electroless nickel containing 4 percent phosphorus deposited on aluminum was shifted to a compressive stress by such a heat treatment. Figure 22.5 shows changes in stress resulting from heat treatment of electroless nickel on aluminum, brass, steel, and titanium.

Electroless nickel (about 0.5 mil or 12.5 μm) produced with dimethylamine borane as the reducing agent exhibited high tensile stress, depending on the boron content, as follows: (14)

Boron Content, percent	Stress psi	Stress kg/sq mm
0.6	69,000	48
1.3	44,000	31
4.7	17,000	12

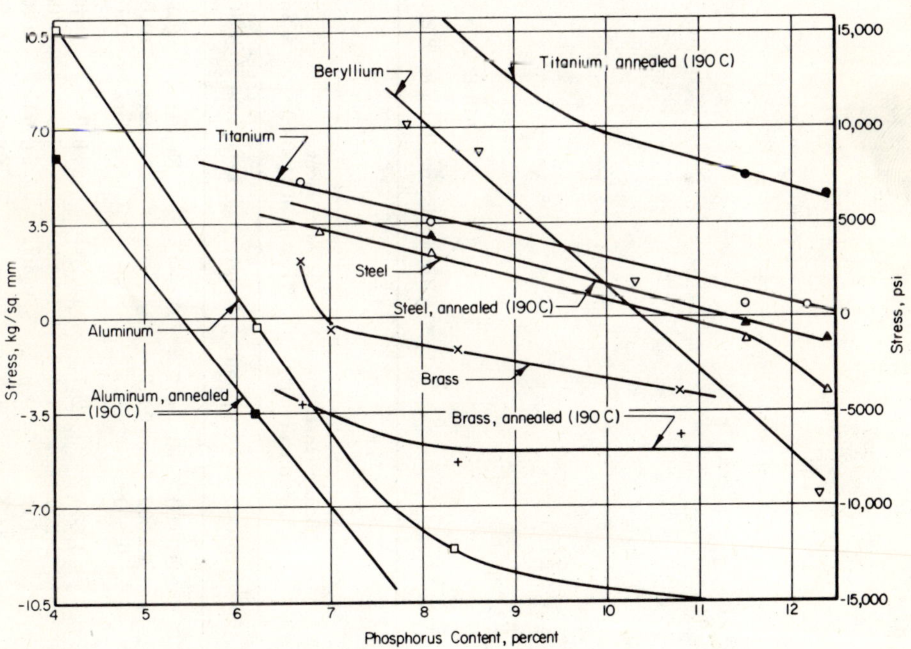

Figure 22.5. Stress in Electroless Nickel as a Function of Phosphorus Content on Metals with a High Expansion Coefficient (Aluminum and Brass) or a Low Expansion Coefficient (Steel, Beryllium, and Titanium). (58)

TABLE 22.5

Internal Stress Data for Electroless Nickel

Phosphorus Content, percent	Stress[a] Before Heat Treatment psi	Stress[a] Before Heat Treatment kg/sq mm	Stress[a] After Heat Treatment (190 C) psi	Stress[a] After Heat Treatment (190 C) kg/sq mm	Solution Composition $NiSO_4$, M	Solution Composition NaH_2PO_2, M	Operating Conditions pH	Operating Conditions Temperature C	Reference
0.03	21,000	14.7	—	No data	0.34[b]	(b)	10.9	90	11
4.5	16,000	11.2	—	No data	0.027	0.19	6.0	82	32
6.0	9,500[c]	6.7	—	No data	0.027	0.19	5.5	82	32
6.9	3,800	2.7	11,200	8.0	0.08[d]	0.23	5.0	82	54
7.0	2,200	1.6	10,400	7.3	0.08[d]	0.23	5.0	88	54
7.2	1,100	0.8	9,400	6.6	0.08[d]	0.23	4.9	94	54
8.0	4,000	3.8	—	No data	0.27	0.19	5.0	82	32

Ca 8	2,700[e]	1.9	9,100	6.4*	(f)	—	—	—	55
Ca 8	20,400[g]	14.7	—	No data	(f)	—	—	—	55
Ca 8	5,000	3.5	9,000	6.5[h]	0.13[i]	0.095	4.5	95	56
8.1	1,900	1.3	9,800	6.9	0.08[d]	0.23	4.5	93	54
8.4	6,300	4.4	20,300	14.2	0.08[d]	0.23	4.5	93	54
Ca 8.5	8,500	6.0	—	No data	0.057[j]	0.17	5.4	90	30
9.0	−2,000	−1.4	—	No data	0.27	0.19	4.5	82	32
10.0	−7,500	−5.3	—	No data	0.27	0.19	4.0	92	32
10.7	−7,800	−5.5	4,200	3.0	0.08[d]	0.23	4.0	97	54
11.6	−12,800	−9.0	0	0.0	0.08[d]	0.23	4.0	94	54
12.2	−10,500	−7.4	−1,200	−0.8	0.08[d]	0.23	4.0	93	54
12.4	−15,400	−10.8	−3,900	−2.7	0.08[d]	0.23	4.0	91	54

[a] A negative sign indicates compressive stress, no sign indicates tensile stress.
[b] This solution contained 0.34 M $Ni(C_2H_3O_2)_2$, 0.06 M Na_4EDTA, 0.8 M $H_2C_2H_2O_3$, and 0.67 M hydrazine.
[c] Stress increased to 17,000 psi (12 kg/sq mm) when the NaH_2PO_2 concentration was increased to 1.1 M.
[d] This solution contained 0.08 M $NiSO_4$, 0.36 M $H_2C_3H_5O_3$, and 0.23 M NaH_2PO_2. Stress values are averages for three measurements.
[e] Minimum of 12 values ranging from 2,700 to 21,000 psi (1.9 to 14.7 kg/sq mm).
[f] Samples of Kanigen Nickel supplied by three commercial sources.
[g] Maximum of 12 values ranging from 2,700 to 21,000 psi (1.9 to 14.7 kg/sq mm).
[h] Stress after aging at 20 C for four weeks.
[i] This solution contained 0.126 M $NiCl_2$, 0.5 M $NH_4C_2H_3O_2$, and 0.095 M NaH_2PO_2.
[j] This solution contained 0.057 M $NiSO_4$, 0.14 M $HC_3H_6O_2$, 0.015 g/l MoO_3, 0.06 g/l hydroxylamine sulfate, and 0.17 M NaH_2PO_2.

TABLE 22.6

Effect of Electroless Nickel Coatings on the Fatigue Strength of Steel

Phosphorus Content, percent	Heat-Treating Temperature, C	Fatigue Strength % Reduction[a]	Cycles	Type of Electroless Nickel Bath	Reference
6 to 15	No heat treatment	30 to 40[b]	—	—	59
6 to 15	190 C—4 hours	2 to 4[c]	—	—	59
6 to 15	400 C	20[c]	—	—	59
Ca 8	No heat treatment	42[d]	5 x 10^6	Acid[e]	30
Ca 8	550 C—2 hours	30	5 x 10^6	Acid[e]	30
8 to 9	No heat treatment	16 to 46[f]	10 x 10^6	Acid[g]	56
9 to 10	No heat treatment	10 to 12[h]	5 x 10^6	Acid[i]	25
10 to 12	No heat treatment	Appreciable	—	Acid	43
No data	400 C—1 hour	40	Above 1 but less than	Acid	60
No data	700 C—induction	None	2 x 10^7	—	60
No data	No heat treatment	4 to 17 (31 to 40)[j]	10 x 10^6	Acid	61
No data	No heat treatment	0 to 12 (15 to 23)[k]	10 x 10^6	—	62
No data	No heat treatment	2[l]	10 x 10^6	Acid	63
No data	400 C—1 hour	46[l]	10 x 10^6	Acid	63
No data	400 C—1 hour	38 to 44	—	—	50
No data	400 C—1 hour	41 to 42[m]	10 x 10^6	Acid	63
No data	400 C—1 hour	5 to 17[n]	10 x 10^6	Alkaline	63
No data	No heat treatment	6 (29)[o]	5 x 10^6	Alkaline	64

[a] Figures in parentheses relate to notched specimens.

[b] The decrease was proportional to the phosphorus content in the nickel coating.

[c] Data for heat-treated specimens shot peened before electroless nickel plating.

[d] The fatigue limit of the bare steel (5 $\times$ 10^6 cycles) was 66,000 psi (46 kg/sq mm).

[e] Solution at about 90 C containing 0.06 M $NiSO_4$, 0.12 M $HC_3H_6O_2$, 0.06 g/l $N_2H_4SO_4$, 0.015 g/l MoO_3, and 0.17 M NaH_2PO_2, with a pH of 5.2 to 5.6.

[f] The nickel coatings were stressed in tension. The reduction in fatigue strength was proportional to the fatigue strength of the steel, as follows:

Fatigue Strength of Steel		Reduction in Fatigue Strength, percent
psi	kg/sq mm	
44,000	31	16
67,000	47	25
97,000	68	46

[g] Solution containing 0.13 M $NiCl_2$, 0.5 M $NaC_2H_3O_2$, and 0.09 M NaH_2PO_2, with a pH of about 4.5 and a temperature of 95 C.

[h] The fatigue limit of the bare steel (5 $\times$ 10^6 cycles) was 36,000 psi (25 kg/sq mm).

[i] Solution containing 0.015 M $NiCl_2$, 0.24 M $NaC_2H_3O_2$, 0.4 M $NH_2CH_2CO_2H$, and 0.28 M NaH_2PO_2, with a pH of about 5.

Effect of Electroless Nickel Coatings on the Fatigue Strength of Steel

Decreases of 10 to 46 percent in the fatigue strength of high-strength steel as a result of plating with electroless nickel were reported by several investigators, as shown in Table 22.6. (25,30,43,59,63) Notched specimens were reduced in fatigue strength at least 15 percent. (61,62,64) Relatively small reductions (12 percent, maximum) were observed for some unnotched steel specimens. (25,62,63) These less harmful results may have been associated with compressively stressed coatings containing more than 10 percent phosphorus, although the phosphorus contents were not reported in all cases.

The fatigue limit (5×10^6 cycles) of tempered steel coated with an electroless nickel applied from an acid bath and continuously sprayed with 3 percent sodium chloride was reduced from about 22,750 to 14,220 psi (16 to 10 kg/sq mm). This reduction was about 37 percent. On the other hand, the fatigue limit of another steel exposed to sulfuric acid solution with a pH of 3.0 was unchanged by plating it with electroless nickel. (30) A 50 percent improvement (from 11,500 to 17,000 psi or 8 to 12 kg/sq mm) was observed in corrosion fatigue resistance to sulfuric acid solution when the coated steel was heat treated to 400 C.

Another report indicated that stress corrosion of AISI 4340 steel was increased by coating the steel with electroless nickel. (65) The study was carried out at 70 percent of the ultimate tensile strength of the steel.

Specimens that were heat treated at 400 C, which precipitates Ni_3P, showed a 20 to 48 percent loss in fatigue strength for steel coated with electroless nickel in an acid bath, in comparison with unplated steel. (50,56,60,63) Even shot peening before plating failed to eliminate the harmful effect of a heat-treated electroless nickel coating on the fatigue strength of steel. (59) On the other hand, steel coated with nickel in an alkaline bath was reduced only 5 to 17 percent in fatigue strength after heat treating at 400 C. (63) The influence of heat treating on stress (which is

---

Footnotes to TABLE 22.6 (Continued)

(j) The fatigue limit of the bare steel (10×10^6 cycles) was 74,000 psi (52 kg/sq mm). The limit for uncoated, notched specimens was 37,000 psi (26 kg/sq mm).

(k) The fatigue limit of the bare steel (10×10^6 cycles) was 71,000 psi (50 kg/sq mm). The limit for uncoated, notched specimens was 37,000 psi (26 kg/sq mm).

(l) The fatigue limit of the bare steel (10×10^6 cycles) was 87,000 psi (61 kg/sq mm), with or without a heat treatment.

(m) The fatigue limit of the bare, heat-treated steel was 91,000 psi (64 kg/sq mm).

(n) The fatigue strength of the bare steel was 87,000 psi (61 kg/sq mm). Fatigue strength was reduced in proportion to coating thickness, as follows:

10 μm	5%
20 μm	10–12%
30 μm	16.5%

(o) The fatigue limit of the bare steel was 46,000 psi (32.4 kg/sq mm). The limit for notched specimens was 31,500 psi (22.2 kg/sq mm).

increased or shifted from a compressive to a tensile condition) may be a significant factor affecting the fatigue strength of electroless nickel-plated steel.

Heat treating at 700 or 800 C eliminated or at least greatly reduced the harmful effect of electroless nickel coatings on the fatigue strength of steel. (56,60) Recrystallization of electroless nickel occurs at these temperatures.

The Corrosion Resistance of Steel Coated with Electroless Nickel Deposits

Electroless nickel-phosphorus alloys are corroded by reagents that attack unalloyed nickel. (6) However, the alloy containing 10 percent phosphorus is corroded at a slower rate by some reagents, such as nitric acid. Table 22.7 includes detailed data on the effects of many chemical reagents. Resistance to chemical attack is improved appreciably by heat treating at 750 C. Improved resistance to corrosion after heat treating in air or nitrogen at about 650 C also was reported elsewhere. (51) By contrast, however, heating in hydrogen is deleterious because of the formation of microcracks which were ascribed to the diffusion of hydrogen into the nickel, combined with internal stress in the coating.

Electroless nickel with a thickness of 1 mil (25 μm), deposited in acid solutions, was better than an equal thickness of nickel electrodeposited in a chloride bath for protecting steel outdoors at Kure Beach, North Carolina and Washington, D.C. (66) These electroless nickel coatings were about equal in corrosion protection to electrodeposited nickel-phosphorus alloys containing 3 or 9 percent phosphorus and much better than electroless nickel coatings deposited in alkaline solutions with a pH of 8.5 to 9.5.

Another report cited good corrosion resistance in air and 600 C steam for steel coated with 1.6-2 mils (40 to 50 μm) of electroless nickel deposited in acid solutions with a pH of 6. (43) The phosphorus content of the coating was in the range of 10 to 12 percent. However, steel coated with nickel containing 4 percent phosphorus, which was deposited in an alkaline bath with a pH of 9, exhibited poor corrosion resistance.

Steel plated with 1 mil (25 μm) of electroless nickel-phosphorus alloy exhibited much better resistance to 20 percent (neutral) salt spray than steel plated with very low-phosphorus electroless nickel reduced with hydrazine or steel electroplated with nickel in a sulfamate solution. (11)

The codeposition of tin (67) or of copper (68) supposedly improved the corrosion resistance of electroless nickel-phosphorus coatings on steel. The tin addition was made to an alkaline solution with a pH of 10.5. (67) To codeposit copper, 0.1 g/l cupric chloride was added to an alkaline bath with a pH of about 9.0. (68)

Structure

Banded or lamellar structures like that in Figure 22.6 are customary for electroless (or electrodeposited) nickel-phosphorus alloys. However, columns perpendi-

TABLE 22.7

Corrosion Resistance of Electroless Nickel (6)

Corrosive Liquid	Temperature	Immersion	Aeration	Test Period, weeks	Penetration, mils per year
		Kanigen as Deposited			
Acetylene bromide	R[(a)]	T[(b)]	No	3	No attack
Amyl alcohol	R	T	No	24	No attack
Benzole	R	T	No	24	No attack
Benzyl acetate	R	T	No	24	No attack
Carbon disulfide	R	T	No	21	No attack
Carbon disulfide and water	R	T	No	21	No attack
Cobalt linoleate	R	T	No	16	No attack
Glucose	R	T	No	16	No attack
Isoamyloctyl orthophosphate, 75 percent	R	T	No	3	No attack
Methyl alcohol	R	T	No	24	No attack
Naphtha, odorless	R	T	No	3	No attack
Petroleum white oil	R	T	No	3	No attack
Photographic: Developer	R	T	No	4	No attack
'Hypo' solution	R	T	No	4	No attack
Refinery brine solution	R	T	No	23	No attack
Sodium: Carbonate, 10 percent	R	T	No	4	No attack
Hydroxide, 10 percent	R	T	No	4	No attack
Acetic acid, 5 percent	R	T	No	4	0.8
Acetic acid, 5 percent	R	T	Yes	4	6.0
Acetone	R	T	No	16	0.003
Acetylene bromide, 1 percent water	R	T	No	3	Slight weight gain
Allyl chloride	R	T	No	24	0.04
Aluminum sulfate, saturated	R	T	No	4	0.3
Ammoniated ammonium:					
Nitrate	R	T	No	16	0.3
Nitrate vapor	R	T	No	16	0.3
Ammonium:					
Chloride saturated, 30 percent NH_3	R	T	No	16	0.2
Hydroxide, 30 percent NH_3	R	T	No	4	2.3
Phosphate, 5 percent	R	T	No	4	0.7
Sulfate, saturated	R	T	No	20	0.05
Sulfate, 5 percent	R	T	Yes	8	1.2
Sulfite, saturated	R	T	No	20	0.02
Thiocyanate	R	T	No	20	0.2
Amyl acetate	R	T	No	24	0.002
Amyl chloride	R	T	No	24	0.01
Aniline hydrochloride, saturated	R	T	No	16	0.5
Beer	R	T	No	4	0.2
Benzyl alcohol	R	T	No	28	0.004
Benzyl chloride	R	T	No	16	1.
Black liquor skimmings	R	T	No	12	0.01
Boraxo, saturated	R	T	No	32	0.1

TABLE 22.7 (Continued)

Corrosive Liquid	Temperature	Immersion	Aeration	Test Period, weeks	Penetration, mils per year
Boric acid	R	T	No	8	0.5
Calcium chloride: 48.5 percent	R	T	No	32	0.008
48.5 percent	R	T	Yes	32	0.04
Carbon tetrachloride	R	T	No	32	0.005
Cetyl alcohol, molten	71 C	T	No	8	0.008
Citric acid, 5 percent	R	T	No	16	0.03
Citric acid, 5 percent	R	T	Yes	8	0.07
Cresylic acid	R	T	No	16	0.002
Detergent solution, 5 percent (Tide)	R	T	No	4	0.04
Dibutyl phthalate	R	T	No	8	0.006
Diphenyl, molten	71 C	T	No	8	0.009
Ethyl alcohol	R	T	No	24	0.007
Fluorophosphoric acid	R	T	No	(12 days)	2.
Formaldehyde	R	T	No	16	0.2
Gasoline	R	T	No	4	0.02
HCl solution: pH 1.5	R	T	No	4	1.
pH 1.5	R	T	Yes	4	5.
pH 2.0	R	T	No	4	0.9
pH 2.0	R	T	Yes	12	3.
pH 2.5	R	T	No	4	0.3
pH 3.0	R	T	No	4	0.1
pH 3.5	R	T	No	4	0.07
pH 3.5	R	T	Yes	4	0.4
pH 4.0	R	T	No	4	0.05
pH 4.0	R	T	Yes	10	0.1
Insectisol	R	T	No	3	0.008
Lactic acid: 45 percent	R	T	No	16	0.1
45 percent	R	T	Yes	4	2.
80 percent	R	T	No	32	0.05
80 percent	R	T	Yes	8	0.7
Lemon juice: Canned	R	T	No	4	0.2
+0.1 percent sodium benzoate	R	T	Yes	(23 days)	0.9
Oleic acid	R	T	No	10	0.008
Orange juice, canned	R	T	No	4	0.01
Petroleum, sour crude	R	T	No	2	0.001
Rosin size: Concentrated	90 C reflux	T	No	3	0.01
50 percent	90 C reflux	T	No	3	0.06
Sodium: Cyanide, 5 percent	R	T	No	4	0.5
Hydrosulfide, 40 percent	R	T	Yes	20	Slight weight gain
Hydroxide, 72 percent	116 C reflux	T	No	16	0.07
Stearic acid	70 C	T	No	4	0.02
Sulfuric acid: 1 percent	R	T	No	4	1.
5 percent	R	T	No	1	1.
Tall oil, crude	R	T	No	12	0.03

TABLE 22.7 (Continued)

Corrosive Liquid	Temperature	Immersion	Aeration	Test Period, weeks	Penetration, mils per year
Tall oil, refined	R	T	No	32	0.02
Tanning solution (Koreon)	R	T	No	3	0.05
Thionyl chloride, anhydrous	R	T	No	3	0.03
Urea, 25 percent	R	T	Yes	16	0.05
Urea, saturated	R	T	Yes	16	0.05
Water, chlorine: 5 ppm Cl_2	R	T	No	4	0.02
10 ppm Cl_2	R	T	No	4	0.01
Water, deionized	R	T	Yes	12	0.02
Water, deionized	49 C reflux	T	No	12	0.01
Water, deionized	82 C reflux	T	No	12	Slight weight gain
Water, distilled	R	T	No	4	0.03
	Heat Treated at 750 C				
Acetic acid: 5 percent	R	T	Yes	20	0.9
10 percent	R	T	No	16	0.2
10 percent	R	T	Yes	8	1.
50 percent	R	T	No	20	0.2
50 percent	R	T	Yes	8	2.
glacial	R	T	No	20	0.02
glacial	R	T	Yes	8	1.
Ammoniated ammonium:					
Nitrate	R	T	No	12	0.03
Nitrate vapor	R	T	No	16	0.3
Ammonium:					
Hydroxide, 30 percent NH_3	R	T	No	29	0.06
Hydroxide, 30 percent NH_3	R	P[c]	No	24	0.2
Nitrate, 63 percent	R	T	No	28	0.1
Nitrate, 63 percent	R	T	No	24	0.01
Sulfate, saturated	R	T	Yes	20	0.02
Sulfate, 5 percent	R	T	Yes	20	0.09
Beer	R	T	No	4	0.008
Calcium chloride: 48.5 percent	R	T	No	32	0.001
48.5 percent	R	T	Yes	20	0.01
Ethylene glycol	Standard corrosion test			3	0.03
Formaldehyde: 37 percent	R	T	No	28	0.007
37 percent	R	P[c]	No	24	0.012
HCl solution: pH 1.5	R	T	Yes	8	3.
pH 2.0	R	T	Yes	8	0.6
pH 2.5	R	T	Yes	12	0.6
pH 3.0	R	T	Yes	12	0.2
pH 3.5	R	T	Yes	32	0.09
pH 4.0	R	T	Yes	32	0.02
Lactic acid: 45 percent	R	T	No	32	0.009
45 percent	R	T	Yes	16	0.8
80 percent	R	T	No	32	0.006
80 percent	R	T	Yes	16	0.2

TABLE 22.7 (Continued)

Corrosive Liquid	Temperature	Immersion	Aeration	Test Period, weeks	Penetration, mils per year
Phosphoric acid, 85 percent	R	T	No	28	0.008
Sodium: Hydrosulfide, 40 percent	R	T	Yes	24	Slight weight gain
Hydroxide, 72 percent	116 C reflux	T	No	24	0.4
Sulfuric acid: 1 percent	R	T	Yes	8	2.
5 percent	R	T	No	16	0.4
5 percent	R	T	Yes	4	2.
Urea, 25 percent	R	T	Yes	16	0.009
Urea, saturated	R	T	Yes	16	0.008
Wine, sherry	Refrigerated	T	No	32	0.004
Zinc ammonium chloride	110 C reflux	T	No	4	2.

[a] R, room temperature, 18 to 30 C.
[b] T, total immersion.
[c] P, partial immersion.

cular to thc bands have been observed microscopically in typical cross sections. The bands in electroless nickel deposited in alkaline solutions are about 10 times as wide as the bands in electroless nickel deposited in acid baths. (29) After heating at 400 C, the precipitation of nickel phosphide is evident in sections like that shown in Figure 22.7. (1,69) Evidence of further precipitation of nickel phosphide and recrystallization is observed during microscopic examination of sections heat treated at 800 C (Figure 22.8) Recrystallization and grain growth also occur at 500 C (69) or 600 C after two hours. (29) Grain growth is more pronounced in the low-phosphorus alloys, in comparison with the alloys containing more than 7 percent phosphorus.

Crystallinity is not observed with X-ray diffraction of electroless nickel until it has been heat treated. However, a microcrystalline structure, prior to heat treatment, has been resolved by electron diffraction. (28,29,70) The alloy is considered to be a supersaturated solid solution of nickel phosphide in nickel. (29, 69) A grain size of 0.02 mil (5000 angstroms) was reported for electroless nickel deposited in an ammoniacal bath with a pH of 9.2. (70) The phosphorus content of this deposit was not given. On the other hand, nickel deposited in an acid bath with a pH of about 5.5, which contained about 8 percent phosphorus, had a grain size of only 0.00024 mil (about 60 angstroms). (30) An alloy deposited from a solution with a pH of 8.2, which contained 9 percent phosphorus, had a grain size of the order of 0.0004 mil (100 angstroms), whereas an alloy containing 5 to 7 percent phosphorus exhibited a larger grain size. (29) Grain refinement with increasing phosphorus content also was reported in another article. (71)

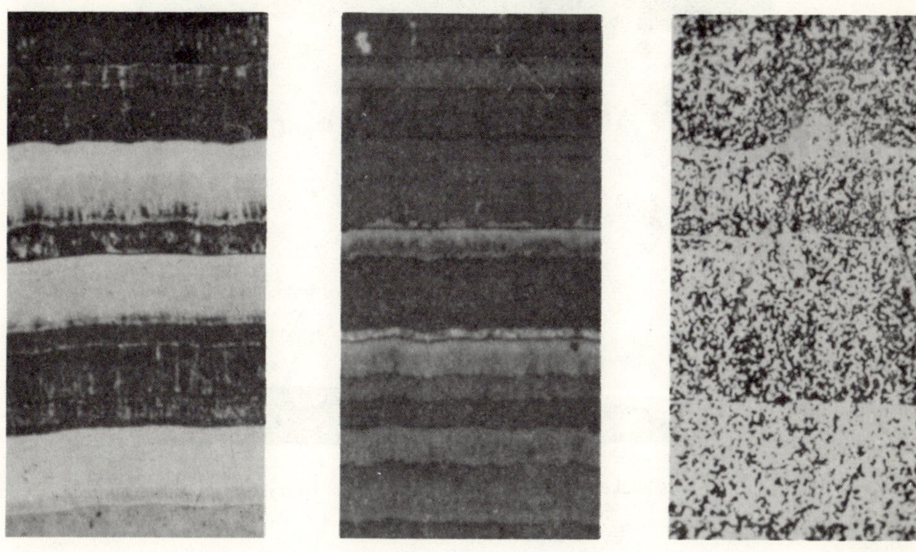

Figure 22.6. Banded Structure of Electroless Nickel-Phosphorus Alloy Deposited in an Acid Solution. (1)

Figure 22.7. Electroless Nickel-Phosphorus Alloy after Heat Treatment at 400 C which Precipitated Ni_3P. (1)

Figure 22.8. Recrystallization of Electroless Nickel-Phosphorus Alloy and Precipitation of Ni_3P after Heat Treatment at 800 C. (1)

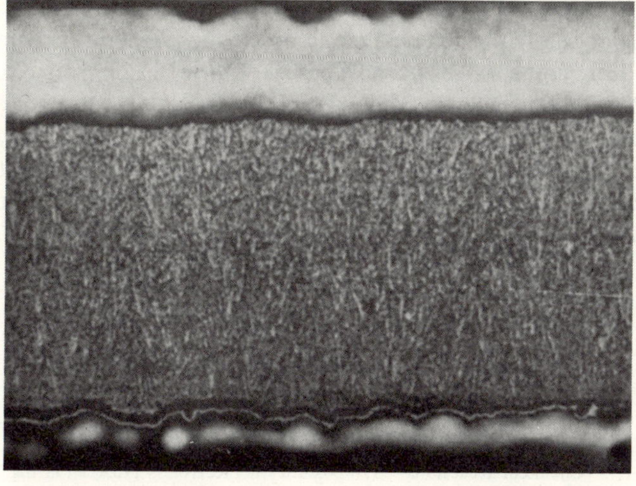

Figure 22.9. Fibrous Structure of Electroless Nickel Reduced with Hydrazine. (11)

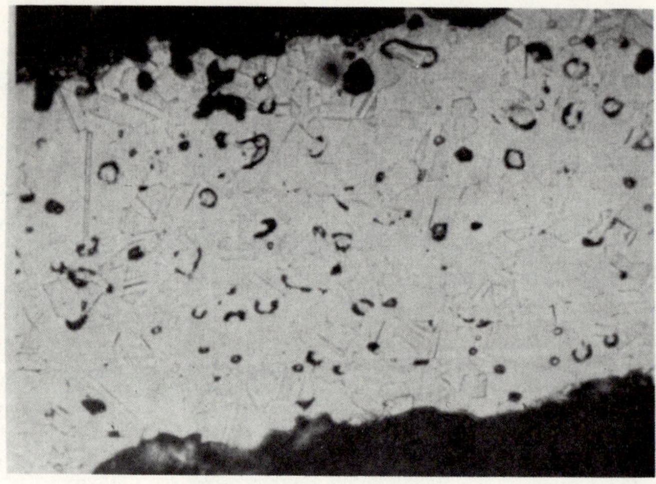

Fig. 22.10. Electroless Nickel Reduced with Hydrazine after Recrystallization at 750 C. (11)

Complete precipitation of nickel phosphide does not occur at temperatures significantly below 400 C (72) when heating times are limited to a few hours. The phosphide has been identified as Ni_3P by several investigators, using electron diffraction (29,59,72) or X-ray diffraction techniques. (30,69) When alloys containing as much as 8.5 or 9.0 percent phosphorus are heat treated (preferably at about 800 C), nickel is dispersed in a Ni_3P matrix phase. (29,69) For alloys containing less than 7 percent phosphorus, Ni_3P is dispersed in the nickel phase. (29) This change in structure at about 7 percent phosphorus is associated with the changes that occur in the strength and ductility of electroless nickel, which are shown in Figure 22.2.

Crystal orientation with the (111) plane of the nickel phase perpendicular to the surface has been noted (29,30) The degree of orientation is increased by heat treatment. (29) The Ni_3P phase also is oriented with its (110) planes parallel with the (111) planes of the nickel.

Unlike the nickel-phosphorus deposits reduced with hypophosphite, electroless nickel reduced with hydrazine exhibited a fibrous structure like that in Figure 22.9. Heat treatment at 750 C produced the structure shown in Figure 22.10.

References

(1) Brenner, A., and Riddell, G., "Deposition of Nickel and Cobalt by Chemical Reduction," Research Paper No. 1835, *Journal Research National Bureau of Standards, 39* (5), 385–395 (1947); *Proceedings Am. Electroplaters' Soc., 34*, 156–170 (1947).

(2) Gutzeit, G., "Chemical Reactions—Symposium on Electroless Nickel Plating," *ASTM Special Technical Publication No. 265*, 3–12 (1959).

(3) Brenner, A., Couch, D. E., and Williams, E. K., "Electrodeposition of Alloys of Phosphorus and Nickel or Cobalt," *Plating, 37*, 36–42, 161–164 (1950).

(4) Jarrett, G. D. R., "Electroless Plating. III. Electroless Nickel Plating," *Industrial Finishing (London), 18* (218), 41,43, 48 (1966).

(5) Gutzeit, G., "Industrial Nickel Coating by Chemical Catalytic Reduction," *Trans. Inst. Metal Finish., 33*, 383–415 (1956).

(6) Gutzeit, G., and Mapp, E. T., "Kanigen Chemical Nickel Plating," *Corrosion Technology, 3* (10), 331–336 (1956).

(7) Fitzgerald-Lee, G., "Chemical Nickel Plating," *Product Finishing (London), 13* (5), 68–70 (1960).

(8) Keil, A., and Berger, B., "Electrical Conductivity of Chemically Deposited Nickel Layers," *Metalloberflaeche, 10* (12), 356–357 (1956).

(9) Private Communication from J. G. Beach, Battelle Memorial Institute, Columbus Laboratories (1970).

(10) Lang, K., "Electroless Nickel Plating by the Nibodur Process," *Galvanotechnik, 56* (6), 347–358 (1965); *Schweizer Maschinenmarkt, 66* (348), 75–81, 83, 85 (1966).

(11) Dini, J. W., and Coronado, P. R., "Thick Nickel Deposits of High Purity by Electroless Methods," *Plating, 54* (4), 385–390 (1967).

(12) Roberts, W. H., "Coating Beryllium With Electroless Nickel," *U.S. Atomic Energy Commission RFP478* (1964), 23 pp.

(13) Ro, B., "On Properties of Electric Conductivity and Corrosion Resistance of Nickel-Phosphorus Alloy Deposits. Studies on Chemical Plating (4)," *Journal of the Metal Finishing Society of Japan, 11* (9), 331–334 (1960).

(14) Mallory, G. O., "The Electroless Nickel-Boron Plating Bath; Effects of Variables on Deposit Properties," *Plating, 58* (4), 319–327 (1971).

(15) Saubestre, E. B., "Electroless Plating Today," *Metal Finishing, 60* (6), 67–73; (7), 49–53; (8), 45–49, 52; (9), 59–63 (1962).

(16) Levy, D. J., "Chemical Plating of Thin Pure Nickel Coatings on Catalytic Metallic Surfaces," U.S. Patent 3,198,659 (August 3, 1965). Assigned to Lockheed Aircraft Corporation.

(17) Levy, D. J., "Thin Nickel Films by Hydrazine Autocatalytic Reduction," *Electrochemical Technology, 1* (1/2), 38–42 (1963).

(18) Alberts, G. S., Wright, R. H., and Parker, C. C., "Effect of NH_3 on Deposition From Alkaline Electroless Nickel and Cobalt Plating Baths," *Journal Electrochem. Soc., 113* (7), 687–690 (1966).

(19) Gorbunova, K. M., Nikiforova, A. A., Polukarov, Yu. M., and Moiseev, V. P.,"Magnetic Properties of Nickel Reduced With Hypophosphite From Alkaline Solutions," *Russian Journal of Physical Chemistry, 38* (6), 854–858 (1964).

(20) Benyon, J. C., "Magnetic Recording Device," U.S. Patent 3,202,538 (August 24, 1965). Assigned to Bunker-Ramo Corporation.

(21) Lawless, G. W., and Fisher, R. D., "Electroless Plating Variables and Coercive Force of Nickel-Cobalt-Phosphorus Films," *Plating, 54* (6), 709–713 (1967).

(22) "Electroless Deposition of Magnetic Coatings," Netherlands Patent 6,512,652 (April 4, 1966); British Application (October 2, 1964). Assigned to International Standard Electric Corporation.

(23) Lawless, G. W., "Stability of Chemically Deposited Magnetic Thin Films With Time and Temperature," *Journal Electrochem. Soc., 115* (6), 620–621 (1968).

(24) Heritage, R. J., and Walker, M. T., "Chemically Deposited Ni-Co Layers as High-Speed Storage Elements," *Journal of Electronics and Control, 7*, 543–552 (1959).

(25) Gorbunova, K. M., and Nikiforova, A. A., "Physicochemical Principles of Nickel Plating," Translated by the Israel Program for Scientific Translations, Jerusalem, 152–153 (1963). OTS 63-11003.

(26) Olson, B. J., and Teply, R. J., "Ferromagnetic Film Memory Elements," U.S. Patent 3,392,053 (July 9, 1968). Assigned to Sperry Rand Corporation.

(27) Korenev, N. A., "Chemical Deposition of Nickel-Cobalt Films," *Izv. Akad. Nauk SSSR, Ser. Fiz., 29* (4), 650–654 (1965); *Bulletin Acad., Sci. USSR, Physical Series, 29* (4), 651–655 (1965).

(28) Flechon, J., "Laboratory Study of Chemical Nickel Deposits," *Revue du Nickel, 27* (5), 123–125 (1961).

(29) Graham, A. H., Lindsay, R. W., and Read, H. J., "The Structure and Mechanical Properties of Electroless Nickel," *Journal Electrochem. Soc., 112* (4), 401–413 (1965).

(30) Wiegand, H., Heinke, G., and Schwitzgebel, K., "Properties of Chemical Nickel Deposits From the Hypophosphite Bath," *Metalloberflaeche, 22* (10), 304–311 (1968).

(31) Jovanovic, S., and Smith, C. S., "Elastic Modulus of Amorphous Nickel Films," *Journal Applied Physics, 32*, 121–122 (1961).

(32) Baldwin, C., and Such, T. E., "The Plating Rates and Physical Properties of Electroless Nickel/Phosphorus Alloy Deposits," *Trans. Inst. Metal Finish., 46* (2), 73–80 (1968).

(33) Read, H. J., and Whalen, T. J., "The Ductility of Plated Coatings," *Proceedings Am. Electroplaters' Soc., 46*, 318–325 (1959).

(34) Cadorna, L., and Cavalotti, P., "Electroless Plating of Metals. (1) Electroless Nickel Plating From Alkaline Sulfate Solutions," *Electrochimica Metallorum, 1* (1), 93–101 (1966).

(35) Morioka, S., Saegusa, F., Endo, Y., and Nawa, T., "Chemical Deposition of Nickel," *Journal of the Metal Finishing Society of Japan, 11* (4), 131–135 (1960).

(36) Yur'ev, S. F., and Sakharova, E. V., "Chemical Coating of Titanium With an Antifriction Nickel-Phosphorus Alloy," *Metallovedenie Term. Obrabotka Metal., (6)*, 28–33 (1964).

(37) Randin, J. P., and Hintermann, H. E., "Electroless Nickel Deposited at Controlled pH; Mechanical Properties as a Function of Phosphorus Content," *Plating, 54* (5), 523–532 (1967).

(38) Brenner, A., "Electroless Plating Becomes of Age," *Metal Finishing, 52* (11), 68–76 (1954).

(39) Mitani, H., Shoji, K., and Kanbe, T., "Phase Changes in Electroless Ni-P Deposits During Heat Treatment," *Journal of the Metal Finishing Society of Japan, 17* (10), 379–383 (1966).

(40) Kocich, J., Tuleja, S., Cech, J., Chachalak, M., "Heat Treatment of Chemically Deposited Nickel Coatings," *Ved. Pr. Vys. Sk. Tech. Kosiciach, 1*, 143–151 (1968).

(41) Lee, W. G., "Age Hardening of Chemical Nickel Coatings," *Plating, 47* (3), 288–290 (1960).

(42) Domnikov, L., "Chromium and Electroless Nickel Deposits. Hardness at High Temperatures," *Metal Finishing, 60* (1), 67–68, 70 (1962). Lozinsky, M. G., and Mirotvorsky, V. S., "Metallurgy and Fuel," *Bulletin Acad. Sci. USSR, Div. Tech. Sci., No. 3* (1959).

(43) Velemitsina, V. I., and Ryabchenkov, A. V., "Chemical Nickel Plating as a Method for Protecting and Strengthening Parts of Power Equipment," *Tr. Mezhdunar. Kongr. Korroz. Metal., 3rd (Proc. 3rd International Congress on Metallic Corrosion), Moscow, 1966*, 358–359 (1968).

(44) Kaczynski, J., and Hajewska, E., "Electroless Nickel Coatings on Aluminum," *Ochrona Przed Korozja, 9* (3), 57–59 (1966).

(45) Goldenstein, A. W., Rostoker, W., Schossberger, F., and Gutzeit, G., "Structure of Chemically Deposited Nickel," *Journal Electrochem. Soc., 104* (2), 104–110 (1957).

(46) Zusmanovich, G. G., "Effect of Heat Treatment on the Adhesion of Ni-P Coatings to Hardened KhVG Steel," *Metal Science and Heat Treatment of Metals (7-8)*, 364–365 (1961).

(47) Kanbe, T., "Characteristics of Electroless Plated Nickel-Boron Films. I. Phase Changes in Electroless Nickel-Boron Deposits During Heat Treatment," *Kinzoku Hyomen Gijutsu (Journal of the Metal Finishing Society of Japan), 20* (6), 279–283 (1969).

(48) Lozinskiy, M. G., Zusmanovich, G. G., and Mirotvorskiy, V. S., "The Effect of Temperature on the Microhardness of Protective Coatings," *Metal Science and Heat Treatment of Metals (7-8)*, 363–364 (1961); *Metallovedeniyt i Term. Obrabotka Metal. (7-8)*, 37–39 (1961).

(49) Nemoto, K., Kanbe, T., and Maruya, T., "The Study on Hardness of Non-Electrolytically Plated Ni-P Deposits at High Temperatures and the Effects Given by Heat Treatments," *Journal of the Metal Finishing Society of Japan, 16* (3), 106–109 (1965).

(50) Vishenkov, S. A., "Use of Chemical Nickel Plating for Corrosion Protection and Increased Durability of Parts, Part I." Abstract of paper presented at the All-Union Scientific/Technical Conference on Corrosion and Protection of Metals. Metal Coatings and Chemical Treatment Sections, and Lacquer Coatings Section, Moscow, USSR (May 19–24, 1958). *Plating, 46* (2), 158 (1959).

(51) Vojtiskova, A., "Effect of Heat Treatment on the Properties of Chemically Deposited Nickel Coatings," *Koroze Ochr. Mater., 13* 1, 5–6 (1969).

(52) Austen, H. E., and Fisher, R. D., "Internal Stress of Electroless Metal Films on Single Crystal Silicon," *Journal Electrochem. Soc., 116* (2), 185–187 (1969).

(53) Judge, J. S., Morrison, J. R., and Speliotis, D. E., "Flexible Recording Surfaces of Electrodeposited Cobalt-Nickel-Phosphorus," *Plating, 53* (4), 441–450 (1966).

(54) Parker, K., and Shah, H., "The Stress of Electroless Nickel Deposits on Beryllium," *Journal Electrochem. Soc., 117* (8), 1091–1094 (1970).

(55) Shemenski, R. M., Beach J. G., and Maringer, R. E., "Plating Stresses From Electroless Nickel Deposition on Beryllium," *Journal Electrochem. Soc., 116* (3), 402–409 (1969).

(56) Spahn, H., "The Effect of Internal Stress on the Fatigue and Corrosion Fatigue Properties of Electroplated and Chemically Plated Nickel Deposits," *Proceedings of the 6th International Metal Finishing Conference, Trans. Inst. Metal Finish., 42*, 364 (1964). "Corrosion Fatigue of Metallic Materials (VII). Internal Stresses in Electrolytically and Chemically Deposited Nickel Coatings and Their Effect on the Properties of Steel Under Corrosion Fatigue," *Metalloberflaeche, 17* (1), 1–9 (1963). Cf. Spahn, *Technische Rundschau (Switzerland), 58* (27), 33 (1966).

(57) Bartlett, B. C., Cann, L., and Hayward, J. L., "Control of Phosphorus Content and Stresses in Electroless Nickel Deposits," *Plating, 56* (2), 168–171 (1969).

(58) Parker, K., and Shah, H., "Residual Stresses in Electroless Nickel," *Plating, 58* (3), 230–236 (1971).

(59) Jungslager, E. F., "Electroless Deposition of Nickel," *Tijdschr. Oppervlakte Tech. Metalen, 9* (1), 2–9 (1965).

(60) Ovsyankin, V. V., Novikov, V. N., and Ryabchenkov, A. V., "Increasing the Fatigue Strength of Electroless-Nickel-Plated Steel by Surface Induction Heat-Treatment," *Zashchita Metallov (Protection of Metals), 3* (4), 496–498 (1967).

(61) Gazez'yan, L. N., and Romanova, E. N., "Chemical Nickel Plating on Steel Parts," *Aviatsionnaya Promyshlennost', No. 7*, 44 (1956).

(62) Antskaitis, V. A., "Chemical Nickel Plating," *Moscow, Filial Vsesoyuznogo Instituta Nauchnoi i Tekhnicheskoi Informatsii AN SSSR, Topic 13*, No. M-58-368/39 (1958).

(63) Borisov, V. S., and Vishenkov, S. A., "Chemical Nickel Plating," *Moskovskii Dom Nauchno-Tekhnicheskoi Propagandy im. F. E. Dzerzhinskogo 2*, 37 (1958).

(64) Efros, D. I., Sharygina, Z. V., and Muzychuk, N. A., "Chemical Nickel Plating on Machine Parts in Alkaline Solutions," *Gor'kii, Tsentral'noe Byuro Tekhnicheskoi Informatsii* (1958).

(65) Hildebrand, J. F., Turns, E. W., and Nordquist, F. C., "Stress Corrosion Cracking in High Strength Ferrous Alloys," *General Dynamics/Fort Worth Report No. FZM-2690* (November 4, 1963), 41 pp. Contract AF 33(657)-11214. AD 423 387.

(66) deMinger, C. H., and Brenner, A., "Studies on Electroless Nickel Plating," *Plating, 44*, 1297–1305 (1957).

(67) Cavalotti, P., and Salvago, G., "Chemical Reduction of Nickel and Cobalt by Hypophosphite. I. Plating of Alloys," *Electrochimica Metallorum, 3* (1), 23–41 (1968).

(68) Luneckas, A., "Chemical Deposition of a Nickel-Copper-Phosphorus Coating," *Zashchita Metallov (Protection of Metals), 4* (3), 338–340 (1968).

(69) Ziehlke, K. T., Dritt, W. S., and Mahoney, C. H., "Heat Treating Electroless Nickel Coatings," *Metal Progress, 77* (2), 84–87 (1960).

(70) Machu, W., and El-Gendi, S., "Effect of Organic Compounds on the Deposition Rate and Brightness of Chemical Nickel Deposits," *Werkstoffe und Korrosion, 12* (4), 223–230 (1961).

(71) Aoki, K., and Ishibashi, S., "Electroless Nickel Plating. XII. Nickel Deposits From Electroless Nickel Plating Solutions Containing Ethylenediamine," *Kinzoku Hyomen Gijutsu (Journal of the Metal Finishing Society of Japan), 20* (3), 115–122 (1969).

(72) Zusmanovich, G. G., "Effect of Heat Treatment on the Hardness of Nickel-Phosphorus Coatings," *Metal Science and Heat Treatment of Metals, (4)*, 229–231 (1960); *Metallovedeniye i Term. Obrabotka Metall., (4), 48–50 (1960). Brutcher Translation No. HB 5385.*

Chapter 23

High-Speed Plating*

High-rate tin-plating processes were first developed about 30 years ago for applying thin deposits on continuous steel strip. For this application, current densities of 0.3 to 0.55 amp/sq cm (2 to 3.5 amp/sq in) have been used to deposit up to 0.75 μm (0.03 mil) in 2 to 3 seconds. Similar processes for plating zinc on steel strip and wire and copper on steel wire were developed later. The maximum current density used for these applications is approximately 0.775 amp/sq cm (5 amp/sq in), which deposits copper at the rate of about 15 μm/min (0.6 mil/min). Developments in high-speed plating of other metals began about 1963 with the objective of conserving space and reducing costs.

Because the success of any high-speed plating process depends on the properties and characteristics of the deposits, Battelle electrochemists devoted considerable effort to the development of property data during a series of programs on high-speed plating. The property data were developed for electroformed foil produced at current densities from about 0.8 to 8 amp/sq cm (5.2 to 52 amp/sq in) on cathodes separated from anodes by a narrow gap ranging in width from 0.25 to 0.625 cm (0.1 to 0.25 in). To sustain high current densities at a relatively low voltage, solution was pumped through the gap between the anode and cathode at a rate sufficiently high to maintain turbulent solution flow. The solution flow rate was usually in the range of 1.2 to 3.6 m/sec (4 to 12 ft/sec). A lower flow rate was insufficient for depositing dense metal at current densities above 0.5 amp/sq cm (3 amp/sq in).

Most of the property data reviewed in this report were obtained from tensile tests of specimens removed from continuously electroformed foil with a width of 2.5 cm (1.0 in). Figure 23.1 shows the rotating-drum facility used to electroform the foil. The 0.3-meter-diameter drum was rotated at 0.05 to 1.0 rpm to deposit foil at current densities from 0.8 to 8 amp/sq cm (5.2 to 52 amp/sq in). Foil thickness ranged from 70 to 100 μm (3 to 4 mils), as a rule.

Cathode efficiency for copper, iron, lead, nickel, and zinc plating was usually in the range of 96 to 100 percent. The efficiency of chromium deposition depended on the current density, as shown in Figure 23.2. At current densities above 4 amp/sq cm (26 amp/sq in), efficiency was greater than 50 percent. Thus deposition rates were 45 to

* Adapted from a paper prepared by the author, in collaboration with C. H. Layer, and presented at the Symposium on the Properties of Electrodeposits as Materials for Selected Applications, held at Columbus, Ohio, November 14–15, 1973.

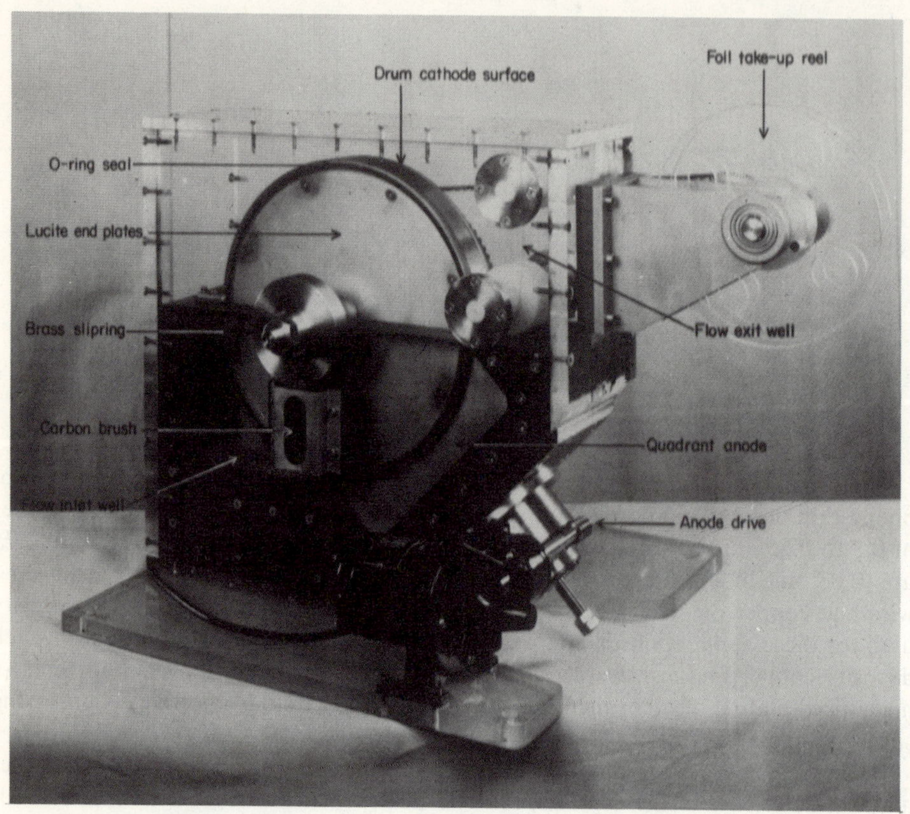

Figure 23.1 The Rotating Drum for Electroforming Foil at High Rates.

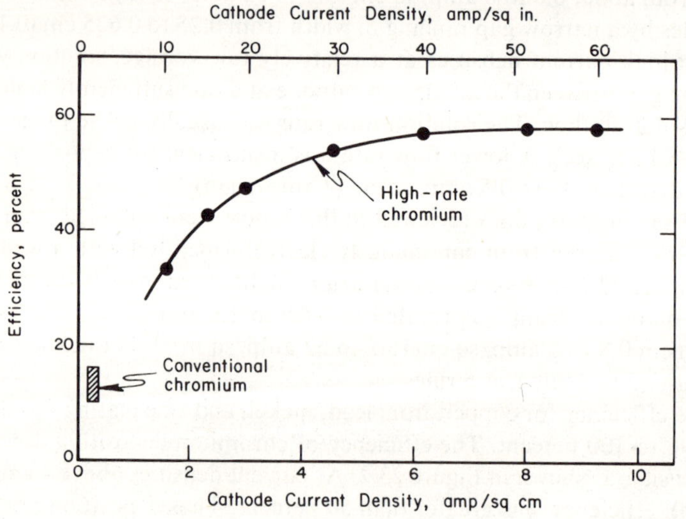

Figure 23.2 Chromium Plating Efficiency as a Function of Current Density for Chromic Acid Solutions at 52 C (125 F) Containing About 3 M CrO_3 and 0.03 M H_2SO_4.

75 times faster for high-speed chromium, relative to conventional chromium plating at 0.3 amp/sq cm (2 amp/sq in), which has an efficiency of 10 to 15 percent.

Density and Electrical Resistivity

Typical values for the density and electrical resistivity of several metals deposited at high rates are listed in Table 23.1. Density and resistivity varied as a function of current density and plating rate in some cases. In the case of copper, density increased with increasing plating rate from 25 to 75 μm/min (1 to 3 mils/min), as shown in Figure 23.3. Resistivity decreased as plating rate was increased to about the same value reported for annealed wrought metal. Changes in plating rate had little effect on density or resistivity of lead and nickel. However, and increase in the plating rate for zinc from 25 to 50 μm/min (1 to 2 mils/min) reduced density and increased resistivity.

The highest density values for high-speed lead and zinc deposits were comparable with the maximum values reported in the literature for the corresponding metals deposited at conventional rates. The density of high-speed copper and nickel was slightly lower than that of their conventional-rate counterparts, but the electrical resistivity of these high-rate metals was similar to that of the corresponding conven-

TABLE 23.1

Density and Electrical Resistivity Data for Electrodeposits Obtained at High Rates

Metal	Plating Rate		Density, g/cu cm		Electrical Resistivity, microhm-cm	
	μm/min	mil/min	High-Rate Deposit	Conventional Deposit (1 μm/min)	High-Rate Deposit	Conventional Deposit (1 μm/min)
Copper[a]	25	1	8.65	8.93, max	2.34	1.70, min
Copper[a]	50	2	8.80	—	1.69	—
Lead[b]	25	1	11.34	11.34, max	21.6	22.90, min
Lead[b]	87.5	3.5	11.22	—	22.1	—
Nickel[c]	25	1	8.69	8.90, max	6.84	7.40, min
Nickel[c]	50	2	8.65	—	6.84	—
Zinc[d]	25	1	7.14	7.20, max	7.6	6.60, min
Zinc[d]	50	2	6.60	—	8.2	—

[a] Fluoborate solution containing 2.0 M Cu, 60 to 65 C.
[b] Fluoborate solution containing 1.75 M Pb (40 C).
[c] Sulfate solution containing 3.8 M Ni, 1.1 M Cl, and 0.3 M H_3BO_3 (60 C).
[d] Sulfate solution containing 1.75 M Zn (75 C).

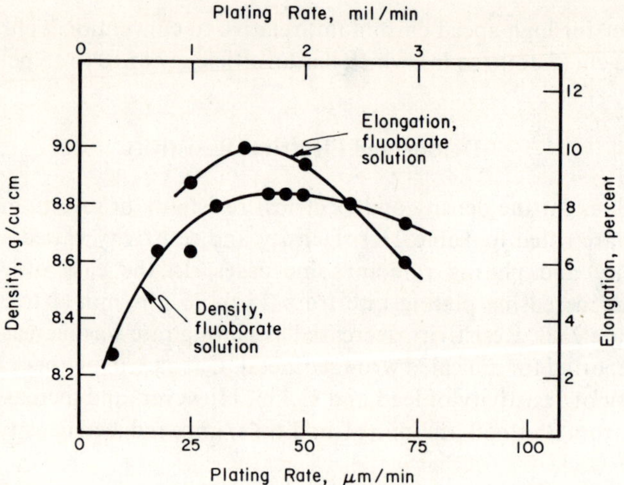

Figure 23.3 Density and Elongation of Copper as a Function of Plating Rate.

tional-rate deposits, indicating a comparable purity for high-rate and conventional-rate copper and nickel.

The oxygen content of high-speed nickel usually was about 400 ppm and was independent of the plating rate. This level of oxygen is approximately the same as the oxygen content of conventional-rate nickel, which ranges from about 300 to 460 ppm in typical, high-purity deposits obtained at current densities ranging from 0.05 to 0.06 amp/sq cm (0.3 to 0.4 amp/sq in). Metallic impurities in high-speed nickel included cobalt (about 1 percent), manganese (0.1 percent), molybdenum (0.1 percent), iron (0.1 percent), copper (0.1 percent), and lead (0.1 percent). These impurities were introduced with nickel carbonate additions for maintaining the nickel content and controlling pH.

High-speed copper deposited at 2.7 amp/sq cm (17.5 amp/sq in) contained 0.004 percent iron, 0.004 percent chromium, and < 0.001 percent nickel, silver, and lead. The iron and chromium contents increased to 0.008 to 0.01 percent, respectively, when the current density was reduced to 0.8 amp/sq cm (5.2 amp/sq in), but an increase in current density to 3.1 amp/sq cm (20 amp/sq in) reduced the iron and chromium contents to only 0.002 percent.

High-speed lead typically contained 0.05 percent tin, 0.02 percent bismuth, 0.02 percent calcium, and < 0.001 percent magnesium. High-speed zinc usually contained 0.15 percent copper, 0.1 percent lead, 0.1 percent cadmium, 0.01 percent nickel, 0.001 percent iron, 300 to 350 ppm oxygen, and 20 to 40 ppm hydrogen.

Strength and Ductility

The tensile and yield strengths of high-speed copper deposited in both fluoborate and sulfate baths increased as plating rate was increased from 25 to 75 μm/min (1 to

3 mils/min). Consequently, the tensile strength of copper electroformed in a sulfate solution at 75 μm/min (3 mils/min) was nearly two times the strength of copper deposited in a similar solution at conventional rates. Tensile and yield strength data for copper are detailed in Figure 23.4. The thickness of the copper foils produced to develop the data in Figure 23.4 was approximately 75 μm (3 mils).

The high strength of the copper electroformed at 75 μm/min (3 mils/min) is attributed to the fine-grained structure of the metal. An etched cross section of a typical deposit from a sulfate bath is shown in Figure 23.5. The thickness of the foil is approximately 90 μm (3.5 mils). Deposits from the fluoborate bath were also fine grained and similar in structure.

Figure 23.3 shows variations in the ductility of high-rate electroformed copper foil as a function of plating rate. The elongation of high-rate copper deposited at 37 μm/min (1.5 mils/min) was 10 percent, but the stronger foil deposited at 75 μm/min (3 mils/min) exhibited a ductility of 6 percent. Foils electroformed in the fluoborate solution were slightly more ductile than deposits obtained in a sulfate bath with about the same copper concentration.

The tensile strength of high-rate nickel was as high as 134 kg/sq mm (190,000 psi) when the plating rate was adjusted to 8.7 μm/min (0.35 mil/min) in a sulfate solution. This nickel is more than three times stronger than nickel deposited at conven-

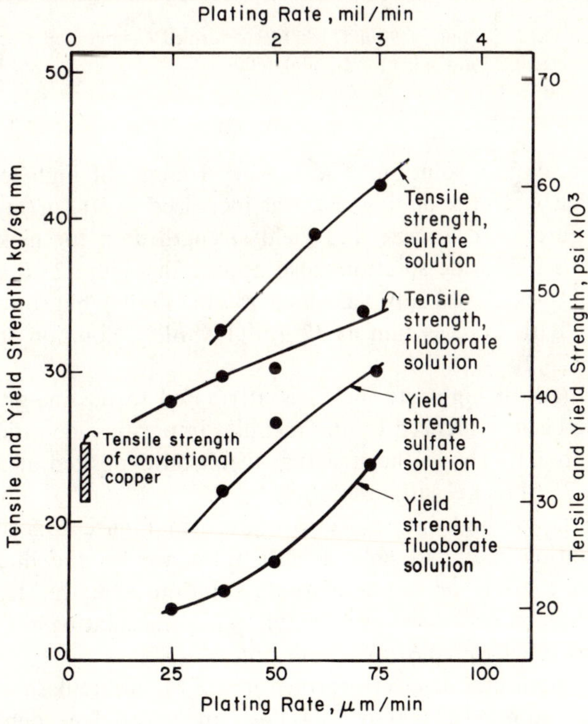

Figure 23.4 Tensile and Yield Strength of High-Rate Copper as a Function of Plating Rate.

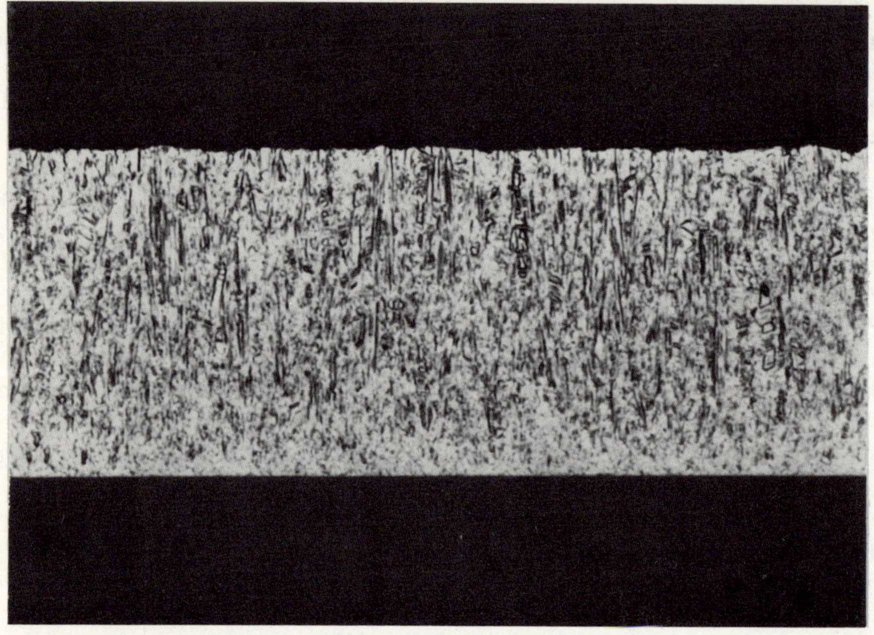

Figure 23.5 Etched Cross Section of 90-μ-Thick Electroformed Copper Foil Deposited in a Sulfate Bath at Approximately 75 μm/min (3 mils/min). (500×)

tional rates in a similar solution. The tensile strength of high-speed nickel was reduced somewhat when the plating rate was increased to 50 μm/min (2 mils/min), as shown in Figure 23.6. Tensile and yield strength data for nickel deposited in sulfate-chloride and chloride solutions also appear in Figure 23.6 as a function of plating rate. Foil thickness usually was about 75 μm (3 mils), but some of the deposits from the sulfate bath were as thin as 37 μm (1.5 mils). Solution compositions are detailed in Table 23.2.

The high strength of high-rate nickel is attributed to its fine-grained structure, which may have been influenced by the metallic impurities identified previously in this report. Figure 23.7 shows the structure of nickel deposited at 30 μm/min (1.2 mils/min) in a 1.75 M nickel sulfate solution.

Elongation data for high-rate nickel are given in Figure 23.8 as a function of plating rate. Plating rate had only a little influence on the ductility of nickel electroformed in chloride or sulfate-chloride solutions, but the ductility of nickel from sulfate solutions decreased appreciably when the plating rate was increased from 17 to 37 μm/min (from 0.67 to 1.5 mils/min).

The tensile strength of zinc electroformed in a 2 M sulfate bath with a pH of 4.0 was 14 to 15.5 kg/sq mm (20,000 to 22,000 psi) and was independent of plating rate from 25 to 70 μm/min (1 to 2.8 mils/min). Yield strength ranged from 13 to 14 kg/sq mm (18,000 to 20,000 psi) and elongation from 3 to 5 percent.

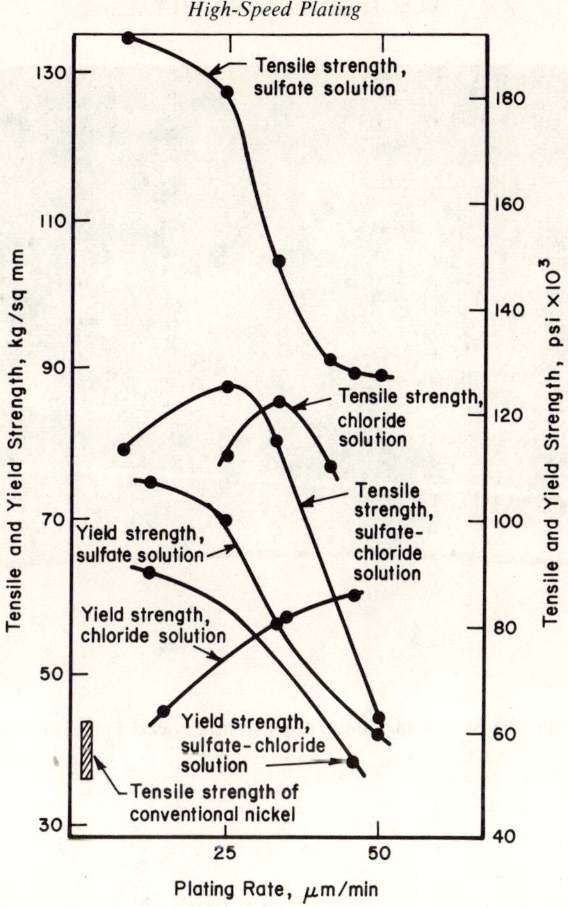

Figure 23.6 Tensile and Yield Strengths of High-Rate Nickel as a Function of Plating Rate.

TABLE 23.2

Nickel Solution Compositions

Constituent	Concentration: Sulfate Solution	Chloride Solution	Sulfate-Chloride Solution
Nickel sulfate, M	1.75	—	1.75
Nickel chloride, M	—	2 to 3	1.0
Boric acid, M	0.3	—	0.33
pH	3.0 ± 0.2	2.0	2.7 ± 0.3
Temperature, C	60	90	60–70

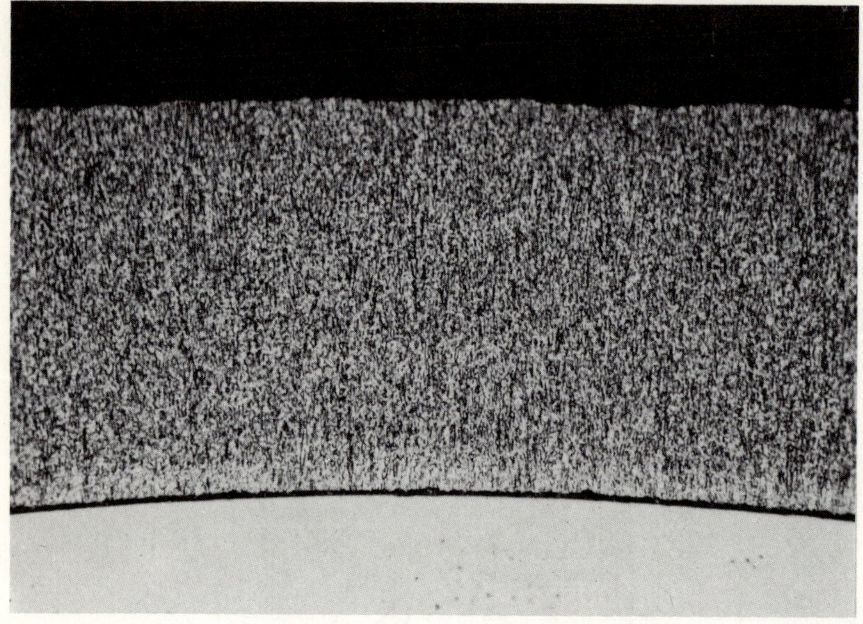

Figure 23.7 Etched Cross Section of 150-μm (6-mil)-Thick Nickel Foil Electroformed at 30 μm/min (1.2 mils/min). (350×)

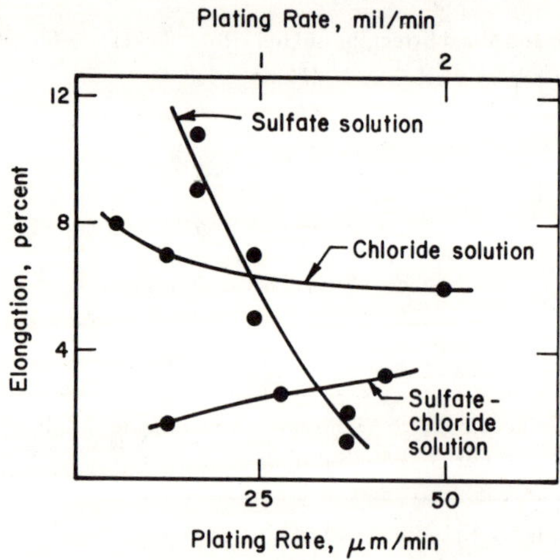

Figure 23.8 Ductility of High-Rate Nickel as a Function of Plating Rate.

High-speed zinc was approximately 30 percent stronger than the highest-strength conventionally deposited zinc and two to three times as strong as low-strength conventional-rate zinc. The higher strength values for high-speed zinc probably were due to a finer grain structure, which may have been influenced by the metallic impurity levels listed previously in this report.

Hardness

Hardness data for high-speed and conventional deposits are compared in Table 23.3. High-speed and conventional nickel span the same range. High-speed chromium and zinc are slightly softer than the corresponding conventional deposits. High-speed copper and nickel-phosphorus alloy can be harder than their conventional-rate counterparts, depending on operating conditions.

The pH of 1 M iron chloride solutions affected the hardness of high-speed iron. Hardness was increased by increasing pH from 2.6 to 3.6, reducing the flow rate, or increasing the current density. Increasing the pH of nickel solutions also increased hardness. Unexpectedly, increasing the current density from about 0.8 amp/sq cm (5 amp/sq in) to 2 to 2.5 amp/sq cm (13 or 16 amp/sq in) decreased the hardness of nickel deposited in sulfate and sulfate-chloride solutions. On the other hand, the hardness of deposits from the chloride solution was not changed as a result of varying current density from 0.3 to 7.75 amp/sq cm (from 2 to 50 amp/sq in). The hardness of all deposits from the chloride bath was approximately 525 kg/sq mm, which was greater than the values measured for nickel deposited in sulfate or sulfate-chloride solutions.

TABLE 23.3

Hardness Data for Metals Deposited at High and Low Rates

Metal	Hardness, kg/sq mm	
	High-Rate Deposits	Conventional-Rate Deposits[a]
Chromium (100-g load)	700 to 825	300 to 1200
Copper (100-g load)	110 to 140	50 to 100
Iron (100-g load)	480 to 750	120 to 640
Nickel (100-g load)	180 to 525	140 to 580
Nickel-phosphorus alloy	800 to 814	500 to 750
Zinc (25-g load)	37 to 43	48 to 80

[a] Deposits obtained with no additives for increasing hardness.

Conclusions

Copper, nickel, and zinc deposited at a high rate above 25 μm/min (1 mil/min) exhibited higher strengths than the corresponding metal deposited in similar solutions at conventional rates below 2.5 μm/min (0.1 mil/min), due to grain-size refinement. The density of the high-rate deposits was slightly lower than maximum values reported for conventional deposits, but comparable with the range normally obtained for conventional electrodeposits. The good density of the high-speed deposits, their high strength, and good smoothness should encourage more industrial use of high-speed plating to reduce space requirements and costs.

Index

WILLIAM H. SAFRANEK is Manager of the Electrochemical Engineering Technology Section, Batelle, Columbus Laboratories. A former president of the American Electroplaters' Society, he has received both Silver and Gold Medal Awards and the Proctor Memorial Leadership Award from the Society. He is a member of the following scientific and professional societies: American Electroplaters' Society, American Society for Metals, Electrochemical Society, American Chemical Society and American Society for Testing and Materials. Mr. Safranek received his B.S. degree in Chemistry from the University of Chicago and has authored or coauthored more than seventy-five articles or chapters in books on electrodeposition.